Technological Innovation:
A Critical Review of Current Knowledge

Patrick Kelly
Melvin Kranzberg
Editors

Frederick A. Rossini
Norman R. Baker

Fred A. Tarpley Jr.
Morris Mitzner

Department of Social Sciences
Georgia Institute of Technology

with chapters by

Simon Kuznets

James R. Bright
Charles F. Douds
John D. Eveland
Thomas P. Hughes
Edwin Mansfield

Everett M. Rogers
Nathan Rosenberg
Richard S. Rosenbloom
Albert H. Rubenstein
W. Paul Strassmann

San Francisco Press, Inc.
547 Howard Street, San Francisco, California 94105

Copyright © 1978 by San Francisco Press, Inc.
547 Howard Street, San Francisco, CA 94105
Printed in the U.S.A.

Library of Congress Catalog Card 77-71749
ISBN 0-911302-34-4

PREFACE: THE GEORGIA TECH INNOVATION PROJECT

Technological innovation has fueled the American economy in the past and has helped elevate the American standard of living to one of the highest in the world. But we cannot rest on past technical achievements to continue to stimulate economic activity and to meet the many — and growing — needs of our own citizenry and the rest of the world. Britain's decline from industrial leadership during the past century demonstrates the danger of smugness and the folly of failing to maintain an innovative thrust. The requirements of our own national wellbeing and the desperate needs of the world's burgeoning population make it imperative that we continue to innovate and, indeed, that we accelerate the pace of innovation. At the same time, the recognition that certain technical innovations have unforeseen and deleterious consequences forces us to strive for newer and better ways of making and doing things.

Although it might be argued that many of the problems we now face — economic, environmental, social, etc. — in some measure derive from unanticipated and harmful results of past technical achievements, that does not lessen our present dependence on technology to assist us in overcoming those problems. In fact, it often increases that dependence, while increasing our responsibility to anticipate and avoid undesirable consequences. For technological innovation perforce remains a dominant problem-solving response in our American society.

If we wish technology to contribute to the achievement of our national goals and objectives, we must obtain a clearer understanding of that complex and significant activity known as the innovation process. The first step in so doing is to determine and critically assess the present knowledge and understanding of the process of technological innovation — and that is the reason for this study.

* * *

This state-of-the-art study of our knowledge of the innovation process has been supported by the National Science Foundation's Office of National R & D Assessment, under Grant DA-39269.* Because research and development (R & D) play a major role in fostering innovation in today's world and because our Federal government has been the major source of R & D funds, it is not surprising that the National R & D Assessment Program of the National Science Foundation has taken the leadership in stimulating research on innovation. That Program is interested, among other things, in questions involving the effectiveness of R & D in bringing about innovation; the obstacles or barriers to innovation in the present organization and application of R & D; and what mechanisms, devices, or institutional arrangements might further stimulate the process of innovation. Other questions raised by the National R & D Assessment Program include the following. Just where in the R & D process should the government concentrate its efforts? What types of innovation and in what fields are our R & D efforts most likely to be productive in terms of new products, greater employment, a finer "quality of life" for our citizens?

Our present knowledge of the technologically innovative response to societal problems is such that integrative, cumulative research is difficult, management of the innovaton process is largely by trial and error, and policy making is extremely problematic. Explication of our present understanding of innovation should be therefore useful beyond the immediate purpose of Federal decision making. It should furnish basic information for decision makers in industry as well as in government; it should serve as a useful guide to managers in R & D; and it should provide a foundation and starting point for further scholarly research. All three groups — policy makers, managers, and scholars — have needs that require attention. We have tried to keep all three in mind in this critical review of current knowledge.

* * *

The research and writing of Part 1 of this work and the overall responsibility for the project were shared by an interdisciplinary research team of six scholars. The members of this Georgia Tech Innovation Project group are:

Dr. Patrick Kelly, head of the Department of Social Sciences, whose special interests are epistemology and the philosophy of technology.

Dr. Melvin Kranzberg, Callaway Professor of the History of Technology, who is a pioneer in this field and editor of *Technology and Culture*.

Dr. Frederick A. Rossini, Department of Social Sciences, whose background includes a doctorate in physics, postdoctoral work in the philosophy of science, and involvement in a NASA technology assessment program.

Dr. Norman R. Baker, School of Industrial and Systems Engineering, whose specialty is management science, especially R & D management. (Now at University of Cincinnati.)

Dr. Fred A. Tarpley, College of Industrial Management, who is an economist specializing in the economics of regulation.

Dr. Morris Mitzner, Department of Social Sciences, who is a sociologist specializing in informal organizational behavior.

*The views and recommendations contained in this study are those of the authors alone and do not necessarily reflect the official position of the National Science Foundation.

This group began its work with a fairly clear-cut division of labor in the usual multidisciplinary mode. However, the level of cooperation was so high that specific credit for Chaps. 1-5 is impossible to assign; in the course of things the project became truly interdisciplinary.

Because the innovation process is itself so complex and the literature concerning it so disparate and diffuse, it was apparent that additional expertise was needed. Hence, nine outstanding research scholars were asked to prepare complementary state-of-the-art assessments of specialized aspects of the innovation process. Their commissioned papers, prepared expressly for this project, comprise Chaps. 6-14. The nine consultants:

Dr. James R. Bright, associate dean of the Graduate School of Business, University of Texas-Austin, a pioneer in technological forecasting. (Now with Industrial Management Center, Inc.)

Dr. Thomas P. Hughes, Department of the History and Sociology of Science, University of Pennsylvania, a well-known historian of technology and author of the prize-winning *Elmer Sperry: Inventor and Engineer*.

Dr. Simon Kuznets, emeritus professor of economics, Harvard University, a Nobel laureate in economics.

Dr. Edwin Mansfield, Wharton School, University of Pennsylvania, the author of many studies linking innovation with the economics of firms and industries.

Dr. Everett Rogers, University of Michigan, a sociologist and a specialist in communications who is the author of several classical studies on the diffusion of innovations.

Dr. Nathan Rosenberg, Stanford University, a former editor of the *Journal of Economic History* and the author of works dealing with the application and transfer of technical innovations.

Dr. Richard S. Rosenbloom, Sarnoff Professor, Harvard Graduate School of Business Administration, the author of studies dealing with the "spinoff" of technical innovations and of decision making in business enterprises.

Dr. Albert Rubenstein, head of the Program of Research on the Management of Research and Development at Northwestern University, who has directed many investigations of R & D organizations, stressing the flow of information.

Dr. W. Paul Strassmann, professor of economics at Michigan State University, the author of books dealing with risk and innovation in 19th and 20th Century American industry.

* * *

Shortly after the project began, we recognized that our literature-based review and integration of current knowledge also required the critical perspective of those actively engaged in the innovation process — the "real-world" doers, administrators, and users of scholarly output. We were fortunate to obtain the services of the following individuals on our Industry Advisory Panel: Dr. Herman Bieber, senior research associate, Exxon Research and Engineering Co.; Dr. W. Gale Cutler, director of corporate research, Whirlpool Corp.; Dr. Edward E. David, former Science Advisor to the President of the United States, now vice president of Gould Corp.; Dr. Jacob E. Goldman, senior vice president, research and development, Xerox Corp.; Dr. William E. Hanford, vice president, research and development, chemicals, Olin Corp.; Dr. Walter R. Hibbard, former vice president, technical services, Owens-Corning Fiberglass Corp., then deputy administrator of the Federal Energy Administration, and now University Professor at Virginia Polytechnic Institute; Dr. Roland W. Schmitt, Research and Development manager, physical science and engineering, General Electric Co.; and Dr. Julian D. Tebo, retired director of technical information services, Bell Laboratories. These men provided us with invaluable insights and correctives for a study that might have been too academically oriented had it not been for this contact with the actualities of the innovation process.

* * *

The project had an analytic and an assessment/integration phase, each of which had two components. The analytic phase involved the reading, classification, and coding of a large body of the recent research literature on the innovation process, and an in-depth abstracting of the most significant subset of this literature. This phase involved the examination of some 4000 pieces of literature (books, technical reports, government documents, and articles from some 200 professional journals), a computerized coding of approximately 2000 of them, and the preparation of detailed abstracts of more than 300 of the most significant ones.

The assessment and integration phase concerned the quality and adequacy of current knowledge of the process, on both the theoretical and empirical levels. As indicated above, the overall state-of-the-art assessment (Chaps. 1-5) was conducted by the Georgia Tech Innovation Project group, and the complementary assessment of selected aspects (Chaps. 6-14) was conducted by the nine consultants.

The report submitted to the National Science Foundation consisted of four volumes, the first three of which form the basis of the present work. Volume 1 of the NSF report was the overall assessment by the Georgia Tech Project Group; volume 2 contained the papers of the scholarly consultants, dealing with selected aspects of the innovation process; volume 3 was the project bibliography, including cross classification by selected categories; and volume 4, the abstracts of the major literature items in the field. An executive summary was also provided. In the present version, volumes 1 and 2 form Parts 1 and 2, respectively, and volume 3 is the basis of the bibliography starting on p. 363 (cross indexed to show page numbers on which the references are mentioned).

* * *

As might be expected from its scope, this project has involved many individuals besides the researchers who actually participated in the project at Georgia Tech and the outside consultants who prepared the specialized assessment papers. In addition to our Industry Advisory Panel, there are many others who provided advice and support and whose contribution to the successful completion of the project must not be overlooked. Foremost among them are the program officers of the National Science Foundation's National R & D Assessment Program: Mr. Leonard Lederman; Dr. Alden Bean; and most importantly, Ms. Mary Ellen Mogee. Indeed, Ms. Mogee shepherded the project through its many phases from inception to completion; she was inquiring but sympathetic, skeptical yet supportive, demanding but understanding.

Other members of the National R & D Assessment Program who provided help at various stages of the project included Dr. Andrew Pettifor, Dr. J. David Roessner, and Ms. Eleanor Thomas.

We were also favored with the assistance and advice from participants in companion projects, including Dr. Charles Douds of Northwestern University and Dr. Edward Wood of the Stanford Research Institute (now SRI International); and from Dr. Charles Susskind, who as consulting editor to San Francisco Press, Inc., supervised the preparation of the present version for publication and, with his assistant Martha Zybkow, prepared the prefatory Argument (p. vi) to establish and support our points of view.

* * *

An essential requirement of any study of this magnitude is the cooperation of conscientious and dedicated people. For their many and varied contributions we are deeply grateful to: Dr. Graham Roberts and the staff of the Georgia Tech library, who generously provided working space and continuous professional assistance, and endured the clatter of our computer terminal; Ms. Frances Kaiser, a first-rate professional, who served the project so ably while "on loan" from the Georgia Tech library and thereafter; Charles Reed and his staff at the Georgia Tech Computer Center, who say they are there to serve and mean it; Jay Norman and Dave Howell, who managed the day-to-day operation of the literature search so ably; Russell Zimmerman, Taylor Little, Lu Ann Sims, Mary Martin, Mark Clark, Fara Eslami, Mary Nelson, Carlos Seminario, L. O. Cox, and Duncan Wood, who searched and searched and searched — with care and good results; Dr. Michael McKinney of the Social Sciences Faculty and his assistant, Ross Herbert, who developed the computer program for the cross-classified displays of the literature base and saw the task through to completion; Mrs. Anita Bryant, Mrs. Wynette Little, Mrs. Ann Singleton, Miss Frances Smith, and Mr. Paul Burkitt, who typed countless drafts with accuracy, good humor, and patience; and finally the administrative offices of Georgia Tech for their help and understanding in the conduct of this complex and extended task.

Above all, we are grateful to the many scholars who preceded us, and whose work provided us with the knowledge and information about the innovation process that served as the starting point and the source material for our own researches and studies. Just as most innovations build on previous innovations, so our study of the innovation process builds on a vast body of research by those who have gone before us. Our major contribution, perhaps, has been to sort out and juxtapose an enormous volume of literature, to integrate the maze of hitherto disparate investigations, and to offer hypotheses and guidelines for future research and action. In the course of this work, we have learned much about the process of innovation, and we hope that this knowledge will be useful to scholars, managers of innovative efforts, and policy makers concerned with stimulating the innovation process.

Patrick Kelly

Melvin Kranzberg

Atlanta, Georgia, 1978

TABLE OF CONTENTS

PREFACE: THE GEORGIA TECH INNOVATION PROJECT, *iii*
Patrick Kelly and Melvin Kranzberg

THE ARGUMENT, *ix*
Charles Susskind and Martha Zybkow

PART I: THE ECOLOGY OF INNOVATION

1. THE ECOLOGY OF INNOVATION, *1*
2. THE WORLD OUTSIDE, *18*
3. THE PROCESS OF INNOVATION: THE ORGANIZATIONAL AND INDIVIDUAL CONTEXTS, *47*
4. THE DIFFUSION OF INNOVATION, *119*
5. OVERVIEW AND PROSPECTS, *151*

PART II: ASPECTS OF TECHNOLOGICAL INNOVATION, *165*

6. INVENTORS: THE PROBLEMS THEY CHOOSE, THE IDEAS THEY HAVE, AND THE INVENTIONS THEY MAKE, *166*
Thomas P. Hughes

7. TECHNOLOGICAL INNOVATION AND NATURAL RESOURCES: THE NIGGARDLINESS OF NATURE RECONSIDERED, *183*
Nathan Rosenberg

8. THE ECONOMICS OF INDUSTRIAL INNOVATION: MAJOR QUESTIONS, STATE OF THE ART, AND NEEDED RESEARCH, *199*
Edwin Mansfield

9. TECHNOLOGICAL INNOVATION IN FIRMS AND INDUSTRIES: AN ASSESSMENT OF THE STATE OF THE ART, *215*
Richard S. Rosenbloom

10. REVIEW AND ASSESSMENT OF THE METHODOLOGY USED TO STUDY THE BEHAVIORAL ASPECTS OF THE INNOVATION PROCESS, *231*
Charles F. Douds and Albert H. Rubenstein

11. ASSESSING THE KNOWLEDGE OF INNOVATIONS IN NEGLECTED SECTORS: THE CASE OF RESIDENTIAL CONSTRUCTION, *263*
W. Paul Strassmann

12. DIFFUSION OF INNOVATIONS PERSPECTIVES ON NATIONAL R & D ASSESSMENT: COMMUNICATION AND INNOVATION IN ORGANIZATIONS, *275*
Everett M. Rogers with John Dudley Eveland

13. TECHNOLOGY FORECASTING LITERATURE: EMERGENCE AND IMPACT ON TECHNOLOGICAL INNOVATION, *299*
James R. Bright

14. TECHNOLOGICAL INNOVATIONS AND ECONOMIC GROWTH, *335*
Simon Kuznets

APPENDIX: SOME PROBLEMS OF ECONOMIC MEASUREMENT, *357*

BIBLIOGRAPHY, *363*

INDEX TO CONTRIBUTORS, *390*

THE ARGUMENT
Charles Susskind and Martha Zybkow

PART I

1. ECOLOGY OF INNOVATION

Technological innovation is an ecological process that spans a range of activities from the initial idea through development, production, and diffusion; it is a complex response to either need or opportunity, requiring creativity and resulting in the introduction of novelty. Exogenous elements such as human wants, social values, and the economic structure affect the nature and rate of innovation itself; for, like any creative endeavor, it arises from the interaction between individuals and the socio-cultural environment. Because innovation brings about changes in how people work and live, and ultimately how they think and act, prediction of the higher-order effects of new technology before its application is important.

Aspects of the innovative process are of interest to a spectrum of disciplines. Researchers in economics, sociology, and anthropology, for example, continue to develop models of the process in an effort to understand it better. Although focused studies with limited emphases increase our insights, they cannot provide a complete understanding of the entire process. It is tempting to simplify innovation into a linear process beginning with invention and culminating in dissemination. But such a model cannot explain the interaction of components and ignores the reciprocal relationships between structure and function. A holistic, "ecological" approach that transcends the bounds of disciplines is better suited to the concepts involved.

The conceptualization of innovation as part of a larger, dynamic system is vital to its comprehension, but we must focus on the most relevant forces relating to technology; structuring devices must be imposed to create a manageable framework on which to base investigation. In this study, the authors order the components of innovation by the context in which the process takes place into "indigenous" factors (from problem definition to production and diffusion) and those that comprise the larger "exogenous" system (e.g., values, institutions). Once this classification is made we can relate the parts, see the relationship of function and structure, and better understand the nature and effect of their integration. Finally, the subject is divided into two phases, innovation and diffusion, as an aid to organization. This classification differs from the classical division of innovation into individualistic (heroic) and deterministic, as well as from the opportunity-need-genius triad of R. B. Dixon or the linear approach of A. P. Usher.

2. THE WORLD OUTSIDE

The three classical elements of the exogenous system affecting innovation are values (individual or group preferences), endowments (inputs to innovation: knowledge, capital, labor, energy, and materials), and institutions (human organizations that develop to serve goals defined by societal preference).

Values. Linking the role of values to innovation is important because capitalism played a crucial part in the development of technical innovation as a central activity of modern Western society. The degree of societal support for research and education reflects the values supporting scientific and technical knowledge. Current studies show that there is growing public awareness of the broader implications of innovation. Concern over the harmful side effects of new technology has given rise to public-interest groups. Changing values are also being "institutionalized," for example in environmental legislation, pollution regulation, and (in the USA) the creation of the Environmental Protection Agency and the Office of Technology Assessment. This desire to evaluate and control the effects of technology may change the rate and kind of innovation in the future.

Endowments. Interaction of science and technology, which has increased the pace of innovation in the past, is the result of the high value placed on the fruits of innovation. However, most innovation does not arise directly from the application of scientific knowledge. A model of a "pulling and pushing" relationship between science and technology has evolved from current studies relating new discoveries and innovation.

New technological innovation is stimulated by "readiness" as well as knowledge. Existing technology is an important tool for advancement. Because technologies are interdependent, advances in one field create the need for innovation in another. The quick mechanization of British textiles during the Industrial Revolution shows how innovation is spurred as new opportunities are created. Improving the spinning process called for faster weaving and in turn a better, faster way of preparing raw cotton.

Dr. Charles Susskind is Professor of Engineering Science at the University of California in Berkeley; Martha Zybkow was his research assistant during 1977.

Innovation also requires money for new ventures, shifts in the quality of labor, and the supply of specific natural resources. In the market structure, demand stimulates and to some extent directs the incentive for innovation. Natural resources and the supply of scientific knowledge are also important. Growth in knowledge decreases the cost of specific inventions. The substitution of materials and energy sources also aids in increasing productivity and lowering cost. Although the range of isolated substitutions is broad, students of systems dynamics stress that in the world context, most resources are finite and in the long run the substitution model cannot hold up. Limits to growth rest in the limits of "substitutability." This bound is set by human values as well as by natural supplies.

Institutions. In the West, the market mechanism provides incentive and legitimization for decisions. Society highly values the products of current technology; the economic structure ought to support the most efficient use of existing resources. Schumpeter and Galbraith argue that competitive organization is not necessarily the best impetus for innovative activity; as technology changes, another type of organization may prove more suitable. This view leads to an examination of the relationships among firm size, market structure, and innovation. Studies have revealed a positive relationship between firm size and the amount of R & D expenditures (and patent activity). According to Schumpeter and to Galbraith, industries in which large firms predominate are more innovative. Smaller firms engaged in close competition want (and need) to innovate more but have less financial and organizational capacity. Ease of entry into an industry is another component of market structure under study. Expense, lack of patents, and difficulty in competing with established firms impede entry to certain industries. In such industries the pressure to innovate is smaller; yet easy entry may actually discourage experimentation and investigation of new areas. More study of the interaction of market structure and innovation activity is needed.

Although the government theoretically plays a minor role in a free-enterprise economy, the U.S. government has become increasingly involved in fostering new technology. It contributes over one-half the nation's R & D funds and is an important factor in determining in which areas of technology innovation will take place. Regulation and control modify the economic institutions that bear on innovation, directly through the patent system and tax structure and indirectly through regulative bodies. A "map" of political institutions and public policies that would indicate the effects of each on innovation and the overlaps and conflicts among them would aid further study.

3. THE PROCESS OF INNOVATION: THE ORGANIZATION AND INDIVIDUAL CONTEXTS

The institutionalized form of innovation has come to predominate, in part as the result of the belief that creativity can be a deliberate and controlled function of the organization. This assumption makes understanding of the creative process imperative. The firm must be able to read external opportunities and circumstances and utilize this knowledge internally. This linkage of environments is assumed to take place at the policymaking level of the firm.

The Individual Inventor-Entrepreneur. Despite the growth of institutionalized innovation, there are proponents of the individual inventor who argue that the increase in the research force has not been matched by the rate of innovation. However, the Charpie Report (1967), which supported this view by a statistical approach spanning this century, missed the change under way at the time of the study; the individual *was* giving way to the group worker in organized research laboratories.

The role of the independent inventor has been studied from sociological and psychological viewpoints; but for the purpose of the present study, the *context* in which innovation takes place is all-important. The function of the inventor must be linked with the entrepreneurial function, which musters the financial support and the managerial and promotional skill necessary for innovation. The phenomenon of "spin-off" firms specializing in invention and development to the point of application is a product of complex technology, which requires knowledge and management beyond the capabilities of a single individual. These firms perform the role of inventor-entrepreneur.

Organized R & D Within the Laboratory. Organizational structure is a major determinant of the function of R & D within the firm. Structure, strategy, and policy interact to influence the orientation of the laboratory, the characteristics of the projects generated and funded, and the effectiveness of R & D. In the past most R & D was subjected to the same hierarchical control and coordination as the firm's other organization functions. Two models (ideas and information flow, and process-phase) were proposed when study of the R & D process began. A different model emerged in the 1960s as a response to change in organizational development, a model that focused on the two stages (innovation and diffusion). In an effort to improve the original two models, project-dominant and phase-dominant composite models have been developed. Various types of projects fit one or the other model better.

Several orientations can be associated with a firm or laboratory. One set is the basis for the process-phase model (i.e., research, development, engineering, etc.); another is the notion of offensive vs defensive R & D (related to the "leader vs follower" notion of firms). The view held by management regarding the proper balance of the R & D program can directly affect the ideas generated, submitted, and undertaken within the laboratory. Dilemmas arise when the relationship of the organizational structure with the nature of the projects undertaken is examined. Organizational conditions most suitable for initiation are opposite to those suited for implementation. The best organization depends on the environment: "mechanistic" (hierarchical and authoritarian) organizations are more effective under stable and relatively certain conditions, whereas the converse is true for "organic" (adaptive and informal) organizations.

Organized innovative efforts change and control individual judgments and activities. However, pre-project activities of problem definition and idea generation cannot be as directly controlled as later project activities. Social-psychological influences of intra- and intergroup relations are important in this early stage. The researcher's perception of his own freedom, supervisory authority that keeps in touch and allows freedom of action while making the needs of the firm understood, and work groups that share information and ideas and are receptive to outside consultation are elements that enhance creative activity. Although relevant variables are now well identified, relatively little has been written about their interdependence or the indirect influence that management can exert. Just as the recognition of needs and demands in its environment and the input of technological information are crucial to a firm's success, internal appreciation of the firm's needs and information flow within the organization are important. Social networks of influence and information flow that are primarily informal and oral are an important area in which indirect influence can be exerted. Since ideas are submitted only if the originator feels that the needs and means for development will be seen as relevant by management, more should be done to foster understanding of the mutual influences and interactions between management and researchers and among various levels of the organization, as well as among the researchers themselves. The importance of project selection is stated as follows.

Of all the subtopics in the literature of organized innovation, R & D project selection may have received the most attention, as it is a focus of both "idea flow" and "process phase" models. The R & D project selection decision is a process by which an intermittent stream of changes are made in lists of currently active and proposed projects. The project-selection process includes generating alternatives, determining the appropriate time to make a change, collecting data, specifying constraints and criteria, and recycling. The decision is complicated by multiple decision criteria which have no natural, common, underlying measure and whose relative importance varies over time. The decision may require contributions from several different organizational levels which are participating in a hierarchical, diffuse planning and budgeting process. Our present descriptive knowledge has not yet been integrated into a framework which offers promise of improving the process.

How the social-psychological environment affects the performance of scientists and engineers is another important topic. Effective workers were found to be self directed, able to interact with colleagues, doing diversified work, and exhibiting individual work styles. An investigation of the causality of effectiveness revealed that performance precedes favorable organizational conditions; productive researchers were rewarded with better conditions. "Thus performance should be viewed as a cause as well as a result of change." Prevalent in high-performance groups was an open exchange of information and ideas, and good interaction with supervisors.

A final area of consideration is the transfer of R & D to implementation and utilization. This process occurs when technological output becomes an ongoing capacity (production, marketing, sales). Approximately 75% of total innovation costs are spent on implementation. The transfer of R & D results requires a coupling of structure, function, information, and authority. Various organizational forms are used to facilitate this transfer. This is a problem area of great importance but the one for which there is the least information.

4. THE DIFFUSION OF INNOVATION

Research to date on the diffusion process is primarily oriented by sectors that are social systems (such as consumers, agriculture, government, industry). The three basic traditions of approach are the social-psychological, the economic, and the "geographical." Diffusion evidently occurs in an outward-spreading movement, with new adoption taking place mainly near existing adoptions. Personal communications and direct observation are the basic channels for diffusion. This conceptualization emphasizes effective flow of information. The interaction of propagators and adopters implies the existence of a social system.

The economist Mansfield identifies key factors in an innovation's rate of adoption as profitability, uncertainties (complexity, observability, and degree of consistency with existing ideas), extent of commitment, and rate of reduction of initial uncertainty. The findings of his 1970 study of intrafirm research and innovation, in which he investigated the relationship between technological change and economic growth, are of interest in the social-psychological realm. The diffusion process is slowed by lack of knowledge and by resistance to change. Nonuse of new innovation is primarily the result of unprofitability. Better educated owners tend to be early users. Resistance to change entails social, cultural, and psychological components. The effect of this resistance on the propagator's strategy is seen as an important topic for future research.

The "classical diffusion model" remains the major approach to a unified conceptual structure for diffusion. This model concentrates on innovation, communications channels through which it is diffused, the concept of a social system, and time (since the adoption decision is a process rather than an act). However, this approach does not provide an adequate framework. A workable theory should include the characteristics of adopters, of innovations, of propagation mechanisms, and of the aforementioned sectors, which are networks of communication. In previous studies, the functions of social interaction have been stressed over structural characteristics. Research plagued by the other-things-being-equal syndrome has produced a static listing of influences: opinion leaders are identified, early adoption is shown to depend in part on early possession of information, communications and information networks are mapped, and so on. However, in the absence of cross-sectoral studies, sector influences on internal diffusion of innovation and the relative importance of these influences remain poorly understood.

The adopter cannot be severed from the organizational structure or from the informal social system in which he operates. The structure of an organization influences its adoption behavior and reflects the judgments of its leaders.

Individuals within the organization look to peers and superiors for confirmation of their opinions. Adoption can be impeded by resistance to change and undereducated decision makers.

Diffusion can be either adopter-dependent or adopter-independent. The social-psychological and economic-industrial traditions have similar lists of adopter-dependent characteristics. For example, the concept of "relative advantage" extends beyond possible profitability to prestige and uncertainty, and depends on the adopter's perspective. Trialability (the extent to which initial adoption is reversible), the adoption of information (i.e., a better way of doing something), and the proprietary or nonproprietary nature of information are adopter-independent traits. Determination of the effect of these characteristics on adopter behavior merits high research priority.

Propagation mechanism are agents, agencies, and vehicles of communication whose primary function is propagation. They provide or increase observability. Technological information "gatekeepers" within an organization, the mobility of engineers within an industry, professional societies, and a broad range of government agencies and programs serve as propagation mechanisms. Quite likely the diffusion of innovation is primarily an oral process. In mass media, information programs and advertising serve as propagators.

The development of a general theory encompassing the above elements requires new questions to be directed at the integration of various aspects of the diffusion process.

5. OVERVIEW AND PROSPECTS

The first part of this report has attempted to determine what is known and what should be known about the innovation process. The specific objectives were to synthesize a diverse literature, to identify gaps and weaknesses in our understanding, and (indirectly) to advance that understanding. Because the subject embraces a wide range of variables of interest to scholars of many disciplines, emphasis on the interdependence and complementary role of the several fields that have contributed to the research on innovation was an additional objective. The investigation was made at several levels of aggregation. It has identified process phases and facets, integrating and at times extending existing conceptualizations. Implicit in the works examined and underlying the institutionalization of the innovation process is the belief that the process can be deliberately altered (i.e., made more systematic) to aid in the creation and diffusion of innovations. However, increased understanding and control depend on the development of a comprehensive theoretical framework. The authors stress the need for conceptual work to achieve this framework.

PART II

6. INVENTORS: THE PROBLEMS THEY CHOOSE, THE IDEAS THEY HAVE, AND THE INVENTIONS THEY MAKE (T. P. Hughes)

The author distinguishes three early stages of innovation — problem identification, idea-response (formulating ideas that will lead to a solution), and invention (ideas given form) — as exemplified by accounts of three inventors at work and other information drawn from Usher's *History of Mechanical Invention* (1954), Gilfillan's *Sociology of Invention* (1935, 1971), and his own *Elmer Sperry* (1971). This approach reveals the individual's ability to synthesize thoughts and activities in his environment. The work of Edison (1847-1931), Sperry (1860-1930), and Midgley (1889-1944) spans a period in which there was strong support for independent inventors. Although each man represents a different type of inventor — heroic inventor, transitional figure, and industrial research scientist — many similarities appear in the three case histories.

Usher believes that major invention in part results from the merging of new and previous insights. In his thesis on "the emergence of novelty," Usher emphasizes the recognition of an unsatisfactory pattern in perceiving a problem. Tension exists until there is an act of insight that can improve the pattern. Gilfillan's theory of invention focuses on invention as an evolutionary social process; inventions are "stimulated" by nontechnological factors such as population and economic growth or industrial specialization. Hughes analyzes Sperry's personal style of invention and finds it to be rational and pragmatic. As a professional inventor, Sperry deliberately sought problems in fields where both capital availability and an eventual market were likely, and critical problems were apparent. He worked on specific problems, not the creation of whole new systems. He found projects by studying patent claims and recent patent applications, reading professional journals, attending conferences, and in general seeing what other researchers were up to. Hughes defines as "reverse salients" problems that retard the efficiency of a system, technological or social. Identification of a problem is itself an act of insight. Sperry commented, "When I had the facts before me I simply did the obvious thing. I tried to discern the weakest point and strengthen it; often this involved alteration with many ramifications. . . ." Part of the developmental nature of invention is the continual modification of an original idea.

The three case histories are meant to show that these theories (Usher's, Gilfillan's, and Hughes's) should be combined into a single model. The first case is Sperry's marine gyrostabilizer. Sperry first patented an automobile stabilizer but it was not well received. He then shifted his attention to the still expanding ship industry, where conditions were more favorable. He realized that existing stabilizers were slow (incomplete pattern). The stage was set by his previous experimentation with automatic controls. Having found the weakest point, he identified friction and inertia as the causes of difficulty and applied his experience with small sensors and servo motors, which led him to use an electric motor to activate the stabilizing response before the gyro could respond. Modifications were made after testing under simulated conditions and later in sea trials.

Edison's invention of the incandescent lighting system was also a response to a need. According to Gilfillan, invention is "a complex cumulation." Electric generators, transmission, and lamps were being improved. Edison had the proper environment in which to pursue the development of a whole system. His Menlo Park laboratory supplied the needed equipment, staff, and funds. Like Sperry, Edison studied previous work and identified many weak points. He set up economic as well as technological parameters so as to develop a system to compete with gas lighting. His act of insight was a response to defined goals. Utilizing his understanding of Ohm's law and Joule's equation he got the idea for a filament that minimized the current to be carried by the conductors, which reduced the size of expensive copper conductors and energy loss. The system saw first use after revisions for efficiency and practicality.

The third case history, of Midgley's gasoline antiknock additive, involves the interaction of the automobile and petroleum industries — two areas of rapid technological growth in the early 1900s. Efforts to increase the cylinder compression ration were hampered by knocking of the fuel. A reverse salient thus arose from the "interface" of two technologies. Midgley, a staff member of an independent R & D firm, first attempted to define the knock. A search of existing literature led to his hunch that the problem stemmed from fast build-up of heat in the fuel, which caused an explosion to strike the cylinder walls. Finding an additive to make the fuel more volatile was a long process entailing prior knowledge, accident, and trial and error. The search was finally narrowed through methodical testing of chemicals similar in structure to a solvent that reduced knock. Tetraethyl was found to be the most effective additive.

Similarities in these case histories help reveal important components of the innovation process. Since inventors mostly solve existing problems, it is important to see how problems emerge. Change generates incomplete patterns, reverse salients, and imbalances. The author states, "Each interpretation and each case reinforces the conclusion that complex change within even more complex change is the process under way when innovation is occuring." An inventor identifies a weakness in a system and seeks to correct the problem. Among factors affecting an inventor's project selection are his education, experience and talent, anticipation of funding and a market, and availability of R & D facilities. Seeing what other researchers have done allows the inventor to see the focus of prior work and find points that need attention. The generation of a solution is a creative act, difficult to analyze. Idea response is a process of small steps. The cases examined show that analogies are employed, similarities in superficially different problems are seen, and earlier solutions aid in finding new ones.

7. TECHNOLOGICAL INNOVATION AND NATURAL RESOURCES: THE NIGGARDLINESS OF NATURE RECONSIDERED (Nathan Rosenberg)

The purpose here is to examine the relationships between patterns of resource scarcity and the innovative process. The supply of natural resources has long been a concern to economists. Malthus envisioned a bleak future as exponential population growth taxed the capacities of a fixed supply of land for food production. Fixed resources, population growth, and diminishing returns became the foundation of classical economics. Although focus shifted from arable land to other resources needed for an industrializing economy in the late 19th Century, economists continued to be preooccupied with finite supplies. The conservationist view is a modern form of an old theme. With growing concern over underdeveloped countries, the problem of population growth again receives attention.

However, Malthus-like predictions have proved wrong. Classical models fail to recognize the dynamic nature of supply and the capacity for substitution which arise from technological advance. The increasing productivity of the agricultural sector is the clearest example of the greater capacities imparted by innovation. Hayami and Ruttan (1971) have constructed an "induced" innovation model in which agricultural productivity is seen as a response to changing factor and product prices. Implicit here is the connection among all sectors of the economy; advances in the industrial sector offer new potentials to agriculture. This model is an important attempt to relate technological change and adaption to shifting relative resource scarcities.

From the classical approach one would predict rising unit cost of extractive projects (agriculture, minerals, forestry, and fishing), but that is not the case. Technological advance has generated substitutions for scarce resources and diminished the economic importance of raw materials.

A redefinition of the economic meaning of the natural environment is needed. Economic changes and growing technological knowledge continually alter the usefulness of various resources. Sharply rising prices trigger substitution and technological adaption. These are mechanisms of an industrial society that sustain both population and income growth. Within this framework, asking how long a specific resource will last ceases to be an important question.

Environmental pollution has joined resource scarcity as a possible limit to growth. The author states that both concerns "fail fundamentally to take account of the human capacity to adapt and to modify technology in response to changing social and economic needs." He goes on to argue that the price mechanism signals the need for redirecting human skill and facilities in response to both direct and indirect (social) costs. A "neo-Malthusian" approach ignores the evidence of the past 150 years that "human behavior undergoes continuous modification and adjustment as a response to changing patterns of opportunities and constraints thrown up by an industrializing society." Inquiry should extend beyond limited supplies. Questions regarding human behavior, social adaptation, and the nature and the relevance of historical experience for today's world are among the many that should be investigated.

8. THE ECONOMICS OF INDUSTRIAL INNOVATION: MAJOR QUESTIONS, STATE OF THE ART, AND NEEDED RESEARCH (Edwin Mansfield)

Problem areas of industrial innovation interest economists. They concern the nature of the innovation process, the appropriateness of the existing rate of innovation, the determinants of this rate and its relationship to industrial organization, and technological forecasting. Mansfield evaluates existing economic research in these areas, focusing on the individual firm, and suggests what kind of research is now needed.

Various models describe the phases involved in the *innovation process;* the relative cost (and time) of each phase vary among inventions. Market and/or production need seem to be a more frequent inducement for innovation than technological opportunities. Although in many industries formal R & D is responsible for a minority of innovations, a larger percentage of major important innovations seems to emanate from it. Profitability is important in shaping public and private policy. In the early and middle 1960s the rate of return from investment in technological innovation appeared to be high. Current estimates of the profitability of such investment is one of the three key areas in need of further study. An analysis of the sources of innovation and the characteristics of innovators is also needed. Finally, success of initiated R & D projects in various firms requires investigation.

What is the proper *rate of innovation* has received little attention. Various levels within a firm's organization (top management, production, marketing, R & D) have different vested interests in change. Top management (which has the greatest control) often feels that the rate is too slow. A main barrier to faster implementation results from poor exploitation of R & D output by the marketing and production departments. The rate of innovation is also important to society as a whole. Does the market economy facilitate enough R & D? Should more attention be paid to the difference between the private and social costs and benefits of new technology?

The optimal rate of innovation is a political as well as economic question. Although the problem is complex, four types of studies would be of great use: (1) We should attempt to estimate better the social return of various innovative activities. (2) We must examine how firms trade off various kinds of risk (e.g., probability of technical or commercial completion and economic success). (3) Studies concerning the optimal extent and kind of international specialization are needed because capabilities and R & D resources in most countries are limited. (4) We must develop better measures of the rate of innovation.

As to the *determinants of the rate of innovation and its relationship to organization,* they depend on factors that influence profitability of technological change (demand, inputs, proportions, availability, etc.). Among the important factors operating at the firm or industry level: the way R & D is coupled with marketing and production, the types of analytical techniques used to manage innovation, the size and sophistication of the relevant market, and the availability of risk capital. More information is needed by policy makers about these factors. The effect of organization, strategy, and personnel changes on innovative performance, and information regarding venture capital deserve focused study. Firm size and market structure also affect the rate of innovation. Although the traditional idea that efficiency is maximized by competition is disputed by Schumpeter and Galbraith, studies have shown that small firms are important for the initial phases of R & D. Evidence indicates that bigger firms get less innovative output per dollar of R & D than smaller ones. Some industry concentration promotes more rapid innovation but high concentration (a small number of independent sources) does not. Moreover, entry of new firms and a mixture of firm sizes both increase innovation; in highly fragmented industries technological change is hampered. Information about economies of scale, the cost of innovation in various industries, and recent major innovations (and their innovators) in a broad range of industries is needed.

As to *technological forecasting,* studies measuring the track record of various techniques would aid in assessing which method is best under particular circumstances. It seems likely that a clearer understanding of the process of technological change must precede improved forecasting techniques.

9. TECHNOLOGICAL INNOVATION IN FIRMS AND INDUSTRIES: AN ASSESSMENT OF THE STATE OF THE ART (R. S. Rosenbloom)

The concept of strategy may offer a framework that would help explain how organizations and environmental elements influence the rate and character of technological change. These factors "can meet the test of maleability, since the elements of the organizational context are within the control of corporate management, whereas the external factors are influenced both by corporate behavior and public policy." The chapter focuses on issues that pertain to policy makers and administrators.

Studies of particular innovations have led to a basic definition of the innovation process and have identified its elements and their interrelated nature. Although central tendencies have been found and crucial factors identified, no adequate theory has evolved to explain their relative importance of *why* variations occur in innovation among firms and industries.

Investigation at the firm and industry level is helpful in constructing theoretical frameworks. Two tendencies prevail. Economists, examining the implications of firm size and industry market structure, focus on the external environment and address problems of public policy. Behavioral scientists focus on the "inner environment" of the firm that concerns managers. They are interested in the relationship between technology and the structure of organizational systems. A firm's tasks determine the organization best suited to it. The technologies the firm uses may also aid in defining the most effective organizational form; moreover, certain structures may be more or less conducive to technological change. Studies exploring the relationship of organization to environment stress the importance of environmental diversity and uncertainty, structural formality, and managerial orientation. Greater differentiation of units within an organization and a high degree of integration across units are characteristic of firms with good performance.

A synthesis between the approaches of economists and behavioral scientists is needed. The firm constantly interacts with its environment. External forces such as competitor behavior, consumer demand, and public policy help shape the environment; whereas organization structure and the kind of technology used affect internal operation. Top managers can influence both internal and external factors. For example, they can bring the firm into new markets, pursue a financial policy that exposes the firm to uncertainty, stimulate sales through marketing policy, and influence consumers.

Kenneth Andrews (1971) summarizes the main ideas of strategy as "the pattern of major objectives, purposes, or goals, and essential policies and plans for achieving those goals, stated in such a way as to define what business the company is in or is to be in and the kind of company it is or is to be." The Harvard Business School has developed a framework for the study of corporate strategy based on the interaction of the administrative processes of strategy formulation and implementation. What formulates strategy is top management's commitment to set goals, policies, and programs that link the organization's capabilities to perceived opportunities in its environment. The implementation process creates the conditions needed for carrying out a chosen strategy, and in doing so alters the capabilities of the organization. The implications of a particular strategy extend beyond the firm's internal operation to the development of technology in society. The replacement of the American steam locomotive by the General Motors Co.'s diesel is a case in point. GM was willing to diversify and utilize manufacturing, marketing, and service techniques that were successful for its automotive products. This new approach helped to transform rail locomotion to diesel by the late 1940s. Forces outside the innovation process that influenced its outcome (structural context), in the case of GM's diesel development, included its decentralized organization, existence of a central research laboratory, support for development projects, and the oligopolistic structure of the locomotive industry.

Technological change can be viewed as a dependent variable that results from the interaction of structural context and the innovative process. The implementation of strategy means turning abstractions into more concrete forms, which gives shape to organizational structure. Technology is thus not entirely outside the control of managers.

10. REVIEW AND ASSESSMENT OF THE METHODOLOGY USED TO STUDY THE BEHAVIORAL ASPECTS OF THE INNOVATION PROCESS (C. F. Douds and A. H. Rubenstein)

Studies of the innovation process should lead to the development of theories that will assist managers and policy makers in their decisions and actions. However, a lack of well-developed and consistently applied methodology for such studies hinders the development of adequate theories. In this chapter, several types of methodologies are described and representative examples are appraised to further the development of research strategies. Much of the literature on innovation reflects three basic methodologies (discursive, case studies, field studies); and a fourth type is of particular importance (field experiments).

The *discursive* approach creates awareness of problems and stimulates action by way of exposition and rhetoric. This approach is not organized for verification, but may point to areas where more scientific methods should be applied. *Case studies* may provide information useful for selecting among various options and propositions for testing. A problem with interpreting case studies is that their nature varies widely. It is also difficult to know how representative any given case is. Cases tend to be selected because they are unusual. However, parallel cases with different outcomes can provide valuable insights. The *field-study* approach refers to research structured to permit generalizations beyond the immediate area examined. Inferences are drawn from an observed sample, key variables are identified, and data are collected and analyzed. Many specialized techniques for "instrumentation," "measurement," and analysis have been developed.

A wide range of methodologies is used in field studies, from simple tabular surveys summarizing a quantity of data for the reader's own interpretation to complex designs aiming at definitively establishing cause-and-effect relationships. Most studies fall between these extremes. Well-designed studies provide a clear statement of their objective — often to test one or more explicit propositions — and a design that minimizes alternative explanations or extraneous factors. Most field-study methodologies inherently cannot establish cause-and-effect relationships. However, among some types of variables such relationships can be established with "longitudinal" studies, which extend through time and make repeated measurements on the focal units (people, ideas, projects, etc.).

Field experiments are appropriate when cause-and-effect relationships are sought and there are clear alternative explanations for what is observed. These experiments can be natural events that are observed or deliberately induced changes (administrative experiments). Independent variables are manipulated to illuminate their relationship to dependent variables. Considerable care must be taken to establish the validity of the results.

Research investigations must be initially well constructed as well as properly assessed. For example, reliability and "generalizability" must be considered in designing a survey. In a field study, utility and credibility must be taken into account. Considerations underlying field research include formulation of explicit questions posed, clear definition of variables and what is being measured, reliability and validity of measures, and the basis for confidence in the results. These considerations also apply to longitudinal studies and field experiments.

11. ASSESSING THE KNOWLEDGE OF INNOVATIONS IN NEGLECTED SECTORS: THE CASE OF RESIDENTIAL CONSTRUCTION (W. P. Strassmann)

Assessing technological change in neglected sectors such as public services and health care is difficult. Not only is their output qualitative rather than quantitative, but these fields are poorly understood in all respects. Residential construction is a "halfway house," providing both physical objects and a flow of services. Barriers among materials producers, designers, builders, and owners are responsible for the inadequate amount and direction of innovation in this sector.

During the Industrial Revolution theories and tests improved types of construction after they were already in use. Utilization of new materials such as iron and reinforced concrete also began with practice and was followed by theory. As the production processes changed, housing technology was also transformed. For example, efficient mechanical saws reduced the cost of thin boards and resulted in the invention of balloon framing. Among factors that restricted experimentation were the separation of building from the fabrication of building materials, the insulation of contributing occupations (leading into increasing specialization), and restrictive desire to curb unsafe building and fraud. More complex building techniques required specialization. Although buildings became safer and cheaper to produce, a rigid system of organization dampened new innovation.

The breakdown of this system is now under way. Government-sponsored building research has pioneered new techniques, materials, and processes. Today a major impediment in overhauling construction methods is the failure to consider "externalities" such as handy transportation facilities, public utilities, and individual needs. "Technological change in housing must raise variety and quality, as sought by the occupants, and not merely lower cost, as defined by engineers." A recent study assessed the role that large American corporations might play in providing housing through mass production and found that institutional constraints limited technological response to the housing problem. Since housing is a stock, durability and maintenance are important areas for R & D. However, innovations must take into account who the occupants will be and where they are.

12. DIFFUSION OF INNOVATIONS PERSPECTIVES ON NATIONAL R & D ASSESSMENT: COMMUNICATIONS AND INNOVATION IN ORGANIZATIONS (E. M. Rogers with J. D. Eveland)

Organizations are an important part of modern society. Understanding the innovation process within them is crucial to understanding technological advance. A new "micro" approach to the study of innovation within organizations is proposed. The classical diffusion model is of limited value becasue of its implicit assumptions. It identifies the principal characteristics of innovation — relative advantage, compatibility, complexity, trialability, and observability — perceptual qualities that are easier to assess in an individual than in an organization. Adopter categories, which classify members of a social system, are based on the relative time at which adoption occurs; but they can be only applied to successful innovation. A measure for judging the value of proposed innovation is needed before a general theory is developed that can deal with failures as well as successes. The decision to adopt an innovation may be an individual's independent choice, a collective choice, or an authoritative decision. Problems arise in transferring the diffusion model from the individual level to the level of the organization. The adoption decision and its implementation may be two different subprocesses. Among the questionable assumptions of the classical diffusion model is the desirability of adoption. We must focus on how decisions are made, with what criteria, and with what consequences. The classical model considers cross-sectional data (comparative information gathered at a given time) as sufficient for evaluating organizational innovativeness, an assumption that ignores the continual change within organizations. Another assumption, that decisions are made by individuals, has led to studies of diffusion among organizations but not within them. We need a research style to facilitate a merger of organizational theory with the diffusion model. Large-survey studies on organizational innovativeness have identified variables but have not revealed how diffusion and innovation occur.

Organizational studies have suffered because of the type of data gathered. Usually only one executive is queried as a respondent per organization. The use of retrospective data leads to a correlation of "yesterday's innovativeness with today's independent variables;" the past is reinterpreted in terms of the present. Incorporating the sources of published and agency records, personal interviews, and observations and inspection in the same study would provide needed information about the process of innovation in an organization.

The relative influence of the personal characteristics of executives and the structural characteristics of the organization are yet to be fully explored. Multivariate statistical methods should be used to evaluate their complementary relationship. Size is most commonly indicated as a factor in innovativeness; it is easily measured and is probably an indicator of other factors that promote innovation, such as total resources and organizational structure.

The authors suggest that it should be used as a control variable only, and not a catch-all for intermediate variables that are harder to measure.

Consideration of structural variables by researchers is a recent and healthy trend. The authors classify these independent variables as individual (leader), internal-structural, or external-structural characteristics. All contribute to organizational innovativeness. The ratio of innovations brought in from outside an organization to those created within it is an important area for study.

Finally, the delineation of the innovation process into five stages requires interpretation and modification. For example, the need-identification phase in turn has several stages. We need research on the innovation process through observation and interviews with people at various hierarchical levels of a small number of organizations. Research should help organizations be more responsive to their needs and to appropriate change. The innovation process is only a part of a larger range of organizational activity. An approach that recognize organizational effectiveness as "the ultimate dependent variable" of innovation should be undertaken.

13. TECHNOLOGY FORECASTING LITERATURE: EMERGENCE AND IMPACT ON TECHNOLOGICAL INNOVATION (J. R. Bright)

The importance of technology forecasting (TF) has grown with the increasing impact of technology on our environment. An increasingly rapid succession of improvements requires huge commitment of resources. Growing concern over the environment has prompted government to require impact studies of its major technological programs and has given rise to technology assessment. But assessment requires anticipation. Traditional forecasting, which relies on expert opinion, is no longer adequate with the growing complexity and interaction of technologies.

Technology forecasting is a qualified, logical prediction. Its objective nature and purpose distinguishes it from opinion and prophecy. The possibility of forecasting is supported by several rationales. General consistency or pattern in technological change provides a basis for trend extrapolation from past data. By choosing appropriate parameters one can extend past time series. Such projections can predict future levels of performance and capability. Technology is a response to needs and opportunities. Normative (goal-oriented) forecasting has arisen from growing willingness to provide public support for technology leading to desirable social goals. Careful consideration of future needs reveals the kind of technology that is desired and will probably be developed. Finally, by monitoring the process of innovation signals can be identified. Since innovation is a long process, most technological innovations are visible before they are widely available or applied.

To avoid confusion and undue criticism from illogical expectations, the limits of technological forecasting must be made clear. It provides information about technology — not management options, the future economic climate, or prophecies about future "breakthroughs." The demand for TF education has grown in part as a result of the techno-economic climate of the 1970s. Traumas like the energy crisis, environmental problems, and trouble with major products such as the Concorde, as well as the success of firms like Xerox and IBM, have increased institutional interest in TF as an aid in planning and management of technological projects. A TF Program forces explicit consideration of the technological future and leads to a better understanding of the forces that affect the technology in one's field.

Statements about the future (prophecy, speculation, science fiction) are abundant and have at times been very influential. The most common method of TF has been the opinion of experts, which is however colored by individual values, knowledge, and assumptions. (The author tabulates the "significant aspects" of TF literature under four rubrics: level of impact or emergence predicted, objective of the predictor, qualifications of the predictor relative to the prediction, and the rationale or methodology used.) Reports of government committees evaluating technological proposals and possibilities embody elements of TF and often influence decision makers. Gilfillan was the first serious student of methods of technical prediction, proposing in 1911 the study of successful predictors through learning their methodology. The first attempt at formal technological forecasting was the U.S. government's report *Technological Trends and National Policy* (1937), which was concerned "not so much with technology as with planning directed at the social effects of using that technology." In that report, Gilfillan provided the nucleus of the concept of systematic technological forecasting recognizing the need to build a database, the use of extrapolation, the principle of monitoring, and the idea of normative forecasting. The report's recommendations were ignored. Not until 1945 was an organizational effort to forecast technology to aid management decisions undertaken, by the U.S. Army Air Corps, which began systematically to examine "the technological future and relate possibilities and realities to the goals of a major organization." In 1967 the U.S. Army began to integrate forecasting and planning into its R & D plans.

An influential survey of the state of the art was prepared by Erich Jantsch for the OECD, *Technological Forecasting in Perspective* (1966). Jantsch found that (1) TF had emerged as a management discipline around 1960 and had shifted from the "informed information" approach to systematic evaluation; (2) "the value of technological forecasting has been proved by its accuracy and usefulness in defining long-range strategies;" and (3) TF was still an "art" with evolving techniques and was undergoing accelerated development.

Although universities have done little to teach or develop technology forecasting, R & D managers in industry and government are "deeply interested and are strong in their support and applications of TF techniques to their problems." The author believes that the National Science Foundation could become "an important element for improving technological innovation" through TF research by "sharpening understanding, clarifying issues and alternatives and by providing more lead time." An appendix reviews TF methods (intuitive forecasting, trend extrapolation, normative forecasting, monitoring, simulation modeling, and cross-impact analysis) and makes recommendations for future research.

14. TECHNOLOGICAL INNOVATION AND ECONOMIC GROWTH (Simon Kuznets)

The author has summarized his findings in nine points, given below largely in his own words.

(1) Technological innovation played a key role in the rise of product and productivity in modern economic growth; and also induced major transformations of conditions of work and life.

(2) These transformations were required to channel new technology effectively, and meant organizational changes in the earlier institutions that governed production. The resulting changes in conditions of work for the active participants were a major element in changing conditions of life. Thus, technological innovations required innovations in social structure and even in prevailing attitudes. They also required adjustment to resulting displacement of resources in earlier, and obsolete, uses.

(3) A major technological innovation involves a sequence of phases stretching over a long life cycle. Of the four phases suggested — preconception, initial application (IA), diffusion (D), and slowdown and obsolescence (SO) — the IA and D phases alone account, on the average, for half a century. The phase sequence, the differences in input-output relations in the several phases, and the phase differences in the interplays between the technological, economic, and social adjustments bear clearly on analysis, policy consideration, and prognosis.

(4) The clustering of even major technological innovations into groups of related changes (stemming from exploitation of one source of power, or from a new industrial material, or from the interrelation of functions within a production process), combined with the interplay between innovations and the social and institutional adjustments to them, lengthens the sequence of distinctive phases and adds to their complexity.

(5) The focus of technological innovations shifts over time from one sector of the economy to another, and creates new sectors. Their immediate impact is always unequal among sectors, and hence among social groups in the economy. This inequality of impact is itself a social and economic problem that requires adjustments.

(6) Because of the combination of conventional economic inputs with required changes in conditions of work and life, and because of the combination of conventional economic outputs with possible nonconventional byproducts of technologically induced economic growth, an adequate quantitative gauge of the *net* contribution of technological innovations to economic growth is still to be established. Current measures of total factor productivity, though possibly reflecting largely technological changes, are either limited to conventional input and output, or involve a variety of *ad hoc,* and not fully tested, assumptions as bases for inclusion of nonconventional inputs or byproducts.

(7) Such a net measure may not be of much value, since a variety of elements, in both inputs and outputs, is needed to give meaning to the comparison. Yet the search for such a net measure helps to focus the analysis. Provisionally, one may justifiably argue that the social valuation of technologically facilitated, modern, economic growth is high and positive, with the critical reaction reflecting responses to temporary lags in adjustment.

(8) Technologically induced economic growth, having been attained, stimulates further technological innovation. A particularly important stimulus is the learning that takes place through mass application of recent new technology and yields new data, new tools, new insights and puzzles to natural science, and helps widen the base provided by the latter to further technological breakthroughs and innovations.

(9) Economic growth hastens the maturity of the older fields by slowing down growth of final demand for their products. It may also limit the conditions for responsive innovative entrepreneurship in the established and modernized fields because of the large scale of the firm, and possible dominance of a few in an oligopolistic or monopolistic situation. Furthermore, the rise in the share of the public sector is a factor, since in its nonmilitary areas it may not be easily responsive to technological innovation. The slowing down of the older sectors, once modernized through technological innovation, helps to shift the focus of innovation to other sectors; such shifts help to maintain a high or increasing pace of technological innovation and of economic growth.

The economic consequences of technological innovation has been a neglected area. Economists have approched problems of economic growth with a short-term orientation in which "technology, institutions, and consumer tastes were all supposed to be given." A lack of data regarding the effects and origins of technological innovation is a consequence of the orientation. Effective policy action requires measurement and analysis of the quantitative aspects of technological impact. Work relating new quantitative data with established measures would be particularly useful. For example, distinguishing aspects of total output, labor force, and capital associated with recent and older technology would help relate technological innovation and economic growth.

PART I: THE ECOLOGY OF INNOVATION

1. THE ECOLOGY OF INNOVATION

I. POPULAR VIEWS OF THE INNOVATION PROCESS

Popular mythology presents only simplistic notions of the innovation process. Yet even this simplistic folklore manifests the complex nature of the process. For instance, we get a half-truth, albeit a very important one, from the comic strips. There the inventor--a somewhat eccentric fellow (after all, he walks around with a symbolic electric bulb suspended over his head)--receives a flash of inspiration (the electric bulb lights up), and lo and behold, an invention has been born. Such cartoons are true, because they depict the importance of individual imagination and ingenuity in innovation; but only in part, because inventive insight is only one in a series of developments necessary for a successful innovation.

"Necessity is the mother of invention" is another half-truth. As we shall see, demand is a strong motive force; but as a full explanation it fails because many needs have not yet given rise to inventions, and many innovations arose from other causes.[1] Besides, in many cases inventions require additional inventions in order to make the original one effective. Thus one could perhaps turn the adage around: "Invention is the mother of necessity." Yet there is also an important truth in the "necessity" explanation, for it forces us to consider the social needs and human wants that help formulate the problems toward which inventors direct their attention.

Another old saw that stresses the same point is the saying attributed to Ralph Waldo Emerson, "If a man . . . make a better mouse-trap than his neighbour, tho' he build his house in the woods, the world will make a beaten path to his door." Although there are many "better mousetraps" that never become successful innovations, Emerson's reputed statement is important because it focuses attention on the need to link together social needs with inventive activity.

The "captains of industry" are also often extolled in popular folklore as the moving force in the innovation process--linking social needs with technical responses in bold and imaginative syntheses. And, indeed, without the entrepreneurial contribution of organizational knowhow, capital, production and marketing capabilities, etc., an invention is stillborn. The well-known "captains of industry," however, have largely been replaced today by faceless organizations; the entrepreneur has become an entrepreneurial function. Further, an idea--an invention--must exist before the entrepreneur can develop it into an innovation. A half-truth "writ large" in folklore is still a half-truth.

Another notion popular since the days of Sir Francis Bacon's *New Atlantis* is that technological innovations derive from scientific discoveries. If this generalization can be said to have any validity, it would hold only for relatively modern times--and even then, as we shall see, the situation is much more complicated than a simple causal relationship between science and technology. Nevertheless, this concept exerts great power--underlying, for instance, the argument for government support of basic research on the grounds that such research ultimately achieves utility.

Still another popular idea is that innovation can come about on command. This idea is based on the myth of technological virtuosity and the "talent and money" syndrome. Nurtured on a diet of "success" stories, the public retains great faith in the ability of technology to meet every challenge put to it. Although some recent scoffers might doubt that technology always triumphs, and although some humanists might decry that triumph (they plaintively, but properly, ask to what human ends and purposes), the general public seems, on balance, to regard technology as beneficent, or at least is not willing to forego the advantages it provides (Metlay 1971, 1972; Taviss 1972). To that is added the belief that "throwing money at problems" will induce technological innovation by the simple expedient of investing talent and money. Crash programs, such as the atomic-bomb and synthetic-rubber projects during World War II and the space exploration program, provide historical examples.

This idea of innovation as induced through the application of talent and money has come to the fore during the current "energy crisis." "Project Independence," which, in its original form, postulated an investment in technological development to make the United States independent of outside energy sources by 1980, was based on a combination of several myths: necessity as the mother of invention, the invincibility of technology, and the efficacy of government financial support.

Although it can rightly be argued that such problems will not be solved without an investment in talent and money, and that technology has in the past responded on command with some spectacular success, the simplistic cause-effect relationship implicit in these notions wilts before the complexities they fail to address. Which technical approach (or combination of approaches) should the money be thrown at? What kinds of talents are needed and at which points in the innovation process? Will just any organizational structure do? What social environmental, political, and human price will there be? Are all the second- and later-order consequences also desirable and, if not, are the undesirable ones acceptable? What, in fact, is the "track record" of "command innovation"; that is, have the failures as well as the successes been recorded? In brief, these apparently simple notions turn out to hide a maze of complex questions.

Yet such popular views of innovation turn out to be extraordinarily influential because there seems to be something inherently attractive about simple explanations for complex phenomena. Even scholars engaged in innovation research are not entirely immune to this attraction. To serve as a corrective for the half-truths that might emerge from such one-sided approaches, we must endeavor to investigate the innovation process in all its ramifications. We are thus brought squarely to face the question: What do we know about innovation?

II. DEFINITIONS, DISTINCTIONS, AND USAGES

Researchers concerned with the process of technological innovation belong to a wide variety of disciplines and typically view the process from these diverse perspectives. As a result there is little uniformity in their definitions and conceptual distinctions regarding the innovation process. For example, some economists reserve the word "innovation" for the application of an "invention," that is, its entrance on the marketplace. Thus the economist Joseph Schumpeter (1939) regarded invention, the discovery of a new tool or technique, as the initial event; and innovation, the implementation of the new tool or technique, as the final event. Mansfield (1968a, p. 83) also defines an innovation as the first application of an invention. To the U. S. Patent Office, the interest is in the initial event--when rights are assigned to an individual--and there is no concern with the eventual application. As we shall see, cultural anthropologists, management scientists, and other specialists have different notions of what constitutes innovation and what are the various elements comprising it.

Scholars also differ in their use of the terms "invention" and "discovery." According to Forbes (1958, p. 5), this difference is tied to a distinction between science and technology: scientific discovery usually recognizes or observes for the first time some natural object or phenomenon, whereas an invention is the creation of something technologically new that had not existed before. For example, man "discovered" fire, but had to "invent" means to start fire and use it for lighting and heating. But such a distinction can be misleading, for sometimes what we call inventions might also be classified as discoveries, such as Perkin's "discovery" of aniline dye--which was at the same time the "invention" of synthetic dyes.

In order to accommodate this wide range of specialized concerns, we have interpreted the term "process of technological innovation" very broadly--as embracing the full range of activities from the initial problem definition and idea generation through research and development, engineering, production, and diffusion of new technical devices, processes, and products. It is tempting to identify these elements as phases in a linear sequential process. But, as we shall see later, this proves to be a speciously attractive but unworkable view that simply does not fit the facts.

Implicit in both the popular myths and the scholarly literature we studied are three major points concerning the innovation process: (1) it is a response to either a need or an opportunity; that is, it is context dependent; (2) it depends on creative effort and, if successful, results in the introduction of novelty; and (3) it brings about or induces the need for further change.

Although most innovations are need induced, such needs do not necessarily express themselves specifically. For example, there might be a general need for quicker transportation, but that does not necessarily dictate that this need be met by supersonic planes. Furthermore, sometimes wants do not make themselves felt until after the innovation appears. For example, there is a general desire to make oneself attractive to members of the opposite sex, and the makers of underarm deodorants try to persuade consumers that their product fills this general want and does it better than alternative means, such as perfumes or soap and water. The "social innovators" of Madison Avenue have proved especially skillful in turning generalized human desires into demands for specific products. In any case, an innovation promises to be in some way "better" than existing means to the same end, perhaps by offering better quality at the same cost, the same quality at lower cost, faster production, etc.

In other cases the stimulus event for an innovative effort may be a newly developed technical capability for which applications are then sought. The development of the transistor, for instance, with the potential for miniaturization it represented, led to a whole "family" of innovations. Some data suggest that major advances are more likely to arise from a new technical capability, whereas smaller, incremental improvements are more likely to be need induced. In either case the innovation process is environmentally dependent from its very outset: that is, it is a response.

It should be noted that innovation is not the only response to felt needs or wants. Another response (a "nonresponse") is simply to ignore the problem, to "make do"; we tend to seek a way--often technical (the "technological fix")--to solve a problem or fulfill a want. Some traditional Eastern religions and cultural systems, however, deny or dismiss such problems or needs rather than resolve them. A possible explanation is that those cultures view reality as primarily spiritual in nature and hence deny the importance or necessity of changing the material conditions of life.

Another type of response is not to innovate but simply to apply more of the technology that is already employed. A typical example in 20th Century America has been to build more freeways to resolve the problem of already clogged freeways.

Even when the response to a perceived need or want is to innovate, the innovation may not be a technological one. There are, for example, social innovations--"new methods of inducing human beings to compete and cooperate in the social process" (Kuznets 1962, p. 19)--such as the Peace Corps or advertising; these innovations are primarily social in nature and only incidentally technological. There are also economic innovations, which do not necessarily have technology as their major component, such as the introduction of double-entry bookkeeping in Renaissance Italy; or marketing innovations, such as the supermarket or installment buying. Political innovations would include new types of bureaucratic organization or new political systems, such as the corporate state and "people's democracies." Our investigation, however, is restricted to technological innovations, and will consider other forms only as this focus requires.

The elements of creativity and novelty also require comment. Whether the initial stimulus is a need or a technical opportunity, creativity is required--both in the initial idea for a response that might work, and at many steps in developing that response into a successful innovation. A host of contextual variables can facilitate or inhibit creative responses, or guide them along certain lines. Especially important in this regard is the fact that those from whom new ideas for technical innovations are typically expected to come--those in the R & D laboratory--are usually engaged in projects that are already under way. It is not at all clear that the environments most conducive to each are identical.

An innovative effort, if successful, results in the introduction of novelty. The patent law has something to say about novelty, but that is for determining legal priority of invention, not for defining the nature of innovation. To the Patent Office, an invention must be new and useful, and it must represent something more than a trivial improvement; it usually means "a contribution over and above the exercise of mechanical skill," and in 1880 the U. S. Supreme Court used the expression "a flash of thought" to describe an essential attribute of invention, thereby stressing the role of creativity.

For our purposes, the notion of novelty requires further explication so as to avoid a semantic confusion that often appears in the literature. Something can be new only in relation to some frame of reference. In the innovation phase, the frame of reference is the current state of the art. In the diffusion phase, however, an innovation is new relative to the particular unit that is adopting it. It is this latter frame of reference to which Hodgen (1952, p. 45) refers when he speaks of a technical innovation "as having taken place when a tool, a device, a skill or a technique, however unknown or well known elsewhere, is adopted by an individual in a particular community and is regarded as new by the members of that community." Novelty, therefore, characterizes an innovation not only upon its first introduction into use anywhere, but also repeatedly as it is subsequently adopted for use in other contexts. The one refers to "new under the sun," the other to "new under this roof."

Finally, whereas the social impact of technological innovation, that is, induced change, is only tangential to the scope of our investigation--our focus being on the process itself--some thinkers single out a purposive element in the process, that of effecting physical-environmental and social change through innovation. Tannenbaum (1970) has stated, "Technological innovation is the novel application of physical knowledge and technique to make premeditated changes in the physical aspects of the environment," and Peter Drucker (1967a, 1967b) regards innovation as a conscious attempt to bring about, through technology, a change in the way man lives.

Although such statements implying the larger impact of innovation are correct, they neglect two major points in stressing its purposive nature. The first is that most innovations do not come about by such transcendental considerations of social change, but rather through more mundane calculations on the part of businessmen regarding profits, resource factors, costs, etc., or by technologists pursuing their own visions of technical efficiency, or by governments responding to various pressures, and the like. Second--and this is a point of major social concern today--innovations may bring about changes other than those envisaged by the innovators. Indeed, the unforeseen changes brought about by the large-scale application of such innovations as the automobile or DDT have given rise to the new art of Technology Assessment, which attempts to study the second- and higher-order effects of technological applications before they are applied (Hetman 1973).

Nevertheless, whether motivated by narrow or broad considerations, whether consisting of a new mix of elements or the insertion of something old into a new context, innovation brings about changes (in varying measure) in what things people make and do, how they make and do them, how they work and live--and ultimately in how they think and act. Is it any wonder that we seek to understand the innovative process, both in theory and practice, in order to help us comprehend the present and, indeed, in order to make the future?

III. CLASSICAL THEORIES OF INNOVATION

If we attempt to categorize the classical theories of innovation, we find that they polarize around two positions: deterministic and individualistic ("heroic," or "great man") theories (Rae 1967, p. 326). The determinist explanation holds that the innovation occurs when the conditions are "right," and it stresses the role of social and other forces, principally military and economic, in bringing about technological change. The "heroic" theory stresses the role of the individual and plays down the influence of external pressures.

These two major theses are not mutually exclusive; they are really matters of emphasis. No scholar has ever claimed that the individual innovator runs his own race entirely unbridled by any external pressures, and even the most ardent adherent of the deterministic school has recognized that the manifold and diverse forces operate through individuals. This accounts for the development of sophisticated, composite theories, embodying elements of both the deterministic and individualistic schools of thought.

A wide variety of exogenous elements--human wants, social needs, economic demands, military requirements, geographical and climatic constraints, and the like--can help determine the rate and nature of innovative activity, and these can express themselves in many different ways. For example, economic requirements might differ in relation to different resource factors, depending on labor costs, capital, fuel, materials, and the like. Such other external factors as geography and climate have been employed to explain why certain civilizations seem to be more innovative than others, and there are those who would ascribe technological creativity only to certain ethnic groups.[2]

The importance of exogenous sociocultural factors is evident from the fact that different societies seem to be especially fruitful in innovations at different times in history. For example, only by reference to sociocultural elements can we explain why China in the Middle Ages displayed technological genius which far outshone that of the contemporaneous medieval West. Similarly, we must rely on a sociocultural explanation of why China failed to undergo an Industrial Revolution, and why the West took the leadership in both science and technology from the 17th Century onward.[3]

Sociocultural factors are also held to be responsible for multiple inventions. Given the widespread diffusion of technological knowledge in the modern world, the similarity of technical problems, and the apparently universal potentialities of the human mind, it is not surprising that many inventors find the same or similar solutions to technological problems at about the same time.[4] Indeed, the great amount of patent litigation over priorities would be proof of that. Further, the number of patent applications turned down because they lack "novelty" would indicate how often the human mind arrives at solutions for technical problems that others have already thought of--and already patented.

Yet, despite the importance of exogenous social factors, we cannot do without the human element in innovation. Human beings define the problems, have the ideas, perform the creative acts of producing a device, do the research and development, and decide upon the application and diffusion of innovations.

Although the heroic theory of innovation does not deny the stimulus of economic needs and the influence of sociocultural conditions, it emphasizes the role of the individual hero in bringing about innovation. For example, scholars of the deterministic school would argue that 18th Century Britain was "ready" for the steam engine, in terms of both economic need and level of technology, so that if James Watt had not invented the steam engine, someone else would have. Proponents of the heroic school, however, would claim that the characteristics and personality of Watt were primarily responsible for the steam engine as it actually came into being.

In his *Lives of the Engineers* (1861-62), a three-volume collection of biographies, Samuel Smiles attributed the landmark inventions and great engineering feats of the Industrial Revolution to heroic individuals. Not surprisingly, he found these men possessed of the standard Victorian virtues of his own time: self-discipline, self-help, devotion to duty, integrity, and perseverance. But such an enumeration of traits, like later and more scientific studies of technical creativity, merely transfer to another level the argument of individualism versus social determinism. After all, it can be argued, these virtues of the heroic Victorian inventors derived from the sociocultural forces of the time--and we are back to the old "nature versus nurture" argument still waged by psychologists, educators, and sociologists on such new battlegrounds as the efficacy of school busing or the validity of IQ tests for ghetto children.

Both the individualistic and social deterministic theories of innovation may be introduced at yet another level, with reference to the entrepreneurial function, which was so greatly stressed by Schumpeter (1939, 1:85-86). It is the energetic entrepreneur who is willing to take risks, who amasses capital to finance the invention, who sees the idea through to actual production and introduction to the marketplace. He is the man who links together the social needs, frequently expressed in terms of profit potentialities, with the creative ideas of the inventor, thereby coupling the marketplace with the invention.

Sometimes the inventor and the entrepreneur are combined in the same individual, as in the case of Edison (Josephson 1959) or Elmer Sperry (Hughes 1971), but more often they are different persons. For example, James Watt possessed remarkable technical ability but lacked capital and business acumen; they were supplied by Matthew Boulton, who became the driving force for the successful introduction of Watt's steam engine. In contemporary business organizations, some of the entrepreneurial functions are carried on by the R & D manager, who brings together scientific knowledge, technical expertise, and knowledge of the marketplace and of economic constraints in an effort to produce profitable innovations (Morton 1971). Other entrepreneurial functions might be carried out by different people or by parts of a large modern corporation. Whether or not the inventor and the entrepreneur are the same person, different people, research teams, or corporations, it is the entrepreneurial *function* that marshals the resources--ideas, talent, technology, money--thus transforming the invention into a utilizable innovation.

In addition to stressing the importance of both individuals and the social environment in innovation, much is often made of the role of happenstance in history. In the case of technical innovations, the intervention of chance is called "serendipity," a term coined by Horace Walpole from the Persian fairy tale, the Three Princes of Serendip (Ceylon), whose heroes often unexpectedly discovered something valuable. Two famous historical incidents illustrate serendipity at work: Charles Goodyear's invention of the vulcanization process, and W. H. Perkin's discovery of aniline dye.

These and other cases of "accidental" discoveries or inventions are not quite so accidental as they might seem. On closer inspection they provide proof that innovation does not occur in haphazard fashion. In virtually every case of serendipitous invention, we find that the inventors were aware of the needs and problems, that they had already conducted persistent and careful searches for what they wanted, and that they were acute and perceptive enough to recognize when a happy accident gave them their answer. In other words, they could appreciate the significance of a chance occurrence and utilize it for practical purposes. Most innovative advances come as a cumulative result of answers to a series of closely directed questions; chance or accidental observations come as a bonus to the perceptive researcher who has already done his "homework." As Louis Pasteur observed, "Chance favors only the prepared minds."

It is apparent, then, that no monistic theory--social-deterministic, individualistic-heroic, or even blind chance--can serve by itself to explain the complexities of the innovative process. It is not begging the question but simply stating a fact to say that innovation, like virtually every other creative activity, derives from the interaction between individuals and the sociocultural environment.

In his pioneer study of the organization of cultures, the anthropologist Dixon (1928) postulated a triad of factors in the background of every cultural innovation: opportunity, need, and genius. Almost four decades later, a study of the conditions fostering successful R & D, conducted by the Arthur D. Little Company (1965) for the Department of Defense, found a similar triad behind innovatory weapon systems: a clearly understood need; relevant ideas, information, insight, and experience; and the men and money to push through the job. Here were Dixon's anthropological factors translated into the context of the modern R & D laboratory.

One of the most influential theories of innovation, comprehending both the individualistic and the sociological points of view, was first presented by Abbott Payson Usher in the 1920s. Usher's theory, drawn from *gestalt* psychology, regards innovation as a social process consisting of acts of insight of different degrees of importance and at many levels of perception and thought. This theory considers innovation as a four-step sequence: (1) the perception of a problem, meaning the recognition of a social need and of the problems involved in its fulfillment; (2) the setting of the stage, involving the existence of a body of technical knowledge and of technological and financial capabilities; (3) the act of insight by which the essential solution of the problem is found; and (4) critical revision, in which the newly perceived relations are thoroughly mastered and effectively worked into the entire context of which they are a part. Today, we would comprehend Usher's "critical revision" step in the "development" part of modern "research and development," and we also realize that Usher's four steps do not always occur in so neat a linear pattern.

Even so, Usher's theory is by no means complete. He was primarily concerned with the "act of insight" in the inventive phase, probably because of the emphasis *gestalt* psychology places on the "Eureka!" or "Aha!" phenomenon. Usher helps tell us how some inventions occur, but not how inventions are translated into true innovations. He neglects both the risky economic decisions involved in the application and diffusion of inventions, and the feedback among the various steps. Above all, he fails to provide a detailed analysis of his crucial fourth stage of critical revision.

More recently, investigators have concentrated on this developmental stage neglected by Usher and other early theorists. At the same time, the development during the past half-century of a science-based technology has enabled them to augment the earlier phases of Usher's sequence by reference to the contributions of science to the innovation process. As a result, terms such as "basic research" and "applied science" enter increasingly into the literature. Indeed, in the popular mind the model of innovation emerges as a direct line from basic scientific discovery, through applied research and development, to the innovation--and even some scholars view the innovative process in those terms. Yet, as we shall see (Chap. 2), the relations between science and technological innovation are too varying and complex to justify such a simplistic model.

Of major significance to our understanding of the innovative process has been the detailed analysis of the crucial developmental stage to which early models had merely given passing attention. In analyzing successful product innovations, Mansfield (1971, pp. 114-18) employed the following categories: applied research; preparation of project requirements and basic specifications for prototype or pilot-plant design, construction, and testing; production planning, tooling, construction, and installation of manufacturing facilities; manufacturing start-up; and marketing start-up. These categories were similar to those derived from their industrial experience by the members of the Charpie panel (1967): research--advanced development--basic invention; engineering and designing the project; tooling--manufacturing--engineering; manufacturing start-up; and marketing start-up. In these and similar studies, Usher's final stage has been extended and broken down into four or more separate categories of activities.

Such sophisticated and variegated investigations by later scholars have provided us with a finer-grained analysis and sharpened our insights into selected elements of the innovation process. Nevertheless, these studies, primarily because of their concentration on certain aspects, suffer from defects that seriously inhibit their ability to provide us with an overall model and complete understanding of the entire process of innovation.

IV. DEFICIENCIES IN INNOVATION STUDIES

It is tempting to compare the literature on innovation to Aesop's fable of the blind men examining the elephant: Each scholar concentrates on investigating but one aspect of the problem's elephantine proportions and believes that his limited knowledge describes the entire animal. But that analogy is overdrawn. For the most part the scholars who have written about innovation are not Aesopian blind men; almost without exception, they have recognized the major contours of the giant with which they are dealing. They choose to concentrate upon selected aspects because of their own research predispositions, which are in turn directed by their disciplinary interests.

A truer analogy would involve scholars working on a gigantic jigsaw puzzle depicting individual figures moving about in a scene involving other people, houses, and a larger landscape. All those fitting together this puzzle are aware of the general picture; but the puzzle is divided into many different-sized pieces. So some researchers are piecing together the lineaments of the individual human faces; others are more concerned with putting together the houses in which the people live and work; and still others are engaged in sketching in the background of the trees, hills, roads, sky, and clouds. To carry the metaphor further, the problem confronting a state-of-the-art study is to fit the individual pieces into the framework of a single picture, and in the process to determine what pieces of the puzzle might still be missing. But what makes the puzzle more difficult is that its contours are constantly changing over time and that changes in one element of the picture bring about changes in other elements, which in turn transform the original elements. One need not be the king of Siam to say, "'Tis a puzzlement."

Although some of the puzzle solvers are scholars of great breadth--for example, Joseph Schumpeter--it is inevitable that they view the innovation process through the lens of the specialized discipline in which they have been trained or that their emphasis corresponds to their particular enthusiasms. Usher's contact with the *gestalt* psychologists is one example; Schumpeter's emphasis on the role of the entrepreneur reflects his economic concerns. Sociologists, anthropologists, organization theorists, etc., all exhibit their own special predispositions in studying innovation.

The most active investigators of innovation have been economists. But economics is not a monolithic field, and all its practitioners do not focus their attention on the same problems. Some economists are concerned primarily with microeconomics--what goes on within an individual firm or a single industry--but the process of innovation might differ from firm to firm or from industry to industry. Others are macroeconomists, involved with the impact of innovation on the national or world economies. Still others specialize in different factor endowments--labor, raw materials, transportation, capital, and the like--and their relations to innovation. And there are those who are fascinated by the changing role of the marketplace in the motivation and acceptance of innovation.

Despite the diversity of approaches, economics alone cannot give a comprehensive picture of the innovation process. Economists such as Solow (1957, p. 312) and Myers (1965, p. 91) naturally think of an innovation in terms of its economic function and impact. Indeed, Simon Kuznets (Chap. 14) insists that unless an invention is regarded as economically useful, it will never become an innovation. To Kuznets, as to most other economists, economic elements are the major external and internal factors affecting innovative activity--although, as might be expected, the economists differ among themselves about the significance of the various economic parameters. Furthermore, even their excellent empirical studies on certain innovations in specific industries--machine tools, transportation, and agriculture--might be too limited to provide generalizations applicable to innovative activity in all fields.

When it comes to a subject like the diffusion of innovation, economists naturally focus on the economic variables, talking in terms of return on investment, labor or capital intensiveness, resource endowments, and the like. These are only part of the story, however (Kranzberg 1967a, pp. 30-31; Rosenberg 1970.) Anthropologists and sociologists treating the same topic focus on elements of sociocultural resistance to change and the interactions of different cultures, whereas geographers might concentrate upon spatial patterns of diffusion (Chap. 4).

The concerns of sociologists are as broad--and as restrictive--as those of the economists dealing with innovations. In his now-classic study of *The Sociology of Invention* (1935), Gilfillan lists "thirty-eight social principles of invention" under such rubrics as the nature of invention; changes evoking invention; the rate of growth and life cycle of an invention; factors fostering, retarding, and locating invention; principles of change; inventors and other classes; tendencies in the craft; and effects of invention. Interestingly enough, the same categories might serve equally for a book entitled *The Economics of Invention*; although an economist would deal with the economic parameters of the topics, Gilfillan and other sociologists perforce focus on the social elements.

Social psychologists may also be concerned with the impact of behavioral patterns on inventiveness within research and development laboratories, and the impact of interpersonal dynamics within the innovating organization.

Management scientists concentrate on the role of management in decision-making processes within the organization, especially as they relate to innovation. Organizational structure may also be studied in relation to the diffusion of innovations, as are the social networks of scientists and engineers (Ben-David 1971; Schon 1967; Marcson 1960; Dubin 1958; Burns and Stalker 1961; Etzioni 1961; Kornhauser 1962; Crane 1972).

Such narrow disciplinary approaches should not be denigrated. They partake of the specialization and reductionism that have led to major advances in the physical sciences, and they provide data and perceptive insights on many aspects of the innovation process. But there comes a time in the history of the sciences when a holistic approach is necessary. This necessity arises from the accumulation of vast amounts of data whose relationships are not yet clearly understood and--as we would urge in the case of innovation--cannot be understood apart from an integration of the specialized contributions of the several disciplines.

One criticism that is usually leveled at theoretical models in other fields of study--that they do not have an adequate database from real events--also seems to apply to theories and models of the innovation process. Although almost all the theoretical works investigated in our literature survey make reference to some empirical data, the theories and models differ--not only because their authors view the data from the vantage point of different specialities, but also because the vantage point determines, in part, what counts as data.

Another kind of selectivity--that is, reliance on data drawn from a single technical field--might skew the model, as was the case with Usher, whose theory is based on mechanical inventions. Hence the resultant model might be inapplicable to innovations occurring, perhaps, in chemical or electrical fields.

Not only might the database be restrictive in terms of the technological field being considered, but it can also be misleading if it rests upon limited historical experience. An example of misinterpretation derived from too narrow a historical perspective can be seen in the Charpie report (1967), whose implicit model of the innovative process--based on historical case studies, to be sure--elevated the importance of the individual inventor and recommended measures by the Federal government designed to improve and favor his lot (see Chap. 2). By looking upon the innovatory environment as static rather than dynamic, the Charpie panel neglected well-established, long-term changes in the pattern of innovation and hence overemphasized the significance of the independent inventor.

Another difficulty emerges in utilizing quantitative data to bolster theories of innovation. Because patent statistics are easily available, they have long been used, sometimes providing important insights, as in Jacob Schmookler's theory (1966) of the predominance of the "demand" factor in stimulating inventions. But, as we shall see later (Appendix), a host of questions arise about both the validity and the interpretation of patent data. Measuring the efficacy of R & D laboratories in the innovation process is similarly muddled, as are the numerous problems of measurement associated with the economic variables in the innovation process (Mansfield 1972b). Despite a plethora of statistics, we are not always certain that we are measuring the right thing nor that we are correlating the most meaningful variables.

Another major deficiency in studies of innovation lies in the elements that are not measured or considered at all. These elements include the failures, the innovations that did not "make it." Sometimes they never got beyond the idea stage, or the patenting stage, or the R & D phase, and were never applied. Sometimes they actually were applied, but never succeeded in achieving a measurable economic or technological impact. One reason for the paucity of evidence on unsuccessful innovations might be the reluctance of corporations to chronicle their failures.

Yet, if our understanding of the innovation process is to be at all complete, we must take into consideration failures as well as successes, as did Project Sappho (Achiadelis, Jervis, and Robertson 1971). In order to learn how various factors can add up to a successful innovation, we must also learn what factors brought about the failure of an innovative effort. Unfortunately, the preoccupation with success has obscured and limited our study by failing to provide us with sufficient empirical data regarding unsuccessful innovations.

Indeed, we are so bemused by some of the spectacular successes of recent times--such as the transistor, nylon, the atom bomb--that we tend to draw our lessons from them, even though they may not be typical of the innovation process as a whole. The transistor, for example, might represent a rare combination of scientific discoveries and technological advancements occurring in a particularly favorable environment. Bell Laboratories had ample money, equipment, and personnel to link together many different aspects of the innovation process; and its management had sufficient foresight and understanding to couple these needs together with scientific and technical knowledge and to define its problems properly (Weiner 1973). Not all innovations are conducted under such favorable circumstances, and perhaps in other fields the pattern of interaction among the several phases of innovation might be quite different.

Even the proliferation of case studies of successful inventions does not always add a great deal to our knowledge, for these studies tend to be noncumulative in nature. For example, the Illinois Institute of Technology Research Institute (1969) published its report on Technology in Retrospect and Critical Events in Science (TRACES) which represented a retrospective tracing of key scientific events that had led to five major technological innovations. The 1973 study on Science, Technology, and Innovation, prepared by Battelle Columbus Laboratories for the National Science Foundation, goes over almost the same ground as TRACES, merely adding more case studies. These case studies reinforce the argument for the importance of basic science in technological innovation, but they make few generalizations regarding the actual innovative process itself (although the implications of some of the Battelle studies are carried through in the chapters following).

Lack of empirical data about unsuccessful innovations and the noncumulative nature of case studies deprive us of information that might highlight the special conditions facilitating success as compared with those leading to failure. In a sense we are deprived of "control groups," which might help us prove or disprove the validity of generalizations or theoretical assumptions. The sponsors of Project Sappho have pointed out, "It is not possible to discriminate between alternative hypotheses [to explain successful innovation] unless failure in innovation is also taken into account," and they have attempted to do so in their studies. We need such studies of aborted and unsuccessful innovations in order to help us pinpoint specific factors contributing to success or failure; these studies might form the basis for testing generalizations about the innovative process in different fields of technology or at different levels of industry.

Even if the study of successes were sufficient to provide us with a valid view of the innovation process, without resort to in-depth studies of the failures, our view might still be distorted because most of our concepts would have arisen from consideration either of the great mechanical inventions dating from the Industrial Revolution or of today's science-based developments arising from physics and chemistry. In either case, we have concentrated on physical technology and, by and large, have neglected revolutionary innovations in the life sciences, including agriculture--which might represent different parameters altogether.

Indeed, as Strassmann points out (Chap. 11), entire sectors have been neglected in the literature on innovation. It is not because these sectors have been lacking in technological advance, but because change in them has been difficult to codify and to measure, and "therefore to analyze at an abstract level. They will be those sectors where advances are qualitative, indirect, and somewhat intangible, making it difficult to aggregate and even to compare one case with another." These sectors tend to be those with a large ingredient of social as well as technical innovation, and Strassmann complains that they are "poorly understood in their non-technological aspects as well."

Strassmann's complaint is justified, of course, but we are equally at a loss in explaining many innovations whose technical aspects are extremely well documented and about which a wealth of quantitative data exists. Furthermore, the ignorance of which Strassmann complains might not be due to the subject matter of the innovations themselves in those neglected sectors (he uses housing as his example), but rather to the diffuse nature of the industry he seeks to investigate. In other words, it might be the difficulty of gathering technical data for an industry that is typically small in scale and dispersed (rather than our inability to link together the qualitative with the quantitative aspects) which has prevented researchers from devoting full attention to such sectors.

Although we have mentioned several major deficiencies in extant studies of the innovation process--the dependence on the investigator's discipline, training, and special field of concern; the development of theories and models based on skewed selections or empirical data; the emphasis on successful innovations and the paucity of information about or consideration of unsuccessful innovations or of neglected sectors--we have not yet mentioned the gravest shortcoming of most of the traditional models: their reliance, implicit or explicit, on a linear-sequential analysis of the innovative process. This shortcoming probably derives from the deficiencies discussed above as well as from our propensity to organize our thoughts in logical (linear) categories and to view matters in historical (sequential) patterns. As we were to discover in the course of our investigations, a linear-sequential analysis does not correspond to what goes on in a complex and dynamic ecological system--a model that seems to provide the best means for describing the congeries of activities that make up innovation.

V. BREAKDOWN OF THE LINEAR-SEQUENTIAL APPROACH

When we first began our survey of the innovation literature, it seemed that the process could best be analyzed in terms of phases in a linear and unidirectional sequence. This assumption was derived from a synthesis of the work of Usher, Schumpeter, Machlup, Gilfillan, etc., and our own initial propensities. Such a scheme, we thought, would allow us to focus on the key decision and/or leverage points in the process. Hence, we distinguished five separate functional phases: (1) problem definition and idea generation, (2) invention (i.e., the prototype device or process), (3) research and development, (4) application (meaning first use), and (5) diffusion (introduction into a context other than original application). A simple block diagram would suffice, showing a linear and unidirectional process, going successively from phase one through phase five. The totality constituted the process of innovation.

We were disabused of this linear-sequential notion rather quickly, however. That is, as we moved more deeply into our study, its inadequacy as an analytical framework became more and more obvious. It was a poor representation of the complexities we found.

Nevertheless, the concept of process phases is a valuable and valid structuring device that we shall use--not as identifying discrete elements in a fixed sequence, but as indicating loosely delineated clusters of activities that may overlap temporally and organizationally and that exhibit complex interrelationships.

The most basic distinction is between the phases of the process that consist of the development of a technological device, process, or product--from the initial conception of its possibility to the point of its first introduction--and the subsequent actions that result in the spread of that innovation to other contexts. The former, referred to as the *innovation* phase, is the focus of Chapter 3; the latter or *diffusion* phase is treated in Chapter 4. It is somewhat awkward to use the same term, "innovation," both generically to refer to the whole process and also to refer to one of its phases or elements, but it is a common practice in the literature and one that we have chosen to follow.

The relationship between even these broad phases of innovation and diffusion is complex. Logically, innovation precedes diffusion; i.e., that which does not yet exist cannot be diffused. Logical priority is not the whole story, however. It is not always possible simply to adopt an innovation; often it must also be *adapted* to its new context of use. Such adaptation, although a form of diffusion, also involves the process of innovation, since modifications are required. Thus diffusion may precede and bring into being new innovations, as well as the other way around.

The innovation and diffusion phases themselves may be divided into subphases. Innovation, for instance, includes problem definition and idea generation, research and development, engineering, and production. There is not widespread agreement on these distinctions. Some authors, as we have seen, identify a separate "invention" phase. However, this term is fraught with difficulties; it requires a distinction between invention and discovery. For example, Charles Goodyear "discovered" that rubber when sprinkled with sulfur and heated would not melt and would retain its resiliency; but this accidental discovery is also regarded as the "invention" of the vulcanization process.

It is likely that such difficulties are semantic rather than substantive, but the more basic question is whether or not the identification of an invention phase, even if cleanly drawn and widely agreed on, is of much value in understanding the innovation process. We think not, for several reasons. First, for many people the term "invention" carries the connotation of an "event," when it is really part of a "process." Even if invention is disassociated from this "event" connotation and is viewed as a phase of a larger process, it is still not a very convenient distinction except in the legal, patent-application sense. In terms of the way in which organized innovative efforts are structured (and as we shall see the innovation process has become increasingly institutionalized), the significant benchmarks lie elsewhere than at the point of patent application. A more important decision point, for example, lies at the point at which an idea has been submitted to management for consideration and funding; the process leading up to such submission is an important phase but only a part of what might be called invention.

The process by which ideas thus submitted are reviewed by management and funded, rejected, shelved, or referred elsewhere in the organization is likewise a significant process phase, but one that would be lost or at least blurred by the choice of "invention" as a phase concept. Finally, as we shall see, R & D is an extremely important and equally complex phase, both conceptually and organizationally, only a part of which could be treated by a phase labeled "invention."

Besides the problem of dividing innovation into process phases, there was also the difficulty, in our early attempts to apply the linear five-phase model, of picking out a starting point in the process. In the linear view, needs or opportunities in some portion of the exogenous world come to be recognized and trigger a problem-definition and idea-generation effort. But in many of the cases cited in the literature, the need, opportunity, or even the clearly defined technical problem arises from what the linear view treats as a "later" phase; for example, R & D, engineering, production, or even diffusion. That is, while problem definition and idea generation are at times "front-end" or logically initial activities, they may also be triggered by later phase developments. Problems are defined and ideas are generated in every phase, not just in the initial one. Examples of major innovations arising from an ongoing R & D effort rather than from "beginning at the beginning" are nylon deriving from Carother's basic research in polymers at DuPont (Jewkes, Sawers, and Stillerman 1969; Mueller 1962), and the development of the transistor by Shockley and his associates at Bell Laboratories (Weiner 1973; Jewkes et al. 1969; Nelson 1962).

True, some innovations followed the linear model, such as tetraethyl lead, invented by Thomas Midgley at General Motors in the 1920s (Jewkes et al. 1969, pp. 312-314). Furthermore, some of the most thoroughly documented case studies--Hindsight (Sherwin and Isenson 1966; Isenson 1967), TRACES (Illinois Institute of Technology 1969), Battelle (1973)--had implied a linear development by utilizing "events" as their chief analytical tool, breaking down each innovation into a series of sequential, and sometimes concomitant, occurrences. Even Myers and Marquis (1969, pp. 3-4), who treated innovation as an information-flow process and who recognized explicitly that the "events" do not always occur in a linear sequence, still found it convenient to identify five "stages" in the process of technical innovation: recognition (of technical feasibility and demand); idea generation (creative act of fusing demand and feasibility into a design concept); problem solving; solution; and utilization and diffusion. Such functional analyses, although more sophisticated and fine-grained than Usher's classical four-step model, do not link functions to the organizational structures and thus do not succeed in showing the actual operation of the innovation process.

Thus, in the relationship between invention and R & D, we encountered a problem that has frequently troubled scholars: When is something invented?[5] When, for example, was nylon invented--when Carother's assistant first noticed the fibers, or when the properties of nylon were verified, or when a first sample in the form that it would be used was produced? (Jewkes et al. 1969, pp. 25ff.)

The question of when an invention actually takes place becomes more difficult to answer when we recognize the increasing complexity of technical devices and processes in our own times. For example, many people claim credit for the "invention" of radar. How can honest men disagree on this question, particularly when radar is so recent and when we have ample evidence from those who actually participated in its development? The answer must be that radar, like a good many other modern innovations, is a highly complex aggregate of a series of innovations. It is foolish to argue over who added the last little bit from "pre-radar" or "almost-radar" to create radar, because all of the increments, or the subinventions, were essential to the creation of radar-- and even when applied and diffused, radar was going through further refinement and development requiring whole new groups of innovations (Kranzberg 1963, p. 137). It is clear that innovation is not a singular event but a congeries of events, or, more correctly, a process.

As we reviewed the various theories and models, we began to realize that in almost every major innovation of recent times each functional phase is linked in some way to the others: every phase in our block diagram has lines connecting it to and from every other block in the diagram. Instead of a linear-sequential picture of a neat flow-chart, with single lines going from one block to another, we had a graphic portrayal of a plate of spaghetti and meatballs!

This same kind of scrambled model also appeared when we applied a systems approach to the innovative process and attempted to illustrate it by a diagram. Mottur, for example, found it necessary to use no fewer than ten charts to show the systems and subsystems "in constructing a single systems model of the processes of technological innovation" and to link together "the sixty-one interrelated major functions or steps of the model, together with the more than two-hundred sub-functions and problem areas" (1968, p. 237). Gellman (1973) encountered similar complications and was forced to develop three separate innovation models--one for producer goods, one for consumer goods, and one for the public sector--and he wound up with three plates of spaghetti and meatballs. Mahdavi, employing a "systems technology" approach in his "efficiency investigation" of innovation, found so "many complicated socioeconomic problems which cannot be formulated in a model" that he did not even attempt to develop one; instead he deplored "wasting money on model building and philosophizing" (1972, p. 83).

The complexity of the process patterns we encountered led us finally to realize that the linear-sequential view described only a few cases. For the rest, the arrows led "every which way," with the starting points being various and with each phase being linked to some, many, or all of the others. It was then that we began to consider innovation as a complex, highly interactive ecological system. This ecological view, or "picture preference," made the interactions between phases easier to conceptualize and comprehend.

VI. INNOVATION AS AN ECOLOGICAL PROCESS

From its literal Greek roots, "ecology" would mean the study of "houses," just as Xenophon entitled his study of the management of households the *Oeconomicus*. Since the household is regarded as a living environment, ecology gradually came to mean "the study of the structure and function of nature," with the further understanding that "nature" includes man and the entire living world (Odum 1963, p. 3).

Like many other sciences, in the first stages of its history--in the late 1800s and early 1900s--ecology was primarily descriptive and taxonomic. Early ecologists looked for patterns in the appearance and structure of organisms coming from different environments, and they developed detailed systems for naming and classifying such communities of organisms. But just as the scientific study of biology outgrew its Linnaean taxonomic phase and began studying the functions and evolution of organisms, so did ecology advance beyond naming, classification, and description to a study of function and of the reciprocal relationships between structure and function. From a descriptive science, ecology became a science concerned with the *dynamics* of interrelationships-- the interrelationships of structure and function (Collier et al. 1973). Indeed, this emphasis on *dynamic systems* and the variations among and within them reveals the applicability of the ecological approach to study of the innovation process.

In order to assist their analysis, ecologists conceptualize levels of organization among biological organisms. Beyond the individual organism are populations, groups of individuals of one kind of organism; biotic communities, including the populations of a given area; ecosystems, where the community and the nonliving environment function together as an ecological system; and the biosphere, the portion of the earth in which ecosystems can operate. These groups might correspond to the various levels involved in innovation, from the individual inventor and the indigenous system of the R & D laboratory and the organization, through the exogenous system of the industry, national economy, and society to the global level. Most important to the ecologist--and to the student of innovation--is to regard the environment as a complex system of interacting components.

In order to understand both the distribution and the abundance of organisms through space and time, ecologists learned that the diversity of phenomena affecting a population meant that any fragmented view would be misleading; they saw that a population and its environment must be studied as a single system. Major emphasis would be on the relationship of function to structure. The history of biology shows how misleading it is to separate the two. For example, the ancient Greeks were familiar with the anatomy--the structure--of the human heart and circulatory system, but this knowledge provided little real comprehension until, many centuries later, Harvey worked out the function of the heart and circulatory system. By analogy, study of the structure of an R & D laboratory is meaningless without an investigation of how it functions; similarly, a study of an industry's structure provides us with little information regarding the innovation process within that industry unless we can tie that structure to functioning and, ultimately, to the innovation process.

The importance ecologists attach to the relationships among the environment, biological organisms, and populations highlights the role of social, cultural, economic, and political factors in the process of innovation. Like the social-deterministic view of innovation, at one time ecologists believed that the interactions of organisms with the physical "abiotic" parts of the environment were mainly one-way: the physical influencing the biological (Watt 1966). They have since learned that these are interactive, feedback processes--just as in the case of innovation. Equally pregnant analogies lie in the ecologist's interest in the factors leading to change, particularly such items as adaptive strategy for survival, much as corporations might seek to innovate for competitive reasons rather than for considerations of immediate profit. Indeed, in an even larger sense, innovation might be regarded as a unique social means devised by man in the process of natural selection and adaptation to his environment--a social and physical environment that previous innovative mutations had helped to create!

Equally stimulating to a study of innovation is the ecological emphasis on trophic structure and dynamics, dealing with the transfer and conversion of energy in various forms through the food, or nutritive, process. In innovation, nutrition comes from new information, or new perceptions or combinations of existing information. This information is varied in nature; it includes knowledge of technical capabilities, scientific theories, economic elements, social needs, legal requirements, and the like. Sometimes the information is incomplete or the conclusions derived from it prove faulty; the result is a stillborn invention or an unsuccessful innovation.

Ecological analysis thus points up the diffuse nature of the decision-making process in innovations. Even before an innovation reaches the market, where economic success or failure awaits it, there are many points in the innovation process where decisions might stop, inhibit, or stimulate its progress.[6] The quantity of information, its quality, and its flow path might dictate--at several points within the innovation process--whether or not the process continues or is nipped in the bud.

In short, recognition of its ecological nature provides us with a powerful intellectual construct for analyzing the process of innovation. It makes us aware of the interrelatedness and interactiveness of its elements, and thus forces us to seek a holistic view. Given the fragmentation along disciplinary lines of the innovation research literature, the pursuit of such an analysis is not the easiest task to undertake, but in the long run it should prove the most fruitful.

VII. SCOPE AND PLAN OF THE STUDY

Although conceptualization of the process of innovation as an ecological system evokes parallels and analogies that can assist us in comprehending the innovation process, it has one serious drawback: the universe it describes is so large and so varied in its interactions that it would be virtually impossible to describe the whole and still derive thematic and working principles. Hence we must narrow our focus to the most relevant forces, to a manageable set of data--although we must be constantly aware of the larger context in which these forces operate.

Even a cursory examination of the research literature dealing with the process of technological innovation shows that it is badly fragmented along three dimensions.

1. The complexity of the process has led investigators to seek out manageable pieces--the R & D laboratory, corporate strategy, patent policy, firm size, adopted characteristics, etc.--with the result that we know something about the pieces, but little or nothing of their interrelationships and mutual influence.

2. The pieces themselves are so complex that there is a strong tendency to deal with only a few (typically two) of the relevant variables--e.g., the influence of supervisory behavior on the performance of scientists and engineers in the R & D lab, or the relationship between education and early adoption of an innovation--with the result that we often do not understand how the "pieces of the pieces" are related.

3. Studies of the pieces, or of the "pieces of the pieces," are almost always conducted from some single disciplinary perspective and in terms of the few variables with which that discipline is preoccupied. Since the number of disciplines so involved is quite large and their cooperative interaction minimal, the outcome is yet another fragmentation of the overall effort and largely noncumulative results.

One might respond to this fragmentation--with some justification--by asking, "how could it be otherwise?" As it turns out, quite narrow foci partitioned from the rest of the process--even if artificially--are all that we can handle with methodological rigor. Also, academic disciplines specialize in certain variables to the exclusion of the rest, but that is their strength--for which no equally fecund alternative has been found.

This response, though speciously damning to the ecological perspective we have adopted, in fact underscores the need for it, since it points out that the fragmentation cannot be eliminated. However, its stultifying effect can be overcome; that is, the pieces and the perspectives can be integrated. Such integration is required if understanding of the process is to be advanced much beyond the present state of the art.

However, the ecological model implies a level of integration that is beyond the scope of this investigation, for we are not equally interested in every aspect of the larger systems that embrace the innovation process, or even in every aspect of those immediate contexts to which it is indigenous. With rare exception, the factors that influence the process in both the immediate and the remote environments, exist for reasons and perform functions other than the innovation-specific ones with which we are concerned.

In defining how wide or narrow the scope of an inquiry should be, Samuelson (1947) gave this advice:

Functional requirements hold as of a given environment and milieu. Of course, to designate this environment completely would require specification of the whole universe; therefore, we assume implicitly a matrix of conditions within which our analysis is to take place. . . .A system may be as broad or narrow as we please, depending upon the purpose at hand. . . . The fruitfulness of any theory will hinge upon the degree to which factors relevant to the particular investigation at hand are brought into sharp focus.

We shall try to make the boundaries of our study explicit.

It became clear in the course of our investigation that to integrate what is known about the innovation process would require a conceptual framework of at least three dimensions. The first of these dimensions involves the various levels of aggregation at which the process may be studied. In our survey of the literature we found it useful to partition levels of aggregation into those that constitute the immediate context in which the process takes place ("indigenous system") and those that comprise the larger context ("exogenous system"). The former we take to include the individual researcher, his work and social groups, the R & D laboratory, and the firm or organization. The latter includes the industry of which the firm is a part and the myriad of national and international influences--economic, social, political, institutional, organization, intellectual, cultural, etc. These levels of aggregation, thus partitioned, constitute one of the three structural dimensions of our study.

The second dimension of our conceptual framework involves a juxtaposition--within and across the above levels of aggregation--of the wide range of variables operative in the innovative process. These variables--typically treated in isolation by the several disciplines--range from the prevailing value system of society to the competence and ingenuity of the individual researcher; from the abundance or scarcity of certain material resources to the attitudes, morale, and cooperation with an R & D project team; from governmental tax, patent, and regulatory policies to the organizational structure and managerial support of the R & D laboratory or the leadership style, policies, and competence of the R & D manager; from the nature, size, concentration, and competitive stance of the industry to the financial condition, policies, and strategies of the firm; etc.

By relating structure and function, we counteract the tendency to treat these variables solely along disciplinary lines. In addition to juxtaposing these variables, we shall also offer a number of hypotheses as to the nature and effect of their integration. Such hypotheses cannot be tested in our literature-based study, of course, but are offered as research agendas for future empirical studies.

The third dimension of our ecological framework lies in the concept of process phases. As was noted above, a distinction between the innovation and diffusion phases is basic. The former, which is the process leading to the first introduction of a new product, device, or process, itself admits of phase (or subphase) distinctions. As it turns out, there are two well-established models of these subphases--one reflecting a management-science orientation (with emphasis on idea flow), the other reflecting an economics orientation (hence emphasizing economic factors). We attempt to integrate these two models, showing how differing organizational configurations have affected the nature of the innovations that emerge. For example, one type of organizational structure may be more conducive to incremental innovations; another may favor innovative breakthroughs.

For the diffusion stage, there are many research traditions. Here our efforts have been directed at integrating these traditions in such a way as to delineate the dimensions that a general model or theory of diffusion would have to embrace.

Using our ecological orientation and the three structural dimensions outlined above, we have assessed the empirical and theoretical literature--and attempted to gauge their correspondence--by asking the following four major questions: (1) What is the current state of the art? (2) How good is our current understanding? (3) What are the major gaps and weaknesses in our knowledge, and what further research is needed to fill them? (4) How adequate are the various commonly employed methodologies for the needs of future research? As will become apparent in the following chapters, a host of subquestions are implied in these major ones.

Chapter 2 deals with the elements of the external system--values, endowments, and institutions--affecting innovation and innovators. It discusses aspects of the total social system, entire industries, sectors of the economy, and governmental functions. Finally, it considers the functions and structures of these elements as they interact with one another, with business firms and individuals, and with the innovation system itself.

In Chapter 3, we turn to those levels of aggregation and contexts to which the innovation is indigenous--the firm, the R & D laboratory, the work group, and the individual researchers. We treat the role of the individual inventor-entrepreneur, but our primary focus is on the institutionalized process. The structure/function relationship, emphasized by our ecological approach, turns out to be an especially important integrative mechanism.

Although we consider innovations arising outside the R & D context, most of our study at this point concerns the laboratory, its internal dynamics and linkages to the rest of the organization. Among the behavioral variables and focusing mechanisms that are examined are individual mind sets and traits, supervisory attitudes and leadership styles, the influence of the individual's primary work and social groups, organizational control and coordination mechanisms, patterns of communication and information flow, etc. This chapter ends with the extremely important--but poorly understood--area of interface between R & D and the operating units within the organization.

Chapter 4 deals with the diffusion of technological innovations. Here we again encounter a fragmented picture, in terms of both the disciplines involved (one source identified seven major and six minor diffusion-research traditions) and an overwhelming preoccupation with quite narrow studies relating only two variables (perhaps as many as 95 per cent of the studies fall in this category). Perhaps as a result, the state of the art in diffusion research is not as advanced as it is in innovation studies.

Chapter 4 first surveys the three basic diffusion-research traditions (the geographical, the economic, and the social-psychological), noting both their differences and their implicitly complementary findings. We then examine the so-called classical diffusion model, which represents the best effort to date to see the process as a whole and identify its basic elements. Finally, on the basis of our survey of the empirical research traditions and the classical model, we describe the four dimensions that a general theory of diffusion should embrace. These dimensions, or major categories of influence, are (1) the characteristics of *sectors* of which adopter units are a part, (2) the characteristics of *adopters* themselves (both individuals and organizations), (3) the characteristics of the *innovations* being diffused, and (4) the *propagation mechanisms* for diffusion.

Chapter 5, the final chapter in Part I, summarizes our findings, both methodological and substantive, and highlights our recommendations for future research. The chapter thus forms, in part, a kind of research agenda, emphasizing what we do not--but should--know. As such, our study has to do with the future as much as the present.

NOTES AND REFERENCES

1. There are many inventions that, to an outside observer, seem to have arisen from trivial motives--although the inventors undoubtedly had strong and serious reasons. Stacy V. Jones, patent editor of the *New York Times*, has published an amusing collection, *Inventions Necessity Is Not the Mother Of*, containing such patented items as a coughing cigarette package (an audible warning of the health hazards of the contents), a bicycle seat with needles (to discourage thieves), snoring preventers, parakeet diapers, dimple grinders (for beautification), and a secret communication device for torpedo control (coinvented by composer George Antheil and movie star Hedy Lamarr) (Jones 1973).

2. For example, Lambert (1971, p. 34) claims that certain parasites, which enervate the human mind and body, were prevalent in tropical climes and inhibited technological advance in those areas; Gilfillan (1969, p. 129; 1971, pp. 73-87) provides a racial explanation of inventiveness.

3. Lynn White (1962) points out the great innovativeness of Western medieval technology as compared with the Chinese of the same era. See the dispute on this point between White and Joseph Needham at the Symposium on the History of Science (Oxford University, July 1961) in Crombie (1963, parts 2 and 3).

4. The international diffusion of the capability for invention is remarked on by Alden Whitman (1974). Merton and Barber (1961) have made a study of multiple scientific discoveries, but there is no similar study for technical inventions.

5. This problem also affects the history of scientific discovery, and is exemplified by Thomas Kuhn's discussion of when Roentgen discovered X rays. Kuhn (1962, pp. 57ff.) concluded that the discovery occurred sometime between Roentgen's first observation, which he later interpreted as X rays, and the completion of his observations and analyses for publication.

6. Gold (1969) points to weaknesses in current models of decision-forming and decision-implementing processes in innovation. Because of the changing nature of the factors entering into managerial decision making, he stresses the need for new analytical perspectives and for more data on specific case studies in various fields.

2. THE WORLD OUTSIDE

I. INTRODUCTION

For analytic purposes it is helpful to partition the world into the various systems relevant to innovation. Our concern in this chapter is with the exogenous system--that part of the world not directly involved with innovation. Since innovation is only one of a number of responses to societally generated needs and opportunities, the elements of the exogenous system are not exclusively or even primarily involved with innovation. Our treatment of these elements focuses on their aspects that interface with the innovation system and whose study increases our knowledge of that system.

On the basis of the available literature, we shall treat three classes of entities in the exogenous system: *values*, *endowments*, and *institutions*.

For our purposes, values may be defined as individual or group preferences. Our concern is with two functions of values. First, traditional values serve as legitimizing justifications for existing institutions and procedures, including the innovation system. Second, value changes can either lead to or legitimate modifications in the ongoing system.

The societal value structure is extremely complex, touching on almost every aspect of individuals' relationships to one another and to society. But only a limited set of human values is concerned directly with innovation. Because of the complexity of the value system, there can be conflicts between values involved in innovation and other values in society such as religious, social, and economic values. Values interact with other elements of the system as well as with innovation itself. They interact with endowments, for example, by placing a premium on the development and diffusion of scientific and technical knowledge; they interact with institutions in their role as legitimators or modifiers of institutions.

We define endowments as factors that serve as inputs to innovation. Among these factors are scientific and technical knowledge, capital, human labor, energy, and material resources. Endowments are not necessarily fixed stocks depleted by use. We can consider them as elements that serve a function or satisfy a set of requirements. Thus one endowment can often be substituted for another. Substitutions can be made between elements of the same sort, as when plastic replaces steel in automobile construction or when relativity theory replaces Newtonian dynamics in mechanics. Likewise, substitutions can be made between endowments of different sorts, as when technical knowledge is substituted for labor. Endowments also couple with other elements of the exogenous system. For instance, scientific and technical knowledge is a key element in stimulating value changes, and talented people enable institutions to function effectively.

The institutions we consider in this chapter are those formal and informal human organizations that serve as external determinants and modifiers of the innovation system. We deal with economic, market-related institutions such as firm and industry; we also treat political institutions, which we subdivide into those that innovate directly, and those that regulate and control innovation. Institutions couple strongly with one another as, for example, do the regulated firm and regulatory commission. They also couple with other elements of the exogenous system, as do research establishments with scientific and technical knowledge.

Our treatment of these elements cuts across all levels of analysis and aggregation. In the case of values, for instance, we shall consider the relationship of some broadly held social values to innovation. On the societal level, values legitimate innovation as a response to needs and opportunities. Institutions such as industry, government department, and firm also have values associated with them. Finally, at the lowest level of aggregation, we consider studies dealing with the preferences of individuals relating to specific innovations. Thus, the present chapter includes a range of information from "wisdom," "common knowledge," and speculation, usually found at the higher levels of aggregation, through comparative studies and statistical material, to tightly designed cross-sectional studies at the lower levels.

Our treatment of the exogenous system stresses the linkages of the elements with one another and with those of the innovation system. We shall move from traditional values as legitimations of the innovation system to specific data that indicate current directions of value changes. Then we shall deal with endowments as important inputs to innovation. Finally, we consider how, why, and to what extent economic and political institutions serve as key determinants and modifiers of innovation.

II. VALUES

A. TRADITIONAL VALUES

A complex constellation of values serves to legitimate innovation as a response to societal needs and opportunities as well as to affect the particular institutions and procedures involved in the innovation system.

At the societal level we can contrast the valuation of technological innovation held by various societies. Classical Greek civilization accorded high value to the development of theoretical knowledge for its own sake and the cultivation of the body for athletic and military feats. Although there were some technical innovations in antiquity, the practitioners of the mechanical arts were not generally held in high esteem; indeed, some authorities claim that slavery inhibited technological innovation (Farrington 1949, pp. 8-9, 166-67).

Although medieval Europe developed and utilized significant innovations such as the stirrup, heavy-wheeled plow, water wheel, windmill, and mechanical clock (White 1962), religious values were of the utmost importance and oriented the society toward the "heavenly city." Innovation, as well as everything else in the society, was valued, in great measure, as a means for reaching God. Such a set of choices put spiritual needs above material needs and gave technological innovation a peripheral place in the scheme of things. Systematic development and utilization of technology was thus not consistently pursued.

A more recent illustration of the role of values in shaping social behavior is Max Weber's global hypothesis of the dependence of the rise of capitalism on the Protestant ethic (Weber 1930). Although this hypothesis is not readily testable, it has stimulated much research and informed speculation. It offers a classic example of putative value-institution linkage in the exogenous system, and is germane to our study because capitalism played a crucial role in the development of technological innovation as a central activity of modern western society.

In contemporary society, the global values that affect innovation include beliefs in the beneficence of material progress, the importance of satisfying material needs, the desirability of innovation as a response to social needs and opportunities, and man's lordship over creation (White 1967) which legitimates his exploiting nature at will.

Affecting innovation at lower levels of analysis are the values that stress the benefits of profit maximization and competition, which encourage firms and industries to innovate or to adopt innovations. Protection of private property serves as a value legitimating the patent system. Also, as we shall see, the high valuation placed on scientific and technical knowledge has led to considerable government support for research and development.

Societal preferences have institutional effects. Education is such a societal value. Incarnated in public policy, this value has led to the development of educational institutions and to raising the level of popular education. Scientific and technological education can affect the rate and type of innovative activity. Also, educational level is a factor in the rate of adoption of innovation; evidence exists both for industrial innovations (Mansfield et al. 1971) and agricultural innovations (Rogers with Shoemaker 1971, p. 7; Hayami and Ruttan 1971) which indicates on balance that the early adopters of innovations have more education than later adopters.

One of the truisms of social science is that a social system seeks to maintain itself by resisting certain kinds of change. Yet at the same time major changes do occur. Technological innovations can function either to facilitate or resist major social change.

Often major changes in a society are directly caused by the widespread adoption of innovations such as the automobile and television. A number of researchers explain these changes in terms of the relative advantage of the novelty over the existent as perceived by the adopting person or group (Rogers with Shoemaker 1971, pp. 142, 350ff.). Additionally, when an alien innovation is brought into a social system, it is most readily adopted if it is compatible with that system (Barnett 1953). Introducing the innovation through agencies already existing in that system is also a help in its adoption (Spicer 1953; Bright 1964, p. 130). For example, radio networks and many owners of local radio stations became owners of TV stations. Yet, it took much longer for motion picture firms to become television program producers.

Working in the area of industrial innovations, Mansfield (1971) found that the most important component of relative advantage is economic gain (see also Griliches 1957). Low initial economic uncertainty or a high rate of reduction of the initial economic uncertainty constitute advantages as does a low level of economic commitment in adopting the innovation (Mansfield 1968b, p. 88). Mansfield's work deals with economic factors, and in the organized industrial context these are central. However, in any context relative advantage might be dominated by political, social, and psychological factors as well as economic (Chap. 4).

Technological innovations may also function to resist social change. Thus, innovations dealing with the production and distribution of energy are currently being stimulated in order to maintain our present system of values and institutions related to automobile transportation.

Since values involve the preferences and choices affecting what will be invented and whether and how it will be applied and accepted, they are one of the most important factors involved in the study of innovation. Unfortunately, there is no well-developed value research methodology (see for example, Williams 1967), but useful clarification of some of the basic conceptual issues has been provided by Scriven (1974), Baier (1969), and Rescher (1969). However, values can be approached through the study of behavioral preferences. Thus, we may reconstruct the values underlying the development and transmission of scientific and technical knowledge by reference to the degree and nature of societal support for research and education. In the next section we shall find that values concerning innovation can be investigated by surveying the preferences of individuals regarding the role of technology in society.

B. CURRENT VALUES AND VALUE CHANGES

Current societal values relating to the desirability of technological change and innovation are influenced by recent experiences with the positive and negative consequences of existing innovations. Trends in values, as seen in current evaluations of innovation, are thus partial indicators of the future nature and direction of innovation.

Until 1970 there was apparently little systematic nonmarketing research on public reaction to technological innovation in general and to specific innovations. Since then, at least four studies have been undertaken to remedy this lack; three confined to local areas (Taviss 1972; Metlay 1971; Metlay 1972), and one on a national basis (National Science Foundation 1973).

The purposes and designs of the studies varied considerably. By integrating their results with the values underlying innovation, we can see some directions of value change in this area. And, importantly, we can see how these value changes might impact on institutions affecting innovation.

Interestingly enough, both Metlay studies (1971; 1972, pp. 82-83) showed that attitudes toward technology bore little relationship to the variables of occupation, education, sex, race. However, there was some correlation with political attitude and age (Metlay 1972, pp. 82-83; LaPorte and Metlay 1974, pp. 13ff.). Despite this lack of strong correlation with the usual social variables, the public gave technology a generally high evaluation.

Metlay claimed that these results "indicate that while the population as a whole holds strong positive evaluations of the social effects of technology, we can isolate a minority who believe that presently employed technologies have not made life better, but in fact, reduced the quality of life" (Metlay 1972, p. 85; see also LaPorte and

Metlay 1974, p. 10). Although technology generally was perceived as being able to solve many or most of the problems facing our society, at the same time there was a strong public perception that some technologies, such as nuclear weapons, were harmful (Metlay 1971). Likewise other technologies, though not primarily harmful, were believed to have had some bad effects.

These public perceptions indicate an awareness of an ongoing cycle of innovation. An innovation takes place to meet a need, but it produces, along with the intended effects, some harmful side effects. Even though the initial problem is solved, additional problems arise from this latest innovation. Resolving social problems by technical means or alleviating the harmful consequences of technologies by further technical innovation is called a "technological fix." For example, industrialization creates pollution. Pollution-control technology reduces the level of pollution, but at the same time raises energy requirements. The increased demand for energy causes additional pressure on our energy resources, thereby creating a demand for further innovation to meet the need, and so on, ad infinitum.

One result that comes through very forcefully in this group of opinion surveys is a desire by the public for a significant voice in the control of technology. For example, when asked who controls the decisions on a variety of technologies such as mass rapid transit and space exploration (Metlay 1972, pp. 117-23), the respondents said in each case that the public *has* the least say in decision making. And in each case they claimed that the public *should have* the greatest say. Clearly people feel that their role vis-à-vis technological innovations affecting them has been too passive; they believe that both the good and harmful aspects of technological innovations affect them sufficiently to be directly concerned. This public concern for "participatory technology" manifests itself in public interest groups such as Nader's "Raiders." Changing values toward innovation are beginning to be institutionalized, for example, in recent environmental legislation such as the National Environmental Protection Act (NEPA). Restrictions are beginning to be placed on innovations which increase pollution, and at the same time innovations in pollution control are being encouraged. The technology assessment movement (Hetman 1973; Teich 1972, part 4), a crucial aspect of which is to understand and head off in advance the undesirable consequence of technologies, seems a response to the perception of the harmful side effects of past innovations and technological fixes. The Environmental Protection Agency, established by NEPA, and the new Office of Technology Assessment are themselves social innovations designed to control technological innovations.[1] This relationship is illustrative of the importance of social innovations, and their interaction with innovations in physical technology in our society. It is also an example of "fixing" social problems caused by innovations with other innovations.

Although such developments offer no positive alternatives to innovation, they have a directive effect on the innovation system. They tend to encourage innovation designed to alleviate certain harmful side effects of technology, as in pollution control. At the same time they may attempt to direct innovation toward implementation of those systems which minimize social cost.

The current "energy crisis" is illustrative of the hierarchy of social preferences. With the impending and actual scarcity of certain forms of energy, especially crude petroleum, a conflict in preferences arose between having almost unlimited quantities of cheap energy and having a relatively pollution-free environment. The Alaska pipeline, almost killed earlier by environmentalists, was resurrected when the "energy crisis" proved to be a greater threat to cherished values and habits. Society, it would seem, perfers cheap energy to clean environment, but environmental considerations are at least examined before the tradeoff is made.

There has been change, as witnessed by the growing numbers of technology assessments, in the direction of wishing to appraise technology more closely and not necessarily preferring an immediate "quick and dirty" technological fix. This tendency seems to indicate a sophisticated public perception of technology that will presumably continue in the future as we continue to rely on technological innovation as a primary response to societally generated needs and opportunities. Technology assessments, together with the desire to minimize harmful side effects of innovations, may indeed change the pace and the manner in which innovation takes place in our society.

The initial thrust may be to slow the rate at which innovations with harmful side effects are introduced--some opponents have called it "technology arrestment"--and to direct innovative activity to minimizing these harmful side effects. This is a way in which the value system, working through institutions, places limits on certain innovations while stimulating others.

But present actions affecting the future ideally should be based on the values which hold when the results of these actions occur, not on today's values. Unfortunately, the methodology of value forecasting scarcely exists (Taviss 1973; Rescher 1969). Thus the continued development of value research methodology, including value forecasting, is quite important in charting the future course of innovations in various fields.

III. ENDOWMENTS

A. SCIENTIFIC AND TECHNICAL KNOWLEDGE

The high value placed on scientific and technological knowledge in American society is manifest in public policy. We seek to institutionalize diffusion of existing knowledge through public education and the development of new knowledge by Federal sponsorship of research and development. Knowledge exists both disembodied in the scientific and technical literature and in other information systems, and embodied in the members of society and in technological artifacts. A large portion of disembodied scientific and technical knowledge is public and can be freely accessed by anyone who has the capacity to understand and use it. Scientific and technological knowledge, both embodied and disembodied, serves as a necessary and crucial cognitive input to innovation. Its substitution for other endowments such as labor has enabled the pace of innovation to increase dramatically.

1. *Relationship Between Scientific and Technical Knowledge and Innovation*

Although it is clear that some technical knowledge is necessary as a starting point for innovation, the role of scientific knowledge is less certain. There is a widespread belief that basic scientific discoveries lead eventually to technological application. Alan Watermann (1965), the first head of the National Science Foundation, claimed that "statistical evidence" (which he did not adduce) existed "that most of the body of science ultimately achieves practical utility." Since this claim plays a central role in decisions relating to public support of basic scientific research and to other aspects of innovation, we must try to determine, insofar as we can, the relationship between scientific knowledge and technological innovation.

This relationship has varied from complete independence to strong interdependence, depending on a variety of factors. In classical Greece there was a deep interest in theoretical science without a corresponding interest in the application of that knowledge to further technological innovation (Farrington 1949). This is the prototype of science without technology.

At the other extreme lie some of the mechanical inventions of the early Industrial Revolution. Inventors such as Hargreaves and Arkwright were primarily tinkerers with no systematic scientific background. Although the science of mechanics had recently emerged with basic contributions by Newton and refinements by other English and continental scientists, this body of knowledge was not explicitly used in the development of these early textile machines. Here we have technology with no direct link to science.

Indeed, many historians claim that as a general rule the technological system generates much of its own knowledge, without recourse to science. Price has pointed out that science and technology each have their own separate cumulating structures, and that "only in special and traumatic cases involving the breaking of a paradigm can there be a direct flow from the research front of science to that of technology and vice versa" (Price 1965; see also Wertime 1962; Smith 1960; Gille 1967).

At the same time there are some recent historical examples of technologies firmly rooted in and developed from science, the outstanding one being 20th Century nuclear technologies. It is extremely unlikely that the technologies for production of nuclear energy could have been developed without the basic scientific framework originating in Einstein's theory of relativity. However, some caution is appropriate here. The idea that matter could be converted into energy and the precise relationship between matter and energy came from Einstein's theory of 1905, but the material devices that allow the actual transformation of matter into energy are not a product of that theory. Although these devices are based on other scientific theories (e.g., the use of liquid metal coolants in nuclear reactors requires knowledge of the properties of liquid metals), they are also based on existing technological knowledge not derived from scientific theory.

As well as initially generating technologies, science may also assist a technology already partially developed. A good example of this mode of interaction is the development of crease-resistant fabrics in the cotton textile industry (Jewkes et al. 1969, p. 245). This development was the result of the establishment of a research group in a firm and in an industry which had hitherto conducted very little scientific research.

Just as science can generate technology, so can technological improvements stimulate scientific discoveries. An excellent example is the development of the science of thermodynamics arising directly from the interest in the relationships among heat, work, and energy aroused by Watt's steam engine (Cardwell 1971). Another example is the development of the cyclotron and its effect on physics research. Invented by E. O. Lawrence in the 1930s at Berkeley, the cyclotron was the first of a long series of machines developed for the acceleration of elementary particles. The use of particle accelerators led to the discovery of many elementary particles and the transuranium elements as well as other important properties of matter (Jewkes et al. 1969, p. 248).

At present there exists no comprehensive theory to account for the complex range of relationships between science and technology. This broad spectrum includes, at one end, scientists who pursue investigations of their own choosing with little technical apparatus and no concern about eventual applications. At the other end are those engineers who are unconcerned about the latest scientific findings and use only the textbook science they learned in school. Then there is a large middle ground where science and technology come together in a variety of relationships, with scientists using technology in order to do their scientific experiments and engineers putting the latest scientific findings to work in the most sophisticated fields of technology (Kranzberg 1967b; 1967c; 1968; Molella and Reingold 1973).

From what has been said, it should be obvious that any simple linear cause-and-effect model of science-technology relationships simply does not fit the facts (Price and Bass 1969). Yet there remain serious students of innovation who cling to the view that technological innovation proceeds linearly from basic science (Peck and Scherer 1962; Marschak 1962). The complexity of the situation is present in the historical dispute about how much and in what way the conceptual science of the 17th Century Scientific Revolution was shaped by the contemporaneous technology (Zilsel 1945; Merton 1938; Hall 1963), just as there are arguments about the degree of science involved in James Watt's steam engine and many other of the landmark inventions of the Industrial Revolution (Bernal 1953; Kerker 1961; Musson and Robinson 1969). Complicating our understanding of these matters was the development of engineering science in the 19th Century which employed the instruments, experimentation, and measurement adopted by science during the Scientific Revolution, but carried on by a different community with its own social system (Layton 1971).

We are still at an early stage in our understanding of these intricate interrelationships between science and innovation. Researchers such as Jewkes et al. (1969) claim that "it is not known whether there is a necessary connection between the growth of scientific knowledge and the growth of technology and invention, or, if there is a connection."

However, useful beginnings have been made to model the process. Kranzberg has suggested (1971) that major technological breakthroughs (e.g., nuclear energy, transistors) derive directly from nonmission-directed scientific discoveries, whereas mission-directed research is in response to specific technical needs and tends to incremental innovations.

Kranzberg also advanced a push-pull scenario in which a scientific discovery attracts the attention of an engineer who can visualize its application (or an entrepreneur who can foresee its profitability). In the course of the technological development of the basic science, it proves necessary to obtain some additional basic knowledge if any workable items are to result. The result is a push from technology directed toward science to obtain basic knowledge to further the development process. But developing one item of knowledge pushes toward some other mission-directed research. In this way science and technology interact, pulling and pushing one another into new discoveries and innovations.

Empirical studies have also been undertaken in an effort to understand these connections. Project Hindsight (Sherwin and Isenson 1966; Isenson 1967), which examined the relationship between science and technology utilized in weapons systems, was initiated in 1964 to establish the effectiveness of the $10 billion invested by the DOD in basic and applied research since 1945, and to determine which, if any, management practices were conducive to high payoffs. Researchers "dissected" each of 20 weapons systems to identify each contribution from post-1945 science and technology that was important to the improvement of system performance or cost reduction; each significant contribution was termed an Event. Events were divided into Science Events and Technology Events, with Science Events being subdivided into Undirected Science and Applied Science. Of the 710 Events identified, only 9 per cent were classified as Science Events, 8.7 per cent as Applied Science, and only 0.3 per cent as Undirected Science.

After the publication of this result, so contrary to the conventional wisdom that technology depended on science, the National Science Foundation, the agency responsible for funding basic science, reacted in 1967 by sponsoring an investigation by the IIT Research Institute to study systematically the role of basic scientific research in technological innovation. That report, TRACES (Illinois Institute of Technology 1969), identified the genealogy of key events that had led to five major technological innovations of social and economic significance: birth-control pills, the electron microscope, videotape recording, ceramic metallic materials, and matrix isolation. Instead of setting a backward time limit of 1945, TRACES went back more than a century in studying the scientific roots of certain innovations. Dividing its key events into nonmission-oriented research, mission-oriented research, and development and application, TRACES attributed 70 per cent to nonmission research, 20 per cent to mission-oriented research, and 10 per cent to development and application. The number of nonmission research events peaked significantly between the twentieth and thirtieth year prior to an innovation.

Interactions of Science and Technology in the Innovative Process, a study by Battelle (1973) for the National Science Foundation, was a lineal descendent of TRACES. Considering three of the TRACES cases as well as five others such as hybrid grains and the heart pacemaker, it attempted to identify important common factors in these innovations such as need, opportunity, technical gatekeepers, and political factors. Of the "significant" events identified in the development of each innovation, 70 per cent were considered research events, divided about evenly between nonmission- and mission-oriented research. The remaining 30 per cent were termed development and nontechnical events. In addition, the researchers designated the most important of the significant events as "decisive" and found that 15 per cent of these were nonmission-oriented research, 45 per cent were mission-oriented research, and 40 per cent were developmental and nontechnical. This was a somewhat different mix from TRACES. The decisive events tended to be closer in time to the point of innovation than those which were only classified as significant.

These three studies--Hindsight, TRACES and Battelle--are not so contradictory as they might seem on first reading. Had Project Hindsight looked further back in time or investigated a wider field than weapon systems, its conclusions dealing with scientific events might have been closer to those of TRACES. On the other hand, TRACES might have come up with a different mix of significant events had it chosen a different set of innovations, particularly some involving mechanical rather than chemical, biological, and electronic technologies. Designating as "decisive" those events usually specific to a single innovation might account for the difference between TRACES and the Battelle study.

More important, all three studies indicate a role for scientific research in technological innovation, with mission-oriented research becoming increasingly important as the final innovation approaches. The difficulties in finding direct connections between science and innovations in these studies accentuate the important finding that most innovations do not arise from the *direct* application of basic scientific knowledge (Achiadelis et al. 1971; Myers and Marquis 1969; Langrish et al. 1972).

As a result, some studies have concentrated on the *indirect* contributions of science to innovation or contributions occurring in conjunction with other factors. Thus, Gibbons and Johnston (1970) and Rosenberg (1974) claim that scientific knowledge acts in conjunction with market demands. Gibbons and Johnston also found that scientific knowledge played an important role in the solution of technical problems arising in the course of innovation directed activities. These studies, together with the work of Langrish et al. (1972), indicate that scientific knowledge makes up a part of the environment in which innovation takes place and that the scientific state-of-the-art may facilitate or impede innovation. Langrish et al. (1972) point out three ways in which science acts indirectly to facilitate innovation:

> First, curiosity oriented science, practiced largely in academic institutions, provides techniques of investigation. Second, it also provides people trained in using these techniques as well as in scientific ways of thought in general. . . . Third, science enters innovation already embodied in technological form. It may be relatively rare for a piece of curiosity oriented research to generate a piece of new technology, but once this process has occurred, the technology can be used over and over again and developed into more advanced technology (p. 40).

Further research on the role of scientific knowledge in technological innovation is obviously needed. To indicate and justify the kinds of research we consider appropriate, we present two brief, suggestive sketches.

The principal alternative to applying scientific knowledge to innovation is a combination of rules of thumb, personal experience, and trial and error. What does scientific knowledge offer the technologist? Science provides a theory, model, or picture cast in terms of intellectual constructs of how a certain part of the world works. It also provides experimental results couched in terms of theory, which is usually highly abstract and says nothing about any specific device. If the technologist can get the scientific theory into a workable form, he can use it to provide knowledge of processes to be incorporated in an innovation. In the case of mechanical technologies, for example, the workings of devices can be immediately evident to the senses and thus usually require no intermediate picture. On the other hand, the internal workings of the nucleus are not immediately accessible, so theory enables the technologist to grasp them. A cursory inspection of the case studies in mechanical innovations in *The Sources of Invention* (Jewkes et al. 1969) indicates that a high proportion of them had no scientific component. On the other hand the critical role of science in electronics innovations such as the transistor (Nelson 1962; Weiner 1973), and chemical innovations such as nylon (Mueller 1962), is well documented. Therefore, possible systematic differences in the relationship between scientific knowledge and technological innovation in different fields of science and technology should be investigated.

One of the findings of TRACES which was supported by *Interactions of Science and Technology in the Innovation Process* (Battelle 1973) was that most basic scientific events important to an innovation took place well before applied science and development events, which may in part explain the absence of significant basic science in Hindsight's relatively brief 20-year time frame. Such a conclusion implies a progressive articulation of basic scientific theories and experimental findings into forms easily accessible to the technologist. It would seem useful to study this "translation process" in order to learn about the mechanisms used in reshaping and diffusing scientific results to the technological community. Brittain (1970) has shown how it required a science-engineer intermediary, Oliver Heaviside, to "translate" James Clerk Maxwell's theories of electrodynamics into a form in which they could be used by engineers. Similar investigations could provide a variety of case studies dealing with the process of technology's use of science and might thus serve as the basis for framing hypotheses about the translation process.

The question of the role of scientific knowledge in technological innovation is of extreme importance, for society supports basic science generously in part because it believes that science eventually leads to innovation. Our theoretical understanding of the connection is virtually nil, for the research bearing on this question consists almost exclusively of case studies at varying levels of detail. Further studies, of the kind we indicated above, are necessary to lay the groundwork which may lead to a deeper theoretical understanding of the role of science in innovation.

2. *Configurations of Technological Knowledge*

From the consideration of the role of scientific knowledge in technological innovation, we now turn to configurations of technological knowledge which can facilitate or impede innovation.

"Technological readiness" refers to the availability of necessary support technologies in order that an innovation be developed and utilized. For example, the Russian I. I. Polzunov (1729-1766) invented a steam engine in 1766. After a few months of operation, his boiler sprang a leak. Unfortunately, the level of technological knowledge in 18th Century Russia was so low that no one could repair it. Thus, Polzunov's invention never became an innovation (Zvorikine et al. 1962, pp. 138-39), and we rightly attribute the steam engine to James Watt, not to Polzunov.

Technical skill is only one element in "readiness." Another element, support technologies, is especially important in the case of large and complex technological systems. Introducing a V/STOL (Vertical/Short Take-off and Landing) aircraft system requires not only innovation in aircraft technology, but also adequate technical support in airport design, pollution abatement, and ground traffic flow technologies, to name only a few. In fact, considerable innovation may be necessary to design the support system and make it operational. With only the aircraft, there can be no system and hence no innovation. As we shall see, "readiness," in the form of support systems, is also essential for diffusion of innovations; e.g., the "green revolution" depends not only on the creation of new hybrid grains, but also on transportation and storage facilities, fertilizer supplies, etc.

As well as requiring support to be effective, technological innovations create pressures which stimulate further innovation. Because technologies are so interdependent, changes in one element of a group of coupled technologies can create "imbalances" or "bottlenecks" that necessitate changes in other technologies. For example, more powerful automobile engines required more powerful braking systems. Improvements in the cutting tools in lathes, with the use of high-speed alloy steels, required accommodations in other parts of the machine for control, lubrication, and disposal of waste material. In Chap. 6 Hughes (1971) refers to these as "reverse salients," using the military image of technology advancing on a broad front with a certain sector, the "reverse salient," lagging.

The concept of imbalance is not confined to changes induced in a single kind of tool or device. Important imbalances can exist between different elements of related **technologies.**

A sequence of imbalances of this kind is to be found in the textile industry in the 18th Century. Richard Kay's invention of the "flying shuttle" speeded up the weaving process, upsetting the usual ratio of four spinners to one weaver; either there had to be many more spinners to supply a weaver with sufficient thread or yarn, or else spinning had to be similarly quickened by innovations in that field. A series of inventions by James Hargreaves, Richard Cartwright, and Samuel Crompton speeded up the spinning process. Then Cartwright set about mechanizing the weaving operation in order to take full advantage of the now-abundant yarn produced by the new machines. The result was the power loom. These machines lowered the price, and hence created a large new market for cotton textiles. Another bottleneck developed in the supply of raw cotton, where the chief difficulty lay in the amount of labor involved in picking the seeds from the bolls. This **problem** was solved by Eli Whitney's invention of the cotton gin, which more than tripled the amount of seed-free cotton which could be produced per man per day. Thus, innovations in one field produced a need for inventions in other related fields (Mantoux 1961, part II, Chaps. 1 and 2).

Our knowledge of technical opportunities and bottlenecks can be used as a valuable indicator of what future technical developments may and/or should take place. That is, they may serve as inputs to technological forecasts. Technological forecasts (Bright 1972; Chap. 13 below) are of two main types: descriptive forecasts predict what will be the case; normative forecasts are goal directed (i.e., what *should* take place). The perspective and interests of the forecaster determine which type of forecast will be used. Although technology forecasting is becoming widely used, it is important to note that its method at present consists in great part of a "bag of tricks," such as Delphi and trend extrapolation, whose ultimate justification awaits further research (Bright 1972; Chap. 13 below).

B. RESOURCE ENDOWMENTS

Besides scientific and technical knowledge, innovation requires other inputs. These include capital, labor, with its embodied knowledge and talent, and material resources and energy. The economics literature refers to these as "factors." Most scientific and much technical knowledge is public and can be used freely and without diminution by anyone who can access it. The resource endowments considered here must be obtained for innovation and production from what are called "factor markets." In this section we shall consider how changes in factors and factor markets affect innovation.

1. *Capital*

Because the innovative process is often quite costly, the ability to secure adequate funds is an important determinant of whether a new product or process will be introduced. Financing an innovation is in some ways similar to the financing of any capital improvement. There are, however, essential differences. New products and processes often involve a degree of uncertainty which is not associated with duplicating existing facilities. For firms with adequate internal sources of cash flow (retained profits and depreciation) investment in new products and processes can be made with little or no recourse to the market. The situation is different for firms without strong internal cash flow. They need access to external sources of capital, so they must persuade these sources of the soundness of their investing in an innovative product or process.

The literature on venture capital in respect to innovation is scattered and mostly of the "wisdom" or "how to do it" variety (for a useful treatment see Mueller 1971). The Charpie Report (1967), though emphasizing the importance of venture capital for small firms, recognized that published data were insufficient even to answer the question of whether there was an adequacy of venture capital. This report could only estimate that $3 billion of potentially available funds existed in 1967. Mansfield (Chap. 8 below) also recognizes this lack of information and calls for the development of data regarding the extent of the venture capital available, the terms of availability, investment criteria, and the experience of innovators in obtaining capital backing.

2. *Labor*

Innovation itself is usually not labor intensive, for the quality, or expertise, of the labor involved is more important than the quantity. The personal qualities and institutional forms which lead to effective performance in organized research and development will be discussed in Chap. 3. As mentioned above, education, resulting in the embodiment of scientific and technical knowledge, is an important factor in the development and utilization of innovations. Thus it is understandable why almost all R & D personnel have some formal education in science and technology, often extending to graduate degrees.

3. *Material Resources and Energy*

Although material resources and energy are necessary inputs to innovation, the innovation process itself uses only a small amount of them. Of course, once an innovation becomes widespread, it can result in vast changes in material and energy supplies.

There are two ways to view these resources: (1) as depletable stocks of particular entities (Meadows et al. 1972); or (2) as entities possessing certain specified properties or performing definite functions (Chap. 7 below; Scott 1962). The latter view is based on what may be one of the most significant forms of technological innovation, namely the substitution of one entity for another when the other is no longer physically obtainable or too costly. Not only does substitutability allow production to continue, but it also offers potential for improvement because of different properties of the new substance. Chemical developments in plastics and polymers, for example, have vastly increased the number and variation of substances available for use while reducing dependence on specific naturally occurring substances such as cotton and rubber. Innovation in extractive industries has developed technologies to tap resources that were previously inaccessible or extremely expensive. These developments help explain why the unit cost of extractive products in the United States fell from 1870 to 1957. Of the four main extractive areas--agriculture, fishing, forestry, and mining--only forest products have increased in cost, and even this increase was accompanied by large-scale substitution of nonforest products (Barnett and Morse 1963; Chaps. 7 and 8 below). Over the same period, output of resources as a percentage of the U. S. Gross National Product has declined from 36 per cent in 1870 to 12 per cent in 1954 (Rosenberg 1973, p. 112). Technology, through substitutability, has decreased man's relative dependence on specific resources--although not, of course, on resources as a whole.

C. ENDOWMENTS--SUPPLY AND DEMAND

1. *Demand*

Endowments such as capital, labor, and resources affect innovation through both supply and demand. Demand operates on the side of the product market, the market in which the innovation is to be sold. Surveying eight **studies** dealing with the relative roles of supply and demand in generating innovations,[2] Utterback (1973) found that between 60 and 90 per cent of the innovations considered were generated in response to market needs and demands. The others were generated as response to scientific and technical opportunities, a supply factor.

On the industry level, at least in capital goods industries, Schmookler (1966, Chap. 6) found after extensive research that changes in investment in an industry were correlated with changes in numbers of capital goods patents. This approach tells us nothing about specific innovations, only about the aggregate of innovations in an industry, for he relied on the number of patents as a surrogate measure of innovation. (The problem with this assumption is discussed in the Appendix on measurement.) Both time series and cross-sectional data were advanced in support of his conclusion.

Schmookler's most complete time series study covered the railroad industry for a period of over a century. He found that railroad investments and railroad capital goods patents exhibited great similarities both in long term trends and long swings. The difference between these two curves was that at the lower turning points in major cycles, the upturn in investment occurred earlier than in capital goods patents. Using somewhat less complete data, he obtained similar results in the petroleum refining and building industries. In brief, demand induced or stimulated invention.

His cross-section work consisted in comparisons of capital-goods patents and investment in over 20 industries. He compared the logarithm of investment in 1939 and 1947 with capital-goods patents in the three subsequent years. Very high correlations were obtained between the logarithm of investment and capital-goods patents (Schmookler 1966, Chap. 7).

Schmookler also argued for the demand hypothesis in the case of consumer goods (1966, Chap. 9). In doing so he relied on consumer preference as generating the demand. However, generating a demand artificially by advertising an existing supply does not constitute demand but supply. Lacking any firm data in the consumer goods area, his results are inconclusive.

2. Supply

Looking at the other side of supply-demand, we find that a number of important inventions, such as the transistor (Nelson 1962) and nylon (Mueller 1962), have come out of programs in basic scientific and technical research undertaken by firms in areas of their interest. It is at least arguable that in some cases initial key inventions marked the beginning of the sort of demand cycle that Schmookler studied. Supply of crucial scientific and technical information leads to the innovation (as in the case of the transistor), which creates an investment opportunity that stimulates further innovative activity.

Analogously Hollander (1965; see also Chap. 3 below) found in studying DuPont viscose-rayon plants that subsequent to a major technical improvement generated by corporate R & D efforts, a host of minor improvements by production personnel and suppliers took place. He observed that these improvements tended to decrease in time, i.e., the potential for improvement became saturated.

In a recent paper, Rosenberg (1974) analyzed Schmookler's demand hypothesis (1966) with great care. Schmookler, he pointed out, assumed it is possible to get as many inventions as wanted in any industry at any time at a constant price. However, neither this case, nor the case where no inventions are possible at any price, is significant.

> The perspective which I am suggesting, therefore, states that, as scientific knowledge grows, the cost of successfully undertaking any given, science based invention declines. . . . Thus, the growth of scientific knowledge means a gradual reduction in the cost of specific categories of science-based inventions. The timing of inventions therefore needs to be understood in terms of such shifting supply curves for individual industries, depending upon the knowledge bases upon which inventive activity in that industry can draw (Rosenberg 1974, p. 107).

To Rosenberg the important economic question is: Given the state of the sciences, at what cost can a technological end be attained? This view considers the importance of both supply and demand together with their interaction. It would be interesting to take the matter one step further back to see how, if at all, economic demand acts to generate basic scientific knowledge.

Markets for factors other than knowledge also affect the supply of resource endowments. The firm must, in the appropriate factor market, pay for the labor, capital, and material it uses. Changes in factor prices and price ratios may be a stimulus to innovation by developing substitutes for costly factors. The factors most frequently considered in the literature have been capital and labor, but it would appear from our earlier consideration that material resources and scientific and technical knowledge are equally significant.

Hicks (1963) drew the distinction between innovations designed to offset changes in factor prices, which he referred to as induced, and all other innovations, which he referred to as autonomous. Induced innovation, he claimed, was biased toward labor-saving innovations as opposed to capital-saving innovations. The reason was that the cost of labor in industrial societies was rising faster than the cost of capital. His findings indicated that no such bias existed in autonomous innovation.

This finding contrasts with Salter's (1960) argument that there was no bias toward innovations which saved in any particular factor. "When labor costs rise, any advance that reduces the total cost is welcome, and whether this is achieved by labor saving or capital saving is irrelevant. There is no reason to assume that attention should be concentrated on labor saving techniques unless because of some inherent characteristics of technology, labor saving knowledge is easier to acquire than capital saving knowledge." Indeed a single firm innovating to lower its costs in the face of changing factor prices would only be concerned with minimizing total cost. Whichever path of innovation it chose could not significantly affect the factor markets because of the small size of the single firm relative to the economy. And because of the lack of bias of the innovating firm, an aggregate of innovating firms would not tend to create a macrolevel scarcity in one factor rather than another because the different firms would tend to make savings in different factors.

The basis of induced innovation is that substitutability which we mentioned

previously in connection with resources and energy. It is possible to substitute some
combination of endowments (knowledge, capital, labor and talent, resources and energy)
for another. Although the economic literature on innovations centers on capital and
labor, all the endowments are involved. In the next section we move considerations
of substitution to the societal level.

D. THE LIMITS TO SUBSTITUTABILITY

Historically the pace of technological innovation in Western societies has been
stimulated by the ability of the innovation system to substitute one sort of endowment
for another or one material or form of energy for another (Chap. 7 below). This pattern
has persisted for at least two hundred years. The modes of substitution involved
include the following (Rosenberg 1973, p. 116):

1. raising output per unit input of resources,
2. development of totally new materials,
3. raising the productivity of the extractive process,
4. raising the productivity of the process of exploration and resource discovery,
5. development of techniques for the reuse of scrap and waste material, and
6. development of techniques for the exploitation of lower grade or other more
abundant resources.

In this optimistic point of view promulgated by technological growth economists,
resources take on a dynamic character. Rather than finite and irreplenishable stocks
of particular entities, resources are substances that perform a certain function or
meet a particular set of specifications (Chap. 7 below; Scott 1962). Substitution
possibilities are extensive. These possibilities are considered for a single substitution in isolation rather than for the aggregate of all substitutions or an interacting
set of substitutions. For example, substitution of plastic for metal in the construction of automobiles would be considered largely in its immediate effects on the production of automobiles without adequate consideration for effects on the plastic and
metals industries and on macroparameters of the world system such as employment,
pollution, level of education, etc. This point of view does not explicitly consider
possible boundary conditions on substitutability in a finite system. However, as
Rosenberg is quick to point out (Chap. 7 below), past trends are not necessarily an
indication of what will take place in the future.

An alternative point of view has been developed by students of systems dynamics
(Forrester 1971; Meadows et al. 1972) who look at the future of the world on a global
level. The model of *The Limits to Growth* (Meadows et al. 1972) regards resources as
finite and irreplenishable stocks of particular substances, so that no additional input
is possible in these closed categories. At the present level of development of this
model, technology is dealt with only indirectly and particular substitutions cannot be
dealt with explicitly (see Kranzberg 1973). This model therefore does not allow for
the possibility of important breakthroughs in system balance such as Haber's discovery
of nitrogen-fixing bacteria in the early part of the 20th Century. Nor does it allow
for flexible responses to particular trouble spots. Thus, no matter what happens,
the model predicts disaster. Either food production will plummet, pollution will rise
to unacceptable levels, or some other calamity will overtake the world. One key reason
for this outcome is that certain important initial parameters of the model, which are
variable in the normal course of events, cannot be changed as the system develops over
time (Meadows et al. 1972).

Boyd (1972) argued that the Forrester (1971) *World Dynamics* model, on which *The
Limits to Growth* model is based, is sensitive to changes in assumptions, and he added
a variable representing technology to the model. He assumed, for example, that a fourfold increase in technology over the 1970 level would decrease pollution output and
natural resource input per unit of material standard of living to zero. His revision
of the model along these lines led to a catastrophe-free future. Damon (1973) pointed
out that such assumptions are inappropriate because of thermodynamic considerations,
and thus that the Forrester model's applicability has not been challenged.

We doubt if either the perspective of the growth economists or that of the world system modelers can adequately represent by itself the future of innovation in our world with its potentials and pitfalls. The economists' model finds its confirmation in a period of history when the innovation system involved only a small fraction of the earth's population. Exploitation of the earth's resource potential was limited both in the amount of environmental disruption and in the number of people benefiting from this exploitation. However, with a greater fraction of the earth's population coming within the pale of industrialization and desiring its share of material goods, the implicit assumption that each substitution process is independent of every other begins to lose its realism. The substitutability process would be treated most realistically in a finite world model with interactions among various substitutability processes. Thus, the interesting question becomes not "What are the limits to growth?" but "What are the limits to substitutability?"

Let us now consider some of the areas where these limits may occur. The population of the earth is rapidly increasing. The percentage of that population involved in industrialization, and hence in the innovation system, is also increasing. As a result the per capita use of resources throughout the globe is accelerating. At the same time, because of depletion and substitution of resources, the industrial process is becoming more intensive in matter and energy usage. There are constraints on the total supply and extraction of particular resources; there are also problems in recycling enormous quantities of materials, and in dealing with large amounts of waste and pollution when these materials involve a significant part of the entire system. The rate of social change should, as a minimum, be such as to satisfy widespread expectations, and as a maximum be such that social disruption is not created by the rapid learning of new techniques and the loss of jobs by persons affected by the changes.

A very crucial area in which limits to substitutability arise is in human values. Values of one sort or another can both encourage and hinder innovations and substitutions. Thus, there appear to be both upper and lower limits to the rate at which substitution, the mechanism for innovation, can take place.

Although we certainly cannot at this time answer the question "What are the limits to substitutability?" we have framed the question and have indicated some areas which can be searched to provide whatever answer is available. An understanding of the future of innovation on the societal levels depends on a realistic and fruitful integration of the perspectives of the technological growth economists and the world systems modelers. Both have contributed to our understanding, but neither provides a complete or adequate description of the complex system we inhabit.

IV. INSTITUTIONS

Along with values and endowments, institutions are an important exogenous force acting on the innovation system. Institutions are social organizations that have been developed to secure the goals shown by societal preferences, utilizing the endowments available. They receive their legitimization from societal goals, and their efficiency is judged in terms of how well they meet these goals.

As a pluralistic society the United States possesses a variety of institutions. Economic questions are influenced by two basic sets of institutions: one primarily economic and the other basically political. In terms of the percentage of output the predominant set is the economic. Firms receive signals through the market mechanism, and then decide what to produce and how to produce it. Such decisions involve the allocation of resources for R & D, introduction of new products, or the adoption of a new technique. As was explained above, innovation is stimulated in general by market demand rather than by the availability of scientific and technical knowledge. The market mechanism provides the incentive and legitimizes the decisions made.

The second set of institutions serving as determinants and modifiers of the innovation process is political. Political institutions impinge on the economic in many ways. For one thing, the percentage of goods and services produced by or for the government or regulated by it is increasing. Furthermore, the government as a major purchaser of goods, especially high-technology goods, influences the innovation process of these goods. Additionally, the government has underwritten directly and through its purchases of goods a large portion, in some years a majority, of the country's organized R & D effort.

The government's impact on innovation can be indirect as well as specific. In carrying out its regulatory, fiscal, and other functions the government can either encourage or discourage innovation in one field and act as a powerful disincentive in another, as with standards lowering automobile engine emissions.

A. ECONOMIC INSTITUTIONS

1. *Technological Progressiveness*

Most economic institutions, and especially the market, are not judged solely or even primarily by how they impact on the innovation process. The basic criterion has been the efficiency with which the economy, utilizing existing technology, translates its given stock of resources (endowments) into products and services. This criterion has received central attention for at least two major reasons. The more important has been that society values highly the current production of existing goods and services produced with existing technology. Increasing the efficiency of this process will allow more goods to be produced with the present endowments. Allocation of resources to innovative activity may have the short-run effect of reducing the current level of output although the effect on long-run output may be very salutary.

The second major reason is that the basic economic models for judging efficacy of market structures have reflected an almost single-minded interest in the efficiency of resource allocation under conditions where technology, consumer preferences, and resources are assumed as given and therefore constants (for an extended discussion see Schmookler 1965a and the works he references). Utilizing such models, economists have judged the relative efficiency of various forms of market structure in terms of resource allocation. Given these comparative-statics models where technology is treated as a constant, the competitive market structure proved to be beneficial in terms of efficiency and equity with its large number of buyers and sellers, no barriers to entry, and a standardized product. Under this type of market structure resources were allocated in relation to consumer preferences, and goods were produced in an "efficient manner." Imperfectly competitive market structures, including oligopoly, with its small number of relatively large firms controlling most of the output and with barriers to entry, provided firms with the power necessary to control supply and thus the price they received for their product. Oligopolistic firms could use their economic power to shortcircuit the process whereby resources are allocated in relation to consumer preferences.

The use of models which treated technology, preferences, and endowments as constants meant that the dynamic nature of technological change could not be dealt with by existing tools. To the extent that it was treated at all, technology was regarded as something exogenous to the basic models. Many economists either explicitly or implicitly extended their faith in the competitive market structure to include technological progressiveness (Stigler 1952, p. 1; Adelman 1954, p. 157).

2. *Schumpeter and Galbraith*

A group of economists led initially by Joseph Schumpeter and later by J. K. Galbraith strongly assailed this preoccupation with judging market structure by the measuring rod of static efficiency. Schumpeter vigorously attacked the proposition that the competitively organized market provided the best impetus for innovative activity. In developing his non-Marxian dialectic of "creative destruction," he argued strongly and lucidly for the technological progressiveness of the large corporation in oligopolistically organized industries. For example, Schumpeter (1943, p. 28) stated:

> As soon as we go into details and inquire into the individual items in which progress was most conspicious, the trail leads not to the doors of the firms that work under conditions of comparatively free competition but precisely to the doors of the large concerns--which, as in the case of agricultural machinery, also account for much of the progress in the competitive sector--and a shocking suspicion

dawns upon us that big business may have had more to do with creating that standard of life than with keeping it down.

Schumpeter's disciples, such as Kaplan (1954) and Villard (1958), have expanded on his ideas.

It was left to John Kenneth Galbraith, in his book *American Capitalism: The Concept of Countervailing Power* (1952, p. 861), to add heat if not light to the debate when he too challenged the conventional wisdom, throwing down the gauntlet with the following statement:

> Moreover, a benign Providence who, so far, has loved us for our worries, has made the modern industry of a few large firms an excellent instrument for inducing technical change. It is admirably equipped for financing technical development. Its organization provides strong incentives for undertaking development and for putting it into use.

The Schumpeter-Galbraith hypothesis is first and foremost a challenge to judge the efficiency of economic institutions in terms of how well they perform in providing goods and services through time, not at a single point of time. Under this standard, technology ceases to be a constant and becomes one of the critical variables.

This hypothesis raises a series of questions concerning the relation between the firm as the innovating unit and the market as a determiner and modifier of innovative activity. First, questions arise about the effect of the size of the firm itself. Is the innovative process subject to economies of scale due to the "lumpiness" of capital, the advantages of specialized personnel and equipment, and the ability to spread risk over a larger set of activities? Additional questions naturally follow. Is a certain minimum size needed to engage effectively in innovative activity? Do economies of scale, if they exist, continue for all scales of output, or do diseconomies of scale appear after a certain size is reached?

A second set of questions relates to the effect of market structure on innovation. For example, how does an oligopolistic market structure affect both the firm's ability and willingness to engage in innovation? Since an oligopolistic market structure presupposes barriers to entry into the industry, do these barriers retard or encourage innovative activity? In the discussion below we examine attempts to answer these questions.

3. *Firm Size and Innovation*

The Schumpeter-Galbraith challenge encouraged economists to examine the relationship between firm size and innovation. Unfortunately, reliable measures of innovation do not exist (see Appendix below). For one thing, it is often difficult to distinguish between major innovations and minor improvements in products and processes. In addition, most of the cost and demand data needed for any measure of the cost and impact of particular innovations are considered proprietary information by the firm. Furthermore, at present the methodology does not exist to classify and measure all the direct and indirect effects of an innovation. This lack has led researchers to rely on the data on R & D expenditure or employment as a proxy for the rate of innovation. Hence the debate concerning firm size and innovation has thus been carried on primarily in terms of the relation between firm size and organized R & D activity. But such measures of innovative output suffer from the fundamental difficulty that inputs are being used when what one should like to measure is output. Only to the extent that organized R & D and innovation vary together can the results be used without qualification. This correlation is, of course, unknown.

After the renewed challenge of Galbraith and his disciples in the 1950s and early 1960s a series of studies appeared dealing with various aspects of firm size and market structure and their effects of innovation. Public attention was focused on the relationship when the U. S. Senate Subcommittee on Antitrust and Monopoly held hearings on Concentration, Invention, and Innovation in May and June of 1965. Commenting on the empirical material appearing around that time, Markham (1965) proposed a "threshold" relationship between firm size and innovation. He suggested:

> Up to a certain size, innovational effort increased more than proportional to size; at that size, which varies from industry to industry, the fitted curve has an inflection point and among the largest few firms innovational effort generally does not increase and may decrease with size (p. 329).

That there appears to be some threshold size necessary for effectively engaging in organized R & D has found support in several studies.[3] For example, Freeman has presented evidence that the threshold for effective R & D may be quite high in the electrical equipment industry (1965). Additional empirical work aimed at documenting the existence and level of this threshold for various industries would be extremely useful.

The idea of a threshold is certainly in accord with data collected on organized research. The National Science Foundation's annual survey on Research and Development in Industry (1973) shows that large firms are responsible for the expenditure of most of the R & D funds. In 1971, the 298 firms with over 10 000 employees performed over four-fifths of all industrial research and development. The twenty largest companies in terms of R & D expenditures accounted for 54 per cent of all industrially performed R & D and received 72 per cent of all Federal R & D funds for 1971. Indeed, five industry groups accounted for 81 per cent of all expenditures in 1971 (NSF 1973).

The National Science Foundation (1973, p. 61) data also show that on the average the larger firms, those with over 10 000 employees, have a much higher ratio of R & D funds to net sales (4.2 per cent) than do smaller firms (1.6-2.1 per cent). Despite this general tendency, there is wide variation among industries in terms of R & D expenditures as a percentage of net sales--ranging from 0.4 per cent for "textiles and apparel" to 16.6 per cent for "aircraft and missiles."

Hamberg (1966, pp. 45-49) has shown that this generally positive relationship between firm size and R & D expenditures varies greatly between industries and is sensitive to the size variable used; it appears stronger when number of employees rather than assets or sales is used as the size variable.

Granted that large firms do more R & D than small firms and that in general a positive relationship between firm size and expenditures on R & D exists, does that mean that continued growth in a firm's size will bring an even greater growth in R & D activity? Markham (1965, p. 329) suggested that there was a leveling off and even a reduction in activity as size increased beyond some point. This idea of a crossover point finds other support in the literature. Scherer found that when he divided the economy into four large industry groups,[4] R & D employment per dollar of sales decreased in three out of the four groups after firm sales reached $200 million. (The exception was basic chemicals and drugs.) When individual industries were examined, the same relationship obtained, although there was considerable variation between industries (Scherer 1965a, pp. 1195-98). Indeed, there is evidence that in the primary metals industry there may well be a negative relationship between R & D expenditures and firm size beyond a very small size (Hamberg 1966, p. 59). The evidence tends to support the proposition that there is a generally positive relationship between firm size and R & D expenditure, but that beyond some size this relationship weakens and may actually decline.

Another attempt at measuring the relationship between firm size and innovative activity has been to relate the patents secured to measures of firm size. At first glance this method does seem superior to that previously discussed in that one would be measuring an output, a patented invention, rather than inputs, R & D expenditures or employment. However, certain grave difficulties are encountered in the use of patent statistics. There is some indication that the number of patents may be a better measure of research expenditure than a measure of the number of inventions (Comanor and Scherer 1969). There is also great variability both in terms of the importance and the quality of patents issued, confirmed by the amazingly high rate of successful challenges to patents, which some have asserted is as high as 70 per cent (Kauper 1973).

Scherer found that for the economy as a whole, sales were more concentrated than patents. When he divided the economy into the same four industry groups discussed above, he found that the number of patents per dollar of sales declined after $200 million of sales in three out of four groups, the exception again being chemicals and drugs. The same pattern was present when individual industries were examined one by one. These data show that large firms do not obtain a **disproportionate** share of patents, given their sales (Scherer 1965, pp. 1195-96). They also indicate that the cost per patent pending is higher for larger firms (Sanders et al. 1959, p. 238). What is missing, however, is some measure of the relative importance of the patents obtained by firms of various sizes.

Given the difficulties associated with patent statistics as measures of innovative activity, definitive statements cannot be made. However, patent studies do tend to support some of the conclusions arrived at in studies utilizing R & D expenditures and employment. It does appear that a crossover point exists in terms of firm size and R & D activity and that beyond this point larger firms do not account for a disproportionate share of patents. Indeed, if anything, the contrary may be true.

Hence, there is no unambiguous answer to that part of the Schumpeter-Galbraith hypothesis that argues that large firm size is needed for innovation. Instead, we are left with four basic observations:

1. There is a minimum size in terms of sales needed to engage effectively in organized research.
2. There is a generally positive relationship between firm size and R & D activity.
3. There is a point beyond which increases in sales do not bring forth proportional increases in either R & D activity or patents.
4. There is a great deal of variation **among** industries.

This last point argues against continued attempts to secure a completely general answer to the relationship between firm size and innovation and argues for industry-level studies. It would appear that the returns would be higher for research which attempted to identify threshold and **crossover** points for various industries and types of innovations within particular industries than for work at a more aggregated level.

4. *Concentration and Innovation*

The structure of the market in which a firm sells its product affects the signals the firm receives and thus influences both its willingness and its need to innovate. Schumpeter and Galbraith in their challenge to "conventional wisdom" not only claimed that large firms were conductive to innovation, but also stated that large firms in industries characterized by a small number of firms were more innovative. Here we discuss the effect of these two aspects of market structure on innovation.

Unfortunately, economic theory is of little help in providing *a priori* models for dealing with the effect of concentration of output on innovation. The reason for this lack is readily apparent. A comprehensive theory would have to integrate the complexity of oligopoly theory, where there is mutual interdependence between the firms in the industry, with the complexity associated with the dynamics of technological change.

Attempts to treat this problem theoretically have most often ended simply in lists of influences which affect the willingness of firms to innovate. For example, Scherer (1970) has developed a model to identify the factors determining whether the speed and amount of innovation will be greater for more concentrated or less concentrated industries. He identifies (p. 369) two basic factors. The stimulus factor relates to "the marginal conditions for profit maximization, an increase in the number of sellers . . . being conductive to rapid innovation." The second factor, the *Lebensraum* factor, relates to the requirement that adequate profits must exist, and that beyond some point a large number of firms may discourage innovation. In an industry with a large number of relatively small firms, the desire to innovate may be strong, but the financial and organizational capacity may be lacking. In industries dominated by a small number of large firms, the capacity may exist but the stimulus may be lacking. Thus in any particular industry, the result depends on the interaction among several variables, such as the size of the overall profit potential, the number of actual or potential competitors, the speed with which rivals can be expected to imitate, the degree of benefit that being first confers on the firm in terms of permanent product differentiation, the effect on the profits of present products and processes, and the magnitude of anticipated R & D cost (Scherer 1967b).

Although **additional theoretical work** is needed, the basic question of the relationship between market structure and innovation is an empirical one. The device used to measure the presence of oligopoly is the concentration ratio, which is the percentage of total output controlled by the largest firms in the industry. A four-firm concentration ratio of 60 per cent means that the four largest firms control 60 per cent of the output of the industry. Unfortunately, concentration ratios as a measure of market power are subject to deficiencies such as arbitrary industry classification, failure to include the competitive effects of substitutes or foreign goods, etc. However, they remain a good first approximation of the degree of oligopoly power that exists in the market (Scherer 1970, pp. 52-57).

Attempts to answer the question of whether more concentrated industries are more or less innovative than less concentrated industries have encountered continual problems. For example, in the late 1950s both Stigler and Phillips attempted to measure the relation between concentration and innovative progress in terms of productivity, i.e., output per worker. Stigler (1956) used firm concentration and found a negative correlation between concentration data and productivity; Phillips (1956) used plant concentration data and found a positive correlation. Stigler's choice of data was more appropriate for answering the question of the relation between firm size and technical process, but the seemingly contradictory results do illustrate the difficulty in securing unambiguous anwers.

In several empirical studies employment of scientists and engineers in R & D has been used as a proxy for innovation and has been related to concentration. These studies have shown that the ratio of scientists and engineers employed to total employment is generally higher when significant amounts of concentration exists (Scherer 1967a). Indeed, in Comanor's study (1967, pp. 645-52) the employment of scientists and engineers in R & D was found to be twice as high in concentrated industries than in unconcentrated industries even when the results were adjusted to take care of the relationship between firm size and R & D employment. When four-firm concentration was low (10-14 per cent), firms were found to hire very few scientists and engineers. The maximum percentage occurred with four-firm ratios of 55 per cent, and at both lower and higher ratios the percentage of scientists and engineers was less (Scherer 1967a).

The results of this type of study must be evaluated with care, for the nature of the technology in a particular industry may mean that the cost of hiring large numbers of R & D scientists and engineers may be a barrier to entry rather than concentration facilitating a large R & D effort. When variables representing opportunity for technological progress in an industry were introduced explicitly into the analysis, the importance of concentration in explaining differences in the percentage of scientists and engineers employed in R & D to total employment was reduced significantly (Comanor 1967). It appears that R & D activity is correlated with both concentration and opportunity, that is, with the "supply" or possibility of an innovation that promises profit.

A few very excellent studies on the relation between market structure and innovation have been conducted at the industry level. Mansfield's 1963 study is especially noteworthy and deserves treatment in some detail. In it, he related firm size and industry structure to the number and importance of innovations in three basic industries: iron and steel, petroleum refining, and bituminous coal. He began by surveying officials of trade associations and trade journals and asking them to identify and rank in importance the most important processes and product innovations in their fields for the period 1918 to 1958. Using these same "experts" plus company records and trade publications, Mansfield was able to identify where the first commercial application occurred in 80 per cent of the innovations, or 150 *in toto*. His basic question was whether the number and importance of the innovations carried out by the four largest firms exceeded their relative share of the market. He received a positive answer for the coal and petroleum refining industries, but not for iron and steel, where smaller firms tended to do more than their relative share of innovation. He further noted that:

> The largest four firms seemed to account for a relatively large share of the innovating in cases where (1) the investment required to innovate was large relative

to the size of the potential users, (2) the minimum size of firm required to use the innovations profitably was relatively large, and (3) the average size of the largest four firms was much greater than the average size of all potential users of the innovations (p. 573).

The ethical (prescription) drug industry has also received attention. In this industry the largest firms are not responsible for a disproportionate share of innovations; indeed their share of innovations is less than their share of market sales (Mansfield et al. 1971, p. 167). Substantial diseconomies of scale in R & D seem to exist in the industry, so that "the marginal productivity of professional research staff is inversely related to the size of the firm" (Comanor 1965). It does appear, however, that the economic impact of innovations by the larger firms tends to be greater than for smaller firms (Comanor 1965; Mansfield et al. 1971).

Industry level studies such as that by Mansfield and those dealing with the drug industry offer useful insights into the interaction between market structure and innovative activity. As this body of literature grows, the aspects of the relationship between industry structure and innovation which are susceptible to generalized statements and those aspects which are industry specific should emerge more clearly.

5. *Entry and Innovation*

A second component of market structure which affects signals received by the firm is difficulty or ease of entry into the industry. Where entry barriers are high, existing firms may feel insulated from innovative pressures of potential market entrants. Easy entry into an industry may discourage existing firms from investing in innovative projects, especially if the projects are expensive, risky, and/or easily imitated. Barriers that retard entry include the need for expensive facilities, the lack of basic patents, and difficulty in competing with market-dominating firms. In many cases a new or existing firm entering an industry for the first time faces a large investment and a significant amount of risk and uncertainty.

An inadequate rate of innovation by existing firms in an industry can be overcome by the entrance of new innovating organizations into the industry if the barriers to entry are not too high. Schon (1967) calls this "innovation by invasion." The invaders can be small, independent firms just coming into existence, or established, technologically sophisticated firms well entrenched in other industries, or even foreign firms. For example, the chemical industry "invaded" the textile industry with both finishes and new materials, and the aircraft industry "invaded" the machine-tool industry with numerical control devices (Schon 1967, pp. 139-171). It was Westinghouse and General Electric which developed jet engines, not the existing firms in the aircraft engine business (Nelson et al. 1967, p. 72). Brown (1957), in a study of new firms developed in the postwar period in the Connecticut Valley, found that many were established by former employees of larger companies that were not interested in applying and developing the inventions made under their auspices.

Ease of entry influences the innovativeness of an industry in at least two ways. First, new firms often bring innovation into the industry. Second, the existence of potential competitors and especially potential innovators may cause, and in many cases require, existing firms either to innovate or to imitate quickly in order to protect a dominant market position. For example, Gillette imitated rather quickly on the introduction of stainless-steel blades in the U. S. market in the 1960s. Sperry-Rand's introduction of UNIVAC digital computers was quickly followed by a set of innovations from IBM. There is some indication that whereas dominant firms in some industries may be slow innovators, they are very quick imitators once their position is threatened (Scherer 1970, p. 371).

Comanor found that moderate barriers to entry, in the form of optimal plant size being 4-7 per cent of industry output and/or capital investment being from $20 to $70 million, produced a higher percentage of R & D employment relative to size than did either higher or lower concentration ratios. He attributed this phenomenon to the fact that lower concentration ratios might represent too easy entry, and profits from innovation could be too rapidly competed away, whereas higher concentration ratios represented situations where the firm felt itself insulated from pressures to innovate (1967, pp. 652-61).

The answer to the question of which industry structure is more conducive to innovation is far from clear. It would appear from the fragmentary evidence that, as Markham suggests (1965, p. 325),

> some departure from the state of perfect competition (or the presence of some monopoly) is a necessary concomitant of innovation, but it does not follow that twice this volume of departures, somehow measured, should lead to twice the volume of innovation.

It would also appear that the degree of concentration needed to secure an "adequate" innovative performance depends on the interaction of a number of variables and may differ widely from industry to industry. The identification of these variables and how they interact in particular industries deserves increased research attention. We may hope that patterns will emerge from these studies that will allow cross-industry comparison.

6. *Unanswered Questions*

It is apparent from the previous discussion that the effects of economic institutions as determinants and modifiers is incompletely understood. The Schumpeter-Galbraith hypothesis proposed a generalized relationship among firm size, market structure, and innovation. As the studies detailed above show, a great deal of variation in the magnitude, and in some cases even the sign, exists between these variables and the rate of innovation in various industries. Disaggregation, i.e., dealing with smaller units, would appear to be a wise course to pursue.

This disaggregation can take several general forms. One would be industry-level research to identify the existence of economies and diseconomies of scale in the various phases of the innovation process for particular industries. Another might be determination of the existence and size of the threshold necessary to engage effectively in R & D and the determination of "crossover" points where increases in size bring less than proportional increases in organized R & D effort. Also needed is additional work on the relationship between the scientific and technological base of an industry and innovation. Are firm size and market structure functions of technology or are they based on financial, marketing and other considerations?

Subindustry level studies would also appear to be in order. Innovations within the confines of a given industry differ in terms of the technologies they use and the endowments they require. It may well be that smaller firms have economic and organizational advantages in innovations requiring sophistication and flexibility or catering to specialized needs. Larger firms may do best in areas requiring large production, marketing, and capital resources. Rather than assuming that a market with a large number of small firms or a market with a small number of large firms is optimal for encouraging innovation, it may well be that we need firms of all sizes--each to innovate where it has an advantage. This in turn requires that we examine mechanisms relating to barriers to entry and economic power relationships within an industry, and how they may be modified to allow for the continued viability of small firms while not adversely affecting the incentives for innovation by larger firms.

At the heart of the discussion are the concepts of an "industry" and a "firm." In many cases the formal definition of an industry is arbitrary and does not capture the essence of how the market operates to carry out the goals of society. For example, industries are often defined in terms of a particular technology, product, or even material. Since innovation has the effect of changing the technology, introducing new products, and changing the materials used for certain functions, industry definitions are a product of previous, sometimes obsolete, technology and as such tend to be static. Also, too little attention may be given to the effect of interproduct competition among goods performing the same basic functions that are produced in different countries. For example, structural steel has encountered competition from extruded aluminum, laminated wood, and special types of reinforced concrete. Measuring the effect of industry structure when the concept of an industry is itself subject to change and may be inappropriate for the question at hand can only lead to confusion or error.

Along with the concept of an industry, most studies use the concept of a firm as a point of departure. A firm is basically a legal concept, which may or may not be functional. Many large, highly diversified, and decentralized firms operate in a number of markets, producing a wide variety of goods, utilizing a number of different technologies. For example, Textron was originally in the textile business. Between 1943 and 1964, it acquired 70 companies outside the textile industry. By 1964, Textron was out of the textile industry and in 37 other industrial product areas from helicopters to watch bands (Harris 1965). Perforce its R & D and other innovative activities reflect the diversified and decentralized nature of its operations. To classify this firm as belonging to the industry wherein its sales are highest--as is currently done--is clearly inappropriate. As an important first step in gathering information about firms from a functional point of view, the Federal Trade Commission is beginning to collect product-line information from large conglomerate firms.

B. POLITICAL INSTITUTIONS--DIRECTLY INNOVATIVE

The United States has been traditionally a free-enterprise economy. Theoretically in such an economic environment the government plays a secondary role in fostering technological innovation. In practice, however, throughout American history, the government has developed and fostered institutions part of whose activity has been technological innovation. These institutions either involve areas of direct constitutional concern to the United States government such as defense, or areas of public concern wherein the private sector lacked the resources or the will to make privately funded R & D feasible, such as agriculture.

These governmental institutions follow three organizational patterns. First are organizations doing in-house research and development, such as the National Bureau of Standards. Then there are institutions which fund research in the educational, nonprofit, and private sectors through a system of contracts and grants, exemplified by the National Science Foundation. The third type is a mixture of in-house and outside research funded by grants and contracts, with the institution coordinating all components. The National Aeronautics and Space Administration (NASA), the Energy Research and Development Administration (ERDA--formerly the Atomic Energy Commission, or AEC), and the Department of Defense (DOD) pursue this mode, which perhaps provides the most operating flexibility of the three types.

Before World War II the private sector was the main supporter of R & D. In 1930, for example, private-sector expenditures for R & D were six times those of the Federal government (Markham 1962, p. 592). Before World War II government R & D expenses were concentrated in agriculture and defense, in that order (Mansfield 1968b, p. 118).

However, by 1972 the Federal government accounted for over half the national expenditures for R & D (National Science Foundation 1973, p. 22) while the total national outlay for R & D had grown from $166 million in 1930 (Markham 1962, p. 592) to about $30 billion in 1972 (National Science Foundation 1973, p. 22). By 1960 most of the Federal R & D money went to DOD, NASA, and AEC (Mansfield 1968b, p. 48), a byproduct of concern for defense and the cold war. In 1966 these agencies took up over 80 per cent of the Federal R & D budget, although a report in the 15 February 1974 issue of *Science* indicates that this fraction had since declined to just under 75 per cent in the budget estimate for fiscal year 1975. Since most of the enormous increases in this budget have been in defense and defense-related areas, the justification for these expenditures has been that of national security.

In 1967 the proportion of research and development funded by the Federal government varied from 80 per cent in the aircraft industry and 61 per cent in electrical equipment and communications to 0 per cent in some industries (Mansfield 1968b, pp. 52-54). Mansfield noted that the industries receiving the greatest proportion of Federal support for their R & D efforts are among those having the highest ratio of R & D expenses to sales.

Over the decade 1963-1972 there was an increase in Federal R & D support for civilian technologies in the areas of health, transportation, energy, and environmental protection (National Science Foundation 1973, p. 25). Since the justification for government R & D funds in the civilian sector cannot normally be national security, a new justification is called upon; a dependence on the "general welfare" mentioned in the preamble to the U. S. Constitution seems to be emerging. Because health, transportation, energy, and environment involve us all, many R & D projects in these areas which carry too high a risk for private venture capital are justified in the name of the "general welfare."

With the Federal government contributing over half of the nation's total R & D funds, it is a very important factor in determining those areas of technology where innovative activity will take place (Markham 1962). Similarly, the government directs the training of scientists and technologists to areas of its interest; the research training of graduate students is determined, in part, by where the money is. Thus the direction of Federal research support is a determinant of areas of training for participants in the innovation process and determines in a broad sense those areas in which scientific and technical knowledge will be developed as well as the level of activity undertaken.

Government institutions are involved in the diffusion of innovations as well as their development. A primary example is the cooperative (Federal and State) Agricultural Extension Service through which information about innovations is brought to farmers. This program provides a two-way channel between the agricultural researcher and the farmer (Baker et al. 1963; Knoblauch et al. 1962). This kind of effective feedback mechanism does not exist in NASA's Technology Utilization Program, in which technologies developed in the course of NASA's missions spill over into other sectors of the economy (Doctors 1969). Despite user needs, NASA's innovative activity is perforce constrained by its missions directed at space exploration and aeronautics.

Although education is not usually thought of as an institutional medium for diffusing innovations, information about technological innovations is spread in the classroom. To the extent that this process impresses student scientists and engineers with the desirability of innovation, it has the long-range effect of developing an innovative frame of mind. Of course education also imparts to the student knowledge and techniques subsequently useful in innovation--and, where the emphasis is on scientific research and engineering design, it might be said to "teach" innovation.

Thus we see that the Federal government has financially supported innovative activity by others, and has actually undertaken such activity itself. Although its greatest direct participation will likely continue to be in defense and space related areas, "general welfare" areas such as health, transportation, and energy will probably increase their share of Federal R & D expenditures. We shall deal with its second role, regulating and controlling innovation, in the next section.

C. POLITICAL INSTITUTIONS--REGULATORY AND CONTROL

In addition to government institutions directly involved in the process of innovation there are others which regulate, control, and guide the process by both direct and indirect means. These institutions are primarily of two sorts. Institutions such as the patent system and tax structure directly control aspects of innovation; regulatory bodies--such as the Federal Communications Commission, the Interstate Commerce Commission, and institutions involved in antitrust policy--influence innovation indirectly by their control of firm and industry structure, rates, and accounting practices. Although the primary concern of the latter is not innovation, such controls act to modify the economic institutions involved in innovation. Other government policies not explicitly related to innovation, such as environmental protection, also influence innovation.

The existence of such institutions derives from social and political values which, while preferring the free enterprise system as a vehicle for economic development, insist that it be regulated for the general welfare.

1. *The Patent System*

Article I, Section VIII, of the U. S. Constitution gives Congress the power "to promote the progress of science and useful arts by securing for limited times to authors and inventors the exclusive right to their respective writings and discoveries." In this view the invention is the property of the inventor, and the protection of his property provides an economic stimulus to innovation.[5]

The first Federal Patent Act became law in 1790. All patents granted before 1836 were awarded on application without examination into the merits or novelty of the inventions. By 1836 the patent system was reformed, and patent examiners were appointed to compare the applications with the prior art to determine novelty and usefulness. Over the years the U. S. Patent Office has issued over 3.5 million patents and continues to issue them at the rate of about 1250 a week.

A patent can be obtained from the U. S. Patent Office by filing an application, disclosing the invention, paying the requisite fees, and then waiting an average of two years for it to go through the process of examination and approval. To receive a patent the invention must represent something more than a trivial improvement. The search for novelty by the patent examiners must be made through all existing patents. In order to expedite the search, the patents are filed by subject in some 85 000 subclasses. When awarded, the patent grants to the inventor a 17-year monopoly on his invention which he must be prepared to defend in court at his own expense against infringements.

Since 1836 there have been minor improvements but few major changes in the patent system.[6] Yet during this period the social and economic conditions in the United States altered tremendously--and so did the process of innovation. The agent of invention changed, in the majority of cases, from an independent inventor to an employee of a corporate R & D department. Beginning late in the 19th Century the USA embarked on a course of limiting monopoly through antitrust legislation, eventually forcing, in some cases, licensing arrangements that weakened the monopoly position of the patentee. In the mid-20th Century the Federal government became the country's largest spender on R & D--both at government facilities and by contracts and grants to the private sector. Because of such changes it is not surprising that problems have arisen in the operation of the patent system. These problems include not only the factors mentioned above, but also such elements in the operation of the patent office as the long period between the patent application and its disposition, and the fact that every invention regardless of field or magnitude is treated the same.

In its beginnings, the patent system was designed for the individual inventor. It protects his property while he seeks ways of developing it. The firm, on the other hand, does not need patent protection as much, and in some cases may find it a liability. For example, a company which develops an invention often has the technical capability to bring it to commercial application before any competitor can. Provided the innovation is properly marketed, this head start may more than compensate for the lack of patent protection: if the firm discloses its invention in a patent application, it allows its competitors an information base on which to "invent around" its patent by developing a patentable invention which performs a function similar to the original invention. "Inventing around" is a common practice, for example, in the ethical drug industry.

Government-financed research also presents problems. What should be done with an invention produced in a government research laboratory by a civil servant? Policy varies from agency to agency. NASA makes a cash award to the inventor with the amount depending on the "importance" of the invention, and licenses the invention for private use. NASA's view is that "NASA-owned inventions will best serve the interests of the United States when they are brought to practical application in the shortest time possible. . . . It is the policy of NASA to grant exclusive licenses when such licenses will provide the necessary incentive to the licensee to achieve early practical application of the invention."[7] The licensee is also required to make a substantial capital investment (specified in detail) and to use his best efforts in achieving early practical application of the invention. Inventions made on government contracts require the same sort of policy decision by the contracting agency, with different agencies pursuing different policies.

Because of the great change in organizational, economic and social contexts of the institutions affected by the patent system, various proposals have been advanced for changing it to accord more closely with these transformations in its context.

Polanyi (1944) observed that using knowledge benefits the user, but the patent system operates to restrict the use of knowledge and hence decreases the social benefit arising from innovation. His solution was to have all inventors make public disclosures of their inventions which could be freely used as long as the user submits to the patent regulatory body the information necessary to determine the social value of the invention. The inventor is then given a monetary reward from public money based on the social value of his invention. Of course, the difficulty with this proposal is the problem of arriving at an effective way of measuring the social value of an invention.

Markham (1962) thought that a crucial difficulty with the patent system was that it treated all inventions, whether major breakthroughs or incremental improvements, equally. His solution was to divide inventions into two categories. The first, consisting of a major technological breakthrough, such as catalytic cracking, the transistor, and float glass, would receive monopoly protection for a long period of time; the second category, consisting of incremental improvements, would receive protection for a relatively short time span. The problem in applying this suggestion is drawing the line between the two categories.

Hamberg (1966, Chap. 2) saw two problems with the patent system from the firm's point of view: (1) industry typically performs R & D not under the stimulus of the patent system but under the stimulus of competition, and (2) government-sponsored R & D results in patents granted for "no risk" activity. He offered four options: (1) abolish the patent system; (2) substitute a system of awards, but allow free use of the technologies; (3) issue patents with terms of protection varying according to the risk of the R & D activity involved; and (4) issue no patents for government-sponsored R & D.

These suggested solutions seem to have little likelihood of adoption. Proposed changes in the patent law have in the past been aimed largely at the mechanics of the patenting process rather than at bringing about fundamental changes in its philosophy.

2. *Antitrust Policy*

The first substantive Federal antitrust legislation was the Sherman Antitrust Act of 1890. Since then United States antitrust policy has become a complex of statutory law, court decisions, and internal policies by several divisions of the Federal government. The goals of this policy are many and varied, and in practice sometimes conflicting. Among them are the following: (1) a preference for diffusion of economic power; (2) the prohibition of certain types of business activities either by statute, Federal agencies (e.g., the Federal Trade Commission), or by case law; (3) a continuing belief in the efficacy of free competition in serving the common good; (4) a similar belief concerning the efficacy of competition as a means of securing technological progress; and (5) a belief that ease of entry into industries must be maintained and that barriers which restrict entry must be torn down.

In the attainment of any one of these goals, conflict may arise with another. For example, maximum diffusion of economic power might lead to a less progressive economy because the small economic units may not possess adequate resources to innovate. The two goals bearing most closely on the relationship between antitrust policy and the innovation process are those of progressiveness and ease of entry (see above).

Antitrust policy influences some of the important properties of market institutions, strongly affecting firm size and market structure by limiting the position of any single firm in an entire industry and by insuring competition. For example, antitrust action destroyed Alcoa's monopoly of the primary aluminum industry in the United States, which resulted in an oligopoly more conducive to innovation than the Alcoa monopoly (Peck 1962).

Realizing the potential conflict between the policies of maintaining competition and insuring rapid innovation, the Charpie Report (1967, p. 52) recommended that in the interpretation and administration of the antitrust laws the effect on innovation as well as competition be taken into account and that there should be clearer antitrust guidelines for activities affecting innovation. However, to date there exists no adequate database on the interaction of innovation and competition.

Antitrust policy and the patent system interact. Patent policy encourages the granting of monopolies to inventions, whereas antitrust policy is designed to break up industrial monopolies which may or may not be based on patent protection. Thus these two policies can conflict. Beginning in the 1930s courts have tended rather strongly to resolve this conflict in favor of the antimonopoly policy. For example, Oppenheim (1957) found that between 1941 and 1957 over 100 judgments involving more than 300 antitrust defendants provided for compulsory licensing or the outright assignment of up to 35 000 patents. The judgments indicated that the compulsory licensing of patents was regarded as a means of reducing the monopoly power of a firm. If the firm chose not to patent, relying on superior technical and marketing capabilities to profit from its innovations, then the option of compulsory licensing would not be available. Indeed, Scherer (1958, pp. 124-134) found that the firms subjected to licensing decrees were the leaders in curtailing patenting. Likewise, Schmookler (1966, pp. 32-33) found that large firms concerned with possible antitrust proceedings by the U. S. Department of Justice have curtailed their patenting practices. However, this curtailment does not necessarily mean that they discourage innovation, only that they were reluctant to seek patent protection which might eventually result in their having to share their innovations with competitors through licensing arrangements.

From the evidence in the literature it is obvious that the patent system and antitrust policy are strongly coupled. Investigations dealing with either of these sets of institutions should deal explicitly with the other when areas of overlap are treated.

3. *Tax Policy*

Tax policy affects innovation in two ways. First, taxation provides the funds used by the Federal government in direct support of R & D. Second, tax policy may stimulate or impede the availability of private funds for the innovation process. At present, industrial research and development can be treated as a business expense. If the firm is profitable, the government effectively pays half the cost simply by not taxing that portion of the profits plowed back into R & D. If the firm is not profitable, there is a five-year loss-carryforward provision.

The Charpie Report (1967, pp. 30-41) dealt with the problems of increasing innovativeness by appropriate tax policy. Unfortunately, this report is biased in considering the inventor-entrepreneur and the small firm as being the prime agents of innovation in our society. These agents play a significant role, but they are no longer the main factors in technological innovation nor is their role increasing. However, although the panel's recommendations are suspect, they did indicate some very important considerations involving tax policy. From these considerations we have developed a brief list of questions suggesting areas in tax policy affecting innovation which should be studied.

 a. Should companies of different size, in different industries, and with different commitments to innovation receive the same or different tax treatment in areas of operating affecting innovation?
 b. How large should be the tax credit on R & D expenditures?
 c. How long should be the carryforward period for losses incurred in R & D?
 d. What should be the tax treatment for exploratory expense for innovations not related to present business?
 e. Should innovative activities with different levels of technological and business risks be treated differently?

It may be possible to determine a satisfactory range of tax treatment for promoting innovation. Special attention should be given to the effects of differential treatment of various undertakings involving innovation, as indicated in questions a and e above.

4. *Regulation and Regulatory Agencies*

Over the years, Congress has established various bodies to regulate certain areas of the economy. The first was the Interstate Commerce Commission (ICC), established in 1887 to regulate the railroads. Later its authority was extended to interstate trucking. The commission form has been extended to the regulation of transportation, communication, and energy.

Although the roles and duties of the regulatory bodies vary, they share common practices:

    a. A firm must be licensed by the regulatory agency controlling the activity in which it wishes to engage.

    b. The prices charged, profit received, and investments made by regulated firms are subject to approval by the appropriate regulatory commission.

    c. The firms regulated by the Federal government in their interstate business are also subject to state and local regulation for intrastate activites.

Two interesting questions can be raised about the effect of regulation on innovation. First, how has regulation of various forms affected innovation in various industries? Any answer to this question is complex, for the type of regulation varies considerably from industry to industry. Second, what kinds of regulations, if any, are most effective in fostering technological innovation? The answers to both questions are muddied further because regulation is only one of many factors affecting technological innovation.

Regulation is at times a means of inducing innovation. For example the enforcement of pollution control regulations by the Environmental Protection Agency has the effect of inducing innovations in pollution-control technologies such as the production of "cleaner" internal combustion engines. Also, AEC has developed devices and processes to insure the operating safety of nuclear reactors.

A number of effects of regulation on innovation have been noted which seem to affect various regulated industries. The Averch-Johnson Effect (Averch and Johnson 1962) is the propensity of regulated firms to develop capital-intensive innovations when regulation is based on return on investment. The firms are allowed a profit equal to total depreciated investment multiplied by allowed rate of return. By increasing total investment, total profits also may be allowed to rise. For example, in 1968 the American Telephone and Telegraph Company (AT&T) built an underwater cable to southern Europe rather than use satellite relays at lower cost because cable would contribute to its rate base but relays would not (Kahn 1971, pp. 76-77).

Regulatory lag results from the tendency of a regulatory commission to respond slowly to changes in the operating situation of a regulated firm. Any cost-cutting innovation can produce added profits until the regulatory body acts to adjust the rate structure. Regulatory lag thus acts as an incentive to cost-cutting innovations (Capron 1971, pp. 6-7). However, it can also have the opposite effect for regulatory commissions are usually slow to allow additional firms to compete unless need for additional service can be shown. Thus established firms are sometimes protected from cost-cutting innovators, which effectively screens technological change from that industry; for instance, until 1968 (see below) telephone-line attachments not owned by the phone company were forbidden by the regulatory agency. However, once change enters a regulated industry, it diffuses very rapidly so that the firms may keep up with the profit and service potentials offered by the innovation (Capron 1971, pp. 9-10).

Differently regulated industries behave differently. For example, innovation in common-carrier telecommunications has been uneven. AT&T has maintained a virtual monopoly by having the Federal Communications Commission acquiesce to the "integrity" of its system. For example, until 1968, "foreign" (i.e., not AT&T owned) attachments were prohibited from the telephone network. Competition in microwave service was beaten off until recently by AT&T's offering the service at nonremunerative rates, which were offset by the profits from its other operations (Capron 1971, p. 206; Shepherd 1971).[8] However, AT&T's research arm, Bell Laboratories, is efficient and has produced many significant innovations, the transistor being the outstanding example. Thus while retaining its monopoly and stifling competition, AT&T has innovated. The question--unanswerable--is whether there would have been more and "better" innovation in telecommunications if there had been freer competition.

It is generally thought that ICC has had an inhibiting effect on innovations in the industries it regulates. Gellman (1971) found that ICC especially inhibited railroad innovation. Indeed, it provided a disincentive to innovate by denying carriers lower rates, even though justified by cost-cutting innovations, because of possible ill effects on other carriers. In addition, the ICC's rate structure does not encourage innovation because it is not cost based. Instead it follows the "value of service" principle, according to which high rates are charged for carrying commodities of high value on the ground that such rates will not discourage their shipment. Thus cost-cutting innovations do not lead to lower rates.

Many students of regulation (Capron 1971, p. 221) feel that the innovative activities of regulated industries fall below what could be reasonably expected. Indeed a recent cross-national literature survey on economic regulation and technological innovation came to the same conclusion (Gellman Research Associates 1974). Regulation is usually "technology specific," which means that the regulatory commission may have vested interests in a particular technology and thus erect barriers to innovations affecting that technology. Yet no evidence exists on how a firm or industry could be more innovative if unhampered by regulation, for there are no cases of no regulation or alternative forms of regulation to compare with existing practice in the case of any industry. In this regard it might be fruitful to consider social experiments in regulation by removing or altering regulation from some firms in a regulated industry and using the others as a control group. However, both monopoly (as AT&T) and cross-system compatibility (as in the railroad industry) may stand in the way of such an approach.

The conference on Technological Change in Regulated Industries held at the Brookings Institution in 1969 offered some useful suggestions for research in this area. We have adapted some of the more important in order to apply them to the effect of regulation on innovation (see Capron 1971, pp. 222ff.).

a. Are regulators more effective in fostering innovation by regulating profits, prices, or qualitative performances?

b. How is regulation affected by the organization and technological characteristics of the industry being regulated?

c. Does the behavior of a regulatory commission toward innovations differ according to its size and composition?

d. Does the actual innovative performance of the industry affect the goals and policies of the regulatory agency? If so, do the interactions between industry and agency also vary systematically with the structure of the regulatory system?

What emerges from our consideration of the effect of political institutions on innovation is the truism that many institutions affect innovation. The effects are overlapping and varied, and often, as in the case of patent and antitrust institutions, in conflict. As a first step in understanding the effect of political institutions on innovation, we feel it appropriate that a "map" of institutions and public policies affecting innovation be made. This map would indicate the effect of each institution and policy on innovation and the overlaps and conflicts among them. Such information would help locate specific areas where information is lacking and which need further study. Based on the literature we have surveyed, regulation would appear to be one of these. The "map" would also be useful in coordinating and putting into perspective individual studies dealing with particular policies and institutions as they impact innovation.

V. OUTSIDE THE "BLACK BOX"

In this chapter we have highlighted the exogenous factors which affect innovation. Dividing these factors into values, endowments, and institutions, we discovered, of course, that these elements interacted with one another as well as with the innovation process. Their interactions with innovation were both indirect and direct, general and specific, long-range and immediate, as well as changing over time. In other words, these are dynamic interactions, typical of any ecological system.

Mostly, values make themselves felt indirectly--through institutionalized mechanisms, such as the marketplace or government policies. These institutions too can provide both generalized and specific "signals." The government, for example, can provide indirect incentives to innovation through tax or antitrust policy, and direct incentives through the funding of research and development programs designed to produce innovations meeting specific needs. Changing supplies of endowments such as knowledge, material resources, energy capital, and talented labor modulate the signals which values give institutions as well as providing direct signals to political and economic institutions as in the recent oil shortage.

If some rough generalizations are called for, we might say that by and large the exogenous factors affect, in a general sense, *what* will be innovated and *where* (that is, in what institutional context it will become an innovation). However, the exogenous factors have much less to say about how something will be invented, innovated, and diffused. The *how* question depends chiefly on the indigenous (or endogenous) world of innovation--that is, the actual locale and operation of the innovation process itself.

From the perspective of this chapter, the world of innovation might be considered a "black box." Our consideration of the exogenous system in this chapter represents a view of the world outside the black box, especially the external wires which lead into the black box and those leading from it.

However, we have not yet looked to see what is inside the black box itself. How is the actual process of innovation carried on, and how does the outside world look from within the black box? That is the task of the next chapter, where we investigate the indigenous elements of the innovation process, or the internal environment of innovation, just as we have studied the external environment of innovation in this chapter.

NOTES AND REFERENCES

1. Technology assessments are being funded by a number of Federal agencies. During fiscal year 1974, the National Science Foundation spent over $2 million on assessments, mainly in the energy and health areas. The Department of Transportation is spending $10 million studying "climatic implication of atmospheric pollution." Finally, the Office of Technology Assessment had $1.4 million for assessments in fiscal year 1974 and about $5 million for fiscal year 1975. OTA's interest is in food, energy, the oceans, health, transportation, and materials.

2. Baker et al. 1967; Carter and Williams 1957; Goldhar 1970; Sherwin and Isenson 1967; Langrish 1971; Myers and Marquis 1969; Tannenbaum et al. 1966; Utterback 1969.

3. Scherer 1965a, pp. 1195-1196; Freeman 1965, p. 67; Chap. 8 below.

4. Electrical equipment and communication; basic chemicals and drugs; petroleum, stone-clay-glass, fabricated metal products, machinery, and transportation equipment; and food, tobacco, textiles, apparel, paper, and primary metals.

5. For a thorough treatment of the patent system see Vaughan (1956).

6. Recently introduced legislation, however, will make significant alterations in the patent system if it becomes law (U. S. Senate 1973).

7. Taken from "An exclusive license agreement between NASA and Consultants Unlimited for vision examination apparatus," U. S. Patent No. 3 737 217, 5 June 1973.

8. To be sure, AT&T sees the matter differently, claiming that microwave-service competitors skim off part of AT&T's most profitable operations without being obliged to offer such unprofitable services as low-density rural telephones.

3. THE PROCESS OF INNOVATION: THE ORGANIZATIONAL AND INDIVIDUAL CONTEXTS

I. INTRODUCTION

The preceding chapter discussed the environment exogenous to the innovation process in terms of three pervasive elements: values, institutions, and endowments. Discussion of these exogenous influences anticipated the basic unit of analysis of this chapter, which in the private sector is the firm, and in the public, various governmental laboratories. Within these organizational contexts are three other levels of aggregation or analysis--the R & D laboratory, various work and social groups, and the individual researcher--which scholars have identified as especially important in understanding the innovation process. Social, political, economic, and informational signals from the external environment are picked up by one or more of these levels; interpreted in light of the needs, goals, and capabilities of the organization; and answered by efforts to achieve new technological syntheses.

In the second chapter, the environmental signals were emphasized. The patterns of response to these signals that characterize organized innovation efforts are the focus of the present chapter. In particular, we shall here be concerned with the organizational, social-psychological, and informational variables that facilitate or inhibit the process within the firm.

Several points should be noted about the emphasis we place on organized innovation activities. First, technological innovation has come to be thought of in terms of the R & D laboratory, and with good reason. Yet not all innovations come about as a result of the planned efforts of research personnel by any means. As we shall see, ideas leading to new innovations also arise from organizational units other than the R & D laboratory. Likewise, we should not lose sight of the individual inventor/entrepreneur whose present role, although perhaps not as "heroic" in the public mind as that of an Edison or a Sperry, is not without its significance. We shall try to capture the flavor of these "independents" both historically and in their modern form, the "spinoff firm" phenomenon.

However, our primary concern is with the innovation process in its *institutionalized* form, for that is clearly the dominant shape of technological advance in the present age.

A. THE DEVELOPMENT OF INSTITUTIONALIZED INNOVATION

In *Wealth from Knowledge*, Langrish et al. (1972, p. 14) have noted that "while it does undeniably still make sense in the twentieth century to talk about the 'independent inventor,' ... 'independent innovator' is almost a contradiction in terms." The point here is that bringing an invention to the point of use requires organizational resources:

> For technological innovation to occur, there must be some interaction between a set of ideas and an institution; the ideas must be interpreted in terms of a need of the institution and put into effect by it. Innovation is almost by definition a corporate and collaborative effort, and it is correspondingly difficult to disentangle the roles played by particular individuals. (Langrish et al. 1972, p. 14.)

Related to this point is a second and more basic one which gets to the heart of the matter of the institutionalization of the innovation process. To innovate is to introduce something *new*. But before novelty can be introduced it must be created. It is therefore tempting to view creative activities which produce novelty as preceding, and distinct from, subsequent activities which bring such novelty to the point of use.

However, such a compartmentalized and linear picture is often a misleading simplification. Problems are defined and ideas generated throughout the process as well as at the "front end." In fact, if the ecological perspective is taken seriously, as indeed we think it should be, it is difficult to point to the beginning of an innovation.

Thus, to speak of the institutionalization of the innovation process is to speak of the institutionalization of creativity as well. That this is possible is itself a bold and novel assumption, at least in the extent to which it is currently accepted and its actualization sought. One writer has remarked:

> The historians of the future may well select the development of deliberate creativeness as the most important development of this century. We have passed through the age of random creativeness and are entering an age of deliberate creativeness. (Rossman 1964, p. xii.)

That the creation and introduction of novelty is now recognized and accepted as a part of the mission of so many organizations reflects the extent to which this assumption of deliberate creativeness has taken root. It is reflected also in most of the scholarly literature dealing with the innovation process. However, it is only an assumption at this point, and the state of knowledge is at present insufficient for full realization of its potential implications.

The process of technological innovation, like all other human endeavors, does not take place in a vacuum, but is always embedded in an environment with particular and describable characteristics. Once this fact is recognized, the question follows, "Might not some environments be more conducive to innovative activities than others?" The answer is "yes," though the reasons may escape our grasp more often than not. Nevertheless, once the question is posed, a process of specification can begin, with both speculative-theoretical and empirical-pragmatic elements. If some environments are more conducive to innovation than others, what are the relevant variables? What are their relative degrees of influence, and how in concert do they facilitate or inhibit innovative tendencies? How may these influences be altered, and what are the results of such changes?

Implicit in the above questions is the assumption that a complex environment can be *deliberately* altered with the predictable result of becoming more conducive to the emergence of novelty that can in turn lead to new innovations. That such environments may arise by chance is an historically safe assumption with a fatalistic addendum that is equally safe. That a more fecund environment can be produced by *trial and error* is the half-way station where we now reside, but it is not what the literature, current practice, or this assessment assumes as the ultimate state. The assumption is that the *deliberate* creation and introduction of novelty is possible.

The date of birth of this assumption is not agreed on. In the Rossman quotation above, the 20th Century is suggested. Alfred North Whitehead (1925, p. 91) placed it earlier when he said, "The greatest invention of the 19th Century was the method of invention." Langrish et al. would place it, at least in its speculative form, much earlier--in Francis Bacon's vision of Salomon's House in his *New Atlantis*.

> Here was a national research institute in prototype, lavishly equipped by Bacon's imagination with all the equipment and facilities he could think of that might conceivably be of use. Among the thirty-six fellows of the foundation, there was well-defined division of labor and allocation of tasks. The program was clearly intended to be a corporate one. All the principal inventors were, it is true, to be commemorated by statues--'some of iron, some of silver, some of gold;' but Bacon does seem to have placed his trust more in his system that in exceptional individuals. (Langrish et al. 1972, p. 12.)

Bacon thus dreamed of a movement beyond the fortuitous and seemingly random acts of individual creative genius, and beyond the happenstance development of such novelty to the point of introduction and use.

But the large-scale effort to actualize Bacon's vision of deliberate creativeness and deliberate introduction had to await the development of the R & D laboratory. Even now, with some years of experience behind us, it is clearly a transition still in progress. And even our view of the progress achieved may be distorted by the nature of case studies, which continue to be the dominant scholarly approach. "The retrospective nature...of these sources probably means that the process has been viewed as much more rational and well ordered than it is in fact" (Utterback 1974, p. 625). The important point is that the transition from random to deliberate creativity is far from complete, and even farther from being completely understood. Nevertheless the assumption is present in the literature and, more important, in the institutionalization of the process of innovation. The question then is one of coming to understand the nature of the immediate environmental influences well enough to know how to alter them with predictable results. Control of the innovation process is sought, and Bacon's vision of its institutionalization is taken as the means.

B. INCREASE IN IMPORTANCE OF INSTITUTIONALIZED INNOVATION

A measure of the magnitude of the need to understand the institutionalized innovation process well enough to alter it with predictable results is the extraordinary growth in R & D expenditures. Total U. S. R & D expenditures increased at an average annual rate of 10 per cent during 1953-1970 (Mansfield et al. 1971). Rubenstein (1957, p. 95) summarized the growth in R & D activity along a different dimension. He noted that the 1956 edition of the *Industrial Research Laboratories of the United States* lists 4834 R & D laboratories operated by 4086 companies. "A sizable proportion of these companies were not operating research programs 10 years ago, and a majority of them were not doing so 15 years ago. As for the programs that did exist then most have grown so fast that today they can hardly be recognized." In Hamberg's (1966) words there has been a "research explosion" and "R & D is being conducted on an unparalled scale offering the potential for unprecedented advances in productivity increases." The extent to which such increases are in fact realized depends in large measure on how rapidly there can be a corresponding increase in understanding the informational, organizational, and social-psychological constraints on the process. The magnitude of our present commitment to institutionalized innovation increases the need for a level of understanding that will permit environmental alterations with predictable results.

However, the above assumption--that *deliberate* alteration of the environment of innovation with predictable results is possible--cannot be examined conclusively because the dominant mode of current research permits no such conclusions. The case-study method (which dominates the current research scene), whether restrospective or "real time," can only take us so far, and the same is true for speculation decoupled from a toughminded empiricism. We cannot understand the innovation process sufficiently to alter its environmental determinants and modifiers until myopic piecemeal empiricism and disjointed speculation are coupled with and guided by a greater concern with systematic theory.

C. THE LINKAGES BETWEEN THE EXOGENOUS AND INDIGENOUS ENVIRONMENTS

As we move from the level of the total societal system, through the levels of the innovation-related systems and industrial sectors, and finally to the aggregations of the individual firm and the particular innovation efforts within it, we discover that we know more and more about less and less. Even though our knowledge of innovation within the firm is far from complete, we know it like the proverbial "back of our hand" relative to what we know about the exogenous influences of the higher levels of aggregation. The consequences of this state of affairs appear when we come to ask about the linkages between the firm and the broader contexts in which it is located. Our knowledge here is "by-guess-and-by-golly" speculation, with the exception of the developing literature about what are termed "technological gatekeepers."

In an attempt to clarify these linkages, Richard Rosenbloom presents a very persuasive argument for a closer look at *corporate strategy* as the integrative framework needed to understand the coupling of the firm and its environment.

> The strategy framework is particularly appealing because it integrates the two relevant dimensions. First, the concept of strategy formulation calls for a perspective that cuts across the boundary of the organization, matching capability (an aspect of the organizational context) with opportunity (an aspect of the environmental context)....The strategy framework demands explicit attention to technical, economic, social-political, and behavioral considerations simultaneously, as it embraces factors within the firm and external to it. (Cf. Chap. 9.)

However, the scholars who have been most concerned with corporate strategy have typically paid little attention to technological innovation.

Our assessment of Rosenbloom's suggestion is that it indeed merits high-priority investigation. One consideration is provided by our earlier observation about the present nature of innovation research, which is echoed by Rosenbloom in a summary comment on what he calls the "particular innovation" level of aggregation.

> In summary, research at this level of aggregation, in which the primary unit of analysis is a particular innovation, has contributed a useful body of descriptive findings. In my opinion, extensions of this tradition of research, whether by enlarged scope or more powerful technique, are unlikely to alter that conclusion. For more adequate explanation, we shall have to move the focus of inquiry to higher levels of aggregation, where we can build on available theoretical frameworks pertinent to the behavior of firms and industries. (Cf. Chap. 9.)

The move suggested here, from the descriptive-comparative research mode to one yielding greater explanatory power by virtue of its inclusion of higher levels of aggregation, might be effected by the integrating concept of corporate strategy.

Another consideration can be briefly summarized as follows. The firm for which the generation of technological innovations has substantial survival value must continue to innovate. To do so it must continuously monitor both its external and internal environment. The only organizational unit within the firm that has complete access to (and in fact responsibility for) monitoring internal information is top management. But such information is meaningful only in terms of a frame of reference. And in the final analysis the frame of reference that counts for the future is the environment exogenous to the firm. Internal circumstances, orientation, and capability limit policy options, but not to the point of their unique determination. Choices within that range are weighted by a reading of the circumstance and opportunity characterizing the exogenous environment. This is the task of top corporate management, since no one else within the structure has sufficient command of the internal whole to distinguish signals from noise.

We thus hypothesize that however environmental signals initially enter the organization, the functional linkage between the two is effected at the policy-making level. This hypothesis is of course *a priori* at this point, and the empirical work, even if confirming, will surely reveal patterns of considerable complexity.

D. DIFFERENT CONTEXTS OF ORGANIZED R & D

In our *a priori* attempt to structure our survey of the literature regarding organized R & D, we sought to find differences between the industrial and agricultural sectors; between the private and governmental sectors; and for organized R & D carried on under government auspices, differences between in-house laboratories and contract establishments. We found no literature dealing directly with any such differences. Even in studies that might have been expected to deal tangentially with this topic, the assumption has been that the R & D depends more on the internal characteristics of the research organization itself than on the auspices under which it is conducted.[1] If that is so, the organizational pattern of the R & D effort would be largely independent of whether or not the R & D is industrial or agricultural, government or private, military or civilian, and done in-house or outside.

However, we find it difficult to accept the notion--especially in the case of agriculture[2]--that innovation could be so independent of its auspices when it is so context-dependent on other variables. It would seem worthwhile to check out this assumption, to see whether the innovative effort is truly independent of external context and depends primarily on the nature of the R & D organization itself. To this end, a series of comparative studies focused on this question might reveal the extent to which differing auspices affect the nature, direction, and rate of innovative activity. Such studies might provide vital information in stimulating and guiding innovation in the future.[3]

II. THE INDIVIDUAL INVENTOR/ENTREPRENEUR

A. IMPORTANCE AND TYPES

The process of technological innovation in this century has been characterized by an increasingly strong trend towards the "institutionalization" of all its phases, including that cluster of activities that often bear the generic label "invention." The emergence and rapid growth of the corporate R & D laboratory, especially in the years after World War II, is the major manifestation of this trend, which we earlier referred to as a move from "random" to "deliberate" creativity. However, this phenomenon should not cause us to lose sight of the role of the individual inventor.

Despite the growth of institutionalized innovation in R & D laboratories, some solo work is still being done. Many investigators discount the role of the "lone-wolf" inventor in contemporary innovations, yet from time to time claims are made that the individual inventor still maintains a major role in the innovation process. In the 1960s this view gained support from the Jewkes, Sawers, and Stillerman study of fifty important 20th Century inventions, showing that over half stemmed from independent inventors or small companies.

From there the argument moved to a quantitative basis with the counting of patents--or a pseudo-quantitative basis which allowed differing interpretations.[4] The proponents of the individual inventor argued against those who claimed that group inventive effort produces more innovations by pointing to the fact that the total annual issue of U. S. patents in the 1960s was no greater than 30 or 50 years previously, and that in terms of patents per unit of population, the number was less in 1960 than in 1870, despite an annual rise of 10-20 per cent in R & D expenditures in the half century from 1910 to 1960. They also pointed out that the number of patents had not grown in proportion to the increased number of scientists; in other words, the research force was growing far faster than the number of patents produced by that force.

Their opponents--the believers in the efficacy of R & D laboratories in producing innovation--first discounted the number of patents as a true index of the nature, amount, and quality of innovative efforts. Further more they claimed that current patents, though fewer in proportion to the population, are individually longer and more technical, and a larger percentage of them were being worked than formerly. In addition, an increasing proportion of inventions were being made by government employees or were in weaponry, and so would be less likely to be patented. They also stated that it was becoming more difficult to make patentable inventions as time went on, because of a tendency of the proportion of basic inventions to shrink while that of minor, unpatentable improvements grew.

The high point of the argument in behalf of the independent inventor was reached in the Charpie Report of 1967. This report, the product of a panel of private citizens convened by the U.S. Secretary of Commerce, was officially titled *Technological Innovation: Its Environment and Management* but is usually referred to by the name of the panel's chairman, Robert A. Charpie, then president of Union Carbide Electronics. The thrust of the report was that the government, primarily through tax concessions, must ease the way for the backyard or garret inventor and for the small company. Its major recommendation--a White House conference on Understanding and Improving the Environment for Technological Innovation--has not been taken up, and few of its other recommendations ever took hold; perhaps just as well, because the Charpie Report was based upon a static and hence an unpredictable database.

History is a dynamic process. Individual inventors in 1959 still accounted for 40 per cent of the new mechanical patents, 35 per cent in electricity and electronics, and 30 per cent of new chemical patents; that was still far less than in previous years. The Jewkes, Sawers, and Stillerman study, constantly referred to by the Charpie panel, dealt with inventions back to the turn of the century, when the structure and nature of technological innovation were far different from what they have become. Furthermore, although the original ideas for many of the inventions cited in that study might have come from individual inventors, their actual development had gone nowhere until they were put into the hands of large corporations possessing industrial research laboratories that could develop them into commercially feasible and salable innovations.

By focusing on a limited number of innovations, by looking upon the innovatory environment as static rather than dynamic, and by failing to distinguish among the different elements entering into the innovation process, the Charpie Panel had diagnosed--and prescribed for--a situation which was at least a quarter of a century out of date. At the very time the Charpie panel was carrying on its deliberations, the percentage of significant inventions made by independent inventors, even measured by their beloved patent count, was dropping markedly. The lone inventor was giving way to the group worker in the organized research laboratory.

Nevertheless, the individual inventor cannot be ignored. With such examples drawn from recent history as Edwin Land (Polaroid camera) and Chester Carlson (Xerography), it is obvious that the solo inventor is by no means obsolete and that he can be responsible for major innovations--although Carlson's original invention had to undergo much development, carried on in a structured R & D situation, before it achieved successful application.

In this section we are concerned with individual inventors whose activities have not been carried on in the context of R & D organizational structures. We can distinguish three such types. First, there is the inventor who is fertile in imagination and technical ingenuity and who can produce an inventive idea or even a prototype device, but who lacks the entrepreneurial capacities to carry it through to an innovation. James Watt at an earlier date and Chester Carlson of today's xerography are representative of this type. Second are those whom Thomas Hughes had characterized as inventor-entrepreneurs, who embody the characteristics of both the independent inventor and the entrepreneurial capitalist. In many cases, such as Thomas Edison and Elmer Sperry, they work alone at the beginning of their careers but later establish their own firms, sometimes working within the context of an organized laboratory. And finally there are research scientists and engineers who begin their careers in a corporate context and later establish new technologically based "spinoff" firms.

B. INDIVIDUAL CREATIVITY

Investigation of the role of the individual inventor brings up complex problems of the wellsprings of technical creativity. Creativity has been studied from two methodological viewpoints, which correspond, interestingly enough, to the social-deterministic and individualistic approaches employed for interpreting the innovation process: sociological and psychological. The psychological approach to creativity tends to focus on forces within the individual, concentrating on such factors as intelligence, personality, and attitudes. The sociological approach, though not denying the importance of those elements, claims that they derive from various types of social background and conditioning.[5] In other words, this is the old "nature versus nurture" argument applied to innovative creativity.

One of the earliest scholars who sought to explain creative genius was Francis Galton (1870), who found heredity a primary determinant of eminence. Other pioneer psychologists also found the explanation for creativity in "native genius" (Cattell 1915; Cox 1926).

More recent studies have relegated heredity to a minor role, although not discounting it completely. Ann Roe (1953) showed other factors to be of major importance, such as the intellectual atmosphere of the home, childhood interests, and position in the birth order. Not until 1955 was a conference on the identification of creative scientific talent held, and then scholars placed differing emphasis on various demographic, cultural, religious, and personality attributes (Taylor 1956). Within a few years, however, it was evident that "profiles" of eminent scientists did not necessarily shed much light on the creative process itself (Anderson 1959).

Samuel Smiles had made out his great inventors to be the most reasonable and virtuous of men (Hughes 1966), but some iconoclastic thinkers of the late Victorian and the Edwardian eras were endeavoring to show that creativity resided primarily in certain choleric and splenetic individuals who refused to adjust to the world about them and who did not adopt its values. Not surprisingly, the psychoanalysts, with their emphasis on the neurotic and irrational elements in the human mind and behavior, came forth with theories relating creativity to emotional disturbances.[6] More recent investigators have abandoned the popular cliché of linking creative genius with a light touch of madness; they now tend to view creativity and psychological health not only as compatible, but as mutually supportive (Rogers 1964; Kubie 1958).

Although one body of opinion holds that the creative act is basically the same in every field of endeavor (Coler 1963), other studies distinguish among various types of creativity and link them to various kinds of activities and goals. Some creative individuals (especially composers, expressionist painters, sculptors, and writers), are simply expressing their inner states; others direct their creativity to meet externally defined needs and goals; and a third type cuts across the first two.[7]

Nevertheless, such studies of creativity would seem to distinguish little between an independent inventor and one operating in the context of an organized innovative effort. This failure to distinguish between the two is justified by our analysis, in which the inventor emerges as a function, not as a person. Before the institutionalization of innovation, the individual and function were merged; in the organized R & D effort, the individual is submerged in the function.

Because invention is a function, discussion of the individual inventors characteristics becomes virtually irrelevant. For our purpose it is much more important to look at the environmental context in which innovation takes place, as we do in our later discussion of the organized R & D effort. Yet we can perhaps learn something about the innovation process by briefly focusing on the independent inventors in their different roles.

C. THE INVENTOR *QUA* INVENTOR

Some light may be cast on the role of the inventor *qua* inventor by viewing this function in the light of Usher's *gestalt* theory of invention. As outlined in Chap. 1, that theory predicated a four-step sequence: perception of a problem, setting of the stage, act of insight, and critical revision. Typically, the independent inventor is strong on the first three of these steps, but his critical revision is frequently lacking in the elements necessary to make the invention into a successful innovation.

James Watt's invention of the steam engine might serve as an exemplar of the Usher theory (Scherer 1965b). When given a model of a Newcomen engine to repair in 1763 he soon perceived the problem of inefficiency caused by heat loss in the cylinder's wall. For two years he set the stage by tinkering with the cylinder and trying wooden rather than brass cylinders. Then Watt tells us of the act of insight which occurred to him "on a fine Sabbath afternoon" in 1765 while strolling on the Glasgow green, for it was then that he hit upon the ideas of condensing the steam not in the operating cylinder, as Newcomen had done, but in a separate condensing chamber.

Although Watt conceived his brilliant idea of the separate condenser in 1765, it was not until 1769 that he obtained his first patent, and it was more than a decade later, in 1776, that the first Watt engine was brought into commercial use. What happened during the eleven years between Watt's act of insight and the first successful commercial installation of his engine proves the importance--and difficulty--of the developmental stage in transforming an idea (and a model, made within three days after the Sabbath afternoon walk) into a practicable innovation. Here the question of defining the technical problems and viewing them in their economic context was to prove crucial; that was to be largely the work of Matthew Boulton, not of Watt who had the original idea.

For the fact is that Watt lacked sufficient capital to devote full-time effort to scaling up his model to an efficient and reliable machine, which involved the solution of many additional technical problems. Watt also lacked the requisite managerial and entrepreneurial expertise. Boulton became the driving force making for the successful introduction of Watt's steam engine; he provided the capital and also brought together the market demand with the creative ability of Watt.

Chester Carlson in more recent times exhibits the same inventive imagination as did Watt. But he too lacked both the capital and entrepreneurial skills to transform his basic concept of xerography into commercial application, and was forced to rely on organized R & D establishments for the critical revision and marketing phases of his invention.

It would seem that the contribution of the entrepreneur is sometimes equal to that of the inventor in arriving at an innovation. However, the entrepreneur need not be an individual, for we are really talking about a function, not an individual. That function--which includes risk taking, the provision of capital, the development from idea or prototype to operational status, and the coupling of the marketplace with the inventor's concept[8]--can be and is increasingly performed by a corporate entity. But it is also possible for the inventor himself to have entrepreneurial qualities and to do the entire job from perception of need through development and marketing. That unique individual is the inventor-entrepreneur.

D. THE INVENTOR-ENTREPRENEUR

In his prize-winning biography of Elmer Sperry, Hughes (1971) utilized the concept of the inventor-entrepreneur to explain the process of innovation. He has offered the following definition and characterizing generalizations of the inventor-entrepreneur in the history of American technological development.

> Inventor/entrepreneurs are inventors who preside over the innovative process from its origins as a problem to, at least, the introduction of the invention into use. The usual reason that inventor/entrepreneurs were not simply inventors was that they were determined to have their invention used, and to achieve that they realized they would have to take the initiative not only in the early phases of innovation but in research and development and marketing. In essence they were inventors; in effect, they had to be entrepreneurs. The evidence, however, is that they found their work more satisfying when identifying problems and inventing solutions, not when presiding over and promoting the other phases. The evidence also tends to support the generalization that in America before 1930 most successful inventors were in fact inventor/entrepreneurs. (Cf. Chap. 6.)

The preeminent example of the inventor-entrepreneur--and indeed America's most spectacular and prolific inventor (some 1093 patents in his name)--was Edison (Josephson 1959). Perhaps the best illustration of Edison's ability to bring an innovation to completion--from definition of the problem to profitable application and diffusion-- is electric lighting. Hughes has pointed out that "only the naive inventor assumes that the challenge is to invent an arc lamp, an electrical generator, a streetcar, or an automobile." Edison saw things in their entirety, and one of the major reasons for his success was that he realized that the problem was to develop an electrical lighting *system*, not just to devise an incandescent bulb (Sharlin 1967).

The scientific principles and technical requirements of a viable electric light bulb were known as far back as 1860 and had been tried by many inventors. Practical electric generators were already at work, and there were arc-lighting systems employing generators, transmission lines, and lamps. The technological level was thus at a stage where further steps could be taken. At the same time Edison possessed the capital to undertake the creation of a complex system because he was the owner of a considerable fortune derived from his previous successful inventions. What is equally important was Edison's well-equipped and well-staffed Menlo Park laboratory, which provided him with resource requirements--shop facilities, instruments, a library, specialized personnel, etc.--for a high level of inventive activity. The mere existence of such facilities illustrates the essential entrepreneurial underpinning for this type of innovative project. Above all, Edison defined the problem in large terms and was fully cognizant of the economic constraints involved: he was attempting to develop an entire lighting system that would compete with gas illumination. Edison's achievement was as much a triumph of entrepreneurial ability, managerial expertise, and economic reasoning as of technical ingenuity.

Unlike Edison's work on a whole complex system, Elmer Sperry focused on bottlenecks, or "reverse salients," in rapidly expanding areas. Such areas were chosen because capital was available and there was a probable market for his invention. Hughes writes:

> Identification of critical problems was a professional capability of Sperry's and one crucially important for his survival as an independent inventor. When Sperry's numerous patents are examined and their claims considered, it becomes clear that he did not invent dynamos, arc lights, streetcars, or automobiles, though the title of his patents might lead to such superficial conclusions. In the case his dynamo

patents, he claimed automatic controls; in connection with the arc light, he invented a regulator for the feed of carbons; and when inventing for streetcars he contributed an operation control. His patent claims in these instances and in numerous others show that he solved very specific problem, which can be aptly labeled "critical problems." (Cf. Chap. 6.)

Sperry's pattern, at least until 1910, was to identify such an area in which his special competencies would apply, move in and concentrate on its weakest point, make his contribution, and then move on to something else rather quickly.

Sperry seems to have lost interest in a field after about five years, which suggests to Hughes "that an inrush of inventors, engineers, managers, and corporations brought by capitalization, growing market, and size convinced him that his special characteristics and circumstances could best be employed elsewhere" (Chap. 6). He may well have realized that after a field has experienced a period of rapid change it then enters a "mopping-up" phase in which incremental improvements are worked out. His own talents were less well suited for that phase.

The pattern of Sperry's activities prior to 1910 allows the anticipation of a point to be discussed later in this chapter. One of the crucial elements in the problem definition/idea generation phase of the innovation process, as conducted in the corporate R & D lab, involves the identification of *market needs*. In fact, as we shall see, most innovations are stimulated in this way. That seems to have also been the case in Sperry's career as an independent inventor. How did Sperry go about acquiring need information?

> Sperry's letters, memoranda, notebooks, and other records reveal that he identified critical problems by close study of technical journals, patterns of patent applications, the patents of others, attendance at professional engineering society meetings, conversations, and his intimate knowledge of expanding technological systems. Articles in the technical journals often told Sperry of the interests of other inventors and therefore of problems on which they were working; in weekly reports of patents granted, like those published in the *Scientific American*, he could discern a pattern of concentration on certain problems (for instance a bevy of patents on arc-light regulation). By a close reading of the claims of these patents he could delineate the problems of focused attention more precisely, and the Official *Patent Office Gazette* provided regular summaries of all patents. Sperry regularly attended the sessions of the engineering society meetings, for there he might gather from fresh reports and papers more intelligence pertaining to critical problems on which other inventors were working. (Cf. Chap. 6.)

The typical modern corporation, with all the information-gathering potential at its command, scarcely acquires need information as carefully, systematically, and with such result as did Sperry working alone.

Once a critical problem had been identified, Sperry "tried to discern the weakest point and strengthen it." After 30 years, this "hit and run" pattern of inventing in response to the weakest point in an expanding field and then moving on to another was finally altered with the establishment of the Sperry Gyroscope Co. in 1910. During the early period we can see most clearly another characteristic of Sperry's inventing behavior, and at the same time note two of the characteristics of the innovation process itself that require an entrepreneurial response.

With the idea for an invention in mind (often the problem definition/idea generation phase in our terminology), Sperry would embark on a process in which "invention merged unperceptibly into development," as he successively "scaled up" towards the environment of intended use. The successive test environments "involved more variables, altered parameters, and new factors, and these **revealed the** need for successive modifications of the first idea by revision and new invention." Such scaling up of course requires resources and the organization of various specialized testing activities, which in turn imply entrepreneurial commitments that may be substantial indeed, far beyond the resources of an individual. Such was the case with Sperry's work on a marine gyrostabilizer, the development of which required access to the U. S. Navy's experimental model basin, and ultimately sea trials aboard the destroyer *Worden*. In such cases arrangements have to be developed in which an independent inventor shares the entrepreneurial commitment, or consigns the development of his invention to an organization with more adequate resources.

However, the independent inventor intersects with and in fact depends on the entrepreneurial activities of others in an even more basic sense. We recalled that Sperry sought out the fields that were undergoing rapid change and looked for the reverse salients, or bottlenecks, to further progress within them. Implicit in this approach is a dependence on the vigorous activities of the entrepreneurs within the field who had brought about the very progress that revealed such weak points and would reward an independent for inventions that strengthened them. Thus the independent inventor, if not an entrepreneur himself, is doubly dependent on those who are entrepreneurs: he requires the progress they impel and organize, and ultimately the developmental capabilities they possess.

One further point should be noted in this connection. As Gilfillian (1935) has pointed out, inventions are themselves changes in a system which necessitate further invention. They introduce perturbations into the ecological system to which new adjustments must be made. Although this consideration poses a constraint on the independent inventor, and may ultimately influence the acceptability of his invention, it is not his primary concern. But the concern of the entrepreneur is with the market fate of a larger system, of which the critical problem addressed by the independent inventor is only a part.

E. THE BUSINESS OF INDEPENDENT INVENTING AND DEVELOPING

Mention of Sperry's need to employ resources in his innovations which were greater than even a prosperous inventor could command helps to account for the development of another recent phenomenon: firms specializing in inventions which develop them up to the point of application. Despite their corporate nature, functionally these firms perform the tasks which were previously the province of the independent inventor or of the inventor-entrepreneur (Bass 1962; White 1961; Machlup 1962).

Some examples of this kind of specialized inventive research and development are well known: Arthur D. Little Co., Battelle Memorial Institute, Stanford Research Institute, Denver Research Institute, Midwest Research Institute. Depending on the contracts offered them by government or private corporations, these firms can perform all the functions of invention from perception and definition of the problem through development, up to the point of application. Sometimes, as in the case of Chester Carlson and xerography, the inventor comes to them with the basic idea, and their task becomes the critical revision and development of the concept to the commercially applicable phase. In the case of the Research Corporation of America, the task becomes the weeding out and patenting of the ideas of inventors--typically university professors--and sales and licensing of the patent to companies which will exploit the invention commercially.

Still another type of such independent inventing firms is a number of the high-technology research firms which sprang up about Route 128 on the outskirts of Boston in the 1960s (Lieberman 1968). These firms were sometimes spinoffs of larger corporations, originated by highly talented scientific and technical people who felt constrained within the organizational structures of larger corporations. Their specialty was the "critical problem" which had attracted the attention of men like Sperry at an earlier date. Regardless of their antecedents, organization, and field of concentration--the Route 128 firms specialized in electronics and computer technology--these firms were selling "know-how." As we have pointed out, technology is a form of knowledge, and these firms specialized in particular forms of scientific knowledge and technical expertise requisite for today's complex process of innovation.

Such firms constitute today's "independent inventor." Indeed the term "independent" can lead to misconceptions. Unless invention is regarded as more than an idle and engaging pastime, the "independent inventor" means little more than an "unsalaried professional." But once invention is regarded as a function, as an activity to provide a useful solution to an existing problem, then it must be coupled with the entrepreneurial function, as in the case of Watt-Boulton, or Edison and Sperry.

Yet, there comes a time in today's highly complex and scientifically connected technology when the requisite knowledge and entrepreneurial base go far beyond what a single individual or a small group can muster. Innovation then requires the efforts of numerous individuals and some type of organized endeavor. At that point the individual inventor "becomes" a team, a research institute, or a specialized R & D firm, which is increasingly the case in modern innovations, as indicated by the story of the spinoff firms.

F. THE SPIN-OFF PHENOMENON: NEW TECHNOLOGICALLY BASED FIRMS

The literature related to technical entrepreneurship and the birth of new technologically based firms was summarized in an article by A. C. Cooper (1973, p. 59). He notes that such firms contribute to the growth and vitality of the economy in a variety of ways: they are important sources of innovation, they serve as new sources of competition which both complement and spur the efforts of established firms, they offer new career opportunities, and they are a desirable form of industry for regional economic development. Because of these contributions, it is important to understand the factors that influence the birth of such companies. Their factors can be organized under three general headings: the entrepreneur himself, the established organization for which the entrepreneur had been working (the incubator organization), and external factors.

Cooper's literature survey identified the following characteristics of the individual technical entrepreneur. Founders of new, technologically based firms tend to be in their thirties when starting the firm (Roberts and Wainer 1971; Susbauer 1969); to have at least a B.S. or other first degree, typically in engineering (Roberts 1970, Susbauer 1969); and to be more single-minded in their devotion to careers than are hired executives (Howell 1972). They often form groups to start new companies in order to obtain a more balanced management team and to provide psychological support at a time of high stress and uncertainty (Cooper 1972; Shapero 1971; Susbauer 1967). Studies involving psychological tests have been conducted with a very limited number of respondents. In these studies, entrepreneurs rated higher than average in esthetic and theoretical orientations, leadership orientation, and achievement orientation (McClelland 1971), but lower than average in religious orientation, need for support, need for conformity, and practical-mindedness. Although they did not score high in regard to economic values (Komives 1972), a disproportionately high percentage of founders were from homes where the father was in business for himself (Roberts and Wainer 1971; Shapero 1971).

In most instances, when a founder starts a new company he leaves some existing organizations, called the "incubator organization.". Table 1 summarizes the industry and incubator organization characteristics which Cooper (1973) reported were associated with low and high birthrates of spinoff firms. At the industry level, factors such as rate of technological change, level of investment required, and economics of scale appear to be important characteristics. Similarly, firm size, firm structure, emplyee attributes, and geographic location apparently are important charactistics of the incubator organization (Cooper 1971); Draheim et al. 1966; Forseth 1965).

The incubator organization also influences such considerations as the location of the new firm (Cooper 1972; Susbauer 1969), the nature of the new business (Cooper 1972; Draheim 1972; Lamont 1971), and the motivation of the new entrepreneur (Cooper 1972; Howell 1972; Roberts and Wainer 1971). Clearly, the incubator organization is an important unit of study for the better understanding of the spinoff phenomenon and identification of possible organizational barriers to innovation.

Many of the major complexes of the new, high technology have grown up around universities (e.g., Boston, Palo Alto, Ann Arbor, and Austin). As a result, several observers have suggested that universities play a central role in the development of local entrepreneurship.[9] Not only do entrepreneurs come from local universities, but the universities attract highly educated young persons to an area and provide consulting assistance to the new firm. However, specific patterns vary widely from area to area, and the degree to which universities play a central or essential role in technical entrepreneurship is still largely unknown (Cooper 1973). Other important external factors affecting the foundation of spinoff firms include the availability of venture capital;[10] the economics of location such as transportation, labor, supplies, markets,

TABLE 1.--Industrial and organizational attributes related to birthrate of new firms (Cooper 1973).

Low Birthrate	High Birthrate
Characteristics of Industry	
Slow industry growth Slow technological change Heavy capital investment required Substantial economies of scale	Rapid industry growth Rapid technological change Low capital investment required Minor economies of scale
Characteristics of Established Incubator Organizations	
Large number of employees Organized by function Recruit average technical people Relatively well managed Located in isolated area of little entrepreneurship	Small number of employees Product-decentralized organization Recruit very capable, ambitious people Afflicted with periodic crises Located in area of high entrepreneurship*

NOTE: All the attributes in either column are not necessarily found together not are all required to bring about a given spinoff rate; various combinations may exist.

*Boston, Mass.; Palo Alto, Calif.; Minneapolis-St. Paul; Ann Arbor, Mich.; Erie-Niagara; Buffalo; Austin, Tex.; and Oak Ridge, Tenn., are identified as geographical areas with high rates of new, technologically based firms being founded.

etc. (Cooper 1972; Shapero 1971); and the region's past history with respect to the success or failure of new, technology-based firms (Cooper 1973). When regional factors are favorable, a self-reinforcing process appears to take place. Past entrepreneurial success makes future entrepreneurial efforts more likely and, in time, a high rate of entrepreneurial activity may develop (Draheim et al. 1966; Susbauer 1972).

Cooper's excellent summary and integration of the literature demonstrates that our knowledge of the spinoff firm is quite high when compared with the other aspects of innovation we have surveyed. However, it would be misleading to conclude that this is not an important area for future research. Several basic questions remain unanswered. Are there conditions under which it is better to develop an innovation in the incubator organization than in the spinoff firm? Is it possible that the spinoff organizations excel in certain elements of the innovation process or under certain conditions? To what extent is the spinoff firm of today the counterpart of the individual entrepreneur of yesterday? Is the spinoff process a viable means of initiating a self-reinforcing process by which a geographical area can experience economic development? This remains a potentially rich field for additional research.

G. SECTION SUMMARY

The conclusions to be drawn from the literature summarized in this section are clear. There has been an increasingly strong trend towards the institutionalization of all phases of the innovation process. The individual inventor, although playing a role of diminishing importance overall, continues to make significant contributions. He should not be ignored; indeed, he should be encouraged by the relevant policy makers. Yet it appears that the role traditionally performed by the individual entrepreneur is in the process of being taken over by the spinoff firm. In light of the move toward institutionalization of the innovation process, it is critical to develop a complete understanding of organized innovation, particularly as performed within the research and development laboratory.

III. ORGANIZED R & D: WITHIN THE LABORATORY

This section is addressed to a special, important phase of the innovation process--organized R & D within the laboratory. In one sense, organized R & D can be viewed as a microcosm of the overall innovation process, in that problems are defined, ideas are created, inventions are developed, applications result, and diffusion to other scientific areas and to other organizational functions is accomplished. To the extent that the R & D and innovation processes have common characteristics, the contents of this section become applicable to the process of organized innovation. The reader is made aware of the conjecture that R & D is a microcosm of innovation but strongly cautioned to keep in mind that this is a conjecture.

Our focus is on innovation within the R & D laboratory, with special emphasis on processes, process variables, linkages, and screens. The purpose is to summarize, integrate, and assess the literature which provides understanding of, and insights to, the ways in which innovation is enhanced, constrained, and inhibited within the R & D laboratory. Our first subsection contains background, underlying definitions, and basic concepts to be used throughout the section. Subsequent subsections deal with problem definition and idea generation, R & D project selection and resource allocation, and performance of scientists and engineers within the laboratory. The final section is a summary assessment of research gaps and needs. Throughout, the theme is that organizational structure has a direct impact on function within R & D laboratories.

A. BACKGROUND, DEFINITIONS AND BASIC CONCEPTS

The 1945-1955 period was an important one for industrial and R & D laboratories in the United States. Both the total level of funding and the number of R & D laboratories increased at a rapid rate (Mansfield 1968b; Mansfield et al. 1971; Rubenstein 1957). As a result, the top corporate and R & D management of many firms adopted for R & D the organizational structures and policies that had been used with apparent success to manage the other organizational functions, such as production. Accordingly, the R & D activity was blended into existing organizational hierarchy either as a new division (usually called a "corporate laboratory") or as a new function in an existing division (usually called a "divisional laboratory"). In most organizations, the R & D activity was subjected to the same hierarchical control and coordination mechanisms as the other organizational functions. Thus, the R & D activity was controlled and coordinated by annual operating budgets, periodic performance reviews, and accounting systems; and these mechanisms were frequently adopted within the R & D activity itself.

In the absence of successful R & D organizational examples and of underlying R & D management theories, the implications of the hierarchical structure and the associated control and coordination mechanisms were not well understood. Although the timing and operation of the control and coordination mechanisms are consistent with the organizational administrative process, it is not clear that they are consistent with the innovation process. Indeed we shall argue that an organized innovation process has emerged over time which derives from the control and coordination and mechanisms adopted in the 1945-1955 period; serious dysfunctions have been experienced as a direct consequence because this mechanism does not correspond with many aspects of the innovation process.

Two models of the R & D process were proposed when academicians began studying the R & D process and R & D management. Economists developed process-phase models of R & D; their colleagues from management and industrial engineering proposed idea and information flow models. A literature grew up around each view and a number of important insights resulted. However, neither view led to a full realization of the importance of the underlying hierarchical structures and the associated control and coordination mechanisms.

During the early and middle 1960s, a new construct emerged in the organization-theory literature called "organization development" (Bennis 1969). "Organization development is a response to change, a complex educational strategy intended to change the beliefs, attitudes, values, and structure of organizations so that they can better adapt to a new technology, markets, and challenges, and the dizzying rate of change itself" (Bennis 1969, p. 2). Within the broader organization-development literature a specialized focus emerged,[11] postulating a two-stage innovation process in organizations:

initiation and implementation (Zaltman, Duncan, and Holebeck 1973). The organizational conditions most conducive to innovation at the initiating stage--high complexity, low formalization, and low centralization--are the opposite of the conditions most appropriate at the implementation stage--low complexity, high formalization, and high centralization--yet both phases are necessary for successful organized innovation (Zaltman, Dundan, and Holebeck 1973). Thus, a dilemma exists when an organization faces the need for innovative activity.

A related dilemma can also be deduced from the R & D management literature (Rubenstein 1964; Scott 1973). The organizational structure most conducive to identifying new product and process R & D opportunities and carrying out the R & D effort apparently is the opposite of the organization structure most conducive for managing the resultant new products and processes. Nevertheless, both functions must be accomplished.

This section begins with a description of the process-phase and idea-flow models of the R & D process. Two composite models are then proposed each of which includes both process-phase and idea-flow components. These composite models are used as a springboard for a more detailed description of the previously mentioned dilemmas and for developing the many important relationships between organizational structure and innovation within the organization.

1. *Distinction Based on Characteristics of the R & D Activity--R & D Process-Phase Model*

Numerous authors have used a "process model of R & D" to structure their discussions and analyses.[12] This process-phase model focuses on the level of technical uncertainty in the R & D activity and on the extent to which the activity is directed toward organizational objectives. Typical R & D process-phase definitions employed by these authors are:

a. *Basic research* is concerned only with the extension of the boundaries of knowledge without any technical or commercial objectives in view: it seeks basic principles and relationships.

b. *Applied research* seeks new knowledge having specific technical and commercial applications, typically in the form of new or improved products or processes: it has a specific practical payoff in view.

c. *Development* begins with an artifact or concept which has been shown to be technically feasible but which requires further change due to production or market needs: it attempts to reduce research findings to practice.

d. *Engineering* refines the knowledge and brings it to its first use or market introduction: it is the refining which leads to commercial exploitation or other practical end uses.

When this perspective is taken, the R & D process is usually conceptualized as the flow from basic research to applied research to development to engineering and, eventually, to some end item or information which is useful to the organization. Moreover, Mansfield et al. (1971), and the Commerce Department's Panel on Invention and Innovation (Charpie Report 1967), use an analogous breakdown for modeling the innovation process within the organization; namely, applied research, preparation of project requirements and specifications, prototype or pilot plant, tooling and manufacturing facilities, manufacturing startup, and marketing startup. One view of innovation considered by the Georgia Tech project team was also a process-phase view--problem definition and idea generation, invention, R & D, application, and diffusion (Chap. 1). Use of a process-phase view both for R & D and for innovation lends credence to the conjecture that R & D is a microcosm of innovation, but it also introduces definitional confusion. One must be cautious to discern whether an author is referring to innovation or R & D phases. In this section, the words "process-phase" refer to R & D, not to innovation.

When the process-phase focus of R & D is taken, phase-related questions tend to be raised by the authors. The following is an illustrative, not comprehensive, list of the typical questions posed in the literature:

a. What percentage of total cost (or total elapsed time) is associated with each step of the process?

b. How do these percentages vary over time? Is the cost of one stage increasing more rapidly than that of another step?

c. Are the estimates of total cost (time to completion, probability of technical success) more accurate for projects in one stage than for those in another stage?

d. Are there significant differences by industry (firm, nation) in the funding (cost, time) pattern across stages? Are there shifts over time in patterns across stages?

e. What is the source of funding by stage? Is the source changing?

In addition to being related to the process-phase model of R & D these questions are also influenced by the fact that most of the authors who have taken this view are economists. The economists tend to focus on the risk bearer, or entrepreneur, who is making nonroutinized decisions, and hence they model the R & D process as viewed by this de-decision maker. Mansfield (1968a; 1968b; 1971; 1972b), Marschak et al. (1967), Hamberg (1966) and Kendrick (1961) are just a few of the many economists who have written on R & D with the process-phase model perspective and the risk-bearer focus. (See Mansfield 1972b, for an excellent summary.)

2. *Distinctions Based on Idea or Information-Flow Model*

An alternative view of the R & D process[13] is based on the flow of information as it impacts on the creation and development of ideas and on managerial project selection/resource allocation decisions. An "information-flow model" example of this approach is summarized below. It is noted that Myers and Marquis (1969) and Utterback (1973) adopted this view for modeling the innovation process. Specifically, the innovation process is viewed as beginning with a new idea that involves both technical feasibility and usefulness, problem solving is activated to solve the inherent problems, and finally the idea is introduced commercially in the market or in the firm. Again, this approach supports the conjectured relationship between R & D and innovation, but introduces additional definitional confusion.

A. H. Rubenstein and his colleagues at Northwestern University have developed an idea and information-flow model of R & D in considerable detail (Rubenstein and Hannenberg 1962; Rubenstein 1968) and have utilized it in a number of empirical studies.[14] According to Rubenstein (1964), an "idea" is defined as " a potential proposal for undertaking new technical work which will require the commitment of significant organizational resources such as time, money, energy." The term "potential proposal" becomes a "project" when resources are allocated to it.

Figure 1 is an information-flow model that identifies some of the activities, linkages, and decision points which arise from consideration of how ideas are created and submitted, proposals reviewed, and projects researched and developed in R & D organizations (Baker and Freeland 1972b). Following the primary path (solid line), we find several screens or decision points at which potentially beneficial or useful ideas may be lost. The first two occur before R & D management has an opportunity to exert direct influence or control, at idea creation and idea submission. At the point when the idea is submitted to a reviewer, the idea originator transfers the control of the idea over to R & D management. The proposal is evaluated, typically relative to other proposals, and, if its evaluation is sufficiently positive, resources are assigned and a project is established. This proposal evaluation is usually referred to as the "R & D project selection/resource allocation decision problem." The project is then assigned to selected technical personnel, perhaps including the idea originator, to be researched and developed. However, management typically maintains control and coordination responsibilities, even though they may relinquish the technical responsibilities.

The following questions are illustrative of those posed by the authors who employ the idea and information flow point of view:

a. How does idea generation take place in the R & D laboratory? Are ideas generated but not submitted? Why? How is the project selection decision made?

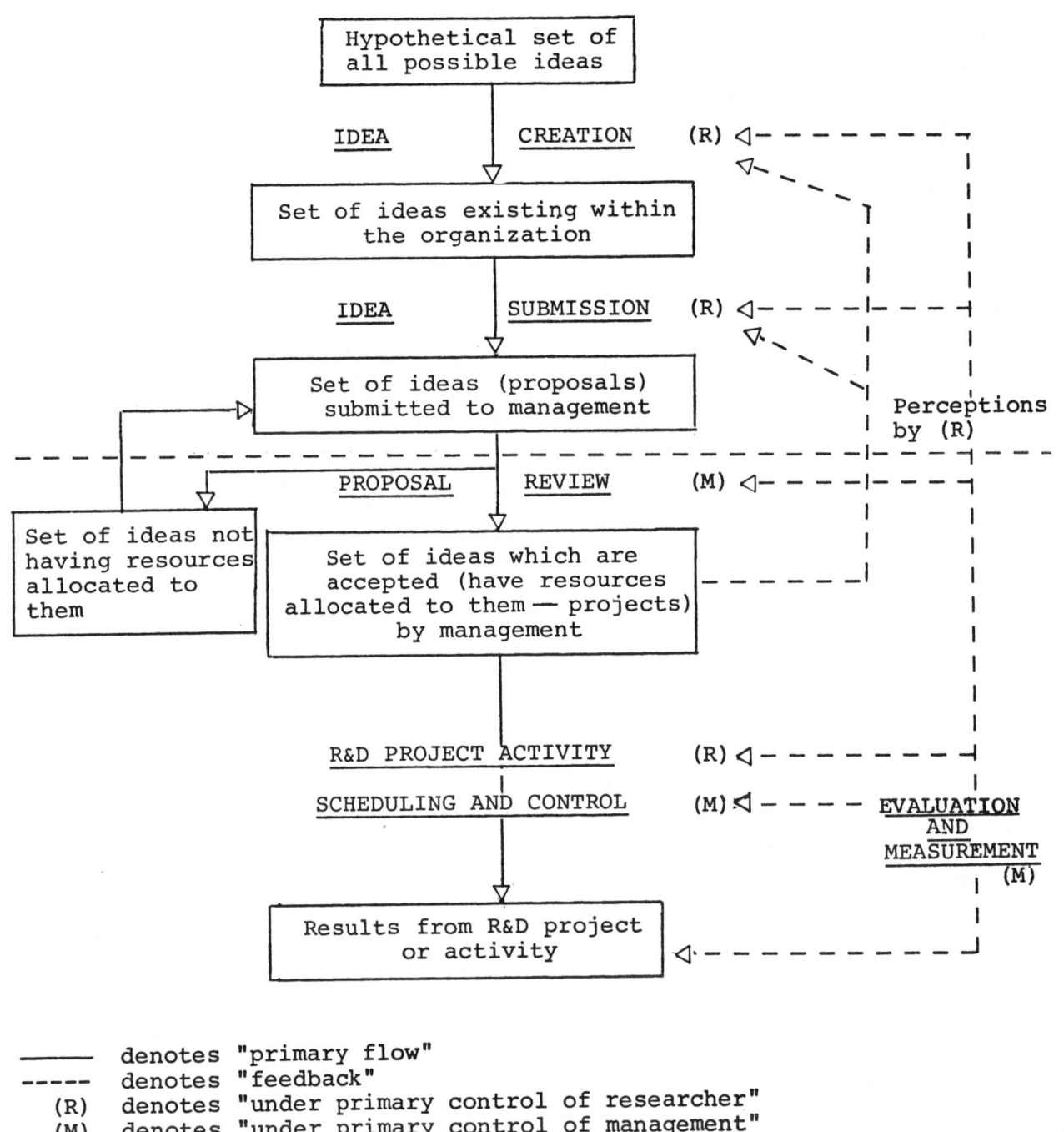

FIG. 1.--Information-flow model of R & D (after Baker and Freeland 1972b).

    b. What information is used at each stage? How is the information obtained? What is its source? How is the information processed? What additional information would be useful?

    c. Which organizational factors are most influential at each stage of the process? Do they function to inhibit, enhance, or constrain the flow of ideas or information? What can be done to improve the existing situation?

    d. What are the impacts of managerial evaluation, control, and coordination behaviors on scientists' and engineers' idea generation and submission behaviors?

    e. What organizational conditions are most conducive for scientists' and engineers' activities in researching and developing projects?

The idea and information flow influence in the above list of questions is obvious. In addition, a number of the authors[15] are on the faculty of schools of industrial engineering or industrial management, and the decision process and organization behavior orientations of such authors also influence the questions. They tend to visualize the R & D process from the view of the organization as an organism which requires information to support current operations and encourage future growth.

3. *Composites of the Process Phase and Information Flow Models*

In the preceding subsections we noted that both the academic-disciplinary background of the author and the point of view adopted focus and limit the questions addressed in the literature. The biases and limitations inherent to each view and author are simultaneously the strengths of the literature. For example, when economists adopt the process-phase orientation, the resulting literature focuses on the reduction of inherent uncertainty, the increase in relevance to organizational objectives, and the accrual of organizational costs as the project flows through the process phases. Thus, attention is called to the project itself, to the impact of the project on organizational resources and their utilization, and to changes in project characteristics. Conversely, when management scientists adopt the idea and information flow orientation, the resulting literature focuses on creation and interaction behaviors, linkages or filters and associated behaviors, and information needs and information-seeking behaviors of the R & D personnel and on the impact of organizational variables and conditions on their behavior. Clearly, both orientations are critical. Any composite of the two viewpoints must maintain the strengths of each as well as overcome their limitations.

In addition, attention must be given to the organizational control and influence aspects of each model. Process-phase literature identifies such organizational control variables as project cost, time to completion, benefit-cost measures, and levels of technical, economic, and commercial risk. This literature thus concentrates on control after a proposal has been submitted, and, most especially, after a project has been funded. It suggests that as a project moves away from the research end of the spectrum, the inherent uncertainty decreases, better estimates of the control variables can be made, and the project thus becomes more amenable to control.

On the other hand, the idea and information flow literature is more likely to concentrate on how managerial behavior and information flow influence the idea generation, idea submission, and problem-solving behaviors of the R & D scientists and engineers. Much of this literature concentrates on presubmission and on organizational conditions that constrain or enhance the idea or information flow. A point common to the two literatures is the project selection/resource allocation decision. It will be important to examine the organizational control and influence implications of the composite models.

One possible composite view would be to structure the R & D activity according to process phase, that is, to organize the project by its phases. Within each process phase organizational needs are identified, ideas are generated, projects are selected, and scientists and engineers are assigned. Project selection and resource allocation take place within a process phase according to the needs of the organization for effort in each phase. The effort in one phase can be related to the organizational needs for effort in another phase. For example, if research uncovers new technological results that could be developed for use by the organization, then a transfer across phases can be accomplished by an appropriately stated need for the development phase. Similarly,

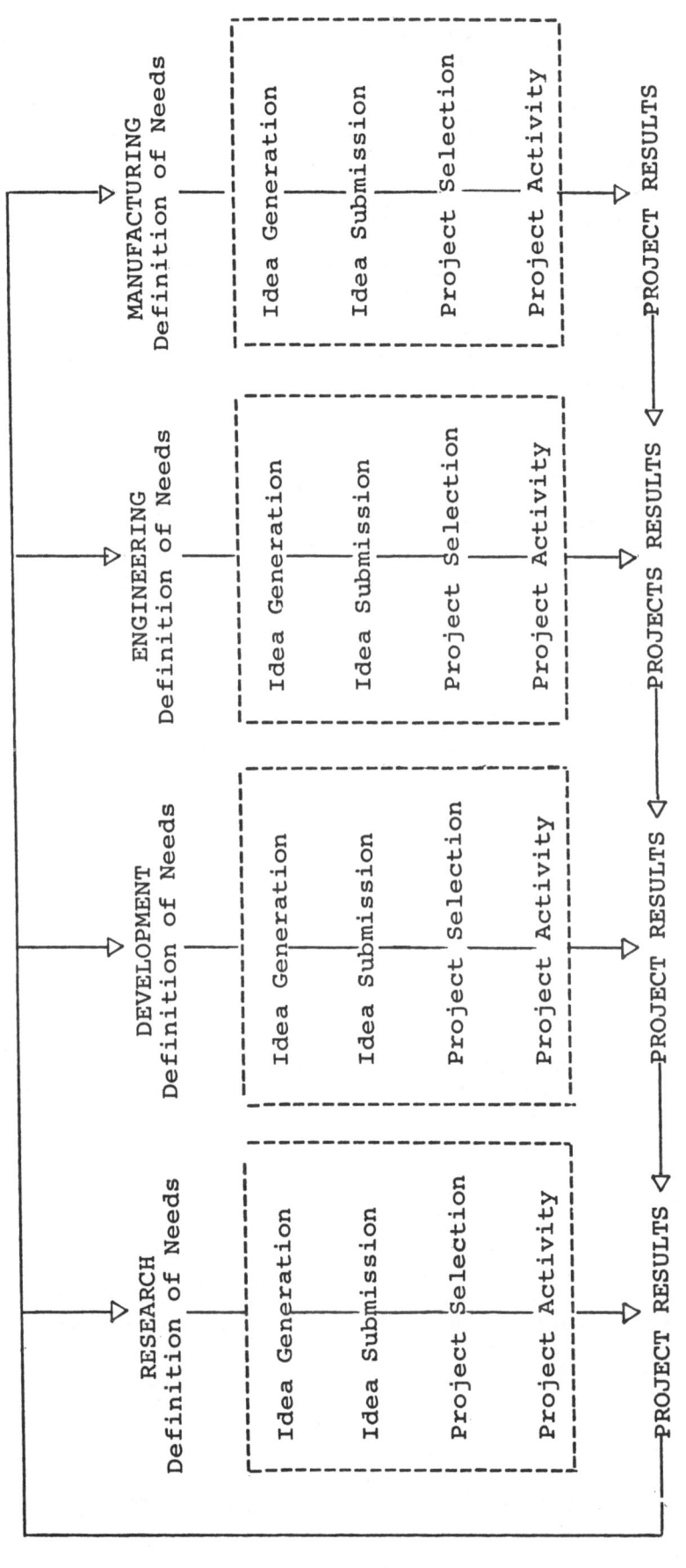

FIG. 2.--Phase-dominant model of the R & D process. The boxes defined by broken lines contain activity conducted within each process phase.

development results could be transferred to engineering and the engineering results then transferred to manufacturing. The flow could also be in the reverse direction; for example, if a development activity is being stymied by inadequate technology, then a research need could be defined and communicated. In order to stress that projects are defined within each phase and that transfer occurs by communicating one phase's results to another phase, the resultant model (Fig. 2) is named the Phase-Dominant Model.

Another composite approach would be to structure the R & D activity according to organizational needs or opportunities, that is, to organize the phases around the project. Ideas are generated, projects are selected, and resources are allocated as described in the idea/information flow literature. A project team is responsible for accomplishing the activities necessary to complete the project and these activities most likely cut across all the R & D process phases. As a whole, the project either progresses through the remaining R & D phases or it is terminated. Thus, resource allocation and scheduling decisions occur within the confines of the project but across process phases. In order to emphasize that results and requirements are not transferred from phase to phase but that the project is viewed as requiring integrated multiple-phase activities, the resultant model (Fig. 3) is called the Project-dominant Model. Table 2 is a summary comparison of the Phase-dominant and Project-dominant Models of the R & D process.

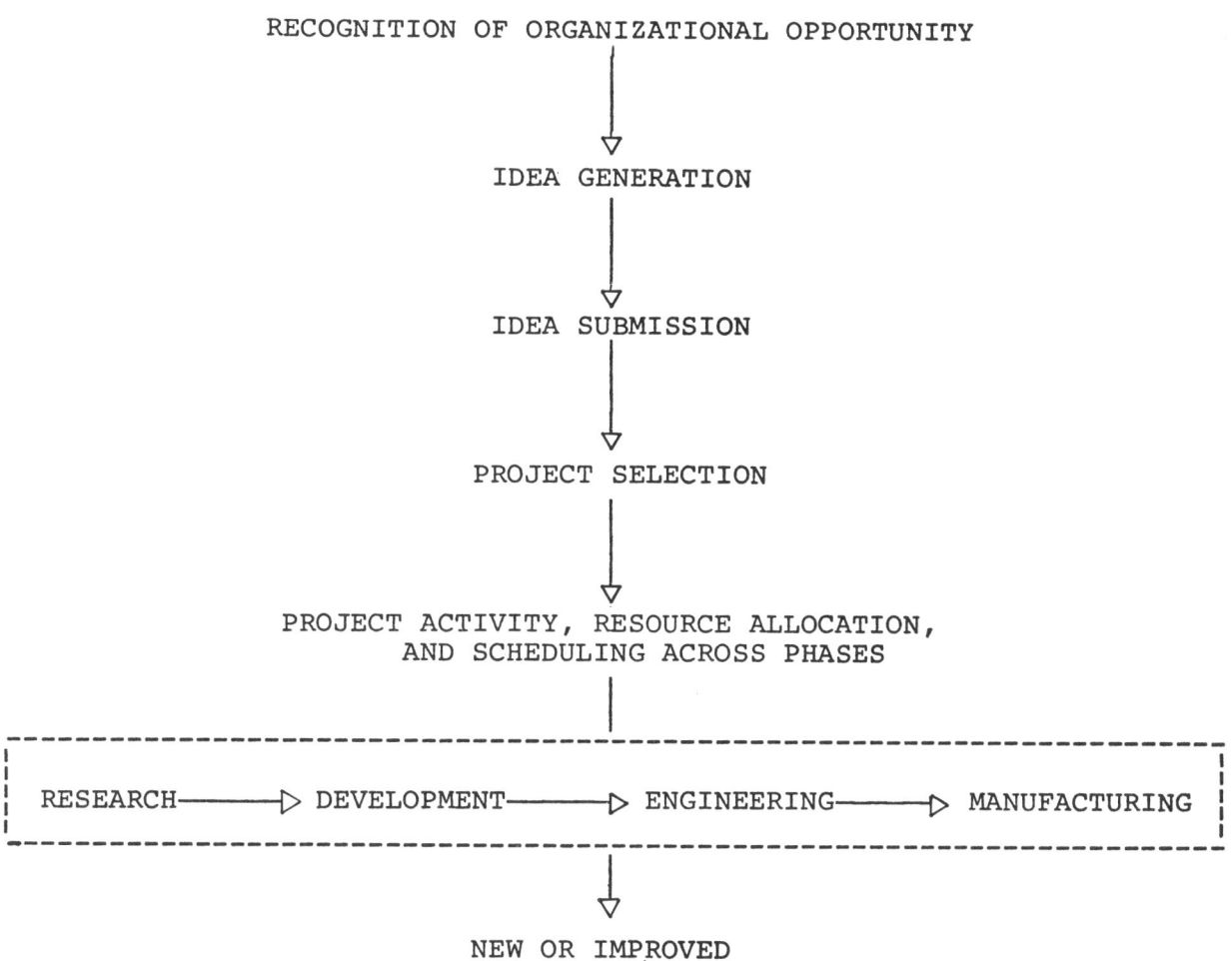

FIG. 3.--Project-dominant model of the R & D process. The box defined by broken lines contains project activity that flows across the R & D phases.

TABLE 2.—Summary comparison of phase-dominant and project-dominant models.

	Phase-dominant Model	Project-dominant Model
Organizational Characteristics		
Structural focus	Projects structured around R & D phases	Projects structured around R & D projects
Structural unit	Divisions within an organizational hierarchy	Project teams that cut across organizational boundaries
Control/coordination	Consistent with hierarchical structure	Inconsistent with hierarchical structure
Information transfer	Project needs and results across phases and organizational units	Needs and results within the project team, but across phases
	Eventual transfer of results to become ongoing activities	Eventual transfer of results to become ongoing activities
Project selection	R & D management--R & D budget	Corporate management--risk capital
Project control	By phase--within each division	By project activity--within team
Top management role	Overview budget summary, yearly review	Continual, detailed project review
Project Characteristics		
Scope	Narrow	Broad
Technical approaches	Inflexible	Flexible
Time horizon	Short	Long
Number of personnel	Few	Several
Scientific disciplines	One, few	Multiple
Process phases	Single	Multiple

The underlying differences between the two models can be illustrated by example. Suppose a firm seeks to add a new camera to its product line. The camera requires new capabilities in the process by which the film is exposed to light, in the paper from which film is made, and in the chemical coatings applied to the paper. Not only are the new capabilities required, but they must be also combined into a single system for use in the new camera. In addition, the housing of the camera must be designed and modifications must be made to the current production process.

A firm with a phase-dominant view of R & D would assign the exposure work to its "light research group," the film-paper work to its "paper research group," and the coating work to its "chemical-coatings research group." The results from each of these research groups would then be transferred to a development team for integration into a single system. If difficulties arose in the development of the system, they would be handled by the development team or, if necessary, the problem would be transferred back to the appropriate research group for resolution. When all the problems had been overcome, the finished system would be transferred to a design engineering team for housing design and to a production engineering team which would specify any necessary modifications to the production process. Eventually, the new camera would be designed and the production process modified, at which time the responsibility for the new product would be transferred to production.

A project-dominant view of R & D would lead the firm to structure the required activity quite differently from that described above. First a project coordinator would be appointed with responsibility for whatever activity is required prior to transfer of the new camera to production. The coordinator would organize a team which collectively has the talent required to develop the new capabilities, to integrate the capabilities into a system, to design the camera housing, and to specify any modifications to the present production process. When all that had been accomplished, the project coordinator would transfer responsibility for the new product to the production department.

A key difference in the two points of view has been illustrated. The Project-dominant Model leads to a single project coordinator who organizes a team to perform all project work. That is quite different from the several organizational administrators, each responsible for his individual portion of the project activity and for its transfer from one organizational unit to another, which follows from the Phase-dominant Model. The Project-dominant Model suggests a matrix organization; the Phase-dominant Model suggests a hierarchical organization. Later we argue that a hierarchical organization, by the very nature of its structure and its control/coordination processes, is led to adopt a phase-dominant view of R & D, perhaps unconsciously.

Our two composite models--the project dominant and the phase dominant--assume greater importance when we realize that the literature distinguishes between (1) R & D projects which are concerned with incremental improvements to existing products or processes or with minor extensions to scientific or technological knowledge, and (2) projects which are concerned with new products or processes or with scientific or technological breakthroughs (e.g., Hollander 1965; Rubenstein 1964; Schwartz 1973). The terms "incremental" as applied to the former and "discontinuous" to the latter are relative terms, of course; they emphasize the degree of their departure from the organization's current products or processes or from the existing level of knowledge in the organization.

Incremental projects tend to be relatively small, single-phase projects that are carried on within an organizational unit, involve one or a few scientific disciplines, and require few R & D personnel to be assigned, and of relatively short duration (up to a year). They are likely to be funded by R & D management out of their annual budgets and to require only the budgetary approval of top corporate management. In fact, except at an overview budget summary level, corporate management may be generally unaware of the incremental projects being conducted within the organization.[16]

Discontinuous projects are of two quite distinct types. One type can be described as the culmination of several incremental projects in which some final technical problem is overcome or some new insight is drawn. Also included are the scientific advances that result from the efforts of the individual scientist or of a small team of scientists. Examples include hybrid corn and other small grains, input-output economic analysis and similar methodological advances, oral contraceptives, etc. The above organizational description for incremental projects appears to be applicable to this first type of discontinuous project, the "cumulative-discontinuous" project. Both incremental and cumulative-discontinuous projects appear to fit well within the phase-dominant view of R & D.

There is another type of discontinuous project, often a new-product, or new-process, project. It can be characterized as a large, multi-year, multiphase project that cuts across organizational boundaries, may involve several scientific disciplines, and requires a substantial R & D effort to carry the project from initial concept through to end item. Examples include the Manhattan project, NASA's manned space effort, the pocket "Instamatic" camera, etc.[17] The decision to undertake such projects often emanates from the highest level of corporate management. Because of the large requirements for organizational resources, top corporate management support must be maintained throughout the life of the project. For reasons that will become clear presently, we refer to these projects as "disruptive-discontinuous" projects (Schwartz 1973). These disruptive-discontinuous projects correspond well with the project-dominant view of R & D.

We realize that the preceding paragraphs are speculative. An important area for future investigation is to evaluate empirically the hypotheses that incremental and cumulative-discontinuous projects can be described by the Phase-dominant Model and that disruptive-discontinuous projects can be described by the Project-dominant Model. To do so one must develop operational definitions for "incremental," "cumulative-discontinuous," and "disruptive-discontinuous," and document case histories for several projects of each type. In order to consider the organizational control and coordination implications of the composite models let us assume for the moment that the hypotheses are true.

The very nature of a hierarchical organization with its annual departmental budgets and responsibilities appears to encompass a Phase-dominant Model of R & D. The small, single-year, single-phase project, either incremental or cumulative-discontinuous, conducted with a single organizational unit by a few researchers and engineers, would seem to fit within the policies and structure of a hierarchical organization. However, the large, multi-year, multiphase project which requires a large number or researchers and engineers and which cuts across organizational units (the Project-dominant Model) would not seem to fit well within a hierarchical organization. Its multi-year, multiphase, multi-organizational unit, and large resource requirement characteristics, would appear to be disruptive to control and coordination mechanisms such as annual departmental budgets and responsibilities. Perhaps that is what Schwartz (1973) means when he suggests that such new product and process ideas are "disruptive to the culture of the firm."

Our hypotheses thus result in questions quite different from those which arose from the noncomposite models. Illustrative questions include:

a. Does the Project-dominant Model accurately describe disruptive-discontinuous projects? Does the Phase-dominant Model accurately describe incremental and cumulative-discontinuous projects?

b. Does the very nature of a hierarchical organization result in a phase-dominant orientation for its R & D activity? Are the incremental and cumulative-discontinuous projects routinely managed within a hierarchical organization? Are disruptive-discontinuous projects difficult for the hierarchical organization to manage? Do project teams with a matrix, rather than hierarchical, organization emerge in order to manage disruptive-discontinuous projects? Do the authorities and responsibilities of the project team cut across organizational boundaries? Are additional organizational problems created? What are they? How can they be resolved? Do the answers to these questions vary across firms, industries, or nations?

c. How are incremental projects funded? How are discontinuous projects funded? At what organizational level are the selection and funding decisions made? What is the role of the top corporate management? top R & D management? the potential users of the project results?

d. Does the way in which organizational needs and opportunities are assessed and the way information is disseminated influence the likelihood of incremental or discontinuous projects to be proposed and undertaken? How?

e. Where are the idea sources for incremental projects? for cumulative-discontinuous projects? for disruptive-discontinuous projects? Are there systematic differences? What are they? Why do they exist? Are the information channels dissimilar for the various types of projects? Why?

f. How do the characteristics of each process phase influence the relationships between organizational variables and conditions and idea generation, idea submission, and project performance behavior by scientists and engineers?

g. Is the project selection/resource allocation decision made differently depending on the process phase? Are different data available depending on the process phase? Do data estimates have different characteristics depending on the phase?

h. Do the models help explain why the bulk of money spent on R & D in industry goes for development, not for research (Hage and Aiken 1970; Mansfield 1968b and 1971; Mansfield et al. 1971)? Why does the contribution of large industrial laboratories tend to be in minor inventions (Hage and Aiken 1970)?

The above questions provide a new focus for empirical research on organized R & D.

Two distinct composite models have thus emerged, each of which includes and subsumes both the R & D process-phase and information-flow models and results in a new level of questioning. Our Project-dominant Model appears to be descriptive of disruptive-discontinuous projects; our Phase-dominant Model appears to be descriptive of incremental and cumulative-discontinuous projects. The models are compatible with different organizational structures--the Project-dominant Model is more consistent with matrix structures; the Phase-dominant Model is more consistent with hierarchical structures. The next subsection develops the organizational structure considerations in more depth.

4. Orientation and Location: Differences Within the Firm[18]

A number of descriptors have been used in discussing the orientation of a firm, or a laboratory, with respect to R & D. One such set of descriptors is the basis for the process-phase model, i.e., research, development, engineering, etc. Another set, common in the R & D management literature (e.g., Hamberg 1966), is the notion of offensive vs defensive R & D. As typically used, offensive R & D is activity performed in order to obtain an advantage over the competition whereas defensive R & D is undertaken in response to some existing or expected action by the competition. Offensive R & D is usually associated with improving the firm's competitive situation, and defensive R & D with maintaining the existing competitive situation. A related concept is the "leader vs follower" notion. Some firms tend to be industry leaders either in technological breakthroughs or in application of technology, whereas other firms in the same industry are satisfied to be followers and to use the breakthroughs of the leaders. Thus, although the distinctions are related, they are different. Leader/follower refers to the R & D activity itself. Offensive/defensive refers to the impact of the R & D effort in the competitive environment.

Gordon (1964) has proposed that the classic process-phase distinctions such as Research and Development should be abandoned, and be replaced by more useful distinctions, such as Urgency and Predictability. Urgency refers to the speed with which the results are required by the potential user of the project results; predictability is the extent to which the steps for obtaining the new knowledge are assumed knowable prior to the actual research (Gordon 1964, p. 3). Based on this distinction, Gordon is able to formulate several interesting speculations regarding the administration of R & D and expected behaviors during research (Gordon 1964a; Baker, Siegman, Larson 1971).

However, it has been argued that the classic process phase distinctions should not be abandoned in favor of urgency and predictability, but that urgency and predictability should be used to develop measures of a laboratory's orientation in respect to the phases (Baker, Siegman, and Larson 1971). In a laboratory oriented toward development, technical personnel and management can be expected to rate highly ideas that are relatively urgent and predictable. Conversely, a research-oriented laboratory should favor ideas that are less urgent and predictable. It would be useful for interorganizational comparative studies to have operational measures for determining a firm's, or a laboratory's, orientation with respect to process phase, offensive/defensive, and leader/follower.

In summary, there are a number of orientations that can be associated with a firm or a laboratory. These orientations often surface as balance considerations; for example, an R & D program should be "properly balanced" along each dimension or orientation (e.g., Hamberg 1966; Mansfield 1972b). The view held by corporate and R & D management regarding "proper balance" can have a direct effect on the characteristics of the ideas generated, submitted, and undertaken within the laboratory. As a result, the location controls used by the organization become important considerations in determining a laboratory's orientation. The answers to questions related to issues such as whether the laboratory is a divisional or corporate laboratory, to whom the laboratory reports, how the laboratory is funded, and where the output of the laboratory is used, indicate the orientation a laboratory is likely to evince.

Let us now turn to an examination of the literature dealing with location and administrative controls. The impact of organizational location and of administrative controls on the orientation of a divisional laboratory has been illustrated by Rubenstein and Radnor (1963). They hypothesized tight control over R & D by divisional general managers, in the absence of direct and continuing counterinfluences by corporate management, with a tendency of the divisional laboratory's R & D program toward:

(a) relatively shorter range programs;
(b) relatively narrower-scope programs in terms of relation to current lines of business, materials, processes, products, and applications of these products;
(c) relatively less flexible programs in terms of the freedom of the researchers to capitalize on the unexpected and follow technical leads that might appear peripheral to the objectives of particular assigned projects.

In the decentralized companies studied, they found pressures in the direction hypothesized and also pressures in the direction of longer range, broader scope, and more flexible R & D programs (Rubenstein and Radnor 1963). Thus conflicting pressures existed.

A gap between theory and behavior was also found in the decentralized organizations studied. Both corporate and divisional management agreed that the orientation should be toward longer range, broader scope, and more flexible R & D programs. However, behavior was much more likely to evince R & D programs of shorter range, narrower scope, and less flexibility. In response to the tendency toward such R & D programs, corporate management acted to change the program orientation of the divisional R & D laboratory by the imposition of performance requirements, the establishment of evaluation criteria, financial controls, methods of funding R & D, and direct influence on the selection of projects and programs (Rubenstein and Radnor 1963). These hierarchical controls serve, to the extent that they succeed, to modify the performance of the divisional R & D laboratory and to coordinate the divisional activities.

Schwartz, in his 1973 Harvard D.B.A. dissertation, presents results that suggest an explanation for the theory-behavior gap. Exploring how firms in high-technology industries make decisions, he found that decisions to innovate required a shared problem conception within the firm and special provisions for risk management. It was necessary for senior managers, middle managers, and lower-level generalists to play differing but complementary roles in defining the projects and in developing the justification required prior to undertaking the project. Technical and marketing risk was "brokered" by assigning various responsibilities to managers. Brokering was a negotiation process in which each manager shaped the proposal by negotiating the terms under which he would lend his support and hence assume a share of the risk. Such justification and risk brokerage functions tended to add a conservative bias to proposals. Thus, even in such high-technology firms as Digital Equipment Corp. and Texas Instruments, the innovations tended to be incremental. Schwartz explains the tendency toward incremental improvements by the justification and risk brokerage functions; our composite models suggest that the hierarchical organizational structure and associated control and coordination mechanisms result in the managers' felt need for the justification and risk brokerage functions.

A 1973 paper by Scott looks at decentralization for managing new products rather than at the decentralization of the R & D program. He states that in the United States and Western Europe the organization decentralized by function is declining and that there is a dramatic rise of the form based on decentralization by product division. This shift is occurring because the divisional structure appears to be the most effective way to manage a strategy of diversification where new profit and growth opportunities have been identified by R & D activity. Scott, it would seem, is more concerned with the utilization of R & D results than with the generation of additional opportunities. Yet, throughout this section we have argued that the divisionalized, decentralized organization is more likely to identify incremental, rather than disruptive-discontinuous, R & D advances. Thus, a paradox results: the organizational form least suitable for identifying disruptive-discontinuous opportunities appears most suitable for managing the resultant innovation.

5. *The "Innovation in Organizations" Literature*

A more general literature, that dealing with innovation in organizations, provides further insight into the paradox identified above.[19] In this literature, an innovation is defined as "any idea, practice, or material artifact perceived to be new by the relevant unit of adoption" (Zaltman, Duncan, and Holbek 1973, p. 10). So the definition is not restricted to technological change, but does include it. Despite this underlying definitional distinction, the literature provides certain concepts that are useful for our discussion.

Wilson (1966) postulated a dilemma apparently faced by organizations when attempting to provide a structure supportive of innovation. The dilemma derives from concepts associated with varying levels of complexity and diversity in an organization's incentive system and task structure. He argued that the greater the complexity and diversity in the incentive system and task structure, the greater the likelihood that participants will both conceive of and propose innovations. However, that same diversity and complexity reduce the proportion of proposals that will be adopted.

Subsequent authors introduced concepts such as formalization (degree of codification of jobs in the organization [Hage and Aiken 1970]), and centralization (the concentration of power and decision making in the hands of a small proportion of individuals [Hage and Aiken 1970]). Zaltman, Duncan, and Holbek (1973) identify an initiation stage (including problem definition, idea generation and submission, and project selection) and an implementation stage. They propose that higher complexity, lower formalization, and lower concentralization are appropriate at the initiation stage, but that lower complexity, higher formalization, and higher centralization are appropriate at the implementation stage. Thus, the organizational conditions most conducive to idea generation and submission are just the opposite of the organizational conditions most conducive for adoption and continued utilization; and yet both activities must be performed--a dilemma strikingly similar to the one which evolved from the integration of the Rubenstein, Radnor, and Scott papers.

The centralization concept identified above is quite different from the concepts "centralized" and "decentralized" as used by Rubenstein. Here centralization refers to the number of people who have organizational power and decision authority, whereas Rubenstein's centralized and decentralized refer to the way in which the organization is structured. Thus, if the divisional general managers exert rigid control over R & D, then that divisional organization would exhibit high centralization with respect to R & D. It is within the high centralization setting that Rubenstein observed the tendency toward shorter range, narrower scope, and less flexible R & D programs. Thus, at least superficially, the two dilemmas appear to yield consistent implications: The organizational conditions and structures most suitable for initiation are opposite to those most suitable for implementation.

Our composite models of the R & D process suggest a refinement to these dilemmas when combined and applied specifically to R & D. An R & D activity structured in accordance with the Phase-dominant Model of R & D will be hierarchical in nature with low complexity, high formalization, and high centralization. It will be satisfactory with respect to the generation and submission of incremental and cumulative-discontinuous R & D projects and will be capable of implementing and managing the output from either incremental or cumulative-discontinuous R & D projects. However, the organization will be unsatisfactory with respect to the generation and submission of disruptive-discontinuous R & D projects.

In order for disruptive-discontinuous projects to be generated and carried out, the organization must be structured in a manner consistent with the Project-dominant Model. It would be a matrix organization and will exhibit high complexity, low formalization, and low centralization. Since all the activities are essential, both organizational forms must exist with the same organization, perhaps within the organization's R & D activity. At this point no satisfactory means of coexistence has been found and thus the dilemma remains.

Several responses to this organizational dilemma have been observed. Among the various strategies are:

a. Set-up a corporate laboratory to do basic long-term, multidisciplinary research.
b. Do some basic research in-house and contract the remainder with a contract research organization.
c. Contract all basic research to a contract research organization.
d. Acquire the research results that have been accomplished elsewhere.
e. Spinoff the in-house research results to a new technologically based firm or to a new company division or subsidiary.

These suggested actions do not completely overcome the dilemma. The dilemma arises at the time when the transfer, internal or external, takes place. This topic is discussed in depth later in this chapter.

6. *Further Insights from the Organizational Development Literature*

In 1961 Burns and Stalker conducted an in-depth study of 19 rayon and electronics firms in England and Scotland. They were primarily interested in the relationships among management practice, organizational behavior, and organizational innovation. Their study found two distinctive types of organization, termed mechanistic and organic. Zaltman, Duncan, and Holbek (1973, p. 171) present a summary description (Table 3) of the mechanistic and organic organizational forms as defined by Burns and Stalker. Basically, Burns and Stalker argue that there is no one best way to organize. When the environment is relatively stable and fairly certain, the mechanistic type is most appropriate; however, when the technical and market environment is changing and unstable, the organic type is more appropriate, especially for gathering and processing information. Note that the mechanistic form could be described as hierarchical (items 4, 6) with low complexity (items 1, 2), high formalization (items 3, 8, 9), and high centralization (items 5, 7, 8); whereas, the organic form is matrix (items 4, 6), with high complexity (items 1, 2), low formalization (items, 3, 8, 9), and low centralization (items 5, 7, 8).

TABLE 3.--Mechanistic and organizational forms (after Zaltman, Duncan, and Holbek 1973, p. 171).

Mechanistic	Organic
1. Tasks are broken into very specialized abstract units	1. Tasks are broken down into subunits, but relation to total task or organization is much more clear
2. Tasks remain rigidly defined	2. There is adjustment and continued redefinition of tasks through interaction of organizational members
3. Specific definition of responsibility that is attached to individual's functional role only	3. Broader acceptance of responsibility and commitment to organization that goes beyond individual's functional role
4. Strict hierarchy of control and authority	4. Less hierarchy of control and authority sanctions derive more from presumed community of interest
5. Formal leader assumed to be omniscient concerning all matters	5. Formal leader not assumed to be omniscient concerning all matters
6. Communication is mainly vertical between superiors and subordinates	6. Communication is lateral, between people of different ranks, and resembles consultation rather than command
7. Content of communication is instructions and decisions issued by superiors	7. Content of communication is information and advice.
8. Loyalty and obedience to organization and superiors is highly valued	8. Commitment to tasks and progress and expansion of the firm is highly valued
9. Importance and prestige attached to identification with organization itself	9. Importance and prestige attached to affiliations and expertise in larger environment

The mechanistic form provides further elaboration of the type of organization that was brought into the R & D laboratory during the 1945-1955 period. According to our earlier arguments the mechanistic form is satisfactory for generating incremental or cumulative-discontinuous R & D projects and for managing the output from either incremental or discontinuous R & D projects. The Burns and Stalker study suggests, however, that the mechanistic form is not satisfactory for managing the output from disruptive-discontinuous projects (Burns and Stalker 1961; Zaltman, Duncan, and Holbek 1973). Their analyses indicate that an organic form is necessary for the generation and submission of disruptive-discontinuous ideas and for the management of output from the disruptive-discontinuous ideas, at least until the environment has become stable and the activities routinized.

7. *Innovation in the Firm, But Not in R & D*

Organization development is concerned with change throughout the organization, not just in change which is initiated by R & D. Thus, the concepts from organization development which are summarized in the earlier sections are applicable, but not restricted, to the R & D activity. One of the important constructs in organization development is that all organizational constituents are responsible for two activities--performing organizational tasks and participating in and being receptive to change (Bennis 1969).

A 1965 study by Hollander provides striking evidence that much innovative effort can be accomplished by organizational constituents other than R & D (Hollander 1965). Hollander analyzed production-cost data during the period 1929-1960 at five Dupont viscose-rayon manufacturing plants and surveyed the personnel employed at each plant.

Defining technical change as "changes in the technique of production of given commodities by specific plants, designed to reduce unit production costs" (p. 23), Hollander pointed out that the changes may be "of a technological nature..., or they may be 'managerial' and consist in improved organization, 'Taylorization,' and the like" (p. 190). A distinction was made between "major" and "minor" technical changes: a change was "major" if its development was considered difficult to accomplish by men skilled in the pertinent arts before the development program, and "minor" if its development was judged a relatively simple process. The cumulative effect of the "minor" changes was far greater than that of "major" changes according to Hollander's data. "Minor" technical change accounted respectively for 46, 79, 80, 83, and 100 percent of the net reduction in unit costs due to technical change at each of the five plants.

"Major" technical changes depended for the most part on formal R & D activity conducted within the DuPont organization. However, the "minor" technical changes tended not to originate in the formal R & D groups. Instead, the larger part of the "minor" technical changes were developed at the plants themselves by personnel intimately concerned with current operations. Some important "minor" changes were also made by equipment manufacturers, suppliers of raw materials, and joint ventures with the equipment manufacturers. Thus, Hollander presents a most interesting finding: "minor" technical change accounted for the greatest percentage of net reduction in unit cost and, for the most part, originated and developed not in R & D but in the production plants.

Hollander qualifies his emphasis on the efficacy of "minor" technical change. The data appear to suggest a "saturation effect" which states that "without some preceding 'major change' the potential stream of 'minor' changes will be exhausted" (p. 205). "Minor" changes can be expected for a considerable time, 10 to 15 years in the DuPont data, after the occurrence of "major" change. Thus, R & D, through "major" changes, improves the process and provides the plant personnel with an output which they further improve by "minor" changes.

It is clear that the "minor" technical changes were incremental in nature; it is not clear which, if any, of the "major" changes would be classified as discontinuous. Hollander provides conclusive evidence that a considerable amount of incremental innovation occurred within five viscose-rayon manufacturing plants of DuPont during the period 1929-1960 and that the activity neither originated in, nor was developed by, formal R & D. Moreover, the evidence is equally conclusive that any discontinuous innovation that occurred was conducted by the formal R & D activity. Discontinuous, or at least "major," innovation is important of itself and, because of the so-called "saturation effect," is also necessary for continued incremental innovation. Thus, Hollander's study indicates the relevance of the organization development literature to incremental innovation and, because of the saturation effect, further supports the need for R & D organizational structures and control and coordination mechanisms which are conducive to disruptive-discontinuous innovation.

8. *The Message*

A consistent theme runs through the literature--organizational structure is a determinant of the function of R & D within organizations. No one model of the R & D process and no one organizational type is most appropriate. The most appropriate model appears to be a function of whether the R & D activity is incremental or discontinuous in orientation. The appropriate organizational form is a function of both the orientation of the R & D project and its stage in the R & D process. Overall the message is clear: organizational structure, strategies, and policies are critical in determining the orientation of a laboratory, the nature of the R & D projects which are generated, submitted, and accepted, and the effectiveness with which projects are conducted and results implemented. Despite the informed speculations presented above, relatively little is known about these important relationships. In a time of increasing international competition, this area is most important and deserves high priority for additional research in innovation.

B. PROBLEM DEFINITION AND IDEA GENERATION WITHIN THE R & D LABORATORY

1. *Preliminary Considerations*

In the above typology of R & D projects, each type is preceded by problem definition. Depending on the nature of the project, the magnitude of resource and top corporate commitment required, and the R & D phase(s) involved, *problem definition* may also be referred to as 'definition of needs' or 'recognition of organizational opportunity.' As has been noted, these preproject activities have been examined most extensively in the management science literature, and with an emphasis on the organizational and social-psychological influences that facilitate or inhibit the flow of ideas and information. In this section we shall survey the several dimensions of these impingements and assess the state of the art understanding of their influence on the patterns of idea and information flow.

It should be noted that the typology of R & D projects offered above has not been employed by those concerned with problem definition and idea generation activities. Hence any systematic differences in the nature of such activities that might result from these distinctions would not be reflected in the literature. For instance, the literature is silent as to possible differences between problem-definition and idea-generation activities leading to incremental projects and those leading to disruptive-discontinuous projects. Likewise, it does not point up differences, if any, between such activities as they precede earlier and later projects in the Phase-dominant Model. Although speculation about systematic differences that might correspond to the types of R & D projects we have distinguished is an attractive possibility, we shall not, by and large, pursue it here. Rather, we shall remain fairly close to the literature's own orientation which, in terms of our typology, deals almost exclusively with problem-definition and idea-generation activities leading to *incremental* projects. That should not be taken to mean that in our judgment no significant differences exist. On the contrary, we think they do--for various organizational, informational, and social-psychological reasons. But these have not been demonstrated (nor has our hypothesized typology of R & D projects), and we shall refrain from such extended speculation.

A more settled distinction concerns the difference between preproject and project activities. This distinction is not blurred by the fact that problems are obviously defined (*refined* is probably a better word) and ideas generated, not only *prior* to the establishment of projects, but during them as well. "Problem definition and idea generation" is simply a convenient, though nonexclusive, label for preproject activities. The differences lie not here, but in at least two other dimensions. First, R & D activities are organized in terms of projects that have been approved by management and to which resources have been committed. Projects thus have a conceptual focus and structure; someone has developed a proposal for solving a particular problem and submitted it to management, which has in turn committed resources to it. It has an organizational existence in that it has been made someone's responsibility, and is thus at or near the top of his official "worry list."

For the individual to whom a project has been assigned there is typically a fairly well-developed set of incentives to carry through on it. In preproject activities, on the other hand, the situation is quite different. There is no project to focus the researcher's efforts. As we shall see, there are focusing mechanisms, but these are tenuous by comparison. Nor are resources typically committed, and individuals charged with defining a particular problem and generating ideas for its solution. To offer a rather weak qualification, many firms say that a certain percentage of the researcher's time is "free" for such ill-defined activities. But, as we shall see, the incentive structure is typically such as to erode this "free" time in favor of the current project.

Second, prior to the establishment of a project, little direct administrative control of a researcher's activity is possible. This being the case, problem definition and idea generation activities seem more susceptible to the influence of localized social-psychological variables than project activities. Thus an organization's success in encouraging the flow of high-quality ideas is usually dependent upon its sensitivity to the social-psychological dimensions of the researcher's environment. On the other side of this coin, the more structured environment of a project is probably "more comfortable" to those individual researchers who prefer to reduce the level of uncertainty through greater structure and focus. Reduction of the initial level of uncertainty associated with a project has been mentioned above as a later R & D phase consequence in economic and corporate terms. What is being suggested here is that the same progression holds for the individual researcher, and that the greatest uncertainty is associated with preproject activities.

Thus whereas problems are certainly defined and ideas generated in both preproject and project activities, the nature and dynamics seem substantially different. Projects have substantially more structure in both conceptual and organizational terms. Someone is put in charge, and progress, or the lack of it, is noted. The organization can directly influence the activity in a variety of ways. Problem definition and idea generation, on the other hand, is a poorly structured and highly unpredictable activity in which the organization has little control, and that indirect. In short, there are many dissimilarities in activities that are speciously similar. From the organization perspective it is the difference between assigning an individual to work on X and trying to determine if he has been thinking about some X or other that it might be worthwhile to work on.

2. *An Influence and Information Flow Model*

Figure 4 indicates the particular problem clusters and major lines of influence and information flow that characterize problem definition and idea generation in organized research and development. Boxes 1-3 represent the problems that cluster around the social-psychological influence elements, and the arrows to and from these boxes indicate the information flow points or filters that are influenced by these social-psychological elements. Boxes 4-13 represent the flow of information from the outside and within the firm. Finally, box 14 links this phase with the R & D phase which follows. The elements of this linkage will be presented in a later section. We shall examine these influences and information flow elements in the order in which they are numbered in Fig. 4, beginning with patterns of supervisory authority.

3. *Supervisory Authority Patterns*

Organized innovative efforts, by definition, channel and control the individual researcher's professional judgments and activities. By becoming a part of an organized R & D effort, he is less free than his independent counterpart to choose his research topics, to approach them as he sees fit, and to change direction unexpectedly in order to follow a new insight or clue. The folklore surrounding this point is that freedom not only enhances creative accomplishment, but is in fact a necessary condition. If that is true, then the need to understand this issue is an urgent one. Unfortunately, few empirical studies bear directly on the topic, perhaps reflecting the conceptual difficulties inherent in terms like "freedom" and "control." Such terms are operationally unclear. They point in so many research directions, though with lack of specificity, that it is not surprising to find that so little has been said about them outside the "wisdom" literature.

One of the directions in which the freedom/control issue points is the role and "style" of the R & D manager. This individual personifies, for better or worse, the

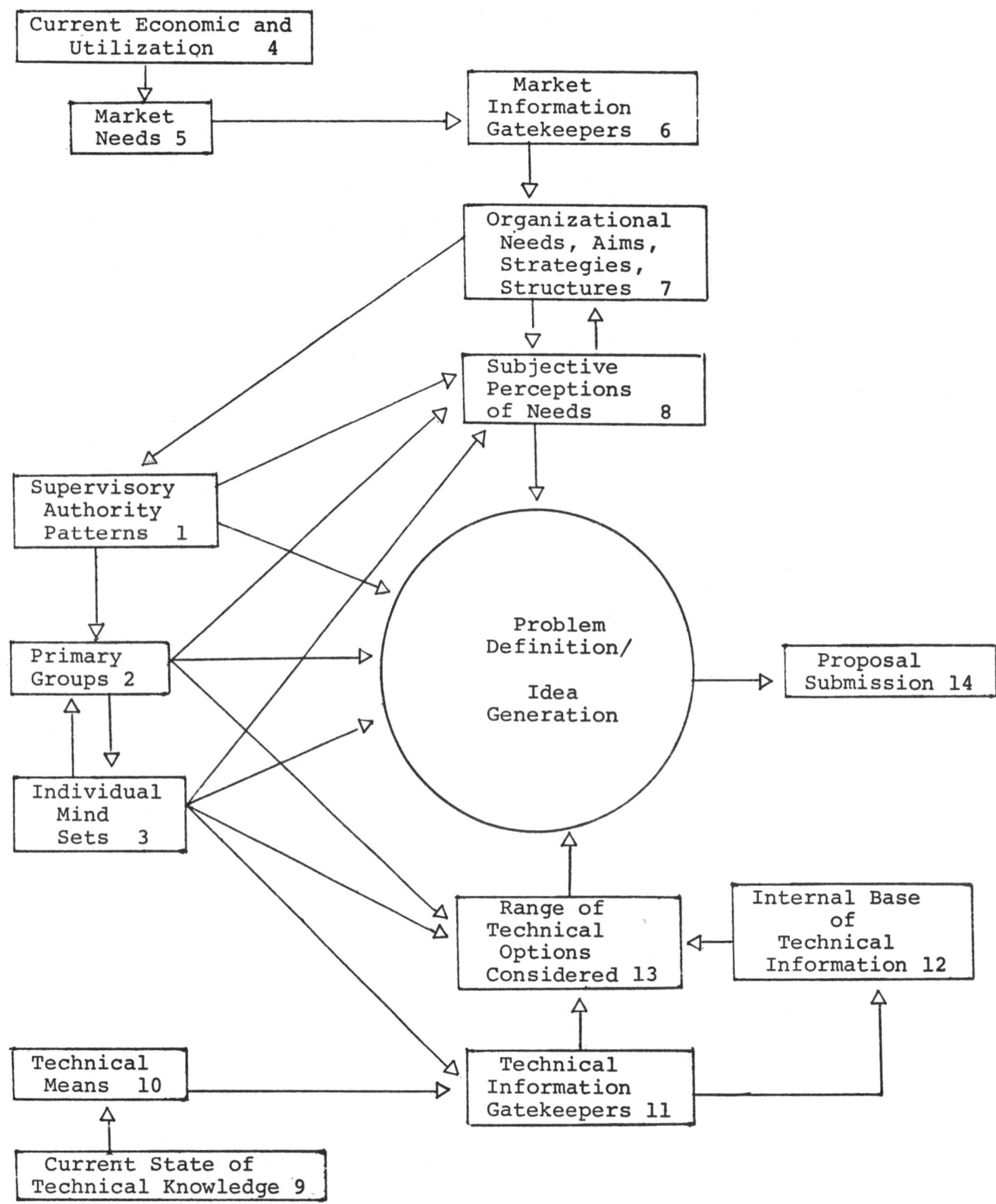

FIG. 4.--Information and influence flow patterns within the problem-definition/idea-generation process.

"control" factor in the enterprise of deliberate creativeness. The manner in which he imposes corporate judgment--to explore certain research areas and not others, to try certain approaches rather than others, and to follow only certain leads which involve a change of direction--determines both the actual limits on the researcher's freedom and, often more important, the researcher's perception of these limits. This observation leads one to posit that different patterns of supervisory authority might differ in impact on the researcher's perception of his freedom and thus upon the frequency and quality of his creative accomplishments. Thus one of the social-psychological influences indicated above (Fig. 4) was labelled Supervisory Authority Patterns. The elements that influence or are influenced by such patterns of supervisory authority are indicated in Fig. 5.

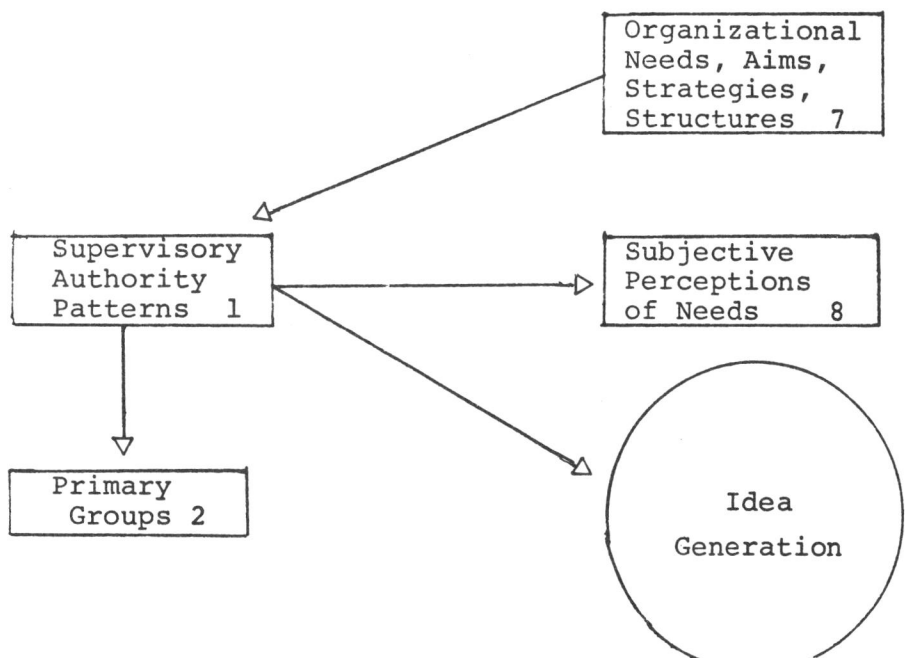

FIG. 5.--Interactions with patterns of supervisory authority.

The unit, or level of analysis, with which we are concerned is that of the individual researcher. One can conjecture that such influences are also operative at the next or higher levels within the organization, and that they also exert an influence on the innovation process at these levels. We shall discuss these influences only at the level of the individual researcher, however, with the implication that *mutatis mutandis* they may also apply at other organizational levels.

In one particularly significant study (Gordon and Marquis 1966), authority patterns in four types of research settings were examined in order to investigate the relationship between freedom and creative accomplishment. The results of 179 research projects conducted in academic and quasi-academic settings (universities, health agencies, hospitals, and medical schools) were assessed by *independent* evaluators in terms of criteria designed to reveal their quality and degree of innovation. Gordon and Marquis hypothesized that maximal research freedom enhances innovativeness only if there is an impetus to innovate. Such impetus can be provided by the "visibility" of the consequences of an individual's research. By "visibility" was meant both the ease with which research consequences may be assessed, and the manner in which they are assessed. The first of these relates to the clarity or obscurity of organizational goals.

> In an organizational setting where the owner of an organization or his representative can accurately evaluate the findings of a project in terms of organizational goals, he can encourage the researcher who shows high probability of solving such problems. As a consequence, the researcher is motivated to seek solutions to difficult but "relevant" problems in preference to less relevant but easier problems. In seeking a solution to the difficult problems, the researcher at times must abandon traditional methods and thinking. This would appear to be as true for the academic as for the non-academic researcher. Kuhn, for instance, has observed that "The novel theory seems a direct response to crises." (Gordon and Marquis 1966, p. 198.)

The clarity of organizational goals and the ability of the supervisor to transmit these goals influence their visibility, and consequently the visibility of research consequences.

Regardless of the ease with which research consequences may be assessed vis-a-vis organizational goals, they are not visible until someone assesses them. That raises directly the question of the pattern or "style" of research management and its influence on innovative behavior. To get at this question the research projects employed in this study were divided into three groups.

> 1. Projects in which the project directors either stated that they had no administrative superior or that they did not discuss their research with their administrative superior. (Low visibility of consequence + freedom.)
> 2. Projects in which project directors had freedom to specify their research procedures and they discussed their research with their administrative superior. (High visibility of consequences + freedom.)
> 3. Projects in which the project directors stated that they had an administrative superior with whom they had discussions and who consistently influenced procedures. (High visibility + limited freedom.) (Gordon and Marquis 1966, p. 199.)

On the hypothesis that *both* high visibility of consequences *and* research freedom are important to creative activity, the second of these three types of authority patterns should be expected to maximize such behavior, and the first and third would minimize it. That proved to be the case: two and one-half times more of the projects rated as most innovative by the independent evaluators were conducted under the ideal authority pattern (type 2) than under either of the nonideal conditions.

> In sum it is not possible to make a blanket statement relating maximal freedom to innovation, but rather maximal freedom is conducive to innovation only when there is an impetus to innovate. It further appears that the institutional settings in which research is conducted--in particular the visibility of the consequences of the research in relation to the goals of the owners of the institution--has a significant effect on inducing innovation. (Gordon and Marquis 1966, p. 201.)

Thus it would seem that creative behavior is more likely to occur where the consequences of such behavior are visible, that is, where the organizational criteria for assessing research results are clear and where the supervisor keeps in touch with what the researcher is doing and how his work is going; but where at the same time the researcher has his freedom. That is, he is not dominated by his superior.

This study was chosen for detailed examination because of the linkage it provides among not only freedom, creative behavior, and different patterns of supervisory authority, but also between these elements and the clarity of organizational goals. On the latter point, there are obviously other informational sources within the environment that also contribute to shaping the researcher's subjective perception of the organization's needs and goals. The primary one would seem to be the immediate supervisor, who does it explicitly through informal discussion, more formal meetings, and implicitly by his decisions concerning ideas submitted. Other dimensions of the concept of "research-result visibility," as a function of the researcher's primary social group within the laboratory, for example, will be examined later.

The typology of supervisory authority patterns implicit in the Gordon and Marquis study and their correlation with innovative behavior finds support in a number of other studies. In summarizing the literature dealing with "leadership styles" Hill reports:

> Most commonly, three general patterns of a supervisor's leadership behavior have been described, though the terminology applied to these patterns has differed somewhat between authors:
>
> (1) nondirective, *permissive*, a laissez-faire, accommodative or abdicative style where the leader relinquishes any influence in setting group goals to the group;
> (2) *democratic*, a participatory, group-centered subordinate-centered, employee-centered, human-relations-oriented style where the supervisor allows and encourages a mutual relationship with subordinates;
> (3) *autocratic*, authoritarian, boss-centered, task centered, production centered, close and punitive style where the supervisor allows his subordinates little or no influence in the setting up of work procedures, while primarily concentrating on achieving task goals. (Hill 1970, p. 11.)

Although the cluster of words associated with each of these three leadership patterns reflects differences of detail and emphasis, there seems to be "... a high concurrence of findings about general leadership patterns across a range of situations" (Hill 1970, p. 11). These three leadership patterns and their correlation with more or less innovative behavior strengthens our confidence in the work of Gordon and Marquis.

The work of Ronken and Lawrence also indicates that the democratic leadership pattern (high visibility of consequences + freedom in Gordon and Marquis' terms) is more effective in stimulating creative behavior than a leadership style that is either too permissive or overly directive (Ronken and Lawrence 1952). One of the problems of a style that is too permissive is that the researcher may interpret being left alone with only minimal contact as a lack of interest in what he is doing and thus a devaluation of his work. Hill also notes this possible interpretation of a leadership style that is overly permissive, and sees it as a "demotivating" influence (Hill 1970, p. 12).

In commenting on the dysfunctional effects of an autocratic style (high visibility + low freedom), Pelz and Andrews state that "...continued direction by the chief will stunt initiative and independence and these are qualities basic to scientific achievement" (Pelz and Andrews 1966, p. 33). And in a perhaps unexpected result, Hill found that researchers perform best (in the sense of the quality and quantity of innovative ideas) under participatory or democratic leadership, but when the leadership pattern was strongly autocratic or overly directive they responded by seeking even greater direction (Hill 1970).

Finally, the studies of Andrews and Farris also support the conclusions offered above concerning the need to "keep in touch," or provide visibility for the consequences of a researcher's work, if freedom is to be effective in stimulating creativity (Andrews and Farris 1967). They found that freedom was unrelated to innovation if the supervisor did not consult with them prior to making decisions concerning their projects. Where freedom was combined with consultation, however, a substantial increase in innovative behavior was observed. A key factor in the effectiveness of such consultation would seem to be the supervisor's own technical competence. In drawing some general conclusions from their research Andrews and Farris say:

> Greatest innovation occurred under supervisors who knew the technical details of their subordinates' work, who could critically evaluate the work, and who could

influence work goals. Thus the widespread practice of including technical competence among the criteria for choosing supervisors seems to be sound. This does not mean that a supervisor should constantly "meddle" in his subordinates' activities. But he should be available, competent in the current "state of the art," actively interested in the project, and informed about it. . .

>What if this kind of structure is not possible, or if a supervisor's technical competence has become obsolete? Again the data were clear: provide substantial freedom for subordinates. Freedom acted as a partial substitute for skilled supervision. But even where subordinates have freedom, the supervisor still makes some kinds of decisions. For freedom to be effective, the data showed that the supervisor must consult with his subordinates before making these decisions. (Andrews and Farris 1967, p. 513.)

Thus the visibility of research consequences, as achieved by an available and competent supervisor who knows what his subordinates are doing, is the intervening variable activating the potential of professional freedom to increase creative accomplishment. However, the influence of this intervening variable appears to depend on the technical competence of the supervisor.

The last point to be mentioned in connection with the "control figure" or supervisor in the R & D context concerns his influence on the researcher's perception of the needs of the firm. Illustrative of this point is an experience related by Hyman.

> I was once talking with the manager of engineering whose company had just lost several million dollars because of marketing a defective machine which later had to be withdrawn from the market. As a result, the company was under tremendous pressure to recover its previous position in a highly competitive market. The manager was worried because, in this market, if you do not look ahead and keep generating new patents you cannot survive very long. Yet his men, eighteen design engineers, had not turned out a patent in the last year or so. Like every one else in the company, their major concern was with current pressures to keep the business out of the red. The managers, in order to change this lack of new patents, first thought in terms of his selection policies. Maybe he had chosen the wrong men. Maybe he should fire some of his present staff and hire new men. One day the thought occured to him: Why not first call the men in and tell them what I want? He called them to a meeting and told them, 'Look men, we need patents, or else we die.' The next month his men presented him with several patent applications. And they have been continuing at that rate ever since. One gets the impression that, essentially, the men just looked at each other and said, 'Well, if that's what he wants why didn't he say so?' (Hyman 1964, p. 70.)

This overly simple and somewhat dramatic little story contains several recurrent themes in the research literature: the dominance of oral communication throughout the process, the phenomenon of the "unsubmitted idea," as well as the role of the supervisor in shaping researchers' subjective perceptions of organizational needs. The former points will be treated later. The latter extends the inventory of ways in which problem definition and idea generation within the laboratory is strongly influenced by the first-line control figure. It is ironic perhaps that one can get the impression from many literature sources that R & D supervisors are but passive functionaries whose role deserves less mention than that of the "product champion" or "top person." But if the evidence is to be believed, their role is a crucial one and their influence substantial.

The next cluster of social-psychological influences to be considered are the researcher's primary work group and his social group within the laboratory.

4. *Primary Groups*

The elements of our information and influence flow model (Fig. 4) that influence or are influenced by primary groups within the R & D laboratory are reproduced in Fig. 6. The term "primary group" has two referents in the laboratory context. Within the formal organizational structure, it refers to *work group* within the laboratory to which a researcher has been assigned for the conduct of a current project. It also refers to the

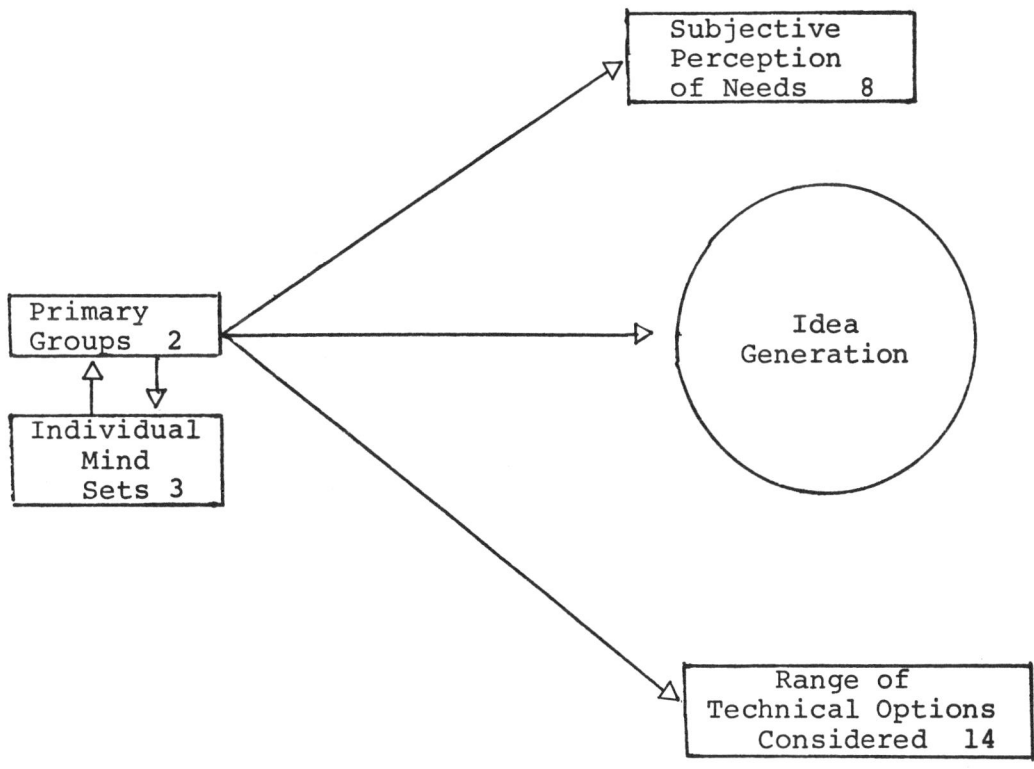

FIG. 6.--Interactions with primary groups within the R & D laboratory.

socialization group to which the researcher belongs in the informal organization of the laboratory by virtue of mutual social choice.

As it turns out, there is very little in the innovation literature dealing with the mediating influence of either primary group on the innovation process. We would claim, however, that this largely neglected area is important in the understanding the patterns of information flow. That is especially true in problem definition/idea generation, owing to the relative absence of mechanisms for direct organization control.

The research community is especially indebted to Thomas J. Allen and his colleagues at MIT for their work on the cluster of issues that relate to the role of the primary group in the innovation process, by bringing certain insights from social psychology to bear on the dynamics of information flow in the R & D laboratory. As with all intellectual efforts in their formative stages, the database is small at present and as many questions are raised as answered. But we find the work to be of both theoretical and practical importance.

Allen and Cohen (1969) have reported the results of a study in which the following concepts were utilized in analyzing the patterns of information flow in two R & D laboratories: (1) the primary work group; (2) the primary social group; (3) a two-step pattern of information flow; and (4) technical-approach attitudes or biases. Unpacking this cluster of interacting elements and determining their relative influence is as complex as it is important. Although the results should be taken as suggestive of directions for future research rather than as definitive, this does not reduce their value as it would in the case of a more incremental research effort. A meaningful direction is a precondition of meaningful results.

In the two R & D laboratories Allen and Cohen studied, the amount of overlap in the composition of the primary work groups and social groups was not statistically significant. As a result their influence on the flow of information was taken as independently measurable. As indicated in Fig. 7, the influence of the work and social groups was measured for three levels of technical information flow. The results demonstrate that for routine technical discussions a researcher's primary social group is at least as important as his primary work group. Indeed, within the work group those individuals whom the researcher seeks out most frequently for technical discussion *may* also be the members of his primary social group--although that cannot be determined from the data. In addition a number of influences are presumably operative in the formation of primary social groups, not all of which would provide a basis for technical discussion choice, so that only a subset of one's primary social group might be sought out for technical discussion.

1. With whom does a researcher engage in routine technical discussion?

2. Who provided the researcher with information that proved crucial in the course of his last completed project?

3. With whom does a researcher discuss an idea for a new project?

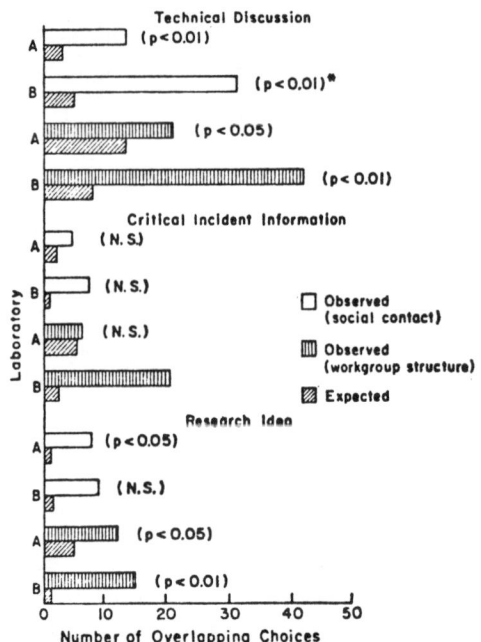

FIG. 7.--Number of overlapping choices between the social contact and work group structure and three networks of technical information flow in R & D laboratories A and B. (After Allen and Cohen 1969, p. 15.) (*Determined by Kolmogorov one-sample test; N.S., difference not statistically significant.)

Nevertheless, the relationship between the primary social group and technical discussion choices is quite strong, and stands out as something requiring explanation. Perhaps such explanation is provided, at least in part, by a factor discussed by Allen and Cohen in an earlier working paper (1966), which was based on the same research. The researchers were asked to indicate their attitudes on each of three rather uncertain technical questions confronting the laboratory in order to test the following hypothesis:

> Technological attitudes, attitudes toward such things as feasibility of particular approaches which are not yet physically testable, will be strongly influenced by

the attitudes held by other members of the primary groups to which the engineer belongs. (Allen and Cohen 1966, p. 7.)

Credit for the formation of this hypothesis was given to Kurt Lewin and others who had suggested that "when an opinion or attitude cannot be tested directly against 'physical reality,' then the individual will resort to a test against 'social reality.' In other words, he will look to his peers for confirmation or disconfirmation and react accordingly" (Allen and Cohen 1966, p. 7).

A fairly strong correlation was found between the technical approaches favored by the individual researcher and the ones favored by those he sought out for technical discussion. Unfortunately, the nature of the data precluded determination of causal direction, i.e., whether technical discussion leads to attitude agreement or whether certain individuals are chosen for discussion on the basis of prior knowledge of agreement. A replication of this study is needed that would permit determination of this causal direction.

Whether formation or reinforcement turns out to be basic function, it may be that the explanation of the strength of the relationship between the primary social group and technical discussion choices lies in this need to "test" one's ideas or position against "social reality." Such confirmation would be sought from those social peers with whom the researcher had established a base for interaction on technical matters.

Let us push this hypothesis one step further. Allen and Cohen made no distinction between "preproject" and "project" activities. As a result they did not determine what the technical discussions were *about*. If they had built in this preproject/project distinction, we hypothesize that for discussions related to preproject activities they would have an even stronger reliance on the primary social group. The argument here is that in the problem definition-idea generation phase the level of technical uncertainty is greater, since it explicitly involves the technical approach to be taken as well as technical problems within a particular approach. Here the "social testing" of one's ideas is crucial, and the Allen and Cohen work should be replicated with that in mind.

To make explicit the practical implications of the influence of the primary social group on the flow of technical information, we should assume the viewpoint of the R & D manager. Whether *formation* or *reinforcement* of technical attitudes turns out to be the more characteristic function of the primary social group, his direct action options are few. Yet they assume added importance for that reason. On the assumption that reinforcement turns out to be more basic, he should be aware of the rather strong evidence that, "engineers, once they have become committed to a particular technical approach, tend to discount very strongly information which would disconfirm their attitude" (Allen and Cohen 1966, p. 21). (The evidence for this phenomenon is developed in our next section.) If in addition the social group influences are strongly reinforcing of particular approaches, then the range of alternatives that receive serious consideration is going to be narrow. Should this restriction prove counterproductive to the laboratory's work, the manager should attempt to introduce more diversity by recruitment or in the formation of work groups. "Merely introducing a single individual with conflicting attitudes should produce sufficient jitter to keep the group aware of other points of view" (Allen and Cohen 1966, p. 21).

If we assume, on the other hand, that the primary group does influence the formation of certain attitudes as well as reinforce already existing ones, a further caution could be offered to the R & D manager. "The possibility that the causal direction is such that interaction leads to agreement, implies that management should periodically rotate their devil's advocates to prevent their capture by the prevailing group attitude" (Allen and Cohen 1966, p. 21).

Another, and perhaps powerful mechanism is available to guard against counterproductive primary group influence in attitude formation, namely the introduction of *diversity* into the laboratory setting by the use of outside consultants, or by the diversification of work assignments. Utterback reports that outside consultants played a crucial role in the generation of ideas for sixteen of the thirty-two new instruments he studied (Utterback 1971b, p. 13). Likewise, Peters in exploring the relationships among consulting, diversity of work assignments, and idea generation in interviews with faculty in four MIT departments, found that 96 per cent of those reporting new ideas engaged in consulting as opposed to 55 per cent of those not reporting ideas (Peters 1968). Such enhancement of idea generation by means of outside consultation is also reported by Morse and Gordon (1968, p. 37). Peters, in the work cited above, also found evidence that diversity in work assignments increases the probability of idea generation; specifically, 70 per cent of those reporting ideas also reported that their work was mixed between research and development, as opposed to 28 per cent of those not reporting ideas. Utterback explains these findings in terms of the need to synthesize information in idea generation. Although this explanation should not be discounted, perhaps a deeper explanation lies in the function of both consulting and varied work assignments in providing more diverse technical perspectives and attitudes than those of the researcher's primary group.

The implications of the work by Allen and Cohen on the influence of the primary group on information flow and attitude formation also need to be explored for organizational levels other than the primary group in the laboratory. For instance, there is the widely held folk wisdom that in any industry some firms are innovation leaders and others are followers. If there is substance to such distinctions, might it not reflect the influence of the dominant attitudes of the primary group at or near the top of such firms? Examples of key attitudes that might be formed or reinforced by such groups would be those taken towards risk taking, the exploitation of technological capibilities, newly recognized market needs, and the like. Such prevailing paradigms or mind sets of top management would be expressed in both broad corporate goals and strategies, and in the disposition decisions regarding particular ideas that are generated in the laboratory. Perhaps related to this point, in negative fashion, is the role of the "product champion" or "top man" as noted by Langrish et al. (1972, p. 72). Such individuals by their own persuasive skill and dogged determination overcome the firm's negative bias towards their idea. What is overcome in these instances, or so we are suggesting, are the attitudes of a primary group which are antithetical to the idea being presented.

Let us turn briefly now to Allen and Cohen's results concerning the primary social and work groups and the flow of "critical incident" information and new research ideas. As indicated in Fig. 7, almost all the relationships here are decidedly weaker than for the technical discussion level, and in fact half are not statistically significant. We would speculate the reason was that the significant informational sources in these instances are the "technological gatekeepers" rather than the primary groups. Allen and Cohen, of course, identified such gatekeepers in their investigation and showed that they were, in fact, the "sociometric stars" of the laboratories. But they did not seem to realize that these gatekeepers might be distributed over *various* work and social groupings, and thus that their influence would not appear significant in data cut along group lines.

We shall discuss the characteristics and influence of technological gatekeepers in a later section, and thus will not get into further details here. But if the leads from social psychology, as applied by Allen and Cohen to the innovative process, are indeed of great potential significance, let us offer one additional speculation. Perhaps the members of primary social groups do not all play an equal role in the development of the group's prevailing *attitudes* towards various technological approaches and questions. Related to this point, Allen and Cohen have hypothesized the existence of a two-step flow for technical information analogous to that discovered by Lazarsfeld and others in the mass communications field.[20]

> It appeared that ideas flow from radio and print to opinion leaders and from them to the remainder of the population.... The intervening variables all involve the individual's social attachments to other people and the character of the opinions and activities which he shares with them. Thus, the response of an individual to a communicated message could not be accounted for without reference to his social environment and to the character of his interpersonal relation.... (Allen and Cohen 1966, p. 3.)

The suggestion here is that perhaps there exists within the laboratory context certain key individuals who perform a *technical attitude formation* role in a fashion analogous to the role of opinion leaders in mass communications. This function may be distinct from the *technical information transfer* role attributed to the technological-information gatekeeper, though we shall later consider the possibility that the same individuals may perform both functions. Such opinion leaders or "technological-attitude gatekeepers" may exist, and operate in either a facilitating or inhibiting way to bias the problem definition and idea generation process towards certain technical alternatives and away from others. The possibility that such individuals exist should be recognized and investigated, since their impact on the process would be quite significant.

We have earlier mentioned the influence of the supervisor in shaping the researcher's perception of the organization's needs. Such perceptions may also be sensitive to more subtle, implicit, and often unintended "communication" from management. The environment which picks up these signals, interprets them (rightly or wrongly), and by subsequent behavior gives them substance, would seem to be the primary social group within the laboratory. Baker and Freeland point to one negative manifestation of such group perceptions, in this case triggered by inadequate management responses to new ideas that have been submitted in the past.

> Thus, expectations regarding organizational rewards for idea flow effort are modified downward and the cycle is ready to repeat. As new employees enter the organization they learn these low expectations from the veterans who have traversed full cycle. In such an environment, it is little wonder that potentially creative employees fail to realize their potential and appear to "go dry" over time. (Baker and Freeland 1972b, p. 111.)

Since the recognition of a need seems to be the precipitating event for most problem definition and idea generation sequences, the influence of the primary group in shaping the researcher's perception of such needs is crucial. As has been the case throughout this section, this too is a little understood phenomenon that merits careful investigation.

In concluding these remarks we should offer a balancing note to what has been a largely negative thrust. The influences of the primary group, and the attitudes they serve to form and/or reinforce, are of course by no means wholly counterproductive to corporate objectives. Whether the directions in which they lead are appropriate or not depends upon many variables in the total environment. Perhaps the influences of primary groups can best be summed up by paraphrasing the line from the nursery rhyme that goes, "When they are good, they are very, very good, and when they are bad they are horrid." Whether "good" or "horrid," such influence would seem to be substantial, yet badly neglected, in the research literature.

5. *Individual Mind Sets*

The concept of "mind set" or "biasing set" is used in the literature to refer to the biasing influence of past experience that an individual brings to his present problem solving activities. Allen and Marquis introduce this concept as follows:

> It is known that the likelihood of finding a solution to a problem may be raised or lowered because the problem solver is set to respond in certain predetermined ways. Prior experience with tools or approaches used in solving similar problems in a certain way may result in a "set" which biases the problem solver and can divert him from consideration of alternative solutions. (1964, p. 158.)

The individual may thus be "set" to transfer information of an approach that he has used successfully in the past to a present problem which he perceives as similar. Such mind sets may have either positive or negative effects on the achievement of a solution. If the transfer of past experience is appropriate to the new situation, it will be a positive factor, but if inappropriate, it may block or delay the discovery of a different and superior solution.

A paradigm illustration of the existence and influence of mind sets has been provided in an experiment conducted by Birch and Rabinowitz.

In this problem the S is required to tie together the free ends of two cords which are suspended from the ceiling to the floor of a corridor. The distance between the two cords is such that the S cannot reach one cord if the other is held. In our arrangement the problem could be solved only if the S would tie a weight to the end of one of the strings and thus convert it into a pendulum which could be set swinging and then be caught on its upswing while the stationary cord was held. The two cords could then be tied together and the problem solved. In our situation only two objects could be used as weights. The first of these objects was an electrical switch and the second, an electrical relay. The conditions of pretest training involved the acquisition of differential prior experience with these objects by our Ss. The pretest training was conducted as follows:

Groups Sw contained 9 Ss who were given the pretest task of completing an electrical circuit on a "bread board" by using a *switch*, which had to be installed if the circuit were to be completed and controllable.

Group R consisted of 10 Ss who received pretest training in the completion of an identical circuit by the use of a *relay*, which is essentially a switch.

Group C, the control group, consisted of 6 engineering students with a wide variety of electrical experience. These Ss were given no pretraining. The Ss in groups R and Sw had little or no experience with electrical wiring.

Shortly after having completed the pretesting tasks, the Ss were presented with the two-cord problem and asked to solve it by using the objects lying before them on a table. Only two objects were present, a *switch* and a *relay*, each identical with the ones used in the pretraining period.

All Ss were individually tested. Upon completing the two-cord problem, the Ss were asked why they had chosen either the switch or the relay as the pendulum weight. (Birch and Rabinowitz 1951, p. 122.)

The control group, who were equally familiar with both switches and relays, chose equally between them as pendulum weights in solving the two-cord problem. Their prior experience was not heavily weighted in favor of one or the other in terms of their utility in a new context in which their function was quite different from normal.

The behavior of the previously naive subjects was strikingly different, however. Of those who had been trained to complete the electrical circuit with a relay, *none* of them used this object as the pendulum weight (i.e., ten of ten used the switch). On the other hand, the subjects who had been trained to use a switch in completing the circuit preponderantly chose the relay as a pendulum (seven of nine). From the combined results of both experimental groups, 17 of the 19 subjects used that object with which they had had no prior experience as the problem-solving tool. Thus there is strong evidence that the nature of the subjects' previous specific experience was influential in determining their problem-solving behavior. Postexperiment interviews reinforced this conclusion.

This study was reported in detail in part because of its intrinsic value as a paradigm, but also to illustrate that the biasing effect of past experience may be negative as well as positive. (For an experiment with strikingly similar results, see Luchins 1942.) Prior experience colors the perceived characteristics or value of an idea, object, or approach in ways that inhibit or enhance its subsequent utilization. The old saying that experience is the best teacher is only a half-truth.

In the information and influence flow diagram presented earlier, the lines of influence indicated for the mind sets to which an individual researcher is subject are given in Fig. 8. For some of these lines of influence there are empirical studies; for others we would offer research suggestions.

Allen and Marquis compared the behavior of eight laboratories in two R & D proposal competitions (1964). They found that mind sets resulting from prior experience do not, by themselves, result in a higher or lower probability of achieving a correct solution. This result may speciously violate a commonsense feeling about the value of experience, but it is hardly surprising when coupled with the reminder that the crucial point about prior experience is not its existence *per se* but rather the appropriateness or inappropriateness of its transfer to a present situation. In other words, when prior experience is appropriate to the present problem the probability of achieving a success-

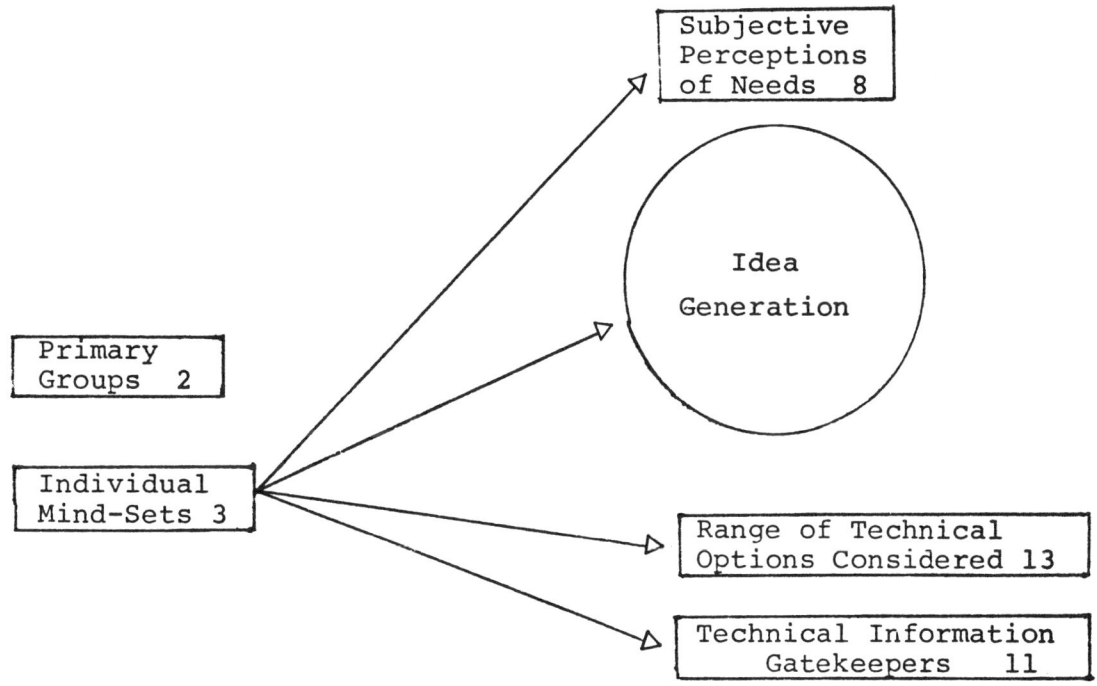

FIG. 8.--Mind sets affecting individual researcher.

ful solution is increased. Thus the biasing set is positive. On the other hand, when the prior experience is not appropriate, the biasing set is negative, i.e., the probability of success is lowered. Prior knowledge was found to have much the same effect as prior experience.

Other studies (Luchins 1942; Maier 1961) have shown that success in overcoming the effect of a negative bias is a function of the number of alternatives considered in the problem-solving process. Allen and Marquis also found this to be the case.

> Of the eight instances in which the laboratory had prior experience with a technique which would be unsuccessful if applied to the present problem four considered no alternative approaches and all four submitted solutions which were evaluated as unsuccessful. In our other instances the laboratory considered two, three, or more alternative approaches, and half of them achieved a successful solution. The additional effort required to search for and compare several alternative approaches is justified by the decreased susceptibility to negative biasing set. (1964 p. 161.)

If one knew *a priori* when an individual's mind set constituted a positive bias and when it was negative, there would, of course, be no problem. One would just look for alternatives in the negative cases. However, an individual is typically not that self-conscious about the nature of his own biases, nor can he know beforehand which approach will prove successful.

Given these constraints, the prescription always to consider several alternatives would seem to be sound. A somewhat subtle complication should be noted, however. If the individual researcher's normal approach to problem solving situations is to consider several alternatives, then by definition the influence of a particular biasing set is not a problem for him in the first place. On the other hand, if he is strongly influenced by a particular set, then this fact will lessen the effect of such advice. Perhaps the point of such advice needs to be reinforced by certain approaches available to management. Several points made earlier about the influence of the primary group should be recalled in this regard. Although primary social groups are not readily open to restructuring, the primary work group is. The individual with a particularly strong mind set might well acquire more flexibility over time if he is a part of a particularly heterogeneous work group, or at least a group that included a "devil's advocate." One might also consider the beneficial effects that have been shown to accrue from a diversity of work assignments.

It is also important in this connection to be aware of experimental work in social psychology on specific problem-solving techniques. Let us mention two. The first involves a comparison of problem-solving behavior under deferred-judgment instructions with such behavior under conditions of concurrent judgment. The former involves posing potential solutions to a problem without evaluation of their quality until several alternatives have been posed. Concurrent-judgment conditions, on the other hand, involve instructions which require only solutions of "good" quality.

The deferred-judgment approach to problem solving is commonly referred to as "brainstorming," and received considerable popular attention and some serious investigation in the late 1950s and early 1960s. Conclusions as to its merit are mixed primarily because experimental conditions were not standardized. Parnes and Meadow report, "Significantly more good solutions were produced under the deferred-judgment instructions than under the concurrent-judgment instructions" (1963, p. 315). Others report much less impressive or mixed results. Although the experimental evidence concerning the deferred-judgment technique is therefore not clearcut, one can assume that it would be of some value in helping the individual or group overcome biasing set influences.

A related technique which might be of even more value in this context is that of "extended effort." A mind set typically leads one to a familiar approach or idea which is transferred to the present situation; then the search stops. There is some experimental evidence, however, that if the conditions are such that one is forced beyond this initial production of ideas with which he is familiar, he will begin to grope for less obvious ones, the quality of which may be higher (Parnes 1964). The advantage of extended-effort instructions in overcoming individual and group biases should be obvious.

The focus in this brief treatment of mind sets has been their influence on the innovation behavior of the researcher in the laboratory. Let us now shift this focus slightly to offer a research suggestion that would seem to merit high priority. In light of the above discussion it is probably a mistake to view technical information gatekeepers as passive conduits in the information-flow network. Relative to their own mind sets and the tenacity with which they are held, they probably also filter such information. The nature and extent of course varies with the individual. But to assume that such a filtering function does not exist is *prima facie* questionable. Since the flow of technical information in the R & D process seems highly dependent on such gatekeepers, this speculative hypothesis should be explored.

This completes our survey of the major influence elements within the R & D laboratory. We shall now shift our focus to the patterns of information flow, beginning with information concerning market needs and opportunities.

6. *The Flow of Information Concerning Needs*

The phrase "problem definition and idea generation" points to two information components on the one hand, and the product of their creative synthesis on the other. The two kinds of information required for the definition of a problem are described by Baker and Freeland as follows:

    1. recognition of an organizational need, problem or opportunity which is perceived to be relevant to organizational objectives; and
    2. recognition of a means or technique by which to satisfy the need, solve the problem, or capitalize on the opportunity. (1972b, p. 197.)

Thus among the necessary conditions for generating an idea is information about both a *need* (problem or opportunity and a means (potential technological capability) for meeting the need.

One question that arises in connection with the above concerns the identification of the "need" component as a need (problem, or opportunity) of the *organization*. Would not an "*organizational* need, problem, or opportunity" be, by definition, "relevant to *organizational* objectives?" The point here is not to introduce a semantic quibble. Nor is it to deny that on occasion the event that stimulated the process to begin was the recognition of a need *within* the organization. But it would seem likely that for the most part the initial stimulating event would be the recognition by someone of a need that lies *outside* the organization, i.e., in the *market.* Baker indicates that in his research this is in fact the case. It was not made explicit in the study mentioned above since his purpose there lay in picking up the process at a later point, "need as perceived by those within the R & D laboratory."

The distinction here is an important one that has received little treatment in the literature. We would suggest that market need information *becomes* organizational need information by a mediated or "two-step" flow process. That is, information about social and market needs flows into the firm through various channels. Unless such channels are unusually anemic, the total set of such communicated needs should far exceed the firm's capacity to respond. This set is then filtered in terms of the firm's capacity (both economic and technical), its corporate strategy and objectives, its aggressiveness and morale, etc. The much smaller subset of recognized needs that pass these filters can then be called organizational needs and subject to the next step, which is their coupling with potential technological capacities for meeting them. This view of a two-step flow of information about needs seems to be supported by Langrish et al. (1972) and by Rosenbloom (Chap. 9).

What Baker has identified as an organizational need, problem, or opportunity would seem (unless it is truly "internal") to be that much smaller set of market needs which has survived the filter process sketched above and is thus compatible with the prevailing characteristics of the firm. However, we recognize that this set of "organizational needs," or perhaps more accurately "firm-compatible market needs," is the product of a filtering process. Otherwise our conceptual structure will not be sufficiently fine-grained to guide empirical investigations on a number of important questions. In the diagram with which we began, the principal elements in the flow of "need" information were displayed as indicated in Fig. 9. This segment of our model is the first we have considered in which the flow of information rather than influence is primary.

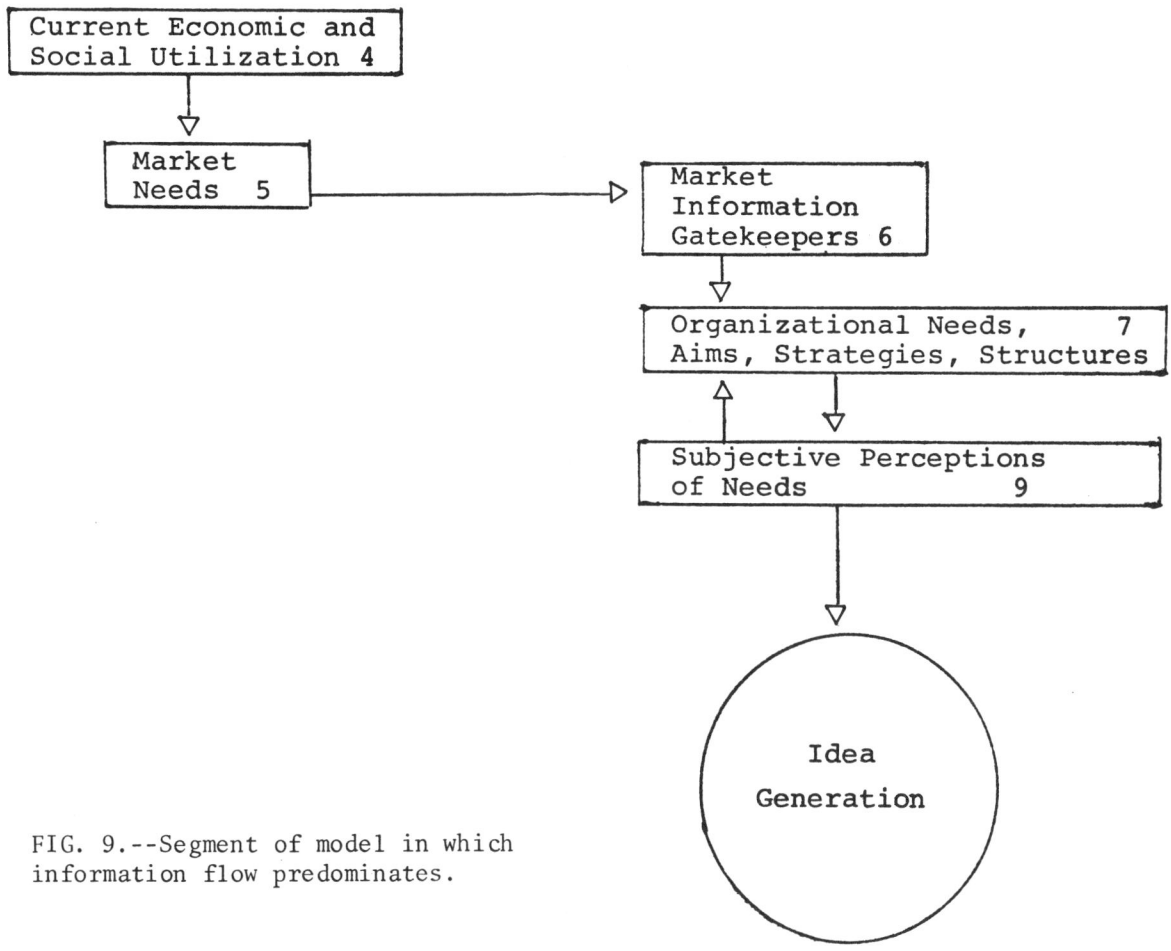

FIG. 9.--Segment of model in which information flow predominates.

Following the terminology suggested by Utterback (1974), the relevant social determinants and modifiers discussed in Chap. 2 are here referred to as Current Economic and Social Utilization. These exogenous environmental influences are of course not the only ones that impinge upon the process. In fact Utterback suggests that the most crucial ones often lie closer to home. "The primary limitations on a firm's effectiveness in innovation appear to be its ability and perhaps aggressiveness in recognizing needs and demand in its external environment" (1971b, p. 81). To this list we would add its lack of information or failure to use the information it has.

At this point the state-of-the-art understanding becomes suprisingly uneven. Some aspects of the flow of market-need information are understood rather well; others are almost opaque. What seems to have been studied most extensively, and about which there is broad agreement, is the event that triggers the innovation process. Sixty-six to ninety per cent of the innovations examined by a number of researchers find their starting point in the recognition of a need. In the remaining cases the stimulating event has been new scientific or technical advances for which an application is then sought. This impressive community of agreement has been discussed in the supply-demand literature (Chap. 2) and has been summarized by Utterback in Table 4.

TABLE 4.--Comparison of studies of the proportions of innovations stimulated by market needs and by technological opportunities.

	Proportion from market, mission, or production needs (per cent)	Proportion from technical opportunities (per cent)	Sample size
Baker et al. (1967)	77	23	303*
Carter and Williams (1957)	73	27	137
Goldhar (1970)	69	31	108
Sherwin and Isenson (1967)	61	34	710†
Langrish (1971)	66	34	84
Myers and Marquis (1969)	78	22	439
Tannenbaum et al. (1966)	90	10	10
Utterback (1969)	75	25	32

*Ideas for new products and processes.

†Research events used in 20 developments.

Thus, the "need-means" pattern seems clearly dominant. That is, in the majority of cases a need first comes to be recognized and stimulates the search for a technical capability which will satisfy it.

Information about market needs seems to come primarily through oral and informal discussion with contacts *outside* the firm.[21] Further, such communication about a need seems to be most often initiated by someone other than the individual who ultimately generates the idea for an innovation. These outside sources of market need information are most often an existing or potential customer (Utterback 1974). The search for market need information (if, indeed, it may be called a search in light of the passivity revealed in the above summary) seems much less structured and deliberate than the subsequent search for matching technical capabilities.

This sketch of the flow pattern for market need information occasions a number of observations, which, it should be noted, are highly speculative. First, if this is even a reasonably complete picture of how a firm comes to know of market needs (which we rather doubt) one must be shaken by its impoverished and haphazard nature. Informal contacts with customers or potential customers is certainly an important channel to the outside world. But if, as is pictured here, most of the contacts that matter are with a researcher in the laboratory, then one must suspect that there are quite rich contacts at other levels within the firm that are not being utilized. Salesmen, for instance, would have much more extensive contact with customers than the researcher and thus should be a better source of market need information. The literature on technological innovation, however, does not deal with this input. Either salesmen do not provide such need-information inputs to the laboratory or their role in this regard has not been investigated by those concerned with the process of technological innovation. The literature on this point is virtually nonexistent, and the authors of this study disagree as to which of these possibilities is the case.

The same point applies to other units within a firm that have "windows" to the outside world, e.g., technical service groups, market research, corporate planning, etc. Certainly these channels of market need information exist, but the research literature on the innovation process does not deal with whatever inputs they make to problem definition in the R & D laboratory. And finally, the literature is silent as to the contribution of top corporate management in the flow of need information, which, on the hypothesized linkage to the external world (with which we began this chapter) would seem to provide a major input.

What is being suggested here is that information about market needs in fact enters the firm via a number of channels and at various levels within the corporate structure.

For the channels that feed directly from the external environment to the researcher in the R & D laboratory, the information-flow pattern seems relatively simple and has been described above. The only caution that should be added to this point is that the researcher himself, and the primary group to which he belongs, provide some filter effect in terms of both their technical receptivity to certain needs rather than others, and their subjective perceptions of corporate needs, goals, and strategies. Were this the only channel of market-need flow, as one could conclude from the innovation literature, this filtering would constitute a serious barrier to innovation, especially in light of the fact that most innovations are need induced.

For the other channels whose existence we hypothesize, the information flow would again involve a two-step flow, from the outside source to a transfer agent within the firm and then to the researcher within the laboratory. This scheme would involve an additional filter or screen which is not only not understood at this point, but has not even been identified in the innovation literature. That such transfer agents or mechanisms do exist, however, is *prima facie* plausible in view of the unlikelihood of the single-channel alternative.

The possible existence of such additional channels of information about market needs raises another possibility that merits investigation. Perhaps there are key individuals at various levels within the firm who, owing to the richness of their external contacts, perform a gatekeeper function. That is, analogous to the *technical* information gatekeepers, there may exist what might be called "*market-need* gatekeepers." This too would be a quite specialized role, requiring a greater than normal "cosmopolitan" orientation, an extensive network of information sources, and unusually broad experience to distinguish market-need signals from the background noise.

The last of these suggested characteristics of our hypothesized market-need gatekeeper clashes in an interesting way with the first. It implies that he must also have a fairly strong "local" orientation, since the distinction between market "signals" and market "noise" is largely a function of the condition, policies, and needs of his firm. That is, since the needs and opportunities of the market are always far greater than any single firm can begin to respond to, the information that is relevant to a firm is that which "fits" its current or projected profile. The rest is noise. The hypothetical market-need gatekeeper must, therefore, be "local" enough to make these distinctions of relevance and at the same time "cosmopolitan" enough to keep on top of the external environment. This consideration alone is enough to warrant investigation of this hypothesized function. Market needs misgauged--whether by an impoverished network or by inadequate measures of relevance--mean ideas (and thus innovative opportunities) lost. Probably no other facet of the problem definition and idea generation phase is so badly neglected and thus so poorly understood. And yet most innovations begin here.

Let us assume, now, that by whatever channels are operative a firm has continuous access to the raw data about market needs. We then need to address explicitly the issue of "relevant" data which was introduced above. At least two frames of reference must be distinguished here: (1) relevance from the corporate perspective as determined by top management and articulated in the decisions of the R & D manager, and (2) relevance in terms of the perceptions of the researcher in the laboratory. We shall comment on these perspectives only briefly since they will receive detailed treatment in the concluding section.

Galbraith (1970) has argued that in dealing with a given external environment and a given set of internal constraints, a firm faces not one but a rich multiplicity of possible strategies. Whatever the innovation strategy adopted by a firm, it becomes the measure of relevance for market-need data. To take only one example, if in a highly competitive market a firm assumes a basically "defensive" posture, reacting to the advances of others by small incremental changes in its product line, the market data considered "relevant" will have certain rather predictable characteristics. If, on the other hand, it adopts a more aggressive, "offensive" stance, seeking to gain a competitive edge rather than just holding its own, different market data will be considered relevant. Therefore, the corporate profile, determined partially by the environment within which it operates, partially by its own internal circumstances, and partially by the aims, strategies and structures it chooses, constitutes a major filter for the market-need data which flow in (or are sought) by the various channels suggested above. The roles played by individual biasing sets, primary groups, and particular leadership styles--as they operate at the higher management levels and influence the nature of this filter--can only be guessed at.

The individual researcher's perception of this corporate profile constitutes another significant filter to which information concerning market needs and opportunities is subject before it becomes an element in the creative process of idea generation. At this point it is more accurate, perhaps, to follow Baker and Freeland and speak of "organizational" rather than "market" needs, since the two have by now become one and the same. The researcher's perception of these needs is constructed from his own experience, and that of his primary group, with management's reception of ideas that have been proposed in the past. As we shall see in the concluding section, there is impressive evidence that these perceptions exert a considerable influence upon idea generation behavior.

7. *The Flow of Technical Information*

The segment of our model respresenting the key elements of technical information flow is repeated in Fig. 10. To begin with the box in the lower left of this figure, Utterback (1974) stressed two crucial environmental boundary conditions for technological innovation. The first, the degree of economic and social utilization of existing technology, was discussed above. The second is the current state of technical knowledge. Since this topic has been treated in Chap. 2, we restrict our consideration of this factor to the characteristics of the flow of technical knowledge within the firm.

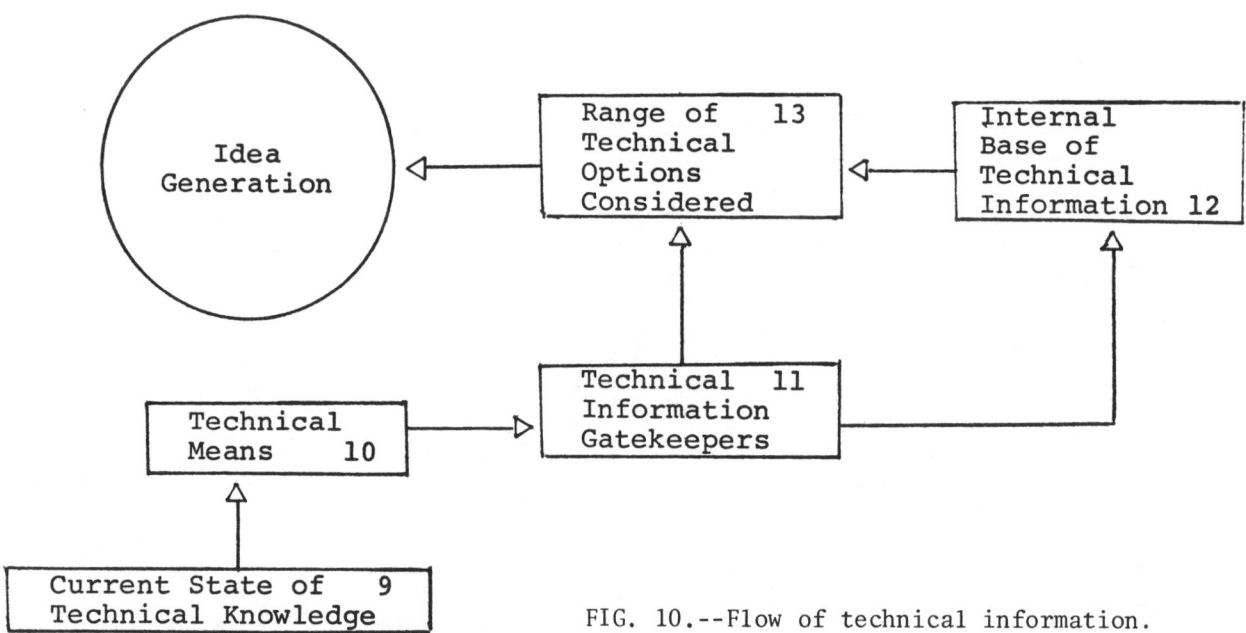

FIG. 10.--Flow of technical information.

The primary means for transferring technological information to the user seems to be oral and informational (Utterback 1969; Marquis and Allen 1967). Publication of previous results is less important than in science, since the utilization of an innovation is more important to its developers than the information about it. Although a journal system exists for technology, it is not cumulative to the degree that scientific literature is (Price 1965). Thus the technological researcher both publishes less and finds his professional literature a less rewarding source of ideas than do scientists.

Marquis and Allen have argued that even in his oral communication the technologist differs from the scientist (1967). He is limited by organizational barriers in the formation of invisible college networks to those demonstrated by Price to exist for scientific communities.

> This organizational identification works in two ways to exclude the technologist from information communication channels outside his organization. First, there are the usual requirements that he work only on problems which are of interest to his employer and, second, that he refrain from early disclosure of his research, to prevent the employer's competitors from profiting from the results. (Marquis and Allen 1967, p. 1053.)

As evidence of the differences in the information flow patterns, Allen compared the frequency with which ideas were brought to the attention of researchers through various channels in seventeen development and two research projects. These results are shown in Table 5. Although it is unfortunate that samples of more comparable size were not available, the comparison of use percentages for the various channels is nevertheless quite impressive. Our basic point that the transfer of technological information is primarily oral is clearly demonstrated, and the relatively smaller contribution of literature sources is evident.

TABLE 5.--Sources of messages resulting in technical ideas considered during the course of nineteen projects.

Channel	Seventeen engineering development projects		Two physics research projects	
	Number of messages produced	Percentage of total	Number of messages produced	Percentage of total
Literature	53	8	18	51
Vendors	101	14	0	0
Customer	132	19	0	0
Other sources external to the laboratory	67	9	0	0
Laboratory technical staff	44	6	1	3
Company research programs	37	5	1	3
Analysis and experimentation	216	31	3	9
Previous personal experience	56	8	7	20

Even the limited reliance on literature sources must be further qualified. Most widely used of all written materials within technology are unpublished or "in-house" technical reports (Marquis and Allen 1967, p. 1055), probably because of the proprietary interest that firms have in the technical information they produce. Technical reports, despite their limited dissemination, serve the need to record such information but restricting its domain of use. Although it is difficult to monitor and control the dissemination of in-house reports beyond the first user, Marquis and Allen report the existence of a norm at the interorganizational level. "There seems to be a rather strong norm against the transfer of another organization's reports beyond the limits of one's own organization" (1967, p. 1057). Such a norm could only operate to limit dissemination to a *third* organization, not in the case of a direct flow from one to another.

Yet the interorganizational flow of unpublished technical reports would seem to be a significant factor in a researcher's effort to stay abreast of his field. This flow may also be largely a two-step process, mediated by the technical information gatekeeper.

To offer a speculation based on the above findings, which have been distilled by Marquis and Allen from a wide range of empirical studies, it may be the case than an "invisible college" also characterizes the flow of technological information, but in a subtler sense than Price has demonstrated for science. The subtlety is introduced by the conflicting needs that firms have to protect their proprietary interest in information on the one hand, and to stay abreast of the state of the art by acquiring information on the other. Conflict between these needs can only be satisfied, and then just partially, by trading one off against the other. That is, a firm must *give* as well as receive in order to maintain a state of the art awareness for itself; but in the giving some proprietary interest must be sacrificed.

We hypothesize that the mechanisms by which the exchange of technical information might take place are quite informal and operate well down in the corporate structure rather than at or near the top. The technical information gatekeepers, who are turned to as resource persons by others in the laboratory, could be primary channels by which a firm *gives* as well as receives technical information.

That such a variation on the invisible-college hypothesis has substance that has not been demonstrated, or even suggested, in the research literature, although one instance has come to our attention in the *engineering* literature (Susskind 1968); we again make a speculative leap beyond what is known. But the phenomenon of interorganizational flow of technical information is clearly recognized, at least implicitly, in the literature. So too is the proprietary interest that organizations have in the information they generate. Likewise, the role of gatekeepers in bringing information into the firm is frequently noted. If the scenario offered is inadequate as an account of the interaction of these known variables, perhaps it will at least serve to stimulate the research necessary to understand it aright.

Before leaving this point, let us add one final consideration. Some items of technical information are *public* rather than proprietary, such as government-held patents and technical reports. But Utterback reports that they "are seldom used in a commercially or socially important application other than the specific one from which the patent or information arose" (1974, p. 623). Although the explanation for this fact is doubtless complex, perhaps the fact that such information is equally accessible to all who are active in this field means that there is no particular advantage to any in its exploitation.

It will be recalled from our earlier discussion that the majority of innovations have as their initial impetus the recognition of a market need. Less commonly a technical capability is first recognized, followed by the search for a need to which it can be applied. In this "means/need" pattern the process begins with a particular technical capacity, and the search is among a range of market needs. What is the nature of certain technical capacities that makes them attractive enough to stimulate the search for a need? Utterback suggests, "Older technical possibilities seldom attract attention spontaneously" (1974, p. 623). On the other hand, a new discovery or technical possibility often attracts the researcher's attention and thus stimulates the search for application. Thus *newness* would seem to be characteristic of technical means that initiate the process. There is also evidence that "the larger technological changes tend to be of the 'discovery push' type" (Langrish et al. 1972, p. 75). That is, innovations stimulated by the recognition of a need are often smaller and more incremental whereas those stimulated by a new discovery or capability are of greater magnitude and constitute greater discontinuities with the past.

One should not automatically assume, however, that these large technological changes have their origin in a scientific discovery. To refine somewhat the material in Chap. 2, we note that Price in fact argues that science and technology develop quite independently of one another, with communication being limited to that which occurs in the education of technologists (Price 1965). Marquis and Allen agree with this general assessment, but offer two refinements. First they note that:

> Occasionally technology encounters a problem blocking its advance, the removal of which requires a fundamental understanding of the scientific basis of the phenomena involved. In this way, science discovers voids in areas which have been bypassed by the research front. Quite frequently, too, difficulties encountered during the research process in a seemingly unrelated area will reveal a gap in the understanding of a basic segment of science. Here again, in a sense, application helps to determine the direction or priorities in scientific investigation. (1967, p. 1057.)

Most cases of technological advance do not seem to be of this sort; they neither require information from the forefront of science nor define a problem there. But when technological advance requires a "gap-filling" contribution from science, "the communication is bilateral, direct, and quite rapid" (1967, p. 1059).

The second enrichment of Price's generalization involves a distinction among technological areas. Some, such as electronics, are probably more closely related to work on the frontiers of science than others, for instance, mechanical technology. However, neither of these refinements disturbs the general point that there seems to be little direct dependence of technological innovation on science. As science builds on earlier science, so technology seems to build on previous technology.

What does that mean in terms of the technological gatekeepers whose role in the flow of information has been anticipated several times above? Allen and Cohen hypothesize the existence and function of such individuals as follows: "There can exists in an R & D laboratory certain key individuals who are capable of effectively bridging the organizational boundary impedance and who provide the most effective entry point for ideas into the lab" (1966, p. 6). Such individuals can be characterized in several ways. First, as opposed to most individuals within the laboratory who have few contacts outside, the gatekeeper's contacts are extensive. There is some evidence that such gatekeepers do not all tend the same gate (or all gates) with equal facility. Some concentrate on the technical literature, others maintain a richer network of informal contacts outside of the organization. Given the dominance of oral communication noted previously, the latter would seem to provide the more fruitful information channels. Allen and Cohen, however, provide a finer-grained distinction which is worthy of note:

> The literature has been shown to provide information which is important for keeping abreast of the state of a technological field, while oral sources are probably better in providing more specific detailed information about particular techniques. Gatekeepers who specialize in knowledge of the state-of-the-art would then tend to expose themselves more to the literature, while those specializing in particular research techniques would interact more with external oral sources. (1969, p. 18.)

This distinction between written and oral sources and specializations may prove valuable, but one might wonder if it is quite so clear cut as it may seem.

Unfortunately, the research literature is silent as to the extent to which the gatekeeper serves, not only as a channel of information from the outside, but also as a *filter* for such information. This point has been raised above as to the influence of the mind sets or biases to which the gatekeeper is subject owing to his past experience. It merits investigation in light of the crucial role that gatekeepers seem to play in the information-flow process.

The final point to be made concerning the gatekeepers has also been anticipated above, but has apparently not been dealt with in the literature. We have hypothesized that gatekeepers may not only bring information into the firm from the outside, but may also share their firm's information with their contacts in other firms. To review the argument here, at least some and perhaps most of a gatekeeper's contacts outside the firm would be those who perform a similar function in other firms. If that is so, then the needs of all parties would be similar and the relationships would by necessity be reciprocal; that is, information would have to be *exchanged*. If one party sought to *receive* only, and never *give*, the informational needs of his contacts would not be met and the relationships would eventually break down. The research agenda suggested by this point rests on the *prima facie* possibility that there exists for technology a communications network, or "invisible college," analogous to that which has been demonstrated for science, but with quite distinctive characteristics owing to the firm's proprietary interest in information.

Mention should be made of another avenue by which technical information enters the firm. Burns has concluded that the transfer of technical information is the work of "agents not agencies," and one of the most efficient forms of this transfer is "on the hoof" (1961, p. 12). That is, an individual researcher joining a firm brings with him not only a certain level of training and technical competence, but also a store of ideas and technical information acquired in his previous work settings. As to the importance of such ideas and information to his new employer, Langrish et al. found that, "in the cases we have studied, the most frequent single mode of technology transfer was by a person joining a firm" (1972, p. 44; also see their Table 7, p. 79).

To speculate again, it may well be that "on the hoof" information transfer is a more important source in technological breakthrough (and discontinuous-for-the-firm) efforts than for incremental innovations. The argument here is that the store of technical information represented by a firm's current R & D personnel is more likely to be adequate in itself, or with routine accretions from outside, for incremental improvements than it is for new and dissimilar activities. One of the immediate implications of a corporate decision to undertake a major breakthrough project may be that it must "gear-up" for it by bringing in new technical personnel with the appropriate specialized knowledge. It would thus be availing itself of information "on the hoof."

Our final topic in this section is the range of technical options considered by an individual researcher or laboratory in generating an idea. This topic in effect forces a recapitulation of much of the foregoing since it points to the discrepancy between the current state of relevant technical knowledge and the subset of such knowledge actually considered in a particular instance. We shall be quite brief in summarizing the influences that contribute to this discrepancy.

The number of technical options actually considered as possible solutions to a recognized market need is a function of a wide range of variables. Of most importance as a filter is the perceived nature of the need itself. A second obvious consideration is the base of technical information within the firm. Since most of the firm's store of relevant technical information was not generated within but enters from the outside, primarily by oral channels, the richness and state of repair of its communications network is a third significant influence. If its gatekeepers are able and if the filters introduced by their own biases are not in a particular instance counterproductive, then the range of technical options considered may be large. If, in addition, new people with relevant experience have entered the laboratory recently then the range could be enlarged even further. If consultants are available at this early stage, that again can be a positive factor.

Anticipating the next section, if the time pressure associated with current project work is not so heavy as to preclude the exploration of a number of possibilities for a new idea, then clearly the number that can be considered will increase. Related to this point, if the firm has established mechanisms for the explicit purpose of encouraging new ideas (see the description of one firm's Idea Generation Groups in Baker et al. 1967), then the range explored could be larger. This should also be the case if the firm's reward structure is such as to recognize and reward effort spent in idea generation activities as well as effort directed to other responsibilities.

In another dimension, the range of technical options considered will be influenced by the primary socialization groups within the laboratory, since these groups serve to channel technical discussion and help either to form or reinforce a researcher's attitudes towards certain technical approaches. And finally, the number of technical options that receive serious consideration is also a function of the mind sets of individuals—especially, it seems, those who enjoy status as opinion leaders.

This check-list of screens or filters that can function to widen the gap between the technical options that exist and those that actually come to be considered in a particular case is on the face of it quite formidable. The actual counterproductive effect of these filters individually and in concert will, of course, vary with the nature of each need that occasions a search, and over time. Minimizing their counterproductive effects and maximizing their productive impact are tasks which accrue to those in supervisory positions within the laboratory.

This completes our review and assessment of the literature on problem definition and idea generation within the R & D laboratory. It remains now to examine what happens to an idea once it exists within the organization.

8. *Post-idea Generation Filters and Linkages*

A significant amount of empirical research has been oriented toward the exploration of creativity. Most of this research[22] has been conducted by psychologists, whose efforts can be categorized as follows: (1) creativity and education, (2) description of creative individuals, (3) identification of creative individuals, (4) prediction of creative performance, (5) early experiences and creativity, and (6) creative techniques.

Several authors including Kuhn and Kaplan (1959) and McPherson (1965) have noted that despite the abundance of literature on creativity, there has been relatively little study of the organizational conditions conducive to creativity in organized R & D. From the relevant work which has been done, initial insights toward identifying organizational factors that influence creative performance in the R & D laboratory and postulating relationships can be drawn. This literature exists in two pieces--that authored by psychologists, sociologists, and management scientists,[23] and that by persons concerned with the management of industrial R & D.[24] In addition, two university groups have been particularly productive in this area, namely, A. H. Rubenstein and his colleagues of the Program of Research on the Management of Research and Development (POMRAD) at Northwestern University[25] (see POMRAD's "Annual Reports") and the late Don Marquis, Tom Allen, and their colleagues at the Massachusetts Institute of Technology.[26] Utterback's excellent summary of the literature dealing with innovation and the diffusion of technology in industry is on the "must" list in this area (Utterback 1974).

The literature dealing with problem definition and idea generation has been assessed above. The purpose of the brief summary above is to set the background for the remainder of this subsection. Relatively little work has been reported which relates organizational variables to creativity; even less has been done on the process by which an idea existing within an organization is communicated for organizational review, i.e., the process by which an idea becomes a candidate for the scarce organizational resources. It is not sufficient for an idea to be generated; it must also be set forth in a proposal, get funded as a project, and be successfully researched and developed before benefit can accrue to the organization. The remainder of this subsection deals with organizational factors influencing idea submission, i.e., the factors determining whether or not an existing idea will be submitted for consideration as a project. These influences are represented in Fig. 11.

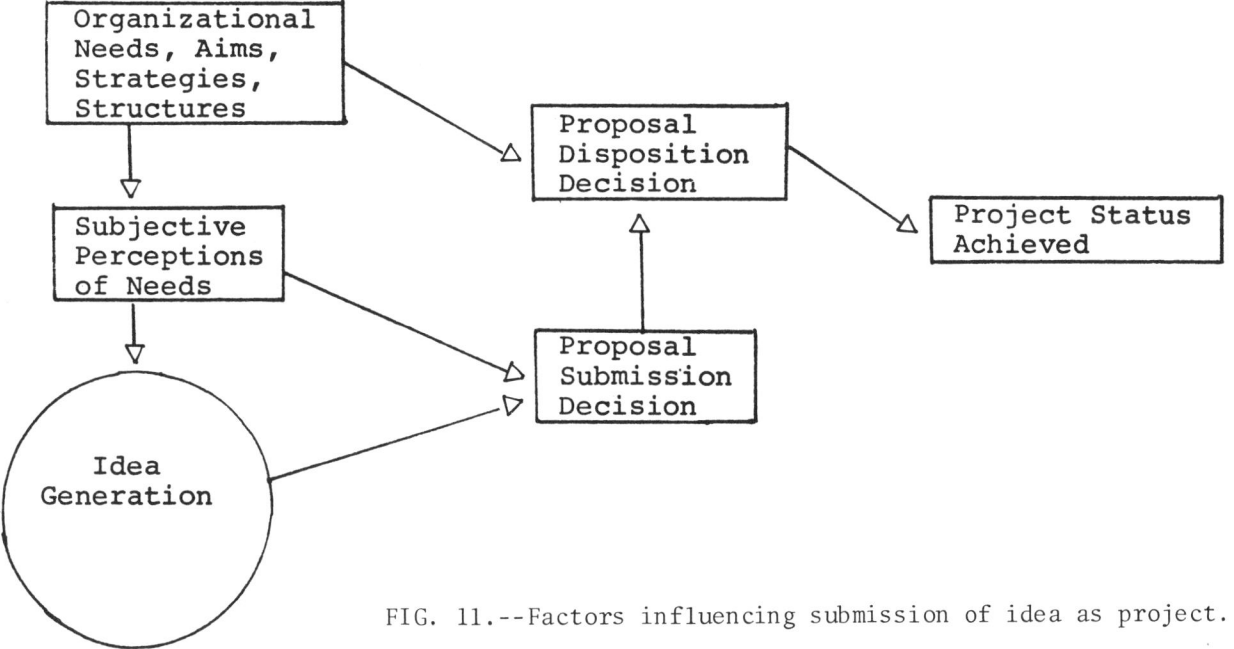

FIG. 11.--Factors influencing submission of idea as project.

Although the literature relevant to idea submission was summarized by Baker (1965) and updated by Siegman, Baker, and Rubenstein (1969), only conjectures and insights can be drawn from it. There are some empirical studies, but they are diverse and of varying methodological soundness; hence, the following summary statements must be viewed with caution.

1. The perception of time pressures associated with the current R & D activity encourages idea generation and development associated with the current work activity, but stifles idea generation and development not associated directly with the current work activity (Jones and Arnold 1962; Kaplan 1960; March and Simon 1958).

2. In order for "free time" to be utilized for idea generation and development not associated directly with the current work activity, the rewards must be viewed as:

a. of the same nature of the rewards for current work activity, but, because of the inherent uncertainty, of greater magnitude; or,

b. valued by the idea originator and of a different nature from the rewards for current work activity.[27]

3. In order for idea generation and submission to continue over time, it is essential that previously submitted ideas be perceived as having been positively and enthusiastically received and evaluated by the organizational reviewers, especially R & D management and supervisors.[28]

4. The perception of organizational goals and needs held by R & D scientists and engineers is influenced by the perceived receptivity of organizational reviewers to previously submitted ideas and by interaction with other R & D scientists and engineers (Avery 1959; Houton 1963).

5. If the organizational reviewers perceive an idea to be relevant, i.e., to satisfy an existing need or solve an existing problem, then they are more likely to receive the idea positively and enthusiastically (Avery 1960; Kaplan 1960).

6. Prior to submission, R & D scientists and engineers tend to screen their ideas according to their perceptions of its relevance (Kornhauser 1962; Morris 1962).

7. Perceptions of organizational needs and problems tend to stimulate idea generation and development congruent with these perceptions, but to stifle idea generation and development not congruent with these perceptions.[29]

Many other independent, dependent, and intervening variables can be conjectured (Utterback 1964); the statements preceding are a logical starting point and summary of the literature prior to 1969.

The primary concern here is with the process by which existing ideas are submitted to organizational reviewers and, most especially, with the constraints and screens in the process. It follows from the summary statements that two screenings occur prior to actual submission--at idea generation and at idea submission. The factors constituting each screen include perceived relevance, perceived time pressures associated with current work, and perceived rewards (or costs) associated with expending effort on "nonrelevant, noncurrent" idea generation and development.

Perceptions of rewards (or costs) and organizational needs and problems held by R & D scientists and engineers, and used by them to determine the relevance of an idea, are based on managerial behavior in reviewing and evaluating submitted ideas and on secondhand feedback from other R & D scientists and engineers in the organization. Note that both of these bases are historical in nature and, accordingly, that R & D scientists may be judging relevance on obsolete grounds. In an environment that changes as rapidly as does R & D, relevance should not be judged entirely on behavior related to yesterday's ideas for solving last year's problems. Thus, timely "need" and "means" information is essential, not only at problem definition and idea generation, but also at idea submission.

For the ideal operation of the screens in the idea flow process "good" ideas should pass through, but "poor" ideas should be filtered out. If "good" ideas are halted, the organization loses their benefits; conversely, if "poor" ideas pass through, the organization must expend already scarce resources to evaluate them. Recent research in one organization demonstrated that both errors were occurring (Baker and Freeland 1972b; Baker, Siegman, and Rubenstein 1967). Ideas were created but not submitted, and, based on both subjective ratings and on subsequent managerial disposition decisions, these unsubmitted ideas were of higher overall quality than those routinely submitted. Perhaps the most startling finding was that 38% of the ideas that eventually achieved project status came from the originally not-submitted category, which contained only 15% of the ideas subsequently reviewed by management. In an unrelated study, Peters and Roberts (1967) found somewhat similar results for a university laboratory. Because the data in this area are scarce and incomplete, the results should be treated with caution. However, these two separate studies indicate that there was a substantial loss in innovative potential due to "good" ideas not being submitted.

In order to obtain further empirical insights into the process of idea submission, Baker (1965; Baker and Freeland 1962b) developed detailed case histories for 45 ideas that were generated, but were not submitted, in an industrial R & D laboratory. Through Baker's intervention, the ideas were subsequently submitted and officially reviewed. As indicated in the preceding paragraph, a number of high-quality ideas eventually became funded projects. This appeared to be an opportune setting for gaining empirical insight into why apparently relevant, timely ideas were not submitted by the idea originator(s).

Case histories constructed prior to the intervention and eventual disposition decision indicate that because of organizational review, scheduling, and reward mechanisms which focus attention on current project activity and because of the uncertainty inherent in the review and reward mechanisms for activities not directly related to current projects, R & D scientists and engineers tend to be cautious in expending time on possibilities not directly related to ongoing work. Ideas are *generated* only if market, mission, or production needs and technical means can be identified. Ideas are *submitted* only if the idea originator believes that the underlying needs and means will be perceived as relevant by management, and if the perceived rewards will be at least equal to the cost of the time involved in creating and developing the idea, i.e., equal to the cost of the time taken from the current project activity. Unfortunately, relevance is time- and reviewer-dependent; e.g., an idea judged irrelevant by a reviewer at one time may be judged relevant by the same, or another, reviewer at another time. Further, since expectations regarding reviewer evaluations are based primarily on reviewer actions on previous ideas, there can be a significant lag between the time reviewers change their evaluation behaviors and the time these behaviors are perceived by the R & D scientists and engineers.

The case histories also provide insights into the impact of managerial (reviewer) behaviors on continued idea generation and submission. Submitted ideas often are not sufficiently developed technically and not sufficiently supported by evidence of relevance so that managment can objectively evaluate them. The lack of completeness is partially explained by the idea originator investing minimal time on idea development because of current project activity pressures, uncertainty regarding organizational reward mechanisms, and lack of knowledge regarding relevant needs and means. The idea may also be submitted in tentative, incomplete form in order to test how management will respond, i.e, to get feedback on its potential relevance. Since management is unable to evaluate objectively the tentative, incomplete idea, it responds in ways perceived as nonrewarding (costly) by the R & D scientists and engineers. Typical responses include "develop in your spare time," "state of art not sufficiently advanced," "too far out," or no response at all.

It is straightforward and "rational" to conclude that the ideas may have been submitted prematurely and that the management response was both expected and correct. However, such management responses can be interpreted otherwise and with severe consequences. If read as rejection, expectations regarding organizational rewards for idea generation and development can be modified downward. Additional uncertainty is introduced regarding the bases on which relevance is judged, i.e., relevant needs and means become even more unclear. As new employees enter the organization, they learn the low expectations and unclear needs and means from the veterans who have experienced the process. Thus, the process builds up, with even less time being expended for idea development prior to submission and fewer ideas submitted. The R & D scientists and engineers, instead of learning that their ideas should be better developed, believe that management is unreceptive to new ideas. Hence they come to believe that ideas should be submitted early, before too much time has been expended on their development and one runs into trouble with respect to current project budgets and deadlines--or not submitted at all.

Although the above somewhat pessimistic characterization is based on a single study within a single organization, it is consistent with the seven summary statements based on the existing literature. It also dramatically illustrates the sensitivity of the innovative environment within an R & D laboratory. If it is a generally applicable description, R & D management should be aware of it in order to understand better the feedback impacts of their review behaviors. Thus, there is need for replicative research which tests the generality of the description for information systems designed to provide timely needs and means (see Baker and Freeland (1972b) for an initial specification), and for better understanding of the behavioral impact of the use of normative budgeting, scheduling, and selection models.

In conclusion, we have stressed throughout this section the rich and subtle network of environmental influences that enhance or inhibit problem definition, idea generation, and idea submission, and in turn hone or blunt a firm's competitive edge in technological innovation. For those attempting to shape this environment there are few reliable rules, but a methodological reminder is appropriate. The technological tradition is strongly oral. In the experience related by Hyman earlier, when the R & D manager finally respected and utilized this tradition he got two quick responses that both ring true. First, his research group said, "Well, if that's what he wants, why didn't he say so?" And second, they produced a wealth of new ideas. Too simple, of course, but instructive.

C. R & D PROJECT SELECTION AND RESOURCE ALLOCATION

Previous sections have discussed idea generation and submission: we now turn attention to idea (project) selection. Rubenstein (1957) and Bradenberg (1966) wrote remarkably similar introductions recounting trends in R & D management which have continued to the present. The post-World War II period was one of optimism, permissiveness, and faith with respect to an industrial R & D revolution. As Rubenstein states, it was a period in which R & D "accomplishment usually came as pleasant, and often complete, surprises to others in the company" (Rubenstein 1957, p. 95). The two authors saw this atmosphere giving way to one in which measurement, control, and evaluation were being imposed on the R & D activity, and by 1964, Baker and Pound (1964) could cite over 80 papers dealing directly with normative models for R & D project selection and resource allocation. However, this statement may be misleading; although the literature reports many models and methods, there is little evidence that the models and methods are being utilized by managers of industrial R & D.[30] The evidence which does exist indicates that classical profitability indices (e.g., rate of return, payback period, and present worth) and simple scoring models, have found some use (Andrew 1954; Baker and Pound 1964; Cochran, Pyle, Green, Clymer, and Bender 1971).

The purpose of this section is to review the current state of knowledge regarding R & D project selection and resource allocation. Our first subsection surveys the descriptive literature. Next, estimation problems are identified and some of the major normative benefit measurement and resource allocation models are described. The section closes with a speculative subsection identifying research opportunities and conjecturing where the literature is headed.

1. *Descriptive Literature: The R & D Project Selection Problem*

Numerous papers contain descriptions of the R & D project selection problem. Most of these are either postulations of how decisions should be made, or based on experience of how decisions are made in the laboratory. Both the *IEEE Transactions on Engineering Management* and *Research Management* have published numerous such papers, as has the new British journal, *R & D Management*. This literature represents a substantial body of information on which to base descriptions of the R & D project selection problem.

Not until 1966 did Brandenberg integrate the literature and conduct an empirical investigation (Brandenberg 1966). The resulting document reports the results of intensive investigation in five companies and preliminary explorations in nine other firms. All of the research activities were company-financed R & D in firms with formal, on-going research departments and budgets. However, the firms varied in size, product-market areas, underlying technologies, and R & D organizational configurations. Despite these differences, a consistent description emerged.

Bradenberg portrayed the R & D project-selection decision in terms of changes made in lists of currently active and proposed projects and of the mechanisms which were used to determine the specific changes which would be made (Brandenberg 1966, p. 17). Typical changes included dropping projects from the list, reprogramming active projects, replacing an active project with a proposed project, initiating a proposed project and reprogramming a current one, or terminating a current project and shifting the resources. Thus, the decision was described in terms of an intermittent stream of investment possibilities, not as a once-a-year decision event (Brandenberg 1966, p. 21).

A detailed process flow model of the project selection decision was constructed. In general, the process consists of six main stages (Brandenberg 1966, pp. 18-20):

1. generating and changing the inventory of project proposals;
2. reviewing the status of current and proposed projects for the purpose of deciding when to make a change;
3. identifying the projects and proposals to be evaluated and compared with respect to change and the criteria, variables, and constraints to be used in the evaluation;
4. evaluating the designated projects and proposals;
5. comparing the alternatives and choosing among them; and
6. recycling to gather additional information, to reformulate criteria, variables, and constraints, and to define entirely new alternatives.

Thus, as observed and reported by Brandenberg, the R & D project selection decision includes generating new alternatives, determining the appropriate time to make a decision, collecting data, specifying constraints and criteria, and recycling, as well as selection of projects and allocation of resources.

One of the important questions we might ask refers to the criteria used by R & D managers during the project selection decision. In industrial R & D, one would immediately identify profit maximization as an important criterion. Although it is important, it is not sufficient by itself. For example, in one organization, Mansfield found that profit maximization accounted for 50% of the variation in the allocation of R & D funds (Mansfield 1968a, Chap. 3). In one industrial laboratory, Baker, Siegman, and Larson (1971) found that urgency (immediacy of the need or opportunity toward which the idea is directed) and predictability (degree of certainty with which the methods and procedures for researching the idea are known) were significantly related to project selection decisions. Rubenstein (1957, p. 97) asked 37 laboratory directors to indicate the criteria used in juding the "results of R & D work" or "progress on R & D projects or program." The general criteria and the number of companies citing each are: perceived future effect on sales volume or revenue (19 companies); perceived future effect on savings in materials, labor, and other costs (17 companies); perceived future effect on profits (13 companies); time and cost characteristics of the technical solution (28 companies); relationship to customer satisfaction (10 companies); and likelihood of technical success with the proposed solutions (16 companies). By 1968, lists of specific criteria had grown to include 25 or more items; however, the only new general criteria were compatibility of project results with existing products or processes and the interest of the R & D personnel in conducting such project activity (Baker, Siegman, and Larson 1971). It is clear that the R & D project selection decision can be characterized has having multiple criteria with no common, underlying measure.

Even if the criteria were known and a measure existed, two significant difficulties remain in the evaluation of proposals and projects. First, there is no known way of generating accurate estimates of the amount a project is likely to contribute with respect to the individual criteria, and this problem is especially acute at the basic research and exploratory development phases. Second, the relative importance of the criteria, and hence the magnitude of contribution expected to accrue from the project, change over time as the project progresses from research to production and as environmental conditions (e.g., market, technology, and resource availability) change. Thus, difficulties abound in estimating and measuring project benefits due to multiple criteria which have no natural, common underlying measure and whose relative importance varies over time.

The project selection decision as described above is imbedded within a more general budget process. In large organizations, the budget process is hierarchical and, despite simultaneous give and take, largely sequential in nature. The hierarchical, sequential nature of the process evolves from the facts that nearly all the organizational forms in which R & D is conducted are hierarchical and that the R & D director is not at the highest level in the organization. Thus, the R & D director typically receives guidance in the form of budgets from the higher organizational levels and passes on guidance in the form of budgets to the lower managerial levels in the R & D organization. In an analogous manner, information detailing how a proposed budget should be allocated flows upward in the organization with each level integrating the input from below. Thus project selection decisions are made at several organizational levels and, at each level, an integration of proposed allocations from lower levels occurs subject to budget guidance from higher levels. Baker et al. (1972, pp. 1-4) and Shumway et al. (1974) describe such a process for one large Federal R & D agency. It should be noted that multi-year planning, as well as fiscal year budgeting, is taking place during this process and that the process may iterate several times.

An alternative but consistent view is offered by Connolly (1972) based on his studies in a large Federal agency different from that studied by Baker et al. (1972). Connolly argues that for several reasons--the lack of an identifiable decision maker, the extended time frame, the importance of structuring and linkage mechanisms--the planning and budgeting of expenditures in a large research laboratory is not conveniently modelled as a decision process. Further, the processes are diffuse in nature, i.e., the processes are temporally diffuse, covering extended periods of time, with indistinct endpoints; are multiperson with decision-making influence and responsibilities diffused across a number of individuals; are geographically diffuse in that the individuals are separated by nontrivial distances; and are diffused across several organizational levels.

As one moves backwards in the R & D process from the output (production) to the input (research), it is likely that the decision problem changes, most likely in the direction of greater diffuseness, less consensus, and more uncertainty. The extreme case of a diffuse decision, then, is a process in which many participants, over an extended period of time, generate a decision in response to some decision problem, working with alternatives which may initially be unclear or unknown, with costs and benefits not reliably estimable, with unclear and/or conflicting preferences, and with modifiable resources and constraints. In the approach proposed by Connolly (1972), the analytic focus for both descriptive and normative studies shifts from the individual decision maker to the decision process as a whole. The concepts of hierarchy and diffuseness in the R & D decision process are new and represent a potentially important area for future research.

In summary, the R & D project selection decision is a process by which an intermittent stream of changes are made to lists of currently active and proposed projects. The project selection decision process includes generating alternatives, determining the appropriate time to make a decision, collecting data, specifying constraints and criteria, and recycling. Evaluation and comparison of projects and proposals is complicated by multiple decision criteria which have no natural, common underlying measure and whose relative importance varies over time. In addition, the R & D project selection decision may be made at several different organizational levels which are participating in a hierarchical, diffuse budgeting and planning process. The empirical support for this description is uneven and noncumulative; considerable empirical work remains before one can safely conclude that the description is complete and valid.

2. *Estimation Problems: Uncertainty, Risk, Cost, Time to Completion, Interactions*

In a 1958 paper, Klein and Meckling (1958) studied 24 post-World War II military developments and commented that uncertainty was an inherent characteristic of the projects which they examined. Marschak et al. (1967, Chap. 3) reported nearly identical conclusions. The recognition of inherent uncertainty in R & D reached such a level of consensus that Mansfield (1971, Chap. 1) defined R & D as an activity aimed at reducing uncertainty. Indeed, the very distinctions underlying the "phase model" of R & D are based on the concept of inherent uncertainty.

Uncertainty in R & D, perhaps in the innovation process as a whole, arises from a number of sources: technical--scientific, engineering, production; market--demand, consumer response, competitive actions; organizational--resistance to change, lack of receptivity by management, organizational conservatism; timing--obsolesence (Bright 1964). Although these sources tend to be present throughout the R & D process, their relative importance most likely varies from basic research through engineering. Existence of these sources of uncertainty results in the R & D project selection decision being substantially more complex than the analogous decisions elsewhere in the organization.

Uncertainty in the R & D process gets translated into risk in the R & D project selection decision process and impacts strongly on the problem of benefit estimation. Throughout the literature, at least three types of risk are noted: technical--that the project will not be technically successfully completed; commercial--that a technically sound project will not be successfully commercialized; and economic--that a technically and commercially sound project will not be economically successful. Mansfield (1968b, Chap. 3; Mansfield et al. 1971) has reported data which confirm that projects do fail as a consequence of each type of risk.

The most common approach for incorporating uncertainty into estimates of benefit is to weight the anticipated benefit by a probability measure, typically intended to include all three types of risk.[31] Several authors have noted the inadequacy of this approach and have attempted to include variance, as well as expected-value, properties in the benefit estimate (Cramer and Smith 1964; Hertz 1964; Hespos and Strassman 1965). As yet no completely acceptable approach has emerged.

Uncertainty also enters into benefit estimation because both R & D organizational objectives and specific project requirements tend to change over time in an unpredictable manner (Marschak, Glennan, and Summers 1967; Rubenstein 1957). As a direct consequence, project requirements and budgets are often initially set in a most flexible fashion based on best predictions of cost, time to completion, and technological advance (Marschak, Glennan, and Summers 1967, Chap. 3). Unfortunately, the initial "best" estimates tend to be subject to large errors, especially at the initiation of the project (Marschak, Glennan, and Summers 1967, Chaps. 3 and 4; Mansfield 1971, Chap. 5). Additional evidence of the difficulty of estimating cost and time to completion exists in the form of rather widely publicized cost and time overruns, both in military (Marschak, Glennan, and Summers 1967, Chap. 4) and industrial R & D (Mansfield 1971, Chaps. 4 and 6). PERT and related scheduling techniques may assist in this area.

The preceding discussion notes that the uncertainty inherent in the R & D process is translated into risk in the R & D project selection decision process and results in difficulties in the processes of estimating project benefits, costs, and time to completion. These difficulties are compounded by several parameter interactions. For example, the time to completion of a project is a function of the rate at which resources are expended (Mansfield, Rapoport, Schnee, Wagner, and Hamburger 1971, Chap. 7; Marschak, Glennan, and Summers 1967, Chap. 2; Scherer 1965c), but only at the expense of an increase in the total cost of the project. Timing is also related to cost in the sense that the largest percentage of the total cost of an innovation apparently occurs when tooling is designed and manufacturing facilities are designed and constructed (Mansfield 1971, Chap. 6; Mansfield, Rapoport, Schnee, Wagner, and Hamburger 1971; Carter and Williams 1957). Finally the probability of technical success is thought to be a function of total cost.

Thus the project selection decision is made even more complex. Not only is there significant uncertainty surrounding the various parameter estimates, but there are also many parameter interrelationships. Hence, there is no logical starting point--in order to determine a good allocation pattern, it is necessary to have estimates of time to completion, likelihood of success, manpower and facility requirements, and the like. However, each of these parameters is a function of the specific resource allocation pattern selected and cannot be estimated until an allocation pattern has been determined.

3. *Normative Models for R & D Project Selection*

The discussion in the two preceding subsections describes the R & D project selection decision from the viewpoints of how the decision is made, the organizational context within which the decision is made, and the environmental factors which make the decision difficult. In this subsection the normative viewpoint of "how the decision *should* be made" will dominate the discussion. No distinction is made between incremental and discontinuous projects. A general normative statement of the R & D project selection/resource allocation problem is:

Given a set of alternatives (projects and proposed projects) which require common scarce resources (such as dollar budgets, manpower, and facilities), determine that allocation of the resources to the alternatives which will maximize the benefit contribution (value) of the resulting program.

Constrained optimization models of the R & D project selection/resource allocation problem consist of two types of mathematical functions:

1. *objective functions* which measure the benefit contribution of an allocation patterns; and

2. *constraint functions* which represent the organizational environment within which the decision must be made (of special importance is the availability of the scarce resources).

The purpose of this subsection is to summarize the state of the art related to normative R & D selection decisions.

A number of papers survey the literature associated with normative R & D project selection models [for example, see Baker and Pound (1964), Cetron, Martino, and Roepcke (1967), and Baker and Freeland (1972a)]. These surveys are periodically updated in the prefatory remarks of numerous papers.[32] Table 6 summarizes the categories used to clas-

TABLE 6.--Summary of literature results.

Authors	Categories Used to Classify Models	Key Topics Covered in the Paper
Baker & Pound (1964)	Decision theory Economic analysis Operations research	Discussion of general descriptive nature of project selection material Characteristics of 30 representative models Brief desctiption of areas of application and data requirements
Cetron, Martino, & Roepcke (1967)	Decision theory Economic analysis Operations research	Features which describe input and output characteristics of the models Ease of use and data requirements discussed Area of applicability identified Table summary for 30 models
Moore & Baker (1969)	Scoring models Risk analysis Economic models Constrained optimization	Brief discussion of each model type Summary of some accepted descriptive insights First empirical data relating output from different model forms
Alboosta & Holzman (1970)	Projected scoring Project index Math programming Utility models Descriptive	Brief discussion of each model type Identification of critical factors not included in most models
Souder (1972)	Linear Nonlinear Zero-one Scoring Profitability index Utility	Scoring models used to evaluate representative models from each class Uses data from 30 actual projects to perform comparative analysis of four models designed to represent main categoreis
Baker & Freeland (1972a)	Benefit measurement Project selection/ resources allocation	Benefit measurement models discussed in detail Numerous research opportunities identified

sify the various models and briefly identifies the content of each paper; it is not intended to represent a complete summary of the papers, but rather to assist the reader who might be interested in investigating the area beyond this presentation. Classification into benefit measurement and project selection/resource allocation models is adopted for use here.

All the benefit measurement methods proposed to date require subjective inputs. Some well-informed respondent (or **group** of respondents) is asked to provide his best judgments regarding characteristics of the project proposals under consideration. Admittedly, the level of subjectivity inherent in a simple rating is considerably higher than in a detailed economic analysis; however, even the economic analysis requires subjective judgment. The benefit measurement methods can be described as systematic procedures for soliciting and integrating subjective and objective benefit data. They can be classified on the basis of the thought processes which these methods impose on the respondents. The classification proposed by Pessemier and Baker (1971) can be summarized as:

1. *comparative models* which require the respondent(s) to compare one proposal to another proposal or to some subset of alternative proposals;[33]
2. *scoring models* which require the respondent(s) to specify the merit of each proposal with respect to each of several project characteristics (criteria) which criterion scores can be aggregated to yield an overall project score;[34] and
3. *benefit contribution models* which require the respondent(s) to tie the projects directly to R & D objectives or to system requirements and benefit is measured in terms of contributions to the objectives or to the requirements.[35]

A number of empirical investigations related to the performance of specific benefit measurement approaches or to assessments of similarities and dissimilarities among the methods have been reported;[36] however, the results are too preliminary to permit general conclusions. In summary, a large number of benefit measurement methods have been proposed, but little is known about their validity when applied within an R & D environment.

Project selection/resource allocation models can be classified along a number of dimensions. The mathematics involved in determining which allocation pattern will maximize benefit contribution range from simple ranking according to estimated (anticipated) benefit, as measured by some benefit measurement approach, to the sophisticated algorithms of linear, nonlinear, dynamic, and integer programming. For constrained optimization models, the benefit measurement methods provide input for the objective function of the model. Linear (Asher 1962; Baker, Shumway, Maher, Souder, and Rubenstein, 1972; Beged-Dov 1965) and nonlinear (Atkinson and Bobis 1969; Hess 1962; Rosen and Souder 1965) have respectively linear and nonlinear objectives functions, linear constraint functions, and continuous-decision variables; however, some allocate R & D personnel or man-hours rather than determine project budgets (Asher 1962; Beged-Dov 1965). Also, stochastic (Charnes and Stedry 1964; Cramer and Smith 1964; Hess 1962) and goal programming (Charnes and Stedry 1964) models have been suggested. Thus, a plethora of R & D project selection/resource allocation models have appeared. Baker and Freeland (1972a) cite over 175 references. However, as the surveys cited earlier report, few of the models have been evaluated empirically and even fewer are being utilized by R & D managers (Baker and Freeland 1972a; Baker and Pound 1964; Mansfield 1971, Chap. 3).

The reviews[37] and the descriptive material in the previous two subsections provide a basis for identifying the limitations inherent in the currently proposed normative models. An integration and summary of these limitations is:

1. inadequate treatment of project and parameter interrelations with respect to both benefit contribution and to resource utilization;
2. inadequate treatment of uncertainty as it impacts on benefit measurement and parameter estimation;
3. inadequate treatment of multiple, interrelated decision criteria which have no common, natural underlying measure;
4. inadequate treatment of the time variant property of the parameters and criteria and the associated problem of continuity in the research program and staff;
5. a restricted view of the problem which (a) portrays a once-a-year investment de-

cision rather than an intermittent stream of investment alternatives, (b) does not include such attributes as timing of the decision, generation of additional alternatives, and recycling, (c) does not recognize the diversity of projects along the spectrum from basic research to engineering, and (d) views the problem as a decision event rather than as a hierarchical, diffuse decision process;

    6. no explicit recognition and incorporation of the importance of individual R & D personnel; and

    7. the inability to establish and maintain balance in the R & D program; e.g., balance between basic and applied research, between offensive and defensive research, between breakthrough and improvement orientations, between in-house and contracted projects, between product- and process oriented projects, and between high-risk/high-payoff and low-risk/moderate payoff projects.

Given these limitations, it is clear why few normative models of the R & D project selection/resource allocation decision have been implemented and use by R & D managers. Most attempted applications have concerned simple scoring or financial index models.[38] As a result of the small number of applications and of the limited types of models which have been applied, it is nearly impossible to assess the potential utility of the normative R & D project selection/resource allocation models.

Souder (1972a, 1972b) has made a promising start in providing approaches for evaluating R & D decision models. Expanding a list of R & D model performance characteristics suggested by Cetron, Martino, and Roepcke (1967) and incorporating responses from a number of R & D managers and management scientists, Souder (1972b) prepared a list of performance criteria and characteristics. This list was used as the basis for a scoring model to determine the "relative suitability" of certain classes of quantitative models. Since the measure of suitability is only a relative measure, it provides little insight into the "absolute suitability" of the models. Two applications of the scoring model have been reported (Souder 1972a), and the results indicate some strengths and limitations of a number of quantitative R & D selection and allocation models which have appeared in the literature. The list of performance criteria and characteristics is also useful as a reference guide during model design. A slightly expanded version of the list is reproduced as Table 7.

Souder (1973) has also provided data which are useful for preliminary assessment of the effectiveness of four resource (budget) allocation models. His four test models are referred to as a "nonlinear model," a "linear model," a "zero-one model," and a "profitability index model" and are intended to be specific representations of broader classes of models. The output from each test model is evaluated against the output from two control models; namely, a "benchmark model" which allocates the budget in a pro-rata fasion (i.e., the budget is distributed proportionately according to the project's maximum annual funding level) and an "ex-post optimal model" which funds at their maximum annual level only those projects that terminated as successes. Perhaps the most interesting result is that, for the specified set of input data, the benchmark model performed as well, and in some instances better, than Souder's test models. In his data, the estimates of the probabilities of technical success undergo significant change over the life of the projects and the benchmark model performs best when the data are least valid. It is impossible to determine whether the result--namely, that the benchmark model is as effective as the mathematical optimization models--is an artifact of the data or is a general result holding for environments with high uncertainty. In any case, that the pro-rata benchmark model is as effective as the mathematical optimization models is an interesting and critical hypothesis which should be systematically tested in subsequent research.

Several papers[39] have been written by individuals who are not academicians, but who are management-science professionals working for an organization and who are attempting to improve the organization's R & D selection and allocation decision process. In all but two possible situations (Bender and Pyle 1972; Nutt 1972) there is little evidence that the models proposed by these practitioners were used, or if used, survived after the departure of a critical management sponsor and/or the model builder. Hence these papers provide little assistance in assessing the potential effectiveness of the proposed models; however, they do suggest some level of dissatisfaction on the part of the R & D management. Two recent studies whose results are just appearing (Baker, Shumway, Maher,

TABLE 7.--R & D project selection models: Performance criteria and their characteristics [after Souder (1972b)].

---

1. Realism Criterion Characteristics

    Model includes:
    Multiple objectives
    Multiple constraints
    Market risk parameter
    Technical risk parameter
    Manpower limits parameter
    Facility limits parameter
    Budget limits parameter
    Premises uncertainty parameter

2. Capability Criterion Characteristics

    Model performs:
    Multiple time period analyses
    Optimization analyses
    Simulation analyses
    Scheduling analysis

3. Flexibility Criterion Characteristics

    Model applicable to:
    Applied projects
    Basic projects
    Priority decisions
    Termination decisions
    Budget allocation applications
    Project funding applications

4. Use Criterion Characteristics

    Model is characterized by:
    Familiar variables
    Discrete variables
    Computer not needed
    Special persons not needed
    Special interpretation not needed
    Low amount of data needed
    Easily obtainable data

5. Cost Criterion Characteristics

    Model has:
    Low set-up costs
    Low personnel costs
    Low computer time
    Low data-collection costs

6. Additional Criteria

    Model considers:
    Competitor efforts
    "Strategic need"
    Project dependencies (value, resources)
    "Flags" for potential problem areas

Souder and Rubenstein 1972; Maher 1972) report on on-site experiments designed to evaluate specifically constructed models. The results reported to date are generally optimistic, but are too incomplete and specific to provide general results. Additional on-site experiments are urgently needed in order to develop general conclusions.

4. *Some Speculation and Conjecture*

Initial research efforts have been reported in the R & D literature and in the relevant methodological literatures which may eventually lead to overcoming some of the limitations identified in the preceding subsection. This subsection contains speculation and conjecture regarding where the R & D project selection literature is headed and identifies a number of fruitful areas for research.

The trend in application appears to be away from "decision models" (models which yield decisions) and toward "decision information system (system which provides information for decision-makers). Three legitimate reasons can be suggested for this trend. First, the existing project selection/resource allocation models are incomplete in the sense that they do not include all the important, relevant aspects of the R & D environment. Second, the decision problem is characterized by multiple criteria, many of which are not readily quantifiable, and the current approaches to quantifying subjective preferences are far from satisfactory. Third, the R & D process is highly uncertain and unpredictable. As a result managers are skeptical of the validity of the numerous input data elements, of the various difficult-to-understand model forms, and of the subsequent allocation recommendations. The general managerial attitude appears to be that the normative models can assist with respect to the routine, preditable activities but are not sufficient to analyze the breakthrough, unpredictable activities which are the most important and difficult to manage.

Response to this attitude appears to be taking the form of constructing scenarios which represent alternative futures, of determining allocation patterns which are expected to produce "good" or "satisfactory" results for several alternative futures, of program or portfolio (rather than individual project) benefit measures. This response can be observed in the work of Dean and Hauser (1967) and Rosen and Souder (1965) in the middle 1960s. More recently, advances in relating project benefit to the feasible levels at which the project may be funded (Alboosta and Holzman 1970; Atkinson and Bobis 1969; Baker, Shumway, Maher, Souder, and Rubenstein 1972) have lead to interactive (i.e., the user can interact with the system) decision systems of the sensitivity analysis type which permit the user to ask a number of "what if" questions (Baker, Shumway, Maher, Souder, and Rubenstein 1972; Cochran, Pyle, Greene, Clymer, and Bender 1971). Although applied in a non-R & D environment, Geoffrion, Dyer, and Feinberg (1971) suggest an interactive approach which does not assume that the benefit function can be quantified and which is specifically designed for a multiple criterion decision problem. These interactive, decision information scenarios offer a promising alternative for coping with the uncertainty inherent in the R & D decision process.

The Charnes and Stedry model (1964) is an interesting integration of the concepts of chance-constrained and goal programming. Goal programming permits the expression of multiple objectives in terms of acceptable goals (e.g., profit, market share, cost in dollars, etc.) and finds allocations which minimize the deviation between the goal levels actually attained and those deemed satisfactory *a priori*. The model also determines the minimum expected short run and long run resource requirements necessary to achieve the specified goal levels and includes explicit consideration of interim adjustments due to research breakthroughs. This models is an important first step in modeling for breakthrough or crisis planning and warrants further attention and development.

Decisions which impact on the allocation of resources to R & D proposals are frequently made at different levels in a hierarchical organization. Each level makes decisions or recommendations based on information from both higher and lower levels. Since each level operates with some autonomy, there is the problem of coordinating the decisions made at the various levels. The resource constraints of any constrained optimization model implicitly recognize this by identifying such factors as total available budget. The Baker et al. model (Baker, Shumway, Maher, Souder, and Rubenstein, 1972) is designed to incorporate constraints which are generated by two hierarchies--an organizational hierarchy and a research effort hierarchy. Thus, constraint intervention as a form of coordination modeled; however, there is no evidence that goal intervention has been modeled or that goal intervention approaches have been applied in the R & D environment (Freeland and Baker 1972).

At a higher level of analysis of hierarchical considerations, it is important to note that resource allocation decisions can be affected by the specific form of the hierarchy, by the nature and content of the information flow, by the way in which coordination is attained, and by the negotiations which result from joint decision-making between officials with different functional responsibilities, e.g., R & D, engineering, production, and marketing. Until recently little opportunity existed for conducting such analyses because of the lack of appropriate mathematical methodology and model structures; e.g., the Dantzig-Wolfe decomposition approach is primarily a computational technique and is not suitable for such analyses (Dantzig and Wolfe 1961; Freeland and Baker 1972) Recent advances, however, appear to offer sufficient mathematical structure so that important general insights into hierarchical decisions can be derived.[40] This research should continue, and its application to the R & D environment should be investigated.

The research opportunities summarized above represent significant areas for advancement in the state of knowledge regarding R & D project selection/resource allocation. In addition to this theoretical and methodological research, companion empirical research should also be encouraged. More studies are required that would support and/or modify the existing descriptions of the R & D project selection/resource allocation decision process and the organizational environment within which it operates, evaluate the prescriptive capabilities of the normative models and the process by which the models are adopted, improve our understanding of and ability to model the many and varied interrelationships, and critically test the implications drawn from the methodological and theoretical research.

D. PERFORMANCE OF SCIENTISTS AND ENGINEERS

The previous section surveyed the existing literature related to R & D project selection/resource allocation from both descriptive and normative viewpoints. One output from the R & D project selection/resource allocation decision process is a list of funded projects, i.e., projects to which organization resources have been allocated. In this section the focus is on the behavioral and organizational variables which influence how well researchers and engineers in the R & D laboratory will perform on their assigned project activities.

1. *A Proposed Scenario of Effective Performance*

One systematic, integrated series of studies exists in the literature. The early work in this portion of the literature is summarized by Pelz and Andrews (1966), who present data from a 1959 study involving 1311 scientists and engineers from 11 laboratories in industry, university, and government. The study was concerned with identifying and understanding the relationship of several organizational factors and a number of output measures such as peer judgments of usefulness, peer judgments of contribution, unpublished reports, published reports, and patents. It is possible using the Pelz-Andrews results to construct a scenario which is descriptive of effective scientists and engineers.

According to the Pelz and Andrews data, effective scientists and engineers are self-directed; they value and seek the freedom to pursue their own ideas. Coordination is provided by a combination of self-direction and allowing several others a voice in shaping their directions. Although self-directed, they also interact vigorously with their colleagues. Their work is diversified; that is, they do not restrict themselves either to application or to pure science, but maintain an interest in both. Accordingly, effective scientists and engineers are not fully in agreement with their organization in terms of their interest and do not necessarily exhibit the behavior necessary for advancement in the organization. Although they tend to be motivated by the same kinds of things as their colleagues, they differ in their work styles and strategies.

One further dimension can be added to the scenario. Andrews and Farris (1972) have studied the relationship between time pressure and scientific performance, and their data strongly suggest that a sense of time pressure can enhance several qualities of scientific performance. In addition to experiencing the most time pressure, effective scientists and engineers also tend to want relatively large amounts of pressure. The critical finding is that desired and actual time pressure should be consistent: either excess or too little time pressure tends to detract from performance. Time pressures come not only from the project work, but also from active communication with colleagues and from administrative duties of a limited nature.

The above scenarios present an internally consistent picture of effective scientists and engineers. However, the reader should remember that the Pelz-Andrews data for both performance and organizational variable data were collected at one point in time. Correlation techniques were utilized for the statistical analysis; hence, causality could not be deduced from the data, although the authors attempted to determine causality by logical argument. In fact, there was reason to suspect that performance preceded the organizational variable for at least some of the relationship. For example, Houton (1963) had rather persuasively argued for the proposition that (1) getting a "good" work assignment is contingent upon demonstrated competence and (2) demonstrating competence is contingent upon having a "good" work assignment; thus, he argued for causality in both directions. In summary, the scenario requires validation in both a methodological and a timeliness sense.

Fortunately, it is not necessary to speculate blindly on the direction of causality. In 1965 Farris (1969a, 1969b) returned to three of the laboratories, all within the electronics industry, to obtain new information from 151 engineers who had participated in the original 1959 Pelz-Andrews studies. Thus, he had data at two points in time and could statistically investigate the direction of causality. Farris concentrated on four measures of output--contribution to the scientific discipline, usefulness to the organization, number of patents, and number of reports--and on six organizational conditions--involvement in work, contact with a relatively large number of colleagues, high influence on work goals, diversity of work activities, high salary, and number of subordinates (1969a, p. 9). His empirical findings are:

1. Engineers who are more involved in their work produce more patents subsequently but, more than that, engineers who are seen as useful and producing more patents, become more involved in their work.

2. Higher performing engineers subsequently receive more influence over their own work goals. Greater influence on work goals was not followed by increased subsequent performance.

3. Engineers who have greater contact tend to (weak finding) perform better subsequently, and high-performing engineers subsequently come into frequent contact with their colleagues.

4. Greater diversity is followed by higher performance, and higher performance is followed by greater diversity of work activities.

5. Engineers who perform well subsequently are paid more, but there is no evidence that those who get paid more subsequently perform better.

6. Engineers with more subordinates (with supervisory level controlled) subsequently perform better but, more than that, engineers who perform well subsequently receive more subordinates (1969a, pp. 13-14).

In summary, the general pattern for the six organizational factors and four measures of performance is for performances to precede the favorable organizational condition, not for the organizational condition to precede performance. That is, effective researchers and scientists were being rewarded for their performance, and some of the organizational rewards assisted them to continue to be effective. Thus, Houton's proposition was clarified and supported by the Farris study.

The predominance of the performance-followed-by-factor relationship was stronger than either Farris or Houton had anticipated. Research should be directed toward more precisely determining the ways in which a person's performance affects his social-psychological working environment. The consequences of performance should be considered explicitly in subsequent theories, so that performance must be treated as both a result and a cause of change in the social-psychological work environment. The implications of this viewpoint for designing and implementing R & D personnel reward systems should be studied.

2. *The Impact of Supervisory and Managerial Behaviors*

Since the impact of supervisory and managerial behaviors is discussed throughout the section on organized R & D, the purpose here is not to repeat that discussion, but rather present what is known specifically about the impact on performance. Farris in his 1972 paper provides an initial framework for understanding the impact of supervisory behavior on group performance in R & D. He employed a system for measuring group performance by a composite score made up of individual measures of how well each group member increased knowledge in his field, extended or refined existing knowledge in his field, contributed to general knowledge in his field, and had been useful in helping his R & D organization carry out its responsibilities. The study was conducted in a NASA research center and focused on 94 nonsupervisory scientists who composed 21 small teams.

The results can be summarized in terms of colleague roles and information flow among group members, between group members and the supervisor, and with persons outside the group. High-performing groups were characterized by (1) technical information and help in thinking occurring among group members; (2) help in thinking, critical evaluations, and administrative help flowing from the supervisor to the group; (3) technical information, help in thinking, critical evaluations, and original ideas flowing from the group to the supervisor; and (4) original ideas from outside the group for administrative help. Conversely, in low-performing groups, the following were observed: (1) discussions of organizational rules and policies followed among the group members; (2) organizational rules and policies and original ideas were passed from the supervisor to the group members; (3) the group members went outside the group for help in thinking and for administrative help; and (4) the supervisor sought help in thinking and original ideas outside the group and received help in thinking and organizational rules and policies. Given the earlier results of performance-followed-by-factor for effective scientists and engineers, one must wonder if this same dominant direction of causality holds for effective groups as well. Farris's data (1972) do not provide the answer.

Farris (1972) notes two surprises in his results--surprises because they are not consistent with other organizational literature. First, none of the several measures in the human relations area were related to group performance. Second, markedly negative relationships were found between the supervisor's performance of such administrative functions as planning and scheduling and his subordinate's performance measures. These surprises as well as the descriptions of roles and information flow are a particularly rich base for additional empirical research.

Some Research Opportunities

The variables and relationships cited by Martino (1973) in his literature survey are drawn from numerous unrelated and nonintegrative studies. His survey indicates the unevenness of the literature dealing with performance; e.g., eight papers cited are associated with the impacting variable "leadership style," but no papers cited as associated with "variables representing characteristics acquired by the researcher." Studying these variables further would obtain more even and in-depth coverage. More importantly, there is need for more complex, integrative studies in order better to understand variable interrelationships.

A scenario describing effective researchers and scientists was constructed from the Pelz-Andrews findings. Farris's subsequent study, although supporting the scenario in general, also raises significant questions regarding directions of causality. In fact, stronger results are found for the hypothesis that performance improvement results in improved organizational conditions than for the other directions, i.e., that improved organizational conditions result in performance improvement. Replications of the Farris study would be beneficial. The consequences of improved performance as well as of improved organizational conditions must be explicitly included in subsequent theories. Maybe these are not causes and effects but both, or perhaps manifestations of something else. Theories should be developed and empirically examined.

The scientific-group-performance research by Farris is in the early stages of development and verification. It is critical that work in this area be expanded since team effort will be required to research the increasingly more complex R & D projects. The question of direction of causality should be investigated for groups in order to determine if the results found for individual scientists and engineers also apply at the group level.

As illustrated in the previous section, the R & D literature abounds with normative project selection and scheduling models. Some literature evaluates the normative models with respect to how well the models relate to the existing descriptive literature and issues included therein. With few exceptions which barely identify the issue (Baker et al. 1972, Souder 1973), no work has been reported which investigates the impact of the adoption and utilization of normative models on subsequent manager, supervisor, researcher, and/or engineer behavior. Research should be initiated to investigate this issue.

IV. TRANSFER FROM R & D: IMPLEMENTATION AND UTILIZATION

The devices and/or knowledge produced by R & D are not innovations until they have been applied. This implementation process is often one of translating an R & D result into a new or improved product which has been successfully introduced into the marketplace or some other environment exogenous to R & D: for example, in the military sector it may entail the adoption of a new guidance system for aircraft. Another common form of implementation involves a new modified process for producing outputs; e.g., the adoption of tape-driven machines to perform functions previously accomplished by operator-controlled machines. In any case, the process by which the technological output from R & D becomes an on-going capability of the operating units, such as production, marketing, and sales, will be referred to as the implementation process.

In order to clarify further what is intended by the term "implementation," let us see what the literature has to offer as definitions. Utterback (1971a, p. 75) considers implementation to include the engineering, tooling, plant start-up, and manufacturing required to bring the solution or invention to its first use or market introduction.

Quinn and Mueller (1963) view the problem as one of transferring three aspects of a potential innovation to the operating units of the organization: (1) information about the technology underlying the innovation; (2) enthusiasm for the innovation; and (3) the authority to use the innovation. The dyadic (two-person) relationship existing between the researcher and the marketer is the aspect of implementation on which Young (1973) concentrates. Perhaps the most unique view is Morton's (1968) who analyzes implementation in terms of the spatial and organizational bonds which facilitate the transfer of information and the spatial and organizational barriers which inhibit the transfer of information: He seeks to maximize feedback and minimize corporate interference.
The common characteristic of these views is that knowledge and/or devices are being transferred across intra-organizational boundaries, from R & D to the operating units of the organization.

Is it important to investigate the implementation process? After interviewing over 200 top operating and research executives in the United States, Quinn and Mueller concluded that "the key problem in research management today is getting research results effectively transferred into operations" (1963, p. 49). That this conclusion is still valid is illustrated by the following 1971 statement by Lawton Hartman, former associate research director at Philco and now a special assistant to the director of the National Science Foundation. "No other problem is so pervasive and potentially mischievous as the failure of top managers and research to communicate with each other. I have never seen it done very well." (Cordtz 1971, p. 108). It seems clear that the implementation process by which R & D results are transferred to the operating units is viewed as an area of critical concern by top operating executives and research management.

One reason for this concern has been identified by Mansfield. The potential innovation leaving the R & D laboratory is entering the phase of its development which is most costly and most time consuming, implementation. From their sample of chemical, mechanical, and electronics firms, Mansfield et al. (1971, p. 118) found that 29.1% of the total innovation expenditures concern prototype or pilot plants, 36.9% involve tooling and manufacturing facilities, and 9.5% involve manufacturing start-up costs. Thus, in the industries surveyed, over 75% of the total innovation expenditures occurred during implementation.

In addition to the economic aspects studied by Mansfield et al., important behavioral and organizational considerations are related to implementation. The organization development literature--particularly the insights drawn from Burns and Stalker (1961), and Wilson (1966), Hage and Aiken (1970), Zaltman et al. (1973), and Scott (1973)--relates organizational structure to organizational context. This literature, summarized earlier (see "Background, Definition, and Basic Concepts" section of this chapter), need not be repeated, but Woodward's (1965) study does deserve special mention. She argues that the technologies utilized in the production process determine the underlying organizational structure. There is one basic structure for firms using mass production technology; another for those which employ a process technology; and still another for firms manufacturing one-of-a-kind articles (p. 36). Transfer would thus seem to require a coupling of structure and function as well as a coupling of information, enthusiasm, and authority.

The most critical transfer points are those where R & D interfaces with the various operating units of the firm. This view is supported by the Panel on Institutional Barriers to Innovation and Diffusion in the Service Sector which reported to a 1973 Engineering Foundation Conference that nearly 25% of the 200 producers' goods innovations examined were blocked by internal management or by organization and staffing (Myers et al. 1973, p. 21). The analysis of liaison agents and coupling relations by Rubenstein, Barth, and Douds (1969), of "systems engineers" as described by Morton (1964), and of the project manager concept by Bean (1968), add further credence to the contention that the transfer of R & D results requires a coupling of structure, function, information, enthusiasm, and authority at the interfaces which R & D has with the operating units.

Quinn and Mueller (1963, pp. 61-62) have identified a number of organizational forms which have been used to facilitate the transfer from R & D, namely:

> *Task-force groups.*--These usually are made up of personnel from research, development, marketing, and manufacturing who are often given total responsibility for ex-

ploiting a new technology. The composition of the group is heavily weighted toward R & D people at first, but it shifts toward operating people as full-scale operations are approached.

Corporate development units.--These, having their own marketing staff and flexible pilot-scale facilities, pick up new research technologies and exploit them. The unit can be a profit center, deriving profits from sale of new products. As the products prove profitable, operating groups want to take them on. Thus, development is constantly forced to seek new technologies from research, and operating resistances are eliminated.

Outside companies.--At times, these are used to entrepreneur new products in specific cases. The research laboratory may take 49% ownership in the new concern formed to exploit the technology. Or it may simply take license revenues. In some cases, large companies have given smaller companies (with special knowledge, facilities, or market access) exclusive rights, under a royalty agreement, to a new technology during a three- to five-year introductory period. When primary demand has been built up, the larger company has the option of continuing the arrangement or introducing its own branded version of the product.

Staff groups at corporate level.--These units serve to coordinate the introduction of new technologies through existing divisional and functional organizations. They are most effective when they either have functional authority over key aspects of line operations or have a budget with which to buy time from line units. Product managers perform this service successfully in some companies.

A top executive with multifunctional line authority.--This executive can effectively force new technologies into operations in small to medium-size companies. In a medium-size consumer products company, the president is also the top technical executive and founding genius. Because of his personal interest and follow-up, new products often move from research to the market in three to six months. He refuses to allow pilot-scale facilities to be built, feeling that they waste time and put less pressure on operating executives than do full-scale facilities.

A research group with a special budget to buy time on operating machines.--This approach is effectively used in flow process industries (such as paper) where (1) the cost of a pilot facility is prohibitive, or (2) the scale of operation vastly affects technical approaches. There are always problem of scheduling these experiments; but if an experiment is successful, research has little trouble in demonstrating its value to operations.

Individual researchers who entrepreneur their ideas through pilot facilities and and into the market.--A pharmaceutical company sets up a profit center for each new product and encourages the researcher, if he has the talent and interest, to follow his idea to commercialization. If successful, he receives a share of the center's profits as additional compensation. Product policy is coordinated by product group managers at corporate level.

Multilevel committee responsibility.--Such committees have been set up in some companies. In fundamental and early applied research, a research committee coordinates the program. In late applied and early developmental stages, coordination moves to an R & D committee. In late development, a new-product committee takes over program progress. Before pilot-scale facilities can be built, the operating committee must approve. A full-scale operation requires executive committee approval. Because decisions tend to be slow and conservative under this system, it must normally be supplemented by one of the other organizations described here.

An entrepreneurial group at corporate level.--Used by several of the companies most sucessfully diversifying through research--produced new products, these groups introduce technologies which are new to the company and which do not logically fit into the organizations of established operating groups. Where they are successful, these entrepreneuring units are headed by a commercially oriented dynamo with a technical background. He has at his disposal a technical group which reduces research ideas to practice, a special budget to build small-scale facilities and underwrite product introduction losses, and a small nucleus of commercially oriented technical men who simultaneously "ride two or three products into the market."

Major difficulties with this approach are (1) finding people with the complex of skills and attitudes necessary to entrepreneur new products; (2) replacing these people as they become committed to products they have "ridden into successful division status;" (3) developing the top-management attitude toward the risk taking such operations must involve.

Although each of the above forms has proved useful in specific instances, none has been shown to have general applicability. The most suitable form varies from firm to firm and from innovation to innovation.

One symptom of inadequate transfer is when developing and newly commercialized products often fail because their design specifications fail to mesh with changing market needs (Young 1973). A number of examples of successes and failures can be found in the literature. Consider, for example, DuPont's seven-year $100 million experiment with "Corfam," which was reported as a successfully developing product in 1954 and again in 1966; but on March 17, 1971 the *Wall Street Journal* reported "DuPont's Corfam leather substitute has been given its final walking papers" (Young 1973, p. 65). This was a clear example of a market failure, for "Company officials attribute Corfam's sluggish sales to the market's indifference to the four features--durability, water repellency, ease of care, and competitive price with leather goods--Dupont thought were its major selling points" (Young 1973, p. 68). Relatively inexpensive foreign imports were also a factor in Corfam's demise. Young (1973, pp. 69-72) offers numerous other examples of innovations which failed due to improper product-design/market-need relationships.

As a result of these observations, Young undertook an empirical study of the researcher-marketer dyad (two-person interaction). Working with 30 firms in the healthcare industry, he found that the most common reason for delay of transfer was improper or incomplete market specification (1973). The researchers and marketers disagreed on such critical areas as the market toward which the product was directed, the current stage of development of the product, the relationship of the product to the firm's overall objectives, and the product's chance of success. Hence, information exchange was found to be the variable with the strongest influence on the dyadic relationship.

Among the specific findings of the Young study are: (1) it is desirable to create an environment in which members of the dyad believe that their requests for information and their answers to questions will be acted upon promptly, even anticipated; (2) the marketer tends to be much more personally committed to the product than the researcher, and this commitment leads to interpersonal conflict between the researcher and the marketer; (3) product managers tended to have responsibility for success and failure, but neither the budget nor the authority to see the product through to completion; and (4) the more risky the venture, the more important is a supportive management attitude in order to reduce personal risk, to encourage new approaches to problem solving, and to provide an atmosphere in which mistakes can be admitted and corrected, rather than covered or ignored. Young has provided an intensive study of the interpersonal relationships inherent to marketer-researcher dyad. Although the study is restricted by its highly selective database, it is well documented and methodologically sound in an area where very little empirical research has been done.

Although the literature is clear that top operating executives and R & D management are concerned about the process by which R & D results are transferred to the operating units of the organization, we were most frustrated in our literature search in this area--as were Grayson (1968, p. ii) and Young (1973, p. 31). In Young's words, "The bulk of the writing has taken one of two forms: a general prescriptive solution or a specific solution" (p. 31). There is almost a complete lack of definitive, cumulative research in the literature--almost, because there are only a few isolated studies such as those cited in this section. Thus, we conclude that the implementation process is of immediate concern to the managers of organized innovation, but that the literature offers little assistance due to the lack of definitive research.

The only other conclusion to be drawn is also largely negative: although a number of organizational forms have proved to be suitable for implementation in specific instances, none has been shown to have general applicability. If a variety of organizations forms will not do the job, then one is led to wonder what other variables are significant. The literature suggests some candidates:

1. at the *interpersonal level*, the Young study identifies information exchange and compatible levels of personal commitment as variables which influence the researcher-marketer dyad;

2. at the *supervisory level*, Young's study suggests that responsibility must be coupled with adequate budget and authority and that a supportive management attitude is important, especially for high risk ventures;

3. at the *organizational level*, it is important that R & D activities be directly related to organizational goals and objectives (Quinn and Mueller 1963; Cordtz 1971; Pessemier 1966) and that careful planning and control be maintained throughout the implementation process (Hill and Hlavacek 1972; Pessemier and Root 1973; Peterson 1967). The prescriptive nature of the literature is apparent, especially at the organizational level of analysis.

Of all the literature areas which we have surveyed there is none for which simultaneously the level of concern is higher and our state of knowledge is lower. There may be an on-going problem because of proprietary information. In any case, definitive, cumulative research is an immediate need. Some additional insights will be forthcoming when the more general diffusion literature is surveyed and assessed in Chapter 4.

V. SUMMARY AND CONCLUSION

Just as a variety of factors in the "external" world--values, endowments, and institutions--influence the innovation process, so too does a wide range of elements in the environment to which the process is indigenous. Interestingly enough, we might have even analyzed the indigenous variables in the same broad terms. The "values" of the firm, of the formal and informal groups within it, and of the individual participants, all strongly influence the nature and thrust of innovative efforts. Similarly, the "endowments"--the technical information and capabilities of individuals and groups, the capital and material resources of the organization, etc.--play a major role. Finally, the "institutions" factor--which has as its analog the structure of the organization--also turns out to be a significant variable in terms of the relation of such structures to their innovative function.

We chose to couch our study of the more immediate environments of innovation in somewhat different terms, however, in order to display more adequately both the broad trends and the fine-grained detail that are to be found in the research literature. To highlight these disparate features of organized innovation, we developed the chapter in terms of: (1) a complex of interactive process phases, (2) several levels of aggregation or analysis, and (3) a wide range of organizational, economic, information, and social-psychological variables.

Perhaps the most significant trend characterizing the process of technological innovation in this century is the increasing institutionalization of all its phases. The assumption underlying this trend is that creativity can be a deliberate and controlled function of the organization. This assumption--though widely held by those involved in innovative efforts as well as those who study the process--can be realized only in part, by trial and error, given the present state of our understanding. In light of the strong institutionalization trend to which this assumption has given rise, an improved understanding of the process has become a matter of some urgency.

A second, recurrent theme to be found in the several literatures concerned with organized innovation is that organizational structure is a major determinant of the function of R & D within the firm. As it turns out, however, no one model of the R & D process and no one organizational structure is the most appropriate; their adequacy depends on the nature of the R & D project and on its stage in the R & D process. The message is clear: structure, strategy, and policy interact to exert significant influence on the orientation of the laboratory, the characteristics of the projects generated and funded, and the overall effectiveness of the R & D effort. There are some informed speculations but relatively little knowledge regarding these interactions.

With regard to the phases of the process, it is important to recognize that the pre-project activities of problem definition, idea generation, and idea submission are not as amenable to direct organizational control as later, "project" activities. Thus, these pre-project phases are unusually dependent on the rich and subtle social-psychological influences of intra- and intergroup relations. There exists a sizable--though fragmented--literature in this area, and the relevant variables have been pretty well identified. Their interrelations and mutual influences are not well understood, however; nor is there wide appreciation for the fact that while management has little *direct* control over these pre-project activities, it can exert *indirect* influence through a variety of mechanisms. The most propitious starting point in utilizing these avenues of indirect influence is the clear recognition that the social networks of influence and information flow are largely informal and oral, and that they help to shape--as well as respond to--organizational "reality." Again, the experience related earlier is instructive in this regard. When the R & D manager whose laboratory was generating few new ideas finally came to respect and utilize the informal social groups and their strongly oral tradition, he got two responses that ring true; "Well if that is what he wants, why didn't he say so?" And second, he got the new ideas he needed. Too simple? Perhaps, but more complex, more formal machinations may miss the point entirely.

Of all the subtopics in the literature of organized innovation, R & D project selection may have received the most attention, as it is a focus of both "idea flow" and "process phase" models. The R & D project selection decision is a process by which an intermittent stream of changes are made in lists of currently active and proposed projects. The project-selection process includes generating alternatives, determining the appropriate time to make a change, collecting data, specifying constraints and criteria, and recycling. The decision is complicated by multiple decision criteria which have no natural, common, underlying measure and whose relative importance varies over time. The decision may require contributions from several different organizational levels which are participating in a hierarchical, diffuse planning and budgeting process. Our present descriptive knowledge has not yet been integrated into a framework which offers promise of improving the process.

Another important literature area addresses how the social-psychological environment is related to the performance of scientists and engineers in scheduled project activities. Recent research indicates that improvement in individual performance are more likely to lead to changes in his working environment than the other way around. Thus performance should be viewed as a cause as well as a result of change. The implications of this viewpoint for designing and implementing organizational changes should be studied.

One additional problem area was discussed in this chapter--the interface of R & D with the operating units of the organization and the effective transfer of R & D project results into operations. With some notable exceptions, namely the Quinn-Mueller survey (1963) and the writings of Morton describing the Western Electric-Bell Laboratory interfaces (1968), the literature on this topic is nearly nonexistent. Our assessment of this problem area is that it is one of utmost importance, but, the one for which there is the least information in the literature. We believe that studies of this area deserve higher priority than some of the proprietary constraints which may have hindered past efforts at study.

In summary, a number of functions have been identified at the individual, informal groups, and organizational levels; idea generation, idea submission, project performance, and transfer. In addition, the influence of a number of structure/function considerations have been discussed, including characteristics of the organization, the structure of R & D within the firm, characteristics of the specific innovation, and the orientation toward R & D which is held by the firm and its members. The overall need now is for a widely accepted conceptional framework for integrating and cumulating across the many fine-grained studies, and for determining the mutual influence of organized innovation and the many boundaries with which it interfaces. The next chapter treats one of the most important of these boundary areas--the diffusion of innovations.

NOTES AND REFERENCES

1. This seems to be the case in the recent work by Pelz and Andrews (1966), and in the numerous works of Rubenstein.
2. The official history of the USDA is Baker et al., 1963. See also Harding, 1947; Wik, 1966. The role of the states can be found in Knoblauch et al., 1962.

3. See Stakman et al., 1967; Hardin, 1969; *The Green Revolution,* Subcommittee on National Security Policy and Scientific Developments, 91st Congress, 1969; GPO, Washinton, D.C., 1970: 38-612.

4. The arguments are summed up in a controversy in *Technology and Culture* 1 (1960): 201-34, consisting of articles by S. C. Gilfillian, Jacob Schmookler, and I. Jordan Kunik.

5. An interesting variation on this theme is the stimulus to creative thought provided by approaching a problem from a different cultural background. Watanabe, 1974, gives examples of contributions of Japanese to Western science, growing out of "the intellectual and spiritual traditions of Japan and that would not likely have been made by Western scientists." Confirming Watanabe's point, Livingston (1974) states that Lee and Yang's destruction of the principle of conservation of parity (1957) might have derived from their prior Chinese cultural upbringing and linguistic tradition, so they approach the problem differently from a scientist "confined to the Western cultural traditions." Livingston claims that such crosscultural exchanges can "result in unusual combinations of thinking" and might "greatly facilitate the effective pursuit of science."

6. Some of this literature is analyzed in Shepard, 1959. See also Stein and Heinze, 1960.

7. Carnegie Corporation 1960.

8. The classic expression of the role of the entrepreneur in innovation is Schumpeter (1939, especially vol. 1, pp. 85-86).

9. Allison 1965; Deutermann 1966; Lamont 1972; Roberts 1972; Susbauer 1972.

10. Baty 1964; Bolton 1972; Cooper 1972; Hodgins 1972; Litvak and Maule 1972.

11. Argyris 1965b; Burns and Stalker 1961; Hage and Aiken 1970; Harvey and Mills 1970; Wilson 1966; Zaltman et al. 1973.

12. Hamberg 1966; Mansfield et al. 1971; Marschak, Glennan, and Summers 1967; Quinn 1959; Roe 1964; Utterback 1973.

13. Allen 1967a; Allen 1971; Baker 1965; Baker and Freeland 1972b; Baker, Siegman, and Larson 1971; Baker, Siegman, and Rubenstein 1969; Utterback 1973.

14. Northwestern University 1964; Baker 1965; Baker and Freeland 1972b; Baker, Siegman, and Larson 1971; Baker, Siegman, and Rubenstein 1967; Siegman, Baker, and Rubenstein 1969.

15. Rubenstein and Hannenberg 1962; Rubenstein 1964, 1968, Myers and Marquis 1969, Allen 1967a, 1971; Baker 1965; Baker et al. 1967; 1971; 1972b; and Utterback 1964, 1969, 1973.

16. Brandenberg 1966; Hollander 1965; Rubenstein 1964; Schwartz 1973.

17. Several of the cases prepared for the National Science Foundation by Battelle (1973) illustrate that innovations can be the combined result of several incremental and discontinuous R & D projects; e.g., the heart pacemaker, electrophotography, organophosphorous insecticides, magnetic ferrites, and video tape recorders.

18. Chapter 12 below by E. M. Rogers and J. D. Eveland is also an assessment of some of the concerns and literature of subsections 4, 5, and 6.

19. E. G. Argyris 1965b; Burns and Stalker 1961; Hage and Aiken 1970; Harvey and Mills 1970; Wilson 1966; Zaltman, Duncan, and Holbek 1973.

20. Lazarsfeld et al. 1948; Katz and Lazarsfeld 1955; Menzel and Katz 1956; Katz 1960.

21. Utterback 1974; Utterback 1971b; Allen and Cohen 1966; Langrish 1971; Myers and Marquis 1969.

22. Summarized well in Haefle 1962; Parnes and Harding 1962; Stein and Heinze 1960; Taylor 1964.

23. Avery 1959; Avery 1960; Baker, Siegman, and Rubenstein 1967; Goldhar 1970; Hillier 1960; Jones and Arnold 1962; Kaplan 1960; Kornhauser 1962; Kuhn and Kaplan 1959; Lehman 1953b; Marcson 1960; Parnes and Harding 1962; Smith 1959.

24. Bralley 1960; Chollar et al. 1958; Gershinowitz 1960; Haefle 1962; Hillier 1960; MacLaurin 1955; McPherson 1965; Morris 1962; Williamson 1960.

25. Northwestern University 1964; Baker 1965; Baker and Freeland 1962b; Baker, Siegman, and Larson 1971; Baker, Siegman, and Rubenstein 1967; Rubenstein 1957; Rubenstein and Hannenberg 1962; Rubenstein 1968; Rubenstein 1964; Siegman, Baker, and Rubenstein 1969; Utterback 1969.

26. Allen 1967a, b; Allen 1969; Allen 1971; Marquis and Allen 1967; Myers and Marquis 1969; Peters and Roberts 1967; Utterback 1969.
27. Jones and Arnold 1962; Kaplan 1960; Marcson 1960; Storer 1962.
28. Jones and Arnold 1962; Kaplan 1960; MacLaurin 1955; Williamson 1960.
29. Bralley 1960; Gershinowitz 1960; Hillier 1960; Utterback 1973.
30. Baker and Freeland 1972a; Baker and Pound 1964; Cetron, Martino, and Roepcke 1967; Souder 1972a, b.
31. Northwestern University 1970; Atkinson and Bobis 1969; Freeland and Baker 1972; Moore and Baker 1969; Motley and Newton 1959.
32. Alboosta and Holzman 1970; Moore and Baker 1969; Souder 1972a, b.
33. Churchman and Ackoff 1954; Eckenrode 1965; Baker and Freeland 1972a; Luce and Raiffa 1957; Pessemier and Teach 1966; Pessemier and Baker 1971.
34. Andrew 1954; Garguilo, Hannoch, Hertz, and Zang 1961; Moore and Baker 1969; Motley and Newton 1959.
35. Ayers 1969; Dean and Nishry 1965; Disman 1962; Hertz 1964; Miller 1970; Nutt 1965; Sigford and Parvin 1965;
36. Dean and Nishry 1965; Eckenrode 1965; Einhorn 1970; Goodwin 1972; Maher 1972; Moore and Baker 1969; Moore and Baker 1969; Pessemier and Baker 1971.
37. Baker and Freeland 1972a; Baker and Pound 1964; Bradenberg 1966; Cetron, Martino, and Roepcke 1967.
38. Baker and Pound 1964; Cochran, Pyle, Greene, Clymer, and Bender 1971; Mansfield 1971, Chap. 3.
39. Asher 1962; Atkinson and Bobis 1969; Nutt 1972; Rosen and Souder 1965.
40. Charnes, Clower, and Kortanek 1967; Freeland and Baker 1972; Ruefli 1971; Weitzman 1970.

4. THE DIFFUSION OF INNOVATION

I. INTRODUCTION

From the ecological perspective, the subsystems within a complex whole are so inextricably interactive and interdependent that the consideration of any one of them in isolation yields only a partial analysis with a large unexplained "residue." Nowhere is this more evident than in diffusion research. Researchers from a wide range of academic orientations have sought, and found, the variables central to their disciplinary approach. Focusing on these variables they have each been able to account, in part, for certain characteristics of the diffusion process. But their success has in each case been only partial; snapshots of part of the action, from a particular vantage point, and with the aid of particular lenses and filters.

The dynamics of the complex whole have largely escaped our grasp. Since the major elements or dimensions are incompletely defined, their interactions and relative influence are largely unknown. As we have seen, studies of the higher levels of aggregation (Chap. 2) often suffer from a lack of direct empirical confirmation, owing largely to the difficultues inherent in gauging the influence of macrolevel variables. An equal and opposite difficulty was encountered in studying organized innovation (Chap. 3), where the level of analysis encourages empirical study, often to the point of sinking the researcher in a morass of microlevel detail. Diffusion research cuts across all these levels of analysis and is thus subject to all the pitfalls encountered earlier. Hence it is a most difficult area in which to seek an adequate empirically based yet theoretically integrated perspective.

In the previous chapters we have emphasized "flows" of information and influence, and stressed the variety of mechanisms and structures that focus, filter, impede, or amplify such flows. This emphasis will be maintained in our examination of the diffusion process. Owing to the unusually fragmented nature of research in this area, however, there is little in the way of an accepted conceptual structure on which a survey and assessment may be built.

Thus we are confronted with a dilemma. On the one hand, we might work our way through the literature, reporting and assessing specific research results without attempting to integrate them, since there is no community of agreement as to the form such a conceptual integration should take. This basically conservative approach is attractive, but would tell us little about what we do not and should know. The assessment of a field requires a view of the whole as well as the pieces. On the other hand, our approach might be to impose an integrating conceptual structure and then try to fit the particular studies into it. But this might prove a procrustean bed, distorting rather than displaying the state of the art.

Dilemmas invite the search for a middle way. Our middle way will be to survey the three major diffusion research traditions which focus on technological innovations; then to summarize and critically assess the classical diffusion model, which represents the most significant attempt to date to cut across the disciplinary barriers; and finally, to identify the dimensions of a general theory of diffusion deriving from the various research traditions and the classical model, and pose a number of questions for research about their linkages and mutual influence.

Before we turn to these tasks, we note that one manifestation of the fragmented nature of diffusion research is the emergence of specialized concepts, the use of which is often unclear and at times conflicting. The growing interest in the diffusion of technological innovations, for instance, has led to widespread use of the term "technology transfer." There seems to be broad agreement that technology transfer is a special case of the broader category of diffusion, but there is little agreement as to what it is. Some researchers (Burns 1969; Doctors 1969; Bar-Zakay 1970) use it to refer to the diffusion of scientific and technical information (as opposed to the diffusion of artifacts or processes). Others (Chakrabarti 1972, p. 11) use "technology transfer" in referring to intersectoral flows, reserving the term "diffusion" for intrasectoral instances. Still others (Spencer and Woroniak 1967) use "diffusion" to "denote a kind of natural process whereby cultural traits (technology being one) flowed from one cultural milieu to another," whereas "'technology transfer' incorporates an additional specific element... i.e., a planned and purposive type of action (on the part of the adopter)." Finally, Brooks (1966) offers quite a broad definition embracing not only the "horizontal" spread of an innovation from context to context but the "vertical" flow of information from basic science to technological applications as well.

Implicit in much of the technology transfer literature (whatever its particular emphasis) is a preoccupation with point-to-point flow (for instance, see Little 1969, p.2). Chakrabarti (1972) thinks this preoccupation reflects a difference among research traditions. Those with a particularly practical, problem-solving concern in the diffusion of technological innovations are more likely to stress a point-to-point transfer, whereas other research interests tend to a broader process view, emphasizing the pattern of spread over time and the variables on which that pattern depends. We would argue that point-to-point transfer cannot be understood apart from the larger context in which it takes place. Thus we shall take technology transfer, whatever its particular referent or emphasis, as a form of diffusion, and thereby subject to the whole range of concerns characterizing diffusion research.

As for the concept of diffusion itself, the following distinctions are important. In the preceding chapter we were concerned with the process by which a new innovation is conceived, developed, and brought to the point of first use. Thus the term "innovation" was used to refer to something that is new relative to the state of the art. The term "innovation" is also used in the diffusion research literature, but with reference to subsequent adoption in contexts *other than* the original one. Here an innovation is "new" relative to the adopting unit; i.e., it is an innovation for that unit even if it has previously enjoyed long use elsewhere. Likewise, the earlier adopters of an innovation within a sector are commonly called *innovators*.

Adoption is the primary process or mechanism by which an innovation is diffused. At times, however, an innovation cannot simply be adopted for use in a new context. It may also require some more or less extensive adaptation--in essence, innovation--to meet the particular conditions of the new applications (Kranzberg 1967a, pp. 30-31). Adaptation, a condition of successful adoption in such instances, is in fact a special case of a state-of-the-art advance which was the focus of Chap. 3. Thus we shall not deal with adaptation here, but only with the variables and influences that impinge on adoption behavior.

With these observations and distinctions before us, we are now ready to examine the major research traditions comprising the diffusion field.

II. THE PRESENT STATE OF THE ART IN DIFFUSION RESEARCH: THE SEVERAL TRADITIONS

Systematic study of the diffusion process, according to Everett Rogers, had its beginning in the late 1930s with the investigation of the diffusion of hybrid seed corn from agricultural scientists to Iowa farmers (Chap. 12 below), particularly in the classical articles of Zvi Griliches. In its early years, most diffusion research was concerned with the agricultural sector, but it soon spread to other fields including education, anthropology, communication, marketing, economics, and medicine. Research traditions grew in each of these specialized areas but with little interchange among them. As Rogers has remarked, "In the early 1960s each of these diffusion traditions operated as a separate invisible college, and the total field of diffusion was relatively unintegrated" (Chap. 12). In terms of Thomas Kuhn's widely read work (1962) on the development of science, diffusion research itself possessed no paradigm, but consisted of schools making use of paradigms from the disciplines listed above. However, recently it has been argued that these boundaries between the diffusion traditions have begun to break down under the realization that the process is a general one, independent of the disciplines, the specific innovations, and the variety of research methods involved in its study. As Rogers and Shoemaker have concluded: "Diffusion research is emerging as a single, integrated body of concepts and generalizations..." (1971, p. 47).

We believe that this conclusion reflects hope and anticipation more than reality. The field is still so fragmented by specialized concerns that even the dimensions of an adequate general theory are poorly demarcated. It is a conceptual cartographer's dream...or nightmare. This judgment is not acrimonious in intent. It is a reflection rather on the fragmented nature of any emerging field, and the enormous range of disparate elements that characterize this one.

What we *know* about the diffusion process tends to be almost wholly sector or adopter-specific (that is, it concerns diffusion among farmers, or physicians, or Colombian peasants, or industrial firms, etc.). What we, by and large, *do not know* is when and to what extent we can generalize from these specific research results. But unless we know what is true generally, and why, specific studies remain forever specific. They are data points without an integrated frame of reference that permits comparison and cumulative development.

In short, the state-of-the-art understanding of the process by which technological innovations are diffused is meager. This condition is not likely to improve until an integrating theory has been developed, around which a community of agreement can begin to coalesce, thus merging the separate research traditions. Our analysis of the diffusion literature leads us to believe that such a synthesis could be achieved with less disruption than Kuhn takes to be characteristic of the emergence of single-paradigm sciences; the various traditions in diffusion research seem, in large measure, implicitly complementary, i.e., their differences are more a matter of emphasis than of substance. We illustrate both the separate and yet implicitly complementary natures of these research traditions in the following section.

A. THE DIFFUSION RESEARCH TRADITIONS

Rogers with Shoemaker (1971) identify seven major and six minor diffusion research traditions in terms of the number of empirical studies conducted in each.[1] The traditions they identify are:

Major Traditions	Minor Traditions
Anthropology	Agricultural Economics
Early Sociology	Geography
Rural Sociology	General Economics
Education	Speech
Medical Sociology	General Sociology
Communication	Psychology
Marketing	

Although identification of these traditions in terms of the number of studies conducted is of historical interest, it sheds little light on their substantive differences and similarities. In terms of assessing the potential for increased interaction among diffusion researchers and ultimately the emergence of a single tradition, it is more important to identify basic differences in approach, i.e., differences in conceptual structure and in the variables that are treated as independent or primary. If this approach is taken, the traditions identified can be reduced to three--the social-psychological, the economic, and the geographical--each characterized by a distinctive conceptual structure.[2] Each of these intellectual traditions offers a distinctive perspective on the diffusion process; however, they are not mutually exclusive; in fact they seem to supplement and complement one another. We shall survey the state of the art in terms of these three traditions, beginning with the geographer's.

1. *Spatial Diffusion: The Geographer's Tradition*

Geographers are concerned with spatial diffusion, specifically the relationship between innovativeness of adopters and their relative position in physical space. Hagerstrand (1968) demonstrated that the diffusion of innovations exhibits three states of growth in the spatial distribution of adopters. In the initial stage adopters are usually concentrated in a small cluster or a set of clusters. In the intermediate stage, expansion takes place in a pattern that indicated that a new adoption is more likely to occur *in the vicinity of existing adoptions than farther away*. This "neighborhood effect" creates an outward movement along a more or less sharply defined frontier, and at the same time the density of adoption behind the frontier continues to grow. A saturation stage may be reached in the central area of dispersal while the frontier is still growing.

If the number of individual adopters are plotted over time, an S-shaped curve, the logistic curve, normally appears. This curve shows a slow take-off stage of varying length, an intermediate stage of more rapid development, and a final stage of declining growth, which approaches a maximum value or ceiling asymptotically. The spatial and temporal aspects of this pattern seem to apply to various categories of innovation adopter units--individuals, villages, cities, and firms (Hagerstrand 1968, pp. 1974-1975).

Hagerstrand feels that personal communication between pairs of individuals and direct observation are the basic channels for the diffusion of innovation. The fact that the spread of innovation from certain centers tends to follow repeatedly the same spatial course implies a hierarchical communication and social influence network with a stable configuration over time (Hagerstrand 1968, p. 176). Thus the concerns of the geographer are linked in complementary fashion to those of the sociologist and social psychologist; in particular, spatial patterns imply the existence of *sectors* or *social systems*.

Hagerstrand has also pioneered in simulation studies of spatial diffusion of innovations (Hagerstrand 1965a).[3] In particular, he has utilized Monte Carlo techniques to simulate the spatial diffusion of farm innovations. In one instance he studied the spatial pattern in the adoption of subsidized pasture improvement innovations in Sweden over a 20-year period, relying on the probability assumptions of "neighborhood effect." Comparing his results with empirical data, he demonstrated that not only does spatial proximity increase the probability of adoption but that simulation techniques and model of S-curve growth are useful predictors of spatial diffusion of innovations (Hagerstrand 1965a). This type of simulation of agricultural innovations has since been repeated by Hanneman (1971) in Colombia with similar results.

Hagerstrand appreciates the limitations of his "neighborhood-effect" model and acknowledges that there are "receptivity factors" which affect the spatial pattern and rate of adoption of innovations. These factors include cost, returns, attitudes, predispositions, and value systems. Although these characteristics are not as amenable to simulation techniques as physical space, they are modifiers of the "neighborhood effect" and hence must be considered in any multivariate theory of the diffusion process (Hagerstrand 1965b, 1968). (Later we shall include these "receptivity factors" among the characteristics of adopters.)

Brown, an urban geographer and economist, is concerned that the Hagerstrand model of spatial diffusion deals only with adopters. Hagerstrand's information flow model, he argues, would suffice only for situations in which there were no active propagators (Brown 1968, 1969). There are, to be sure, communications networks through which information flows from a few early adopters to the larger group. But the basic tenet of Hagerstrand's conceptualization is that adoption is primarily the outcome of a learning process, which implies that only the factors relating to the effective flow of information need be considered.

Brown questions this assumption, arguing that market factors, especially the role of propagators who champion rapid and complete diffusion, must be considered as well as information-flow factors. Market factors would include both the distribution policies of propagators and the shopping behavior of potential adopters which determines the markets at which they trade. Brown and Cox (1971) analyze the differences between situations in which there is a propagator of the innovation with an interest in its rapid and complete diffusion and those where there is not such a person or entity. (We shall emphasize such differences in our later development of the dimensions of a general theory of diffusion.)

Brown distinguishes among macro-, meso-, and microscale diffusion within an urban area (Brown and Cox 1971; and Brown 1972). Macroscale diffusion takes place within an entire urban system, encompassing the processes of diffusion from the propagator of the innovation to intermediate diffusion agencies, including the establishment of the agencies themselves. Mesoscale diffusion takes place within a specified subarea of the whole urban system, encompassing diffusion from the agencies to the population at large. Microscale diffusion consists of diffusion among individuals within a small area or single community. Thus Brown employs the notion of social systems within social systems (which will turn out to be particularly important in our later integrative task). However, we would offer the caution that the boundaries of an "urban system" might prove too diffuse and poorly defined to support investigations of the influence of larger aggregations on adopter behavior.

Brown is particularly interested in identifying conditions that influence spatial aspects of diffusion at the macro- and mesolevels, and the patterns of diffusion generated by these conditions. He hopes that this research will provide a bridge between work in the genre of Myers and Marquis on propagator decisions (1969), and work in the genre of Rogers with Shoemaker on the adoption behavior of individuals (1971). This linkage would seem to be an especially important one for understanding the diffusion process.

To augment the geographer's spatial concern with the consideration of propagators we must turn to the economic tradition in diffusion research.

2. *Diffusion: The Economic Perspective*

Like the geographers, the economists have also brought their perspective to bear on quite diverse sectors. Griliches's work (1957) on hybrid seed corn, for instance, is recognized as a classic. Beginning with data by state and crop reporting districts within states, he fitted logistic growth functions to these data, reducing differences among areas to differences in estimates of three parameters of the logistic or S curve: origins, slopes, and ceilings. His implicit hypothesis was that profit maximization is directly related to the regional development of hybrid corn. Griliches's two major empirical findings were:

a. Differences in the long-run equilibrium use of hybrid corn (ceilings) and in the rates of approach to that equilibrium (slopes) are explained *in part* by differences in the profitability of the shift from open pollinated to hybrid varieties in different parts of the country.

b. The lag in the development of adaptable hybrids for particular areas and the lag in the entry of seed producers into these areas (differences in origins) are explained on the basis of varying profitability of entry ("profitability" being a function of market density) and the cost of developing and marketing the innovation.

Griliches's major conclusion was that the entire process of diffusion, the process of adapting and distributing a particular innovation to different markets, and the rate at which it is adopted was largely guided by expected payoff, "better" areas being entered first. His perspective, although involving adopter behavior, was basically concerned with the reasons for certain propagator decisions, that is, the decision of propagators to enter certain markets or areas.

Mansfield (1966) is also concerned with the relationship of "profitability" or "payoff" to diffusion, though his focus is on the industrial rather than the agricultural sector and on adopter rather than propagator behavior. His perspective was also somewhat more complex, for he viewed an innovation's rate of diffusion as determined, in large part, by four factors:

a. the extent of the economic advantage of the innovation over older methods or products;

b. the extent of the uncertainty associated with using the innnovation when it first appears;

c. the extent of the commitment required to try out the innovation;

d. the rate at which the initial uncertainty regarding the innovation can be reduced (p. 123).

It should be noted that each of these factors influences not only the behavior of potential adopters, but of propagators as well. For instance, the economic advantage offered by an innovation influences whether and how a propagator chooses to market it, as well as the decision of a potential adopter to invest in it. Since each of Mansfield's four factors can thus be viewed from two perspectives (propagator and adopter) they actually represent a total of eight process variables. In addition, the responses of the propagators and potential adopters to each factor influences the perceptions and subsequent actions of the other. For the sake of brevity we shall for the moment follow Mansfield's lead and restrict our attention to the influence of these factors on potential adopters, realizing however that *mutatis mutandis* they apply to propagators as well.

With regard to the first of these factors, which parallels Griliches's "profitability" thesis, Mansfield (1961) offered evidence that more profitable innovations are adopted more rapidly than less profitable ones. Nabseth and Ray (1974) also point to the significance of profitability in an innovation's rate of adoption, but rightly indicate that we know little of its *relative* significance.

> A general conclusions seems to be that profitability is an important factor in explaining the diffusion of new processes, not only in distinguishing between users and non-users, but also in explaining diffusion among firms and within them....But although profitability is a significant variable in most calculations, it is more difficult to say anything about its importance relative to other factors in explaining diffusions of new technology (p. 303).

This echoes our earlier remarks about the currently "static" state of the art which analyzes factors in isolation, and the need for investigation of their linkages and dynamic interaction.

Uncertainties other than economic may also be associated with the adoption of an innovation as indicated by Mansfield's second factor. If potential adopters are very uncertain of an innovation's performance, it tends to spread less rapidly than if they are relatively sure of its potential. As a part of such uncertainty, innovations that are relatively easy to understand seem to be accepted more rapidly than more complicated ones (Mansfield 1966, p. 123). This particular influence on adoption behavior is also emphasized by Rogers and Shoemaker (1971, p. 154) in their generalization that the complexity of an innovation, as perceived by members of a social system, is negatively related to its rate of adoption. It seems to be true in the agricultural sector as well, for Rogers with Shoemaker cite similar results in studies by Kivlin (1960) in the U.S., Singh (1966) in Canada, and Petrini (1966) in Sweden. Petrini found that complexity, along with relative advantage (which would incorporate Mansfield's and Griliches's factor of profitability) explained 71 per cent of the variance in the rate of adoption of innovations among Swedish farmers (Rogers with Shoemaker 1971, p. 154).

Besides complexity, Mansfield includes two other aspects in "extent of uncertainty." First, innovations that have easily observable results, and second, those that are more consistent with existing ideas and beliefs, seem to spread more rapidly than others. Rogers with Shoemaker also support these points, offering the generalizations that the observability and compatibility of an innovation, as perceived by members of a social system, are positively related to its rate of adoption (Rogers with Shoemaker 1971, p. 156). In its support they cite studies by Hruschka and Rheinwald (1965) showing that the more observable innovations demonstrated by German "pilot farmers" diffused more widely than the less visible innovation. Although Mansfield is primarily concerned with diffusion of industrial innovations and Rogers with Shoemaker with agricultural innovations, they are in point-to-point correspondence with reference to the relationship of the perceived attributes of an innovation to its rate of adoption.

Mansfield's third factor emphasizes the extent of commitment required to try out the innovation. He demonstrated (1961, p. 763) that the rate of diffusion of an industrial innovation is inversely related to the size of the investment required to put it into use. In other words, a relatively small initial investment for an innovation is positively related to its rate of adoption. Rogers with Shoemaker (1971, p. 352) make virtually this same point for agricultural innovation, though in somewhat different terms. They say that the "trialability" of an innovation (the degree to which it may be experimented with on a limited basis) is positively related to its rate of adoption. In support of this generalization, they cite studies by Fliegel and Kivlin (1966), Singh (1966), and Fliegel and Kivlin (1968). Again, in this respect, industrial diffusion does not seem to differ from diffusion in agriculture, which supports our point about the implicitly complementary results of the largely separate traditions.

The rate of reduction of the initial uncertainty regarding the innovation's performance--Mansfield's fourth factor--is largely conjecture. Difficulty in obtaining supporting evidence is due to differing intrinsic characteristics of innovations. The nature of some innovations is such that information regarding their performance can be obtained quickly; others require a long time. This is not a wholly independent variable, though; its impact is mitigated by the degree of sophistication and training of potential adopters.

Rogers with Shoemaker allude to the rate of reduction of initial uncertainty regarding the innovation's performance when they discuss what they term the "confirmation function." Citing supporting evidence by Mason (1964) and Francis and Rogers (1962), they indicate there is evidence that a decision to adopt or reject is not the terminal stage in the adoption process (Rogers with Shoemaker 1971, p. 113). Instead, in the confirmation period the adopter seeks reinforcement for the adoption decision he has made, but he may reverse it if he acquires negative or conflicting messages about the innovation.

Using his four factors, Mansfield (1961) constructed and tested a simple multiple correlation model in which the probability that a firm will adopt a new innovation increases with the number of firms in the sector already using it, and the profitability of doing so, but decreases with the size of the investment required. This models appears to be a useful forecasting device for certain aspects of interfirm diffusion of innovations.

In a recent international study of the diffusion of eight innovations, Nabseth and Ray (1974) summarize a wide range of considerations that influence interfirm diffusion and thus might lead to an S-shaped pattern of adoption over time within a sector.

> Considering first the opening phase, where we are looking at a process of adoption which is initially slow but begins to accelerate:
>
> (1) There are usually a few firms willing, and even anxious (having some special compelling reason) to be among the earliest to try out new techniques, which may initially involve considerable uncertainty and risk.
>
> (2) If a few pioneer firms overcome the teething troubles of a new technique, they substantially reduce the risk in the eyes of those who have yet to adopt it.
>
> (3) Good reports of a new technique from entrepreneurs already using it may carry considerably more weight with the large majority of firms than reports in the press or publicity by suppliers.
>
> (4) Modifications to the new technique in the early stages of commercial application may substantially increase the potential range of production to which it can be applied, as well as increasing the superiority of the new method over existing methods in the feasible range of application.
>
> (5) There may be a bunching of the new adoptions as part of a cyclical mechanism of the Schumpeterian type (Schumpeter 1939).
>
> (6) Other factors may also work in the same direction, for example, the age of existing technology to be replaced by the new one.

These factors are all consistent with an acceleration in the diffusion process following a slow beginning. After a time, however, other factors may slow down the process:

> (7) It may transpire that there are areas of production in which the new technique is, after all, not very suitable; probably the most promising areas will have been exploited first.
>
> (8) The very success of the new technique in its early stages may stimulate some firms to improve their existing methods of production for fear of being driven out of the industry, or for other reasons (Nabseth and Ray 1974, p. 9).

As with the other studies being considered in this section, Nabseth and Ray bring together considerations of various kinds with little attempt to categorize them by type. (We shall attempt to correct this methodological weakness at a later point.)

Mansfield (1963a) is also concerned with the rate of diffusion *within* a particular firm. He defines intrafirm diffusion as the rate at which a particular firm, once it has begun to use a new technique, proceeds to substitute it for older methods. Employing data on the substitution of diesel for steam locomotives, he examined the rate of diffusion in the various firms in the railroad industry. "Once they had begun to dieselize" was operationally defined as 10-percent achievement of dieselization, where substitution is considered fairly complete when 90-percent dieselization had taken place in a firm. Nine years was required, on the average, to increase a firm's stock of diesels from 10 to 90 per cent of its number of locomotives. This finding, pertaining to only one innovation, provides little more than a data point for understanding intrafirm diffusion in general. The relevant variables would seem to be multiple and complex. The rate of reduction of the initial uncertainty pertaining to an innovation's performance, for instance, may vary from one innovation to another and from one adopter to another, as may the condition of, and capital that is tied up in, existing technology.[4]

Mansfield's work on intrafirm diesel diffusion is similar in structure to the one he used (1961) to represent the rate of interfirm diffusion. This similarity suggests a certain degree of unity and similarity between the two diffusion processes (interfirm and intrafirm), at least with respect to the variables under consideration. Further efforts should be made, using data for other innovations to test this sort of econometric model of the rate of diffusion of an innovation. However, something will have to be done about operationalizing the concept of "rate of reduction of the initial uncertainty," which will involve achieving greater clarity about the characteristics of innovations, adopters, and their mutual influence.

Mansfield's most comprehensive work on the diffusion of industrial innovations is *Industrial Research and Technological Innovation: An Econometric Analysis* (1968a). In it he investigated four large national industries: iron and steel, petroleum, coal, and railroads. Secondary data were utilized to analyze the adoption of 150 innovations by firms in these industries from 1919 to 1958, and a hypothesized regression model was tested for fit with the actual data.

A number of significant findings emerged from this massive undertaking. First, the length of time a firm waits before using a new technique tends to be inversely related to its size and the profitability of its investment in the innovation (p. 205). Second, 20 years or more are often required for all the major firms in an industry to adopt an innovation. Nabseth and Ray report similar results (1974). Third, the number of firms adopting is positively related to the rate of adoption, which is in keeping with the previously described S-curve of growth prior to the saturation stage. Finally, "the personality attributes, interest, training, and other characteristics of top and middle management may play a very important role in determining how quickly a firm introduces an innovation" (p. 205). However, Mansfield's study revealed no close relationship between a firm's rate of growth and the rate at which it adopts an innovation (p. 203).

From his earlier studies, Mansfield (1963b) expected to find that larger firms would introduce innovations more quickly than smaller firms because the larger have greater financial resources and more extensive engineering and technical resources. They can pioneer more cheaply and with less risk. This finding was substantiated in Mansfield's 1968 work and in Mansfield et al. (1971). Interestingly enough, firms with younger top management did not adopt an innovation sooner than firms with older top management, which is consistent with Rogers's analysis of farm innovations (Rogers with Shoemaker 1971, p. 185). However, Mansfield and Rogers with Shoemaker (pp. 354-356) find that earlier adopters are more highly educated than later adopters.

In one of his most recent collaborative works, Mansfield et al. (1971) looked at intrafirm research and innovation, in order to enhance understanding of the relationship of technological change to economic growth. Dealing with the diffusion of numerical control in the tool-and-die industry, the authors, using secondary data, found that:

  a. The diffusion process is slowed by lack of knowledge and resistance to change.
  b. The primary reason given for nonuse was that the innovation would be unprofitable.
  c. Both the larger firms and firms with more highly educated owners tend to be early users.

These findings contain substantial noneconomic dimensions and thus are interesting not only in themselves, but also for the evidence they afford of the efforts of those in one diffusion research tradition, the economic, to bridge the gap to another, the social-psychological.

Similar linkages between economic and social psychological variables were suggested by Nabseth and Ray (1974), particularly as regards profitability calculations and management attitudes.

> One general conclusion seems to be that calculating the profitability of a new process is more difficult than is usually acknowledged in studies on the subject. For some processes, for instance numerically controlled machines and continuous casting, profitability turned out very difficult to calculate *ex post*,

and even more difficult *ex ante*. This does not mean that firms do not try to estimate the relative advantage of a new process, but rather that their calculations are very subjective. Of course, new investments always involve uncertainties, but they are probably greater in the introduction of new technologies than, for instance, in straightforward replacement. It follows that profitability calculations for new processes are very much linked with management attitudes, especially when experience of the technology is scarce and perhaps contradictory....The very presentation of a profitability calculation therefore involves subjective elements, with some managements stressing the risks and uncertainties, others the benefits that can be gained (p. 302).

Thus they take the determination of utility, not as a purely economic calculation but as containing social-psychological dimensions as well. Suggestions of the possibility of such linkages, whether arising from the recognition of unexplained "residues" or other sources, represent a positive signal that integration of the several research traditions can be achieved.

3. *The Social-psychological Tradition in Diffusion Research*

As noted above, Mansfield found that the process of diffusion was slowed by resistance to change. Barnett, an anthropologist, has also dealt with this topic in a highly speculative social-psychological work (1953). A new idea, Barnett argued, must be compatible with the norms of the social system, and the adopter must perceive relative personal advantage before he adopts. But as Rogers with Shoemaker point out, changes occur in the meaning of an innovation and the use to which it is put even as diffusion proceeeds (Rogers with Shoemaker 1971, p. 169). If the adopter cannot satisfy himself on the score of compatibility and relative advantage, he resists change. There is, then, the problem of "merchandizing" the innovation in such a way that the potential adopter can maximize his future expectations.

Mansfield et al. (1971) established quantitatively that resistance to change (rejection of numerical control in the tool-and-die industry) within the firm (which is one kind of social system) did exist, but could only *partially* explain the resistance on the basis of profit maximization. Rogers with Shoemaker, having offered the generalization that the relative advantage of a new idea as perceived by members of a social system is positively related to its rate of adoption, supported this claim with data from 29 studies (1971, pp. 142, 350-351).

What is needed is additional research into the nature of "relative advantage" and "resistance to change"--beyond the economic aspect--in order to determine the relative influence of social, cultural, and psychological components.

Two separate studies heuristically illuminate social, cultural, and perhaps psychological components of resistance to change in the diffusion of technological innovation. Spicer (1953) studied fifteen crosscultural cases of successful and unsuccessful attempts to introduce new ideas and methods in agriculture, industry, and medicine. He concludes that the change agent, coming from another culture, must be careful to operate through existing channels of communication and cooperation and must understand the relationship of his position as an outsider to the members of a given social system. He must attempt to map the linkage of customs and to make sure that his product will be perceived as compatible. Above all, he must avoid ethnocentrism--making judgments in terms of standards, norms, and values of his own culture.

Bright (1964), studying resistance to innovation as a problem of management, came up with findings similar to Spicer's. To Bright, resistance is associated with the degree to which institutions and individuals feel themselves threatened by change. Where only slight changes in work behavior and habit are required, there is less resistance. Resistance is heightened, however, when the innovator is contemptuous of other members in the social system and of existing work routines. It should be noted that whereas Spicer and Bright are concerned with the *adopter's* resistance to change, they deal with it in terms of its implications for *propagator* strategy and behavior. The interactions of these roles is an important topic for future research.

In this connection (and in anticipation of a later point), it should be noted that the *dogmatism scale* developed by Rokeach (1960) is widely accepted as a reliable indicator of the extent to which an individual has a tolerance for new and unfamiliar situations. Not to be confounded with *attitudes*, which have a specific object or referent, dogmatism refers to an individual *trait* or characteristic way of behaving when confronted with the new and unfamiliar. Those who are "high dogmatic" have more difficulty (i.e., feel more anxiety, engage in more avoidance behavior) adjusting to a new or unfamiliar situation than those who are "low dogmatic." Dogmatism seems unrelated to either education or intelligence, and individuals differing markedly in their dogmatism levels may perform equally well in situations that are not new and unfamiliar.

The relation of this trait to the "resistance to change" phenomenon in the diffusion process is that of an intervening variable. That which may be, or has recently been, adopted is new *in that context,* so that its impact on the *status quo* is uncertain. Those who are high dogmatic will tend to be more uncomfortable, anxious, and threatened by the prospect and thus more resistant to the change. Although not a diffusion-research-specific instrument, the dogmatism scale has significant potential for contributing understanding in this area.

The second finding of Mansfield et al. (1971) was that "the primary reason given for nonuse was that the innovations would be unprofitable." Mansfield has also used the term "relative advantage" (in its economic sense) in this connection. Psychologically, relative advantage is a function of selective perception. It has been established by Bruner and others that perception is a highly subjective phenomenon (Bruner and Postman 1947). Each individual perceives a new idea, process, or product in terms of his own past experience and technical competence, his current needs, and his future expectations. Because this perception differs from one individual to another, both the propagator and potential adopter of an innovation subject purely economic and other intrinsic aspects to selective perception, thereby affecting the rate of adoption. When Mansfield et al. (1971) said that the primary reason given for nonuse of an innovation was that it would be unprofitable, they were saying in effect that there were other reasons (of selective relative advantage or disadvantage) left unstated. That relative advantage is a major reason given for adoption of innovation, but that it means different things to different groups, is supported by Rogers with Shoemaker in their analysis of the perceived attributes of 286 innovations and their rate of adoption (Rogers with Shoemaker 1971, pp. 140-141).

Nelson and Phelps (1966), as economists, are interested in the relationship of "human capital" (in the form of advanced education) to technological diffusion and economic growth. Attempting to explain why earlier adopters of innovations, in both industry and agriculture, have more education, they suggest that in a technologically progressive or dynamic economy, management is a function requiring adaptation to change and that the more educated a manager is, the quicker he is to introduce new techniques of production. Since educated people make good innovators, they argue, education speeds up the process of technological diffusion. By way of example, they contend that the remarkable innovative thrust of the American agricultural sector is due to the greater education of today's farmer, which has increased his ability to understand and evaluate the information on new products and processes disseminated by the Department of Agriculture, the county agent, farm journals, radio, and seed and equipment companies. Rogers found that the majority of studies support the view that those who know of an innovation early have more education and more exposure to mass media channels of communication than those who learn of it later (Rogers 1962).

Nelson and Phelps view education as an investment in people, i.e., that educated people are bearers of human capital, and account for technological progressiveness of society. Hayami and Ruttan (1971) tend to support the analysis of Nelson and Phelps. They assume that significant productivity growth cannot be brought about by reallocation of resources in traditional agricultural systems, but becomes available by diffusion of technological changes. They distinguish three types of technological diffusion: material, design, and (what is especially germane here) capacity (Hayami and Ruttan 1971). Investments in research and education (human capital) provide a basis for diffusion, technical change, and productivity growth in agriculture. After studying the history of agricultural innovation and diffusion in the United States and Japan, and then examining developing countries currently experiencing enhanced agricultural production, Hayami and

Ruttan conclude that one of the major factors in a country's capacity to adopt agricultural innovations from elsewhere, as well as to diffuse indigenous technology, is investment in public education. A literate population is an endowment, increased by public education.

Hayami and Ruttan also indicate how diffusion of technological innovations can overcome deficiencies in material endowments. In Japan, the scarce factor in agricultural output was land; in the United States it was labor. Several innovations made up for these these endowment scarcities. Mechanical technology, by increasing the power available per worker, facilitated increases in the land area that could be worked by a labor force of a given size in the U.S. New biological and chemical technology increased the efficiency of the process of solar energy conversion by plant life and was, in effect, a substitute for additional land in Japan. These widely diffused technological innovations relieved the constraint on production imposed by limited land area in Japan and that imposed by an expensive labor force in the United States.

Underlying not only the social-psychological tradition of diffusion research, but implicitly at least, the other traditions as well, is an agreement on the importance of social networks in the diffusion of innovations. An important early study by Coleman, Katz, and Menzel (1957) may be taken as representative of the best research evidence on this topic. They focused on the ongoing social processes that finally led to the widespread adoption of a new drug (euphemistically called "gammamyn") by physicians in four cities. They were concerned with the effectiveness of the networks of interpersonal relations at each stage of the diffusion process. Structured interviews were conducted with the physicians to determine the sociometric choice of colleagues and the patterned network of their interpersonal relations. They then undertook a systematic search of the prescription records of local pharmacies to determine the month in which each physician first used the drug in the 15-month period following its release.

The physicians were divided into two major groups. Doctors who were more profession-oriented (i.e., were more in communication with their colleagues) adopted the drug earlier than those who were patient-oriented (i.e., more isolated from their colleagues). Adoption of the drug by the profession-oriented physicians followed the logistic or S-shaped curve, since the physicians were themselves propagators, whereas that of the more isolated or patient-oriented doctors, who were not propagators, was exponential. Coleman et al. (1957) also found that the rate of adoption by the patient-oriented doctors differed between those practicing alone and those in partnerships, with the latter adopting earlier. However, both groups of patient-oriented phyisicans followed the exponential curve, thus lending support to Hagerstrand's assumption that the S-shaped curve implies a stable communications and influence network.

This study by Coleman et al. (1957) is a fine example of the importance of social networks in the diffusion process as well as an indication of the usefulness of mathematical formulations. (We shall treat the concept of social networks in more detail in a later section.)

This concludes our brief overview of the three major traditions of diffusion research. To offer a preliminary assessment to be explicated in subsequent sections, the fragmentation and conceptual jumble of the field, as revealed in these pages, constitutes a major barrier to its further development. The most pressing need is the development of a unified conceptual structure, in terms of which the several traditions and their implicitly complementary specialized concerns can be integrated, and in terms of which a cumulative base of research results can begin to develop. There exists one outstanding example of just such an effort to bring the field together--the "classical diffusion model." This model deserves close examination.

III. THE CLASSICAL DIFFUSION MODEL

The so-called classical diffusion model represents the attempt to synthesize some 2400 diffusion research publications carried out under the direction of Everett Rogers at the Diffusion Documents Center at Michigan State University, and can be found in Rogers with Shoemaker (1971).[5] Four elements, which are taken to be central to the study of the diffusion process, form the basis for the model: "(1) *an innovation* (2) *communicated* via certain *channels*, (3) to members of a *social system*, (4) who adopt it over a period of *time*" (see Chap. 12 below).

We can question the adequacy with which these elements are conceptually structured and their importance in understanding diffusion. Rogers himself levels criticisms against his own, earlier model in Chapter 12 below.

The first element identified in the classical model as central to the conduct of diffusion research is an *"innovation,"* which is taken to include ideas, practices, or objects "perceived as new by the relevant unit of adoption."[6] We agree that the innovation itself is central to an understanding of the diffusion process; yet the point here is not the truism that without an innovation there would be nothing to adopt, and thus no diffusion. Rather, as we have seen in the last section, adoption behavior is influenced in significant ways by the particular characteristics of an innovation. Its "trialability" is a significant influence in the adoption decision, as is its complexity, its relative advantage, its associated uncertainty, etc.

Yet the conceptual structure erected by the diffusion researchers to explain the influence of an innovation's characteristics on adoption behavior is inadequately refined. We would urge a distinction between the characteristics whose influence is independent of the nature and circumstance of the potential adopter and those whose influence depends on certain adopter characteristics. The former we call "adopter-independent" characteristics, and we present the argument in behalf of this distinction in the following section. We shall also re-examine the characteristics identified in the research results summarized above and suggest two additional characteristics that previous studies have tended to neglect. With regard to the first element of the classical model, we concur its centrality, but believe that it requires further refinement before it can support needed research.

The second element of the classical model concerns the *communications* channels through which information about an innovation is diffused. In emphasizing the importance of this point, Rogers with Eveland suggest that diffusion should be considered "...a subset of communications research that is concerned with new ideas" (see Chap. 12 below). Whether or not communication takes place, and if so, its effect on the potential adopter depends in large measure on the social relationship between the source and the receiver of the information. Further, the communication channel employed (whether impersonal mass media channels or personal communications) is a function of the purpose of the communication (information dissemination or persuasion), the size of the audience, and the relationships among its members.

Although we agree with this emphasis on the flow of information and influence, we do not feel that the conceptual structure supporting this element of the classical model is sufficiently strong to permit an integration of the several research traditions, and thus the emergency of a cumulative research mode. What is required for a more adequate conceptual structure is the close coupling of the channels of communication and influence with the various social systems involved in the diffusion process. The channels cannot be adequately understood apart from internal and interactive dynamics of such systems. Rogers does not deny such coupling, but his conceptual structure does not deal with them adequately. Where a potential adopter is an organization, for instance, its *internal* social system may be quite complex, reflecting the interaction of the formal organizational structure, the informal hierarchy, and mutual-choice social groupings. In addition, such an organizational adopter is a part of a larger social system composed of similar adopters and other related units. The patterns of communication and influence and the resulting adoption behavior reflect these levels of complexity as well, and cannot be understood apart from them. In the following section we offer an alternative structure for this element.

The third element identified by the classical model as central to diffusion research is the *social system*. This is perhaps the most basic concept in diffusion research, and as such bears a heavy load of multidimensional complexity. Rogers with Eveland discuss the concept in the following terms:

> A social system is defined as a collectivity of individuals, or units, who are functionally differentiated and engaged in collective problem-solving with respect to a common goal. The members or units of a social system may be individuals, informal groups, complex organizations, or sub-systems....The defining feature of such a

system is the interaction between its component elements; it is not necessary for such interactions to be consciously purposive for a system to exist, although the existence of common objectives simplifies the detection of interactions for the analyst. It should be noted that the term "structure" has multiple acceptable meanings; the communication model discussed here emphasized *communication structure*, which is not necessarily equivalent to either the role structure or the formal authority structure, although it may correlate highly with them. The terminology of "open systems theory" with its emphasis on the permeability of system boundaries, allows the development of very flexible and useful analytical models. (Rogers with Eveland, Chap. 12 below.)

We call attention to the various dimensions of complexity inherent in this "open-systems theory," for it will prove useful to have them disengaged for later detailed analysis. In particular, the following characteristics of a social system should be noted:

a. A social system may be of various sizes, presumably with a lower bound of two individuals and an indefinite upper boundary.

b. The units comprising a social system may themselves be social systems as well as individual persons.

c. As a result, social systems may exist at various levels of aggregation and thus overlap or be imbedded within one another, i.e., social systems within social systems.

d. A social system may embrace different types of "structure" singly or in combination; i.e., formal authority structure, and/or role structure, and/or communications structure.

3. As a result of its structure(s) a social system may be, and usually is, hierarchically ordered.

The concept of a "social system" is thus much better developed than the two elements of the classical model discussed earlier. We would suggest that the integration of the several research traditions might be well served by distinctions among the relevant social systems in terms of levels of aggregation and their roles in the diffusion process. Social systems may (1) function as adopters, (2) be composed of adopter units, (3) be part of a larger whole which is an adopter, or (4) function as propagators. The importance of these several types of social systems lies in their influence upon *adopter behavior*. This being the case, it is important to distinguish them, as we attempt to do in the following section.

The final element identified by the classical model is *time*. Rogers offers three reasons for this emphasis on the temporal dimension: (1) the adoption decision is a process not an act; (2) the point in time at which a unit adopts, relative to others, provides a measure of the *innovativeness* of that unit; and (3) the rate at which an innovation is adopted is a function of the number of adopters in a given period of time. We agree with the first of these points, of course, but view it as a given or presupposition, not as requiring a major role in the explanatory structure. The second "reason" in fact defines a *measure* of innovativeness; the third is a definition of a *rate*. Thus they are methodological rather than substantive considerations. An adequate general theory or model of the diffusion process requires the provision of an explanatory structure adequate to *account for* "time to adoption" and "rate of adoption" differences, i.e., they are *dependent* rather than independent variables. Hence we question the need for the emphasis time receives in the classical model.

In Chap. 12 below, Rogers with Eveland offer their own critique of the classical model, though in terms of certain of its assumptions rather than a point-by-point analysis. One criticism deserves particular mention. They note that of the more than 2400 diffusion research publications, only some 373 are concerned with organizational adopters. Further, most of the subset dealing with organizations treats them *as if* they were individuals; i.e., the influence of processes *within* the organization tends to be ignored. Thus they argue that the classical model is largely predicated on the assumption that adoption decisions are made by individuals rather than organizations. We agree with Rogers's own critique of his earlier model and feel that the correction of this bias can be made by the development of appropriate distinctions regarding the role of social systems.

In concluding our examination and assessment of the so-called classical diffusion model, we must reiterate that this model was introduced as an attempt to integrate several separate research traditions. Although we view it as a significant step in that direction, for the reasons indicated, we do not believe that it constitutes an adequate conceptual framework for unifying the field. In the following section we attempt to take the next step in the direction of an adequate general theory, though we do not offer such a theory. Rather, relying on the findings and special concerns of the various research traditions and the partial synthesis effected by the classical model, we indicate the *dimensions* of an adequate general theory. Although this is a more limited and modest task than developing such a theory, it is one that is clearly in line with our assessment focus and a necessary preliminary in any case. After all, surveyors are needed to plot the terrain before the roadbuilders and architects can erect the structures.

IV. THE DIMENSIONS OF A GENERAL THEORY OF DIFFUSION

By their emphases and preoccupations, researchers in the several traditions have identified four major sets of influences operative in the diffusion process. The classical model, although structured rather differently, also reflects a concern with these same influence sets. Thus the "dimensions of a general theory" to be identified in this section should be viewed as emerging from, rather than imposed on, the literature base. As with any new conceptual structure, however, differences of perspective result in somewhat different questions being raised. The long-term value of the structure we offer lies in its explanatory and heuristic power concerning the linkages, interdependence and relative influence of its elements. The questions raised cannot be answered *a priori* in this study, but constitute a fecund "research agenda" for the future.

What we take to be the four dimensions of an adequate general theory of diffusion are identified in Fig. 1 as impinging on the decision process by which an innovation comes to be adopted. Our first dimension involves the characteristics of "sectors." Sectors are social systems, the constitutents of which are adopter units (individuals or organizations) plus certain other units with whom the adopters regularly interact. (In a sense the sector is, for the adopter, the counterpart of the exogenous system with which the entire innovation process interacts.) The second dimension of influence on adopter behavior involves the *characteristics of adopters* themselves. The third concerns the *characteristics of innovations*, and with some refinement parallels the corresponding element in the classical model. The final dimension concerns the various *propagation mechanisms* that are often active in the diffusion process. We shall discuss each of these dimension in turn, noting not only the support they derive from the diffusion literature, but also certain parallels with the process of R & D discussed earlier.

A. SECTOR CHARACTERISTICS

Innovations are adopted by particular individuals or organizations. However, the behavior of such adopter units cannot be understood in isolation from the larger context of which they are a part. Though their descriptions and emphases vary, diffusion researchers have long recognized the influence of such larger systems on adopter behavior. Rogers with Eveland have used the term "social systems" (see Chap. 12 below) to refer to such aggregations, in order to emphasize their role as social influences and communications mechanisms. In light of our concern with the process of technological innovation and thus with industrial firms, it is tempting to refer to these larger aggregates as "industries." Adopter units are not always firms, however; moreover, there is good reason to consider these larger aggregates of influence as also including units other than potential adopters. Therefore we have chosen the more neutral term, "sectors."

In previous chapters the term "sector" has been used to refer to a functionally differentiated but ecologically interactive whole, composed of the firms in a particular industry and various related organizations that provide a service to, regulate, or promote the activities of that industry: university research, governmental agencies (funding and regulatory), professional and trade associations, etc. On this structural view alone (and considering only the cases in which the adopter units are firms), sectors

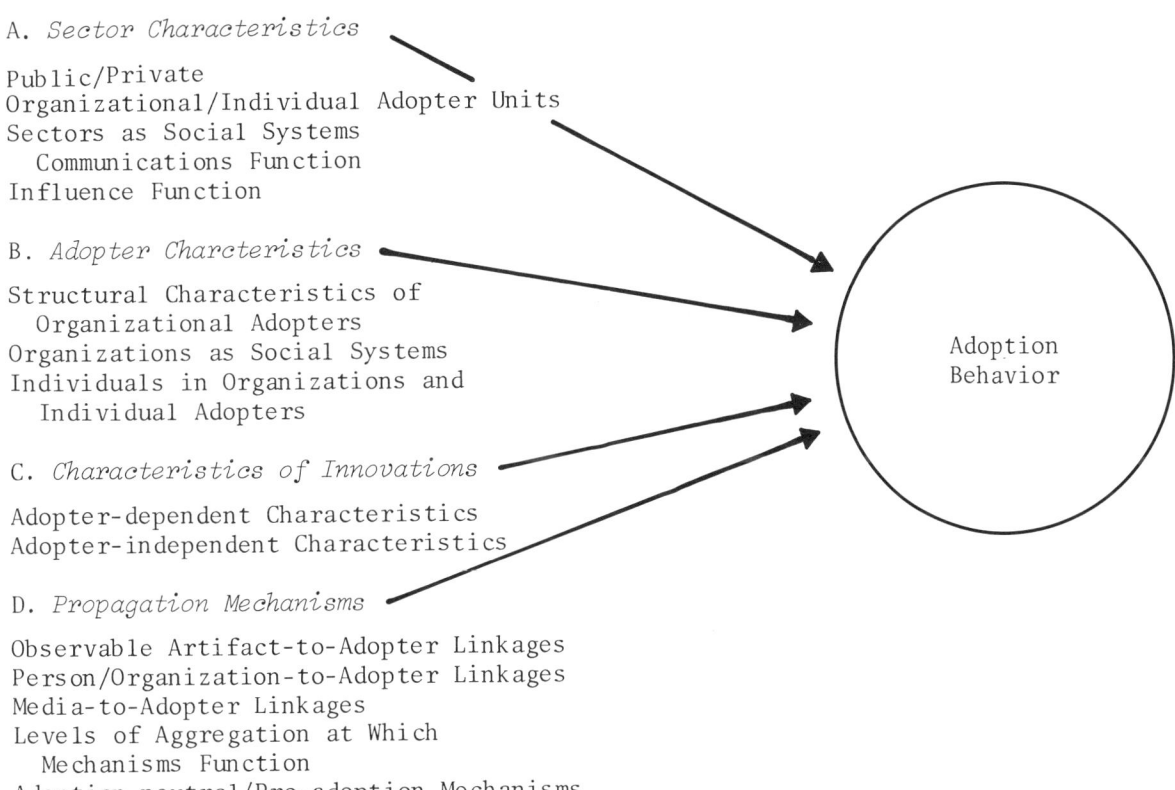

FIG. 1.—Dimensions of a general theory of diffusion.

vary widely not only along the dimensions of number and size of potential adopter units, concentration of resources, high versus low technology, and highly competitive versus stable distribution of market shares, but also in the type and extent of participation by the related service, regulatory, and promotional organizations.

However, this list hardly exhausts the complexity that must be embraced by the concept of a sectors, or the litany of their potentially relevant characteristics. In addition to sectors in which the adopter units are firms, there are also sectors in the public domain, such as those composed of state, county, or municipal governments, each with its own set of related organizations. There are also sectors such as agriculture, in which the adopter units are largely individual farmers rather than organizations. And finally, by far the largest class of adopter can only very loosely be identified with a sector at all, i.e., the consuming public. Recently there are signs that nascent consumer groups are beginning to constitute organized communities of influence and communication akin to the sectors described above. Consumers also participate in, and are influenced in their adoption behavior by, larger and better organized units of aggregation. The complexity of this situation, though, may be such as to preclude its coverage by a general theory of diffusion, at least until the state of the art is much more advanced.

In addition to their inherent structural features, sectors must also be viewed as *communications networks* and as *mechanisms of social interaction and influence*. As such they are a primary means by which potential adopters learn of an innovation, its success in contexts similar to their own, and the actions that have been taken concerning it by significant "others" (i.e., those viewed by the potential adopter as pace setters, major competitors, opinion leaders, etc.). Diffusion researchers have heretofore tended to stress these communications and social influence *functions* of a potential adopter's sector more than the sector's structural characteristics.

This emphasis is to be expected if a researcher takes the sector's structure as his frame of reference, rather than as a process variable. He can thus assume (1) the similarities of activities in which the potential adopters in that sector are engaged; (2) the similarity of problems and opportunities with which they are confronted; and (3) the communications and influence patterns which in fact characterize the sector. In short, if the structure of a particular sector is taken as the frame of reference--as a *given* rather than a possible process variable--then the way in which that sector *functions* will, of course, receive the lion's share of the researcher's attention. This seems to be the state of affairs in most diffusion research, with the work of Mansfield (1968a) and Nabseth and Ray (1974) being among the exceptions.

Our ecological approach, relating structure to function, causes us to raise a number of questions concerning the influence of sector characteristics on adoption behavior:

1. In what ways is the influence of sectors in which the adopter units are individuals (e.g., farmers) similar to that of sectors composed of organizations? In what ways is such sectoral influence different?

2. Is the influence of sectors composed of private organizational units (i.e., firms) different from sectors composed of public organizations (e.g., state governments)? How is it different? Similar?

3. Is the influence of a sector in which there are strong indigenous propagator units (e.g., county agents in the agricultural sector) different from sectors in which the major propagator mechanisms are external? What are the differences and how significant are they?

4. What difference does it make if the exogenous propagator mechanisms are very weak to nonexistent (as seems to be the case for sectors in the public domain), as opposed to sectors where there are active exogenous propagators?

5. Is the influence of a sector on adopter behavior different where there is strong Federal governmental input in the form of funding and/or regulation?

6. Is the strength of a sector's trade and/or professional associations a significant factor in its influence on adopter behavior?

7. What are the types of formal communication channels within sectors, and what is their relative influence upon the diffusion process?

8. What are the effects of the following sector variables on its influence on the adopter behavior of its components:

   a. number of adopter units?
   b. concentration?
   c. high or low technology oriented?
   d. spatially compact or diffuse?
   e. R & D intensive or not?
   f. capital vs consumer goods orientation?
   g. mobility of skilled individuals among sector units?
   h. labor intensive vs capital intensive?
   i. field of technology?

These questions concerning the effect of a sector's structural characteristics on its influence on adopter behavior cannot be adequately answered by the sector-specific studies that currently dominate the diffusion-research literature. Cross-sectoral studies might not show all the above questions to be significant; some would probably turn out to be significant, others would not. In addition, research on the effect of the structural characteristics of sectors on their adopter-behavior influence would doubtless reveal significant variables that have been omitted from the above list.

The basic point here is that in the absence of more cross-sectoral studies only very limited progress can be made in understanding the influence of sectors on the diffusion of innovations within them. Until it is known which sector characteristics influence adoption behavior and which do not, the generalizability of sector-specific results cannot be determined. Generalizability of results is, of course, the key to an integrated and cumulative database, which in turn, permits the rapid development of a field.

One further point should be noted about the above list of questions. In addition to their focus on sector characteristics, several of the questions also concern the linkage of sector influence to other influences in the process, e.g., adopter and propagator characteristics (questions 3, 4, and 5). The diffusion process results from the dynamic interaction of these dimensions, so it is the *relative* influence of each, in concert with the others, that must be investigated.

As we have seen, there is impressive evidence that sectors function as social systems.

In such studies as Coleman, Katz, and Menzel (1957) on the diffusion of a new drug (gammamyn) among physicians, and in Hagerstrand (1968) and Mansfield (1968b), the rate of adoption was found to be influenced by the number of adopters at a point in time. This transient sector characteristic was taken as reflecting the existence and influence of a stable and enduring social system among the adopters. But as was demonstrated very clearly in the gammamyn study, the influence of adoption trends within the sector was greater for the well-integrated active participant than for his more isolated counterpart. The influence of the social system is thus not an independent variable, but depends to some extent on the characteristics of the adopter. For the individual adopter (physicians, farmers, etc.), certain personality traits, educational or cultural background, etc., may strongly influence his participation or nonparticipation in the sector's social system. For the organizational adopter, management strategies, attitude towards risk taking, the significance attached to prestige, etc., may be decisive. The characteristics of particular innovations themselves and the activity of propagators also help determine the influence of the number of adopters at a point in time.

The sector's social influence extends byond the number of adopters at a point in time. Particularly in the early phases of diffusion, *who* the adopters are may be more important than their number. One would expect some members of a social system to be more influential than others, and thus their adoption to carry more weight with the group than that of another, less influential, member. Walker (1969), for instance, reports evidence that if an innovation was first adopted by other than an opinion leader, it spread slowly if at all.

But there is troublesome complexity here also, for opinion leadership may result from a variety of considerations, some of which may have little to do with innovativeness. For instance, a firm may enjoy considerable status within a sector by virtue of the market share it controls, the high quality of its current products, or simply because it was an early pioneer in the field. However, these same factors may cause the firm to feel its status threatened by an innovation and thus to resist it strongly.

The present state of the art does not permit a clearcut evaluation of the role of opinion leaders in fostering or hampering diffusion. A distinction made by Becker (1970) illustrates the complexities involved. He found that for "low-risk" health innovations (such as measles immunizations, viewed as amenable to easy acceptance), the opinion leaders among the 95 health departments studied were the earliest adopters. For more risky innovations (e.g., diabetes screening), however, the earliest adopters were not the recognized opinion leaders but rather those well down the prestige hierarchy. Rogers with Shoemaker (1971) report comparable findings among individual adopters. This same phenomenon was also noted in Chap. 3 with regard to the initial development and marketing of high-risk innovations, with the industry leaders often seeming content to be quick imitators rather than innovators.

A pattern found throughout the diffusion-research literature emerges from these considerations. It is what might be called the "other-things-being-equal" syndrome. Other things being equal, the number of adopters at a point in time is of substantial influence on the behavior of the remaining potential adopters; or, other things being equal, early adoption by an opinion leader is of substantial influence in future adoption behavior within the sector.

Rogers with Shoemaker (1971) have reference to this characteristic of diffusion research when they say, "Our generalizations deal almost entirely with pairs of concepts, whereas the real nature of diffusion is certainly a cobweb of interrelationships among numerous variables" (p. 93). In fact, they report that 95% of the empirical generalizations in the diffusion research literature through 1968 were of the bivariate type. A particularly good example is the 275 studies supporting, and 127 not supporting, the generalization that "Early adopters have higher social status than later adopters" (p. 375). The Becker (1970) study reported above seems to be in the "not supporting" category, but the distinction it makes between characteristics of innovations and the influence of these characteristics on the adoption behavior of opinion leaders is an important one. The "other-things-being-equal" approach, by studying the effects of only one independent variable at a time on adoption behavior, results in only a static listing of influences whose relative weight is unknown.

This approach also characterizes the results reported earlier by Rogers with Shoemaker

(1971) and Mansfield (1963a, 1968b, 1971) that the larger units within a sector tend to be earlier adopters than smaller ones. That this is the case for the agricultural and industrial sector examined is an important result, but of even more importance would be a determination of the factors that yield this result. Many of them would presumably be concomitants of size--greater capital resources, greater capacity to absorb risk, more technically able personnel, etc. But in certain cases it may be a characteristic of the innovation rather than the adopter that favors the large firm over the small. For example, Nabseth and Ray (1974) report the following concerning the adoption of the float glass process:

> The huge output of even the smallest float plant in comparison with the market size is a highly important factor. One float line alone, if utilized at a rate anywhere near normal capacity, makes more glass than the whole consumption of a country the size of Austria. This is important in relation to the behavior of firms in smaller markets, or indeed smaller independent companies (p. 211).

That is not to argue that bivariate results are totally without value, but their explanatory power is not very great because other influences are operative.

As has been noted, a sector's social system is a mechanism for the flow of information as well as influence. Rogers with Shoemaker (1971) offer a number of bivariate generalizations about the centrality of early adopters in a sector's communications network. They tend to be more cosmopolitan, have more contact with agents of change, have greater exposure to the mass media and interpersonal channels of communication, seek more information, and hence have greater knowledge of innovations than later adopters (pp. 367-374). We would strongly caution against drawing simplistic, cause-effect conclusions from such data, for we suspect that early adoption rests upon a more complex set of conditions than simply being a sector's gatekeeper and outstanding communicator. The early possession of information may well be a necessary but not a sufficient condition of early adoption.

Becker (1970) in fact challenges the assumption that a unit's centrality in the sector's information flow network is the *cause* of early adoption. In fact, he urges just the opposite, i.e., that early adoption is more likely to be the *cause* for an adopter's early possession of information than its effect. "A desire to maintain or increase prestige (tempered by the risks of adoption) motivates the profession to seek...innovations" (Becker 1970). On the other hand, Rogers with Eveland (Chap. 12 below) points out that Becker's data may be inadequate to permit the determination of causal direction in this matter. Nevertheless, Becker has rightly questioned the widespread assumption (also inadequately supported) the the causal arrow runs in the opposite direction.

Thus opinion leaders have been identified, communications and information-flow networks mapped, and early to late adopters categorized in many diffusion studies; yet the relationships among these sector characteristics are far from clear. It seems that those within the sector who adopted an innovation at an early point in time constitute an important influence on the behavior of the rest of the sector and thus upon an innovation's rate of diffusion. But why these things are so--why certain units adopt early, why the number of prior adoptions is important, and why the influence of adopters varies--are complex matters that await full explanation.

Beyond these few generalizations there is little empirical evidence regarding the influence of sectors on adopter behavior. This lack reflects the point made earlier, that the sector is most often treated as a fixed site in which studies are made, and not as a variable. Thus sector characteristics have been little examined for their influence on the process. The major recommendation we would offer is that sector-specific research be balanced by more cross-sectoral studies, and in particular that the kinds of comparative questions raised earlier be given high research priority. Until this deficiency is overcome, the category of what-we-don't-but-should-know about the diffusion process will remain embarrassingly large.

B. ADOPTER CHARACTERISTICS

Sectoral influences on adopter behavior, as discussed above, depend on particular characteristics of adopters, and vice versa. Such interdependencies cannot always be determined *a priori*, so that diffusion research must be an iterative process. The more we know about the characteristics of a sector to which an adopter unit belongs, the better we are able to understand that unit, its characteristics, and behavior. Conversely, the more we know about the characteristics of a particular unit, the better we can understand that unit's behavior as a component of the sector.

We shall deal first with organizational adopters, and their characteristics as formal structures, and as social systems; then we shall look at the individual, as a participant in an organizational adopter unit and as an adopter unit in his or her own right.

1. *Structural Characteristics of Organizational Adopters*

Rogers with Eveland speak of the need to merge the diffusion research tradition with organizational behavior research (see Chap. 12 below). This study affords a unique opportunity to sketch some of the questions and hypotheses that might result from such a merger, since a large portion of Chap. 3 was devoted to assessing the impact of organizational structure on the R & D phase of the innovation process. If we assume a parallel between the R & D activity and the process by which organizations come to adopt an innovation, then a number of the points made in the preceding chapter are relevant here.

First, as we have seen in Chap. 3, a number of literature sources converge on the organizational structure/function dilemma. Briefly stated the situation is as follows. There seem to be no one "best" structure for conducting all of an organization's responsibilities in all environments. The structure best suited for "business as usual" is hierarchical and characterized by low complexity, high formalization, and high centralization. However, this "mechanistic" structure is apparently poorly suited for responding to the need for significant changes which would require new ideas and major innovations in the organization's operation, service, or product line. Such pressures on the organization require a more "organic" or matrix structure characterized by high complexity, low formalization, and low centralization.

Except for the most stable of sectoral environments, an organization is called on both to "mind the store" and to respond in innovative ways to needs and opportunities. Thus an organization is subject to conflicting pressures, the responses to which have different structural requirements. One organization adopts a more hierarchical form and continues to do well what it has been doing, but finds change difficult. The other, with a more organic form, has and initiates new ideas with relative ease, but may have difficulty in implementation and the maintenance of stability.

A sector characteristic mentioned earlier, which may find its *partial* explanation in this contrast of the structural characteristics of organizations, is that of the "influential" or "opinion leader."[7] We would hypothesize that the "leader" organization in terms of early adoption would have a less rigidly hierarchical and more organic structure than the followers or late adopters in the sector. The strength of this correlation should be higher in the sectors that are more highly competitive or for some reason are experiencing a period of rapid change. One would expect the more rigid, less open hierarchical structure to be a decided disadvantage in such environments. However, that is speculation, as we know of no published diffusion research on this point.

A further point on the influence of organizational structure deserves mention: Richard Rosenbloom's thesis (see Chap. 9 below) that the primary linkage between the organization (firm) and the larger contexts of which it is a part is to be found in the concept of corporate strategy. The particular structure of an organization, which in turn influences its adoption behavior, must in some significant measure reflect the strategies pursued by those in positions of leadership. In a later section this point will be linked to considerations of the influence of individual characteristics upon organizational-adopter behavior.

2. *Organizations as Social Systems*

In addition to the influence of an organization's formal structural characteristics on its adoption behavior, we must also consider effects occuring from the manner in which that organization functions as a *social system*, which would include many informal and other elements. Because organizational-adopter units have typically been treated as "black boxes" by diffusion researchers, we cannot make this discussion diffusion-specific by referring to an appropriate body of diffusion-research literature. However, many of the social-psychological considerations of importance in organized R & D also seem relevant to organizational-adoption behavior. We briefly sketch the major points here in the hope of stimulating the research needed to demonstrate their applicability to the diffusion process.

The influence of an organization's social system on adoption behavior can be viewed from various perspectives. One of the most fruitful is provided by Kurt Lewin's hypothesis that individuals when confronted with uncertainty as to the actual nature of something seek to reduce such uncertainty by "social-reality testing." That is, they look to others--to "significant others"--for help in formulating an opinion, or for confirmation or disconfirmation of their opinions. (Our frame of reference here is the process of adoption internal to an organization, but this concept of "social reality testing" also has suggestive parallels with relations between adopters.) These "significant others" tend to be the members of his primary social group, and, in particular, the "influentials" at or near the top of that group's social hierarchy. We would add further hypothesis, for which there is some empirical support (Hill 1970), that in a tightly structured, hierarchical, formal organization, there also seems to be a strong tendency to look to the immediate supervisor for guidance.

The guidance sought in such "testing of social reality" may concern a potential or recent adoption either directly or indirectly. In the direct case one may ask a member of his social group what he thinks about the merits of a particular adoption, or may offer an opinion for which he seeks confirmation. He may also check social reality at a higher level by discussing the adoption with his supervisor informally, before formally proposing a course of action. A rebuff may end the matter there. It also influences the individual's perception of organizational goals and/or organizational receptivity to new ideas, which will then influence future behavior (Baker and Freeland 1972b). The particularly strong or self-confident individual who is also deeply committed to an innovation (the term "product champion" is often used here) may not react in this way to a negative response or lack of support. He may, in fact, persist in the face of stiff opposition, and at times succeeds in reshaping the prevailing opinion (Langrish et al. 1972, p. 67). Such behavior, of course, goes beyond the concept of merely "testing" social reality.

In the indirect case, that which is tested is not an idea concerning an innovation, but rather the individual's perception of the environment, as either potentially receptive or unreceptive to a particular idea, or to new ideas in general. In this indirect mode the question is not "what do you think about X?" but rather, "How do you think the boss would react if someone suggested that we adopt X?" Implicit in this indirect form of social reality testing is a greater degree of uncertainty about the receptivity of the organization to new ideas or about the nature of the organization's goals. If this negative impression or uncertainty about goals is reinforced in the social testing process, the idea is likely to go unsubmitted (Baker 1965).

An important point about testing social reality rather than the idea itself (i.e., submitting it as a formal proposal and then seeing how it fares) is that nonexistent barriers can thereby become real. That is, if the social testing proves negative, and the idea is dropped on that basis, it is as if a barrier existed in the formal organization (even if it did not).

Thus, in addition to the actual filters of the *formal* organization through which a proposed adoption must pass (compatibility with organizational goals and strategies; economic feasibility; compatibility with existing practice, product line, or services; etc.), there are also filters within the informal or social system, arising from shared perceptions of what is and is not possible. Such barriers are nonetheless real for residing in the social rather than the formal structure. An idea that is filtered out is lost to the organization, whatever the filter.

It would seem, in the ideal case, that the socially held perception of an organization's goals, strategies, needs, and degree of openness to new ideas would be continuously updated and thus would correspond closely to what is in fact the case. Yet a substantial lag can develop between organization reality and the social perception of it shared by various subgroups (Baker and Freeland 1972b). We hypothesized (in Chap. 3 above) that a major factor in such a lag or gap is the immediate supervisor with whom the members of a social group interact. For better or worse the immediate supervisor, by his decisions, attitudes, communications, etc., represents the organization to those under him. He operationalizes its policies, strategies, and goals, and distributes rewards and sanctions. If in so doing he distorts the need picture, then the perception of the organization held by those under him will also be distorted.

The tone of these remarks concerning the influence of the social structure of an organization on its adoption behavior has been largely negative. That is not to say that the social system cannot serve to facilitate the adoption process. Although we have perhaps emphasized the social "grit" between the organizational cogs, it should also be clear that the social system is also the lubrication without which the cogs would not turn at all.

3. *Individuals in Organizations and Individual Adopters*

Thus far we have dealt with organizational adopters. The basic constituent units of organizations are, of course, individuals. Their influence on an organization's adoption behavior must also be considered. In addition, individuals may themselves be adopters (e.g., farmers). Combining our treatment of these two roles should not prove to be a distorting procedure as long as we keep in mind that in the organizational context individual characteristics are but one influence element among several, whereas in the case of the individual as adopter they are the indigenous whole.

In Chap. 3 we discussed at length the concept of "mind sets," by which is meant the biasing influence of past experience that an individual brings to a present problem-solving or decision-making activity. The crucial point about prior experience is not its its existence *per se* but its appropriateness or inappropriateness in a new situation. If what has been learned in the past is appropriate, then the bias is *positive* and of help in assessing an adoption; if it is inappropriate then the bias is negative and makes an accurate assessment more difficult.

The concept of mind set operationalizes a part of what diffusion researchers mean by "resistance to change." Overcoming the individual bias which favors the old way of doing things may be a formidable barrier to the adoption of a new innovation, especially if the need for change is not perceived as urgent. That raises the question, does *need* or an *opportunity* most frequently serve as the stimulus event for the adoption process? For the R & D process it is clear that most innovations are need-induced (see Chap. 3 above), but we know of no research on this question with regard to diffusion.

The next individual characteristic whose influence on the adoption process we judge to be considerable is the psychological trait referred to earlier as "dogmatism." Dogmatism functions as an intervening variable in the resistance-to-change phenomenon. Individuals who are "high dogmatic" have more difficulty adjusting to new and unfamiliar situations than those who are "low dogmatic." The influence of this trait on an individual's adoption behavior (or his participation in an organizational adoption process) should be obvious. That which has been recently adopted (or is under consideration) is new at least for that context and its impact on the *status quo* is uncertain. Those who are high dogmatic tend to be more uncomfortable, anxious, and threatened by its prospect and thus more resistant to the change. The influence of high dogmatic individuals on organizational adoption depends, of course, on their relative position in the organization's formal and informal hierarchies. The higher their position and/or status the more likely it is that this personality trait will sway the adoption decision. For the individual adopter a high dogmatic orientation may be decisive, i.e., it may well override strong pro-adoption influences.

The final individual characteristic to which we would call attention is the education level of the adopter. As we saw earlier, for both organizational and individual adopters the educational level is of significant influence on adoption behavior. Although the role of this variable is undoubtedly complex, it would seem that its influence is in large measure a function of the technical complexity of innovations.

This completes our analysis of the influence of adopter characteristics on the diffusion process. As before, our assessment of what we know (at least in the sense of cumulative, generalizable knowledge) is that it is meager, relative to what we do not know. In particular, organizational adopters have been badly neglected by the diffusion research community. Correction of this situation should be given a high priority.

C. CHARACTERISTICS OF INNOVATIONS

The characteristics of innovations were identified in the classical model as a key element in the study of the diffusion process. We agree with this assessment, though we would also argue for the importance of a distinction between "adopter-independent" and "adopter-dependent" characteristics of innovations. The former are qualities of the innovation itself, independent of the nature or circumstance of the potential adopter; for the latter, the nature and circumstance of the adopter are intervening variables in the determination of their influence.

The distinction between adopter-independent and adopter-dependent characteristics has not been made in the literature, but it is nevertheless commonly recognized that the characteristics of innovations influence their adoption. In fact, as we have seen, research in quite dissimilar sectors has resulted in similar lists of such characteristics (most of them being, in our terminology, adopter-dependent).

1. *Adopter-dependent Characteristics*

As was noted earlier, Rogers with Shoemaker (1971) in the social-psychological diffusion research tradition and Mansfield (1966) in the economic-industrial tradition have developed quite similar lists of the characteristics of innovations that influence their adoption. Those we take to be wholly, or largely, adopter dependent are compared in the following tabulation.

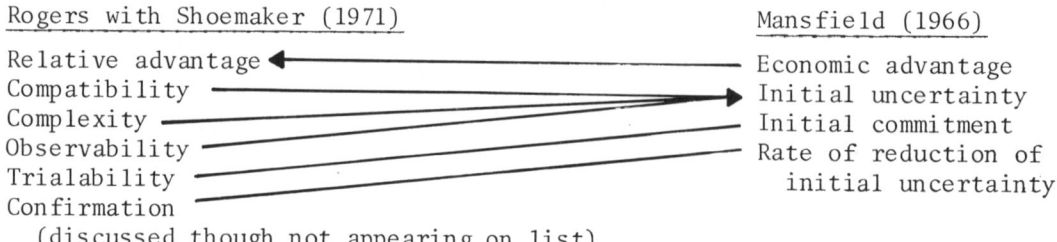

Rogers with Shoemaker's use of the term "relative advantage" includes the economic dimension identified by Mansfield, but also goes beyond it to embrace social-psychological considerations of advantage as well. The influence of these additional aspects is more difficult to assess than the economic because of their nonquantifiability, but if Becker's (1970) argument is correct--that the desire to maintain or increase prestige is more important in early adoption than centrality in the information network (the latter having been traditionally considered a dominant if not decisive factor)--then prestige may play a major role in the adopter's calculation of relative advantage. This would seem to be a potentially important focus for future research.

As we have noted earlier, the second characteristic identified by Mansfield, namely the initial *uncertainty* associated with it, embraces three of the characteristics identified by Rogers--*compatibility*, *complexity*, and *observability*. This congruence of results from quite different sectors and research traditions increases confidence in them. The additional conceptual refinement offered by Rogers will probably provide the better basis for future research. That is, the compatibility of an innovation with the potential adopter's context, and the ease with which the results of its adoption may be observed, differ significantly from one another and from the difficulty an adopter may have in understanding and using the innovation.

Mansfield's third characteristic, the extent of the *initial commitment* required to try out the innovation, is similar to Rogers's "trialability," though the one refers to the "lumpiness" of the capital requirement, whereas the other involves the "lumpiness" of the innovation itself. We shall argue later, however, that whereas these characteristics have a similar influence on the interpretation given here, there is a significant difference between them on another interpretation, under which trialability is adopter independent. One might be inclined to view both as adopter independent, since some innovations (e.g., adoption of the continuous-casting process in steel making) in fact require a substantial initial investment, whereas others (e.g., the application of gibberellic acid in malting) cost little;[8] and that some (again, continuous casting) must be either adopted or not, whereas others (a new hybrid seed) may be tried out on whatever scale one chooses, independent of the characteristics of the adopter. On reflection, it becomes clear that initial commitment and trialability (in the sense of incremental adoption) *are* adopter-dependent, since what counts as a substantial initial commitment depends on what the adopter's resources are like; i.e., a firm with a large number of steel mills can try out continuous casting in only one of them.

The final characteristic identified by Mansfield, the rate at which initial uncertainty can be reduced, also finds a parallel in what Rogers has called the "confirmation function" and fits well into his more detailed structure. Such a rate would clearly seem to be adopter-dependent; i.e., as with the other characteristics identified above, the nature and circumstance of the adopter function as intervening variables in the determination of its influence on the adoption process.

Although it is difficult to be confident that this set of adopter-dependent characteristics of innovations will prove to be fully adequate for future research, they appear to offer a sound basis for proceeding. However, we suggest that the concepts of "relative advantage" and "compatibility" embrace a considerable complexity and will probably have to undergo further refinement as the diffusion-research field develops. Such additional specification may well emerge quickly from an increased interaction of the several research traditions.

2. *Adopter-independent Characteristics*

There remains a second set of characteristics where the nature and circumstance of the adopter seems to make little difference. The first of these characteristics concerns a sense of trialability other than incremental adoption; namely the extent to which the initial adoption (in whatever increment) is reversible at a later time. In view of the world's population problems, vasectomy is a potentially quite important innovation, but one with major drawback that it is virtually irreversible, i.e., it has no trialability at all for the individual adopter. Adoption of the continuous casting of steel, though not completely irreversible, represents a level of commitment that makes it quite difficult to return to the older three-stage casting process (ingot casting, soaking pit, blooming mill). Thus it has low trialability. The same is true for a local government building a new sewerage treatment facility or a farmer adopting a force-fed irrigation system. A great many innovations would seem to possess this adopter-independent characteristic of poor trialability in that they allow little reversibility of the initial commitment. However, since this characteristic has not been specifically identified in the literature, there is no empirical evidence of its degree of impact upon diffusion.

A second adopter-independent characteristic of an innovation is its form. Our primary emphasis has been on the adoption of a new product, device, or process. However, often what is adopted is information rather than an artifact or process. For example, Kottenstette and Rusnak (1973) report the rapid diffusion of the plane-strain fracture-toughness test throughout a number of industrial sectors. What was adopted in this case was basically information as to "best practice" in the selection and testing of metals for various uses.

Whether what is adopted is information or takes the form of an artifact or process is independent of the adopter. The innovation simply has the form it has. Differences in adoption behavior arising from different forms of innovations are unknown at the present. The lack of such knowledge raises again the issue of the generalizability of research results.

Introduction of the matter of the diffusion of information leads into another adopter-independent characteristic, the proprietary or nonproprietary character of information. The plane-strain fracture-toughness test is an example of the latter. It was developed by the American Society for Testing and Materials at the request of the U.S. Department of Defense and the National Academy of Science, and was ultimately incorporated in the ASTM Standards (Kottenstette and Rusnak 1973, p. 104). As such it was public information and was adopted quickly by firms within a number of industrial sectors. Nonproprietary information is not always diffused so rapidly. However, the point is that in such cases there are no proprietary barriers or constraints on its dissemination.

In other cases, however, information may be considered the property of the unit (usually a firm in a private sector) that developed it, and its diffusion is severely constrained. The extraordinary secrecy about the ingredients of Coca-Cola is a paradigm example. Not all barriers are so impenetrable, however. As was noted in our earlier discussion of the technological gatekeeper function, proprietary information does flow between competitors in a sector. There even seems to exist informal norms for restricting, but not preventing, such diffusion (Marquis and Allen 1967, p. 1057).

The influence of this adopter-independent characteristic of proprietary or nonproprietary information on adopter behavior has seldom been an explicit research focus and is thus poorly understood. It would seem that a negative influence would be greater in the proprietary instance, i.e., that only here would barriers exist. There is some evidence that nonproprietary or public information may be of various types, some of which may pose barriers to diffusion while others do not. The plane-strain fracture-toughness test was nonproprietary and spread very rapidly. On the other hand, Utterback has found (1974, p. 623) that government-held patents and technical reports (also nonproprietary or public information) are seldom diffused beyond the source of their original application. The reason for such differences have not been established, but they constitute evidence that even nonproprietary information may possess characteristics that limit its diffusion.

Since the distinction between adopter-dependent and adopter-independent characteristics of innovations has not been a part of the conceptual framework that has guided diffusion research, our list of characteristics may not be exhaustive. We have some confidence in the adequacy of the adopter-dependent list, since the strong research bias towards the adopter perspective would seem likely to have led to their identification. But since adopter-independence so far has not even been recognized as a factor, one must suspect that the examples discussed above do not exhaust the set. Identification of the major adopter-independent characterstics of innovations and empirical determination of their influence on adopter behavior would seem to merit a high research priority.

D. PROPAGATION MECHANISMS

We come finally to the fourth of the major dimensions of influence on diffusion: propagation mechanisms. In its broadest sense the term might embrace every influence serving to diffuse an innovation, including all those discussed above. On a narrower and more useful interpretation, however, reference is restricted to the agents, agencies, and vehicles of communication for which propagation (either in an advocacy or informational sense) is a primary and explicit function. Thus, although the behavior of a particularly influential farmer and the county agent may both be crucial in the rate of adoption of a new hybrid seed within an area, only the latter would be considered a propagator, since the primary function of the former is farming, not the propagation of innovations.

A typology of propagation mechanisms is offered in Table 1, together with examples of each type. The axes of this matrix are (1) the types of propagation mechanisms with which the adopter may be linked, and (2) the levels of aggregation at which the mechanisms function. The types and levels arrayed along these axes are not mutually exclusive, but do reflect important differences that have been recognized in the literature. Griliches's classic work on the diffusion of hybrid corn (1957), for instance, rests on the assumption of active organizational propagators at the intersectoral level. Myers and Marquis (1969) and Brown (1968, 1972) are also concerned with this type and level of propagation. Rogers (1962) and Rogers with Shoemaker (1971), on the other hand, deal primarily with intrasectoral diffusion, in terms of all three types of mechanisms (though with an emphasis on the individual propagator). However, we know of no studies that systematically compare the function and relative influence of all three types of mechanisms across the various levels of aggregation.

TABLE 1.--Typology of propagation mechanisms.

Type of Linkage / Level of aggregation	Observable Artifact to Adopter	Person/Organization to Adopter	Media to Adopter
International	World fairs, shows, exhibitions	International Organizations (e.g., WHO) Private-sector sales (e.g. multinational firms) Licensing arrangements	Mass Media Professional journals Private-sector promotional literature
Intersectoral (External to Adopter's Sector)	Exhibitions, shows	Professional societies (e.g., ASTM, ASME) Federal agencies (e.g., USDC, NASA, EPA, SBA, USDA) Private-sector sales Licensing arrangements	Mass Media Professional journals Private-sector promotional literature USDC Clearinghouse
Intrasectoral (Internal to Adopter's Sector)	Trade shows Observation of adoption by others	Professional societies (e.g., AMA, SAE) Trade associations Agricultural Extension County Agent Private-sector sales Licensing arrangements	Trade journals Private-sector promotional literature
Intra-Adopter (Organization)	Trial adoption	Gatekeeper Product Champion	In-house technical reports

In addition to the distinctions offered in Table 1 the function performed by the various propagation mechanisms may range from a strongly "pro-adoption" mode (i.e., committed to the rapid and complete diffusion of an innovation), to that of a more "adoption-neutral" channel of information. (This distinction is implicit in Rogers with Shoemaker 1971, p. 24.) An example of the former would be a firm selling an innovation in a competitive market; of the latter, the dissemination of information by the Clearinhouse for Federal Scientific and Technical Information of the U.S. Department of Commerce. Although all the propagation mechanisms identified in Table 1 seem to have a dominant orientation as either pro-adoption or adoption-neutral, one can also find instances in which most of them functioned in both modes. Professional societies, for instance, though typically adoption-neutral as between most competing innovations within their domain, can also assume an advocacy role with respect to professional standards or "best practice."

The differential impact of this active/passive distinction on adoption behavior is perhaps substantial, but poorly understood at present; so is the influence of a shift from one mode to the other. The endorsement of a particular fluoride toothpaste by the American Dental Association, for instance, is widely assumed to have had a major impact upon its rate of diffusion. Perceived shifts from straight reporting to advocacy by the mass media often occasion a great outcry, again indicating an assumption, at least, that the influence of shifts in a propagator's normal function is substantial. This topic merits a high research priority.

The typology in Table 1 offers a structure for examining what is known about the influence of various propagator mechanisms on the behavior of potential adopters. As has been the case with the other influences we have examined, however, there are few studies which examine the linkage of propagator with other variables.

1. *Observable Artifact-to-Adopter Linkages*

Because the foremost writers on diffusion are academicians, they are perforce immersed in literature. Hence they sometimes tend to overlook the importance of the actual demonstration of the innovation to the potential user--what Rogers with Shoemaker (1971) have called its observability. Thus the study of artifact-to-adopter diffusion has been largely neglected, yet it would seem to be a powerful diffusion mechanism. Indeed, the efficacy of this technique is known to every primary-school pupil under the descriptive title of Show and Tell. Every technical salesman, wherever practicable, carries samples or demonstrator models with him for the show-and-tell effect (thus coupling visual with oral propagation techniques). Trade shows and exhibitions form part of many trade-organization and professional-society meetings. There might thus be some quantitative data available--attendance figures, or, buried within corporation ledgers, some analysis of sales derived from showing innovations to potential users--for demonstrating empirically the efficacy of this diffusion mechanism. However, we know of no study that presents such data.

However, there is some historical information on the impact of world's fairs and international exhibitions on technological progress and diffusion. Much of this literature centers on the role of such exhibitions in creating a sociocultural climate of acceptance of technological growth. Little attention has been given to the way in which exhibitions fostered the diffusion of specific innovations, although there is the well-known case of Britain's introduction to the American technique of interchangeable parts at the Great Crystal Palace Exhibition of 1851 in London, which led to the dispatch of a commission under the great machine-tool designer Joseph Whitworth to see the "American system of manufactures" in operation. On the basis of the Whitworth commission's report, the British equipped the Enfield Armory with American machine tools, and hence accelerated the diffusion of these innovative devices and processes (Rosenberg 1969).

Such diffusion-specific literature as exists on the observable artifact-to-adopter linkage has been concerned with observability as a characteristic of innovations, rather than with the propagator mechanisms that provide or increase such observability. Thus the mechanisms, as opposed to the observability itself, have hardly been identified to date. The examples offered in Table 1 must be therefore taken as illustrative rather than exhaustive.

2. *Person/Organization-to-Adopter Linkages*

People-to-people transfer of technical information is being increasingly studied. Allen and Cohen's concept of the "technological gatekeeper" (1966), the individual within the laboratory who informally links his colleagues with outside sources of information, is being extended by work in progress on the "informational entrepreneur" carried on at the Georgia Institute of Technology. The Georgia Tech study[9] stresses the passivity of most current information systems (even highly computerized information storage-and-retrieval systems represent nothing more than a fast librarian), and the need for a more active and flexible scheme of information diffusion that would allow for feedback from the user to the generator of information, as in the highly successful Agricultural Cooperative Extension Service.

The mobility of professional engineers can be important in the diffusion of technology (referred to in Chap. 3 as "on-the-hoof" transfer). Gilfillan claims that the mobility of labor, industrial spying, disclosure in patents, and the inability to seal off the research and productive processes tend to erode the differential level of knowledge among firms (Gilfillan 1935). When an engineer moves from one firm to another, or when a scientist or technologist moves from a government laboratory into private industry, or vice versa, he carries with him knowledge and information, much of which may legally be transferred to his new employer. Many unpatented processes and techniques can thus be diffused by the movement of people from firm to firm, from industry to industry, or from country to country. Indeed, in the mid-1960s, the "brain drain" from other nations to the United States was held responsible for the "technological gap" between the United States and other nations. The brain drain, the impact of which was doubtless exaggerated, represented not so much the transfer of actual devices and signs as it did the transfer of technological capabilities.

The role of professional societies as propagators has already been mentioned in the case of the American Society for Testing and Materials reported by Kottensette and Rusnakc (1973). Here the organization itself assumed an advocacy role in the diffusion of "best-practice" information at the intersectoral level. A contrasting professional-society role reported by Coleman, Katz, and Menzel (1957) illustrates not only activity at a different level of aggregation (i.e., the intrasectoral), but also an adoption-neutral propagation function. The physicians most active in the organization were not the earliest adopters of the drug because the society formally advocated it (it did not), but because of their centrality in the sector's social structure. The society was a propagator, however, in the neutral, information-dissemination sense. In light of this result, one might hypothesize that the adoption-neutral propagation function depends for its influence on the existence of social systems. (The two-step flow hypothesis discussed in Chap. 3 also lends support to this view.) This hypothesized linkage merits further investigation.

One might ask at this point whether a linkage might also exist between pro-adoption propagators and the potential adopter's social system. Such data as exist at this point seem to indicate an affirmative answer, though the relationship is somewhat different. For instance, Spicer (1952) and Bright (1964) offer evidence that the influence of pro-adoption propagator varies considerably according to whether such a propagator is perceived as being a member of the potential adopter's social system or not. Thus the linkage in this case is more complicated, with propagator influence depending not only on the existence of a social system as above, but also on the relationship of the propagator to that system.

Thus, in addition to the distictions offered above, it would also seem to be important to differentiate between pro-adoption propagators as internal or external to the potential adopters sector. (In terms of Table 1 this is the distinction between the intrasectoral and intersectoral levels.) The latter may be of several types. For instance, a pro-adoption external propagator may represent a private firm that has marketed an innovation for which there are for a time no competitors offering comparable advantages (e.g., Polaroid Corp.). What differences in adoption behavior might accrue from this circumstance as opposed to one in which there are several competing propagators advocating adoption of highly similar innovations? Or the external propagator might be a professional association to which only some adopters in a sector belong. Would that make a difference in the adoption behavior of members of that sector (independent of the characteristics of the particular innovation under consideration), relative to a sector in which all the potential adopters were members of the association? How would the propagator's behavior differ in these instances? How would the behavior of adopters be different?

Included under the category of external propagators (those functioning at intersectoral level) is a wide range of federal agencies, a representative sample of which are:

- Technology Transfer Program of the Environmental Protection Agency
- NASA's Technology Utilization Office

- Technology Utilization Program of the Small Business Administration[10]
- The Clearinghouse for Federal Scientific and Technical Information, the Office Field Services, the Economic Development Administration, and (during the late 1960s) the State Technical Services Program,[11] all of the U.S. Department of Commerce
- The Agricultural Extension Service of the U.S. Department of Agriculture

The range of propagator activities of the Federal government undertaken by these programs, and others that could be mentioned, are too diverse to be surveyed in this context. Two, however, deserve special mention. The most successful of these efforts has been the Agricultural Cooperative Extension Service; a good deal of the success of American agriculture during the 20th Century is testimony to the effectiveness of this device. The agricultural program, fostered by the Federal government beginning with the Morrill Land Grant Act of 1862, which founded the land-grant colleges for promoting the agricultural and mechanical arts, combines all elements of the innovation process within it. Agricultural research stations, both state and Federal, usually connected with the land-grant agricultural college, generate the discovery or invention; and the Cooperative Extension Service, through county agents, works to see that the innovations thus produced are applied and diffused.

The county agent not only transmits scientific, technical, and economic information (in a form and manner aimed directly at the farmer-user), but he also serves as a means for feedback from the farmer-user to the information-generating source. For example, when the individual farmer is faced with a new problem, the county agent transmits this user need to the information-producing source, which then seeks out or adapts old information or creates new information to provide the answer. In other words, direct feedback from the user helps in the problem-definition phase of the innovation process. The agricultural research station faced with the problem has access to past literature and to laboratories for carrying on experimentation. It can thus perform all the functions associated with R & D, from idea generation through the research and development phase. Then the county agent diffuses it among other potential users. Finally, the county agent serves as an important feedback in the ecology of the innovation process.

During J. Herbert Hollomon's term as Assistant Secretary of Commerce for Science and Technology, he sought to develop an industrial analog of the agricultural extension service. His efforts were not completely successful, because the Congress never approved of the program as a whole nor provided sufficient funding for the parts of which it did approve. However, the State Technical Services, which emerged for a time, did result from Hollomon's work (Nelkin 1971).

Nevertheless, there are nonagricultural and nongovernmental modes of propagation which resemble somewhat the agricultural extension service and which have exhibited some success in the marketplace. One example would be the "detail men" in the pharmaceutical industry, who go directly to the users (doctors) to present them with the literature and samples of new drugs. At the same time, they feed back to their employers the needs of the doctors; although not medical men themselves, they provide the crucial coupling between innovator and user. Yet there seems to be little serious study in the diffusion literature regarding the role of the drug detail men in such two-way diffusion of information (Burkholder 1963).

The "sales engineer," the salesman of technical products, performs a similar diffusion function. He brings information of new technical products and processes directly to the customers, and at the same time serves as a conduit by which his customer's technical needs are transmitted to the manufacturer-innovator whom he represents. To the best of our knowledge, there have been no scholarly studies of the role of technical salesmen as propagators, or as articulators of consumer needs and hence as problem definers for innovators.

A second widely heralded government attempt to diffuse information regarding innovation is the NASA Technology Utilization Programs (Doctors 1969). Through contracts with Midwest Research Institute, and others, NASA has actively sought, in accordance with its mandate, to transfer the scientific and technological "spinoff" of the space effort to civilian economy. Despite good intentions and the expenditure of considerable money and effort, this attempt to diffuse innovation has had mixed results. In an unpublished research study prepared for the NASA Office of Scientific and Technical Information,

Kranzberg and Rossini have indicated flaws in NASA's Technology Utilization Program: it does not allow for feedback, and when user needs cannot affect the information generation source, the information-coupling mechanism is weak. On the other hand, in that part of its mission relating directly to civilian aeronautics, NASA has an excellent record--reaching back to its old days as the National Advisory Committee on Aeronautics, when its innovations were directly in line with the needs of its users.

The defects in the NASA Technology Utilization Program are further proof of the ecological nature of the innovation process. If there is no feedback, no interplay, among the different phases, the innovation process becomes flawed and imperfect. In the case of NASA, civilian user needs could not directly influence the problem-definition and idea-generation phase of the innovation process. Such needs were outside its mission. Instead, NASA's Technology Utilization Program must rely upon serendipity; the innovations arising from the space program might happen to be useful for civilian industry, but it is not the task of NASA to make them so nor to embark upon innovations with that end in mind. In order to apply NASA's technological innovations, the civilian user must frequently redefine his own problems, rather than having the innovation produced in answer to his needs from the outset (Doctors 1969).

An important international propagation mechanism has only recently come to public attention, although it has been around for some time: the multinational corporation.[12] Multinational corporations take on many different organizational forms; some allow a substantial autonomy to their international parts, others maintain a high degree of centralized control. Although there is a vast literature on the diffusion of technology to less developed countries through government and international aid and financial programs (e.g., the World Bank), there are only beginning to be serious studies of multinational corporations. However, most deal with political, financial, and international monetary problems arising from the operations of multinationals, and little attention has yet been paid to their role in the diffusion of innovations.

Unlike international aid and development programs sponsored by governments, which (especially since World War II) have served as devices for the transfer of technology to the less developed countries, the multinational corporation diffuses technology primarily to already industrialized nations. In fact, multinational corporations carry on approximately two-thirds of their activities in industrially developed countries, whose advanced economic level and similar institutional and social structures have facilitated their spread (United Nations 1973). With the exception of the United States, which is primarily a home country, developed market economies serve simultaneously as home and host countries. Although only one-third of the direct investment by multinationals is in developing countries, their presence in those countries is perhaps of greater relative significance than in industrialized nations.

Herman Kahn claims that the multinational corporation "is probably the most efficient social, economic, and political institution ever devised" to accomplish, among other things, the transfer of technology.[13] The Diebold Institute for Public Policy Studies (1973), however, indicates that some critics of multinationals claim that the impact of such transfer is often minimized because: (1) R & D is generally carried out by the home country; (2) the training of host country nationals for R & D work is often neglected; and (3) the technology itself is often closely held and not diffused within the host country.

The report of an ad hoc panel of the Board on Science and Technology for International Development of the National Academy of Science (1973) looked at the prospects for industrialization in low-income countries through the work of multinationals. The panel examined five industrial sectors: automotive and agricultural machinery, chemicals, electronics and communications, pharmaceuticals, and food processing. It pointed out that the multinationals do not engage in or support much research, development, and engineering in the low-income countries, and suggest that they reconsider their position. The panel concluded that the multinationals and the host countries would find it in their mutual self-interest to foster such cooperative and proprietary programs of research, development, and engineering. The panel (1973) recognized the limits of "workability" posed by what might be called the host country's "receptivity threshold." "The development of basic engineering capabilities--the ability to manage quality-control systems, introduce material specifications and standards, maintain tool shops, and establish other production-support activities—normally must precede more ambitious developmental and applied research on product design, new materials, equipment design, and other changes in production or processing techniques. In other words, the logical and chronological sequence is E, D & R (Engineering, Development, and Research) rather than R, D & E." This concept of a "receptivity threshold" is similar to the adopter-dependent characteristics of particular innovations which Rogers has called "compatability."

In terms of Ruttan and Hayami's (1971) three modes of technology transfer (material transfer, design transfer, and capacity transfer), the panel's point is that multinationals should be more concerned with raising the host countries receptivity to advanced technology by the transfer of "capacity." They are, of course, concerned with material transfer, which refers to the diffusion of the technological artifacts themselves, and often design transfer also, as is the case with licensing arrangements.

3. *Media-to-adopter Linkages*

Much of the research of potential relevance to this topic of linkages between the media and adopters has been previously discussed (in Chap. 3) and need only be mentioned here. The flow of technological (as opposed to scientific) information in R & D is a highly oral process, in which literature sources play a decidedly secondary role. We would hypothesize the same oral pattern for the diffusion of technological innovations. Although the role of such propagation mechanisms as professional and trade journals is certainly not negligible, and would vary considerably in terms of certain sector and adopter characteristics, it is probably not so important an influence in adopter behavior as the other types discussed above. However, that is conjecture, and even if the needed research should prove it to be a valid generalization, it would be important to know such exceptions as may exist.

Not all mass media propagation sources are literature based, however. Radio and television, in both informational programs and advertising, also serve adoption-neutral and pro-adoption propagation functions. Television offers the additional advantage of a "show and tell" capacity. Moreover, as has been indicated by the two-step flow hypothesis, the mass media have not the same direct influence on all potential adopters; the social system serves as an intervening variable. Thus Rogers with Shoemaker report that the preponderance of studies show early adopters to have greater exposure to the mass media than later adopters (1971, p. 372), but they also report them as having greater exposure to interpersonal communication channels (1971, p. 374). Not only are the causal influences hard to disengage, even the determination of the causal *direction* is difficult. For instance, Rogers with Shoemaker also report that early adopters have a higher degree of opinion leadership than late adopters within their social system (1971, p. 375). If this is coupled with Becker's (1970) contention that desire for prestige is a more important factor in early adoption than centrality in the communication system, then the relative influence of the mass media and the early adopter's receptivity to it becomes very hard to determine. There is a need for carefully designed research to enable us to begin to sort out these multivariant complexities. The simplistic, "other-things being-equal" approach will simply not serve to advance the state of the art much beyond the present fragmented condition.

This completes the description of what we take to be the four major dimensions of types of variables basic to an adequate general theory of diffusion. As we shall indicate in the following section, the holistic view of the process to which the consideration of its dimensions naturally leads, introduces quite a different level of questioning than has thus far characterized diffusion research.

SUMMARY AND CONCLUSION

At the present stage in its development, diffusion research has little explanatory or predictive power at its command; it has no theoretical structure of its own and offers few empirical generalizations that may be confidently applied to a wide range of diffusion instances. That this is the case should not be surprising in the light of the following considerations:

1. the field is a relatively new one and deals with extremely complex phenomena;

2. it is shared by an unusually large number of research traditions that interact with one another very little;

3. empirical research is dominated by studies that deal with only two variables at a time--"whereas the real nature of diffusion is certainly a cobweb of interrelationships among numerous variables" (Rogers with Shoemaker 1971, pp. 93);

4. it is a field that does not seem to possess even the most rudimentary sort of theoretical structure or paradigm--the so-called "classical model," which represents the best attempt in this direction to date, has only limited value in this regard.

This largely negative characterization of the state of the art in diffusion research is tempered somewhat by the implicit--though unexploited--complementarity of its three major research traditions. We have attempted to draw out and build upon the complementary thrusts of these traditions by gleaning from them the dimensions of a general theory of diffusion. Such a theory, when developed, could serve to overcome the present fragmentation of effort, and provide the basis for achieving research results more commensurate with the complexity of diffusion phenomena.

Identifying the dimensions of a general theory is, of course, just the first step, a static categorization of the variables operative in the diffusion process. How these influences interact with one another, and how in multivariable clusters they impinge on adopter behavior--in short the dynamics of the diffusion process--calls for empirical work that is beyond the scope of this study.

The development of a general theory, which would capture the dynamic interaction of the dimensions we have portrayed statically, requires a new level of questioning. These questions have as their point of departure the concept of types of variables or, as we have called them, process dimensions. This concept leads one to ask about clusters of variables--not some x and y in isolation--and about clusters that embrace all the process dimensions, not just one.

That is not to say that the research design for every investigation can (or should) embrace the total diffusion process in all its complexity. Nor does an ecological view require that. But it does mean that whereas the several process dimensions may be treated as a "quasi-separable" system for purposes of analysis, ultimately they must be integrated. And even in the analytic phase of an investigation an awareness of the interacting whole must be maintained.

Illustrative of the kinds of questions that need to be addressed by diffusion researchers in the analytic or "quasi-separable system" mode, but can only be answered by theory-based cumulative research, are the following:

1. What is the relationship, if any, between public and private sectors on the one hand, and the types of propagation mechanisms that prove most effective in the diffusion of innovations on the other?

2. What is the relationship, if any, between highly aggregated and highly disaggregated sectors, and the existence and influence of pro-adoption propagators? adoption-neutral propagation mechanisms?

3. Are there adopter characteristics that are sector-specific while others are found with similar frequency in widely divergent sectors? With regard to the latter, have they all a direct influence on adoption behavior, or is the influence of some indirect, i.e., dependent on the presence of some other, intervening variable(s)?

4. Does the distinction we have urged between adopter-independent and adopter-dependent characteristics of innovations correlate systematically with certain propagation mechanisms, and the rate at which innovations are diffused?

5. What are the differences/similarities in the meaning and determination of "relative advantage" across diverse sectors (e.g., public/private, high/low competition, etc.)?

6. What is the relationship between an adopter unit's organizational structure and the influence of sector characteristics and/or certain propagation mechanisms on its initial adoption behavior and/or on the rate of intra-organizational diffusion?

7. Which sector, adopter, and innovation characteristics, and which propagation mechanisms are most significant determinants of the effectiveness of Federal participation in the diffusion of technological innovations?

Such questions, though lacking specificity, are illustrative of the kinds of linkage issues that arise once the two-variable generalization approach is seen as merely the first weak step in understanding the process. They can hardly begin to be specified, however, in the absence of a theoretical framework, and cannot be answered outside the context of a "normal" science, cumulatively building a shared knowledge base.

Lest this assessment of the state of the art understanding of the process of diffusion seem unduly pessimistic, let us repeat an observation made earlier. Although the perspectives and emphases of the several diffusion research traditions are substantially different, they are, by and large, not incompatible or in conflict. They are, in fact, implicitly--and occasionally explicitly--complementary. The unexplained "residue" of one is often a major preoccupation of another. Thus, we would venture the judgment that the development of a comprehensive theoretical framework, and the emergence of a single, interactive research community could, in fact, take place rather quickly. We would urge the assignment of the highest priority to the conceptual work necessary to make this a reality.

NOTES AND REFERENCES

1. The exception to this "number of studies" criterion is early sociology as a major tradition "because of its considerable influence on most of the other traditions which develop later" (Rogers with Shoemaker 1971, p. 48).

2. We would group the traditions identified by Rogers with Shoemaker as follows:

The Social-Psychological Tradition	The Economic Tradition	The Geographic Tradition
Anthropology	General Economics	Geography
Early Sociology	Marketing	
Rural Sociology	Agricultural Economics	
Medical Sociology		
Communication		
General Sociology		
Psychology		

Two traditions identified by Rogers with Shoemaker have been omitted in this arrangement: education and speech. They offer a low evaluation of the contribution of education studies to the understanding of diffusion. As Carlson (1968) notes, "Data collection on acceptance has not been characterized by rigor....Given this weak base, it is rather difficult to count on what is known about the diffusion of educational innovations."

3. See also Rogers with Shoemaker (1971, p. 194).

4. Nabseth and Ray (1974, p. 210) report, for instance, that the rate of adoption of the float glass process was very slow for some large European companies "because of existing expensive and relatively new equipment which was using the latest pre-float technology."

5. A brief description of this model may also be found in Chap. 12 below.

6. The point here has been made previously, that, with regard to the process of diffusion, "new" is relative to a potential adopter, whereas in the innovation process something is new relative to the state of the art.

7. These particular terms reveal the bias of the classical model toward the individual adopter, i.e., they both carry more the connotation of evaluation and social-psychological "pecking orders" than offensive/defensive strategies of firms in the marketplace. Their analog in organizational sectors would perhaps be the "leader/follower concept."

8. See Nabseth and Ray (1974) for a discussion of this point.

9. "The Flow of Scientific and Technical Information in the Innovation Process," NSF Grant GN-42061.

10. See Kottenstette and Rusnak (1973) for an analysis of the similarities and differences in these programs.

11. See Arthur D. Little & Co. (1969) for an evaluation of this office.

12. The United Nations has recently adopted the term "transnationals" to refer to these firms in order to reflect that, although they may do business in many countries, typically they are not multinational in terms of ownership and decision making.

13. An undated publication, "If the Rich Stop Aiding the Poor..." United Nations Centre for Economic and Social Information, *Development Forum*.

5. OVERVIEW AND PROSPECTS

I. AN ECOLOGICAL PERSPECTIVE

This assessment of the current understanding of the innovation process was designed to achieve several objectives. One was to synthesize a diverse and scattered literature, so as to make what is known at present about the process more easily available to scholars, students, and researchers and managers actively engaged in technical innovation. A second objective was to identify the gaps and weaknesses in our understanding--what we should but do not know--as inputs to the research priorities of individuals, research groups, and funding agencies. Finally, it was hoped that the very act of assessing the state of the art might serve to advance both the understanding and practive of innovation, so that we may be afforded, in our national and international dependence on technology, greater choice, diversity, and control.

As our investigation proceeded, a fourth, unanticipated objective emerged. A subject matter as complex as technological innovation embraces an unusually wide range of variables, and thus requires for its understanding the cooperative effort of scholars from many disciplines. The unexplained residue of one disciplinary perspective is often a central focus of another. This interdependence has typically not been dealt with in the research literature, with a resulting isolation of the several perspectives and fragmentation of what is known. We became convinced that this fragmentation of effort and results constitutes a major barrier to substantial advances in understanding. Thus it became an additional objective of our study to emphasize the interdependent and complimentary roles of the several research communities that are involved in innovation research, and to point out, wherever possible, linkages that need to be explored.

Initially, we attempted to structure our assessment in terms of a simple but speciously attractive view of process phases following one another in linear fashion: (1) problem definition and idea generation; (2) invention; (3) research and development; (4) application; and (5) diffusion. We were disabused of this simplistic notion rather quickly--it just did not jibe with the facts--and cast about for a more adequate representation that would capture both the intertwined complexity of the process and its dynamics.

A closer analogy, and the one we came to adopt, is that of an ecological system. Viewing the innovation process as an ecological system not only enabled us to develop a more adequate notion of process phases, but also to portray the wide range of immediate and more remote influences and their complex linkages, and the interaction of embedded and overlapping levels of aggregation. We have therefore come to treat an innovation and its diffusion as the outcome of a multiphase but often nonlinear process in which a host of social-psychological, economic, organizational, and informational variables interact throughout several levels of aggregation.

Although we speak of a lake or forest and its inhabitants as constituting an ecological *system*, we recognize that this parlance does not mean that they are completely self-contained or self-sufficient. Such systems are linked to others as a part of a larger whole. In the same way the process of technological innovation affects and is affected by systems and factors exogenous to the immediate contexts in which it takes place. Thus we have not only given the term "process of technological innovation" a very broad interpretation--as embracing the whole range of activities from the initial problem definition and idea generation through to the diffusion of an existing innovation to new contexts--but have also included in our investigation the exogenous influences of societal values, endowments, and institutions.

On the other hand, we were not equally interested in everything about the larger systems which embrace innovation, or even in every aspect of the immediate contexts to which the process is indigenous. With rare exception the factors that influence the innovation process, in both the immediate and remote environments, exist for reasons (and perform functions) other than the innovation-specific ones with which we are concerned. Most Federal agencies whose actions affect innovation, for instance, also perform many other functions that have little or no direct relevance to it. Likewise, even a highly innovative firm may still have as its dominant orientation the continued production of goods for which innovative improvements are not contemplated or called for. Thus while we recognize the other, often dominant, activities in which most of these institutions are also engaged, we have restricted our study to their innovation-specific roles, and have dealt with their other functions only as they impinge on innovation. Given the constraint that we were concerned only with the innovation process, and given an ecological perspective that precluded treating this process as if it were self-contained and independent of all else, what evolved was a conceptual structure of three dimensions. That is, we were concerned simultaneously with the process (1) at various levels of aggregation, (2) as involving a wide range of disciplinary variables, and (3) as amenable to phase distinctions. The first two of these structural dimensions are reflected in Sec. II below, which deals with the environment exogenous to the context of innovation; and all three are interwoven in Secs. III and IV, which treat innovation and diffusion, respectively. Each of these chapters is summarized below.

II. THE WORLD OUTSIDE

Innovation is but one of the possible responses to a perceived need or want, and in the broader sweep of human history, it does not seem to have been man's dominant one. On the contrary, the fatalistic assumption that things are as they have always been and cannot be otherwise, with the deadening and dreamless acceptance that this assumption entails, has surely had a longer and stronger grip on the human mind. For some cultures, and for some manifestations of the religious quest, there has been an alternative to innovation in the face of need or opportunity. In those cultures the material world has been taken as somehow less real, less important than the spiritual, and thus such groups have not been preoccupied with schemes to change it. Some, in fact, have regarded the failure to meet a physical need--save by endurance--as the measure or path of spiritual attainment.

To the extent that such alternative responses to material needs characterize a society's value orientation, the innovative response is not considered, or is viewed as misguided or beside the point. Thus if innovation is to flourish, the cultural "soil" of its basic value orientation must be supportive.

While a change-oriented value system is thus necessary, it does not exhaust the ways in in which values and preferences influence the innovation process. They also operate at every level of aggregation--national economic policy, corporate behavior, and individual responses, for example--to channel, inhibit, facilitate, and focus innovation. The nature of the value systems operative at these several levels is poorly understood; their linkages, hardly at all.

Societal values and the particular preferences they yield are not the only exogenous influence to be considered. There are also societal endowments of various kinds--material resources, labor talent, knowledge--whose nature, scarcity or abundance, cost, distribution, etc., impinge on the process. Likewise, a host of institutions, governmental and private, play a major role--by inadvertence as well as design, and as a result of their structure as well as their function. Our assessment of what is known about the interactions and influence of these elements of the exogenous system--values, endowments, and institutions--is summarized in the following.

A. VALUES

We know that values and preferences both influence innovation and are influenced by it. As recent surveys indicate, the public is still generally positive about technology, but feels some serious uneasiness. Although the public realizes that technology is responsible for much of the material welfare it enjoys, at the same time it holds technology responsible for changes, not always viewed as beneficial, in established life patterns in many areas, including employment and leisure. The uneasiness apparently stems from a feeling that individuals or the public as a whole have little or no control over how technological change affects their lives. In the future this latent frustration may be translated into more structured demands for public control over technology similar to what has occurred with regard to environment and has already led to the technology assessment movement. Thus, we must consider modes of public participation in decisions regarding technological innovations which transcend the present governmental mechanisms.

Unfortunately, on balance, we know little about values from a scientific point of view. It sounds simplistic to say that we need an adequate value-research methodology, but that is in fact what we need. We should be able to determine value systems existing at various levels of aggregation such as the government, firm, and individual, and how they interact. Furthermore, we must be able to determine the ordering of preferences within these systems as well as the linkages and conflicts among values and value systems. Finally, we must learn to forecast trends in value changes so that we can make today's decisions affecting future innovations in terms of these trends as well as present values.

True, considerations of value research methodology go far beyond the limits of our study. Yet they are important in understanding innovation because values act to facilitate or constrain various innovations. We know how to survey individual preferences, and we can observe specific choices made by firms and governments. The problem is to comprehend the relationships between preferences at different levels, and the multifaceted linkages between values and innovation, as well as to understand the interactions of values with institutions and endowments. The alternative to developing value-research methodology is to remain at the present "wisdom level" supplemented by some cross-sectional studies indicating specific preferences.

B. ENDOWMENTS

Fortunately, the state of our knowledge regarding endowments is greater than our knowledge of values. We know more about how endowments affect innovation and how they interact with other elements of the exogenous system.

In the case of the relationship between scientific knowledge and technological innovation we are confronted with collections of case studies which recount, with varying degrees of detail, the "events" leading to an innovation. However, there has been virtually no attempt to dissect the case studies in order to determine differential relationships by field of science and technology. Perhaps the existing data could be culled to determine whether such relationships do in fact exist, and what further studies are appropriate to clarify them.

Scientific knowledge appears to play different roles at the various phases of technological innovation. For example, relativity theory underlies nuclear technology. But more localized scientific investigations, dealing, for example, with the properties of materials and not relativity theory, play a role in improvements in existing nuclear technologies. However, the most fruitful and basic formulations of relativity theory and the theory of matter are not generally accessible to every technologist who is developing specific innovations. A fuller understanding of the various uses of scientific knowledge at different points in the development of an innovation would thus be useful; so would be better understanding of the process of translating a scientific theory from its general and abstract version to a form usable by technologists. Enough case-study data probably exist to enable us to begin to structure the problem and to find sources of relevant information.

An area that has been studied in bits and pieces is the relationship of supply and demand to innovation. Although the balance of evidence indicates that innovations occur largely in response to demand, it might be worth while to examine the origins and types of demand. Demand may be generated from changes in factor markets or market needs. Schmookler's classical studies consider only the dollar measure of demand--in terms of influencing patent activity--without delving into why this money was committed in the fist place. It is tantalizing to speculate that the origin of demand may, in part, be created by the supply of appropriate scientific and technical knowledge. A look at the sources of decision to commit investment capital might thus yield some useful information. Moreover, Rosenberg's recent analysis of Schmookler's work has effectively pointed out that supply and demand factors are highly interactive. We need additional research on this interaction.

One of the most interesting questions raised in our study of endowments is, "What are the limits to substitutibility among endowments in our world?" We have tried to lay the groundwork for structuring an approach to this problem whose solution would give us some grasp of the factors determining the upper and lower limits of the rate of innovation that would preclude disasters such as those predicted by *The Limits to Growth*. The answer to this question might have more effect on the future of innovation than the contributions of both the technological-growth economists--who study the mechanisms of technological development--and systems dynamicists--who model the total context in which technological development takes place.

Throughout its treatment of endowments the literature tends to consider sections of problems in isolation, without seeing how the partial solutions offered might link with the solutions to other parts of the same problem. One advantage of our ecological framework is that it relates the parts to the large whole. Certain endowments are determined and influenced by values and institutions. For example, public and private support of scientific research and education is a manifestation of social values translated through institutions. Public perception concerning pollution and environmental decay set limits on how much coal and oil may be recovered and what technologies may be used in their recovery. The nexus among endowments, values, and institutions is multidirectional and complex. However, this complexity does not lessen its importance.

C. INSTITUTIONS

Researchers studying institutions have again generally not stressed their mutual linkages or their connections with other elements of the exogenous system and with innovation itself.

In discussing the economic institutions which affect the innovation process, economists have been limited in their analyses because of their tradition of treating technology as a given external constant. Until the innovation process becomes an integral part of the economic theories of both firm and industry and until these theories become more dynamic in nature, technological change will continue to receive inadequate attention.

Some attempts have been made to overcome these static suppositions of economists. Schumpeter in the 1930s and Galbraith and his disciples in the 1950s and 1960s hypothesized that large firms in concentrated industries were more innovative than small firms in competitive industries. Although the Schumpeter-Galbraith hypothesis stimulated much empirical research, this research has encountered both definitional and data problems. One major problem is defining and measuring innovation. Economists have utilized R & D expenditures or R & D employment as a surrogate for innovational output, which meant that they utilized an input as an estimate of an output, with all the uncertainties implicit in that procedure. On the other hand, attempts to use patents, an output, as a measure of innovative activity have resulted in a different set of problems and in equally inadequate perspectives.

Attempts to match the Schumpeter-Galbraith hypothesis with existing data show that the original hypothesis is simplistic. The relationship between firm size and measures of R & D activity, such as R & D expenditures or R & D employment, is far from linear. Initially, increases in firm size bring more than proportional increases in R & D activity. But beyond some point, which apparently differs from industry to industry, that is no longer the case. R & D activity flattens out and increases less than proportionally to additional increments in firm size.

The effect of industry concentration on R & D activity is even less well understood. It does appear, however, that some concentration may stimulate innovation--or at least R & D activity--but that too much concentration may lead to opposite results.

Additional empirical work is necessary before the role of firm size and market structure in the innovation process can be more clearly understood. Much of this work should be conducted at the levels of industry, the firm, and the individual innovation. In other words, we require disaggregated information to augment what we now have in aggregated form. One difficulty in conducting research of this type is that data are classified in accordance with schemes based on existing technology. Such classification schemes are inadequate for measuring changes in technology itself and give inadequate attention to interproduct competition. Similarly, the conglomerate firm, which operates in a large number of differing product groups, is often an inappropriate unit for measuring technological advances. Many of the divisions of these conglomerates perform most of the functions usually associated with a single firm, and yet under existing classification schemes the firm belongs only to that industry in which its sales are greatest. The program recently initiated by the Federal Trade Commission to require large firms to report operating results by product line is a first but important step in providing the data necessary to understand more fully this aspect of innovation.

Political institutions also play important roles as modifiers and determinants of innovative activity. In many cases they interact with economic institutions. For example, antitrust policy and patent policy are closely related to questions of firm size, market structure, and technological progressiveness. Understanding the relationship between market structure and general economic performance, including innovativeness, is a first step in securing antitrust and patent policies that will stimulate innovation. The second step is making sure that present laws, policies, and procedures are working in harmony with the values of society. That is not easy; our society is not monolithic and has many different and sometimes contradictory values.

Government policies, such as patent policy, tax and spending policies, and direct regulation of certain industries, all affect the innovation process not only separately but also as they interact with each other. Whereas each individual policy may be consistent with the attainment of some particular goal, the overall effect in terms of general government policy might contain conflicts and contradictions. For example, patent and procurement policy may conflict with antitrust policy. As a general rule, existing studies look at the effects on innovation in terms of only one governmental policy, one single institution, or one narrow and closely related group. As a first step in structuring future research, we recommend that a "map" of all government policies and institutions affecting innovation be made. This "map" would indicate the effect of each entity on innovation as well as showing linkages and conflicts. Not only would this initial territorial survey indicate critical areas for future research, it would also allow researchers in one area to see important links which should be considered in their research design.

The purpose of our study of the world outside the innovation context was to structure that realm so that its aspects most important to innovation are highlighted. We indicated the major linkages between the world at large and innovation taking place in an organized context as well as connections within the exogenous system important for innovation. In short, we have first looked at innovation from the outside, as a closed box; we have located that box and shown its connection to the environment. The consideration of values, endowments, and institutions at various levels of analysis proved a most useful approach in partitioning the exogenous system. These same considerations also prove useful inside the innovation system.

III. THE PROCESS OF INNOVATION: ORGANIZATIONAL AND INDIVIDUAL CONTEXTS

As has been indicated, our ecological perspective led us to structure our investigation of the innovation process itself in terms of process phases, levels of aggregation, and disciplinary variables. The most basic phase distinction is between that part of the process which results in the first introduction of a device, product, or process, and that part which involves the spread of an innovation to other contexts. The former phase we have called *innovation*; the latter, *diffusion*. Although it is somewhat awkward to use the term "innovation" both in a generic sense to refer to the whole process and as referring to one of its phases, it is a common practice in the literature and one we have followed.

A. THE TREND TOWARD ORGANIZED INNOVATION

Technological innovation in this century is characterized by an increasingly strong trend towards the institutionalization of all its phases. This conclusions is not intended to imply that the role of the individual inventor has become negligible. Such examples as Edwin Land (Polaroid camera) and Chester Carlson (Xerography) demonstrate that, although the role of the individual inventor may have diminished, it cannot be ignored. The individual inventor can still claim credit for many major inventions which later become innovations.

However, the relative importance of the individual inventor appears to vary from industry to industry and from technological area to technological area. The reasons for this observed variance are not well understood; indeed, the significant parameters have not even been identified. The unanswered questions include the following: Do independent inventors excel for certain types of innovation or in certain settings? Are independent inventors more likely to invent in some technical fields than in others? Do individual inventors contribute more in high-technology industries than in low technology ones? Not only are the answers to these questions not known, but our state of understanding is such that it is difficult even to determine whether these are the relevant questions.

According to the explanation of Schumpeter, the innovation function must be coupled with the entrepreneurial function if the invention is to evolve into an innovation. In some rare instances, e.g., Sperry and Edison, a single individual performs both as inventor and entrepreneur. More frequently, however, the inventor is not the entrepreneur. Increasingly the entrepreneurial function is being performed by the larger corporations, just as invention is becoming a group process. Corporations possess the resources necessary to develop, produce, and market the budding innovation. Even when the entrepreneurial function is not initiated by the larger corporations, it may expand to become a large corporation which the individual inventor/entrepreneur can no longer control, e.g., Sperry Gyroscope Co.

The so-called spin-off organization (that is, an innovative corporation "spun-off" by a larger one, or, more usually, a group of inventive individuals within a large firm, who, restive under the constraints of a large organization, form their own firm) can be viewed as a modern-day counterpart of the individual entrepreneur. Research on the spin-off phenomenon reveals agreement on the industrial and organizational attributes related to the birth rates of new firms. On the surface it appears that this research area would not warrant further study, but such a conclusion may be premature. Several basic questions remain unanswered. Are there conditions under which it is more effective to develop an invention in the incubator organization? Is it possible that the spin-off organizations excel in certain elements of the innovation process or under certain conditions? To what extent is the spin-off firm of today the counterpart of the individual entrepreneur of yesterday? Is the spin-off process a viable means of initiating a self-reinforcing process by which a geographical area can experience economic development? These interesting unanswered questions remain in the study and understanding of the spin-off phenomenon.

Nowadays the innovation process is represented chiefly by organized R & D--as might be expected, since the institutionalization of innovation has been accomplished primarily by organized R & D, even though innovation also occurs in other organizational units. There is some evidence that R & D is the primary source for major, or discontinuous, innovations of all types, but that non-R & D units may be the primary sources for the minor or incremental process improvements that cumulatively can result in major cost savings. Incremental improvements are made in the major improvement until a saturation (diminishing-returns) point is reached; i.e., until the cost of the improvement exceeds its benefits. At this point another major improvement is required, and the process repeats itself. The role played by R & D and non-R & D (operating) units in organized innovation and the process by which major and minor improvements result and interact are promising areas for research relevant to organized innovation.

Technological innovation does not occur in a vacuum, but is embedded in an environment with particular and describable characteristics. The literature supports the hypothesis that some R & D environments are more conducive than others for organized innovation. Once that is recognized, either as an assumption or as an empirical conclusion, derivative questions arise:

1. If some environments are more conducive than others, is it possible to isolate and define the critical individual and organizational variables?

> *Status:* A number of studies have identified variables in information flow between organizations, information flow within the organization, and performance of scientists and engineers. Although these variables have been identified, their interrelationships have not been determined, nor has their relative importance been measured.

2. Once the relevant variables have been defined, is it possible deliberately to modify an existing R & D environment and make it more conducive to innovation?

Status: In general, the answer to this question remains unknown. A major difficulty is that no acceptable measure exists to relate the environment directly to the innovation, so it is very difficult to evaluate the consequence of the modification. The resource allocation and scheduling literatures of operations research and the creativity literature of applied psychology suggest progress toward approaches that might lead to enhanced innovation. The prescriptive literature assumes that deliberate improvement is possible, but does not test the assumption.

As suggested by the preceding questions, there is a literature which is concerned with description, explanation, and understanding, and another literature oriented to prescription and improvement. In general, the prescriptive literature is not based on the descriptive, explanative literature, but is based on an author's conjecture of how the process should operate, either generally or in specific instances. Our primary emphasis has been on integrating the descriptive literature as a basis for better informed prescriptions.

Various contexts of organized R & D exist, e.g., the industrial, agricultural, and military-space sectors. However, when the focus narrows to the internal R & D activity, these differences seem to disappear and the same questions, approaches, and results appear in the literature. To some extent that uniformity is a consequence of the tendency to study technological innovation as a process operating *within* an organization. The organization is viewed as a study site, not as a process variable. Consequently, process similarities are emphasized and organizational differences are minimized. Comparative studies specifically designed to investigate differences in innovation resulting from the various contexts of technological innovation are certainly justified.

B. FOCUSING MECHANISMS AND THE BEHAVIOR OF SCIENTISTS AND ENGINEERS

Each organization receives input from the external environment which it translates into organizational goals, needs, and capabilities. The existing studies indicate three mechanisms by which this information transfer and translation is accomplished. First, top management and their planning staffs determine corporate policy and strategy--the incarnation of the firm's values--and thus effect a match of opportunity or need with capability. Corporate policy and strategy operate as guidelines and constraints at the lower levels of the organization, and it is within these guidelines and constraints that projects are proposed and evaluated. Second, technological gatekeepers operate as a conduit by which technological information enters the firm. Third, need and opportunity information may be brought into the firm by market-need gatekeepers such as personnel in marketing, technical service, or sales. Thus, at least one formal mechanism, determination of corporate policy and strategy, and two informal agents, technological and market-need gatekeeprs, interact to transfer external data into internal information and translate it into goals, needs, and capabilities.

The literature presents convincing empirical evidence for the existence and role of technological gatekeepers, but the information transfer and translation role attributed to top management, and the existence and role of what we have called "market-need gatekeepers" have not been verified empirically. The process by which environmental input is transferred and translated into internal information for use in organized innovation is thus not well understood or even well documented. In fact, Hughes has provided a more complete description of Sperry's information-gathering behavior (see Chap. 6 below) than exists in the literature for organizations. Such studies would appear to offer a high payoff, due both to the lack of existing knowledge and to the importance of information transfer and translation for organized innovation.

Although relatively little is known regarding the process by which environmental input is transferred and translated into organizational information, there is widespread agreement that information about market needs tends to come from sources outside the firm. The significance has been demonstrated in a number of empirical studies indicating that from 60 to 90% of the ideas which resulted in innovations were initiated by the recognition of market, mission, or production needs. The needs, when coupled with a technological capability which later was coupled with a need or opportunity.

An idea is generated when a reasonably well-defined match is made of an organizational need or opportunity with a technological capability. Corporate policy and strategy from top management and information from technological and market-need gatekeepers are important informational inputs for the individual researcher and engineer. They serve to focus his attention on needs, opportunities, and capabilities, and thus simultaneously broaden and restrict his perceptual base.

Other important factors that serve to focus the individual researcher's effort include:

1. the specific mind set of the individual, which reflects his past experience and the habitual ways of thinking that he brings to his current problem-solving effort;
2. the organizational control and coordination processes--including project selection, budgeting, scheduling, and project-review systems--that restrict individual behavior and provide secondary input regarding corproate policy and strategy;
3. the attitudes and actions of the supervisor in his roles as (a) a middle man in the communication of corporate policy and strategy, (b) a path by which organizational rewards are obtained, (c) a source for critical evaluation of ideas, (d) a provider of resources and administrative help, (e) an implementer of control and coordination mechanisms, and (f) an enforcer of rules and regulations;
4. the role of the primary group, both social and work, as another source of administrative, technical, and market-need information and idea evaluation, most especially as a vehicle for introducing diverse opinions regarding the range of technical options considered and the interpretation of corporate policy and structure.

An overall assessment of the literature dealing with focusing mechanisms reveals that a number of variables have been identified and their general significance has been demonstrated empirically; however, the specific impact of each variable on the process of organized technological innovation, and the variable interrelationships, remain in the category of informed speculation.

The literature on focusing mechanisms deals primarily with idea generation as an important aspect of organized innovation. But in addition to generating ideas, scientists and engineers perform the technical work necessary to move an idea to its fruition as a utilizable innovation. The mechanisms by which the organization rewards its scientists and engineers and by which projects are controlled tend to overemphasize assigned, scheduled project activity at the expense of idea generation and idea submission (the process by which an idea becomes a candidate for the allocation of organizational resources). There is evidence that the "unsubmitted-idea" phenomenon is real and significant. For example, in one organization a number of ideas existed which had not been submitted to management for review and possible funding. Through intervention by an outside research observer, some of these projects were funded and resulted in substantial benefit to the organization. Yet it is unlikely that these ideas would ever have been brought to the attention of management without the observer's intervention. Study of the "unsubmitted-idea" phenomenon should be replicated in other organizational settings; these studies should explicitly consider the type of innovation (incremental vs breakthrough) and the view taken of the R & D process (project dominant vs process dominant), and try to explain what organizational conditions lead to unsubmitted ideas.

A systematic, integrated series of studies have been reported regarding the effectiveness with which scientists and engineers perform the technical work associated with an R & D project. Basically, these studies report that effective scientists and engineers:

1. are largely self-directed, i.e., directed by their own ideas;
2. value and seek the freedom to pursue their own ideas;
3. are coordinated by allowing several others a voice in shaping their directions;
4. interact vigorously with their colleagues;
5. work throughout the phase cycle of the project--i.e., perform research, development, design, etc.

Surprisingly, the literature does not conclude that effectiveness can be enhanced by deliberate modification of the environment to induce these characteristics. Instead, the evidence is quite the contrary. The most recent data are more consistent with the hypothesis that it is effective performance which leads to the above behavioral characteristics than with the opposing hypothesis that the above behavioral characteristics lead to effective performance. Again, we are frustrated in the attempt to determine whether it is possible deliberately to modify an existing R & D environment and make it more conducive to innovation.

The research undertaken in these three areas--(1) focusing mechanisms and their influence on idea generation, (2) idea submission and the "unsubmitted-idea" phenomenon, and (3) individual work style and its relationship to effective performance on assigned tasks--attempts to explain the influence of organizational and individual variables on the specific behavior of individual scientists and engineers. In addition, such factors as individual mind sets, time pressure, interaction with colleagues, and organizational goals are considered in all three topics. Despite these commonalities, the three topics are essentially disjoint in the literature; we have not found a research study that simultaneously studies all three activities and only a handful consider more than one. Research that explicitly studies the three behaviors, their interactions, and their trade-offs should be given high priority.

C. RELATIONSHIP BETWEEN STRUCTURE AND FUNCTION

Because of the concentration on behavioral variables, the literature dealing with the above topics tend to downplay (1) the importance of the specific innovation and its characteristics, (2) the relationship between organizational structure and the function being performed, and (3) the mutual, self-reinforcing relationship between organizational structure and orientation toward R & D. Our view of innovation as an ecological process forced us to consider explicitly the relationship between structure and function, which in turn led to a consideration of the specific innovation and of orientation toward R & D.

We were thus able to identify two R & D orientations. The *phase-dominant* orientation results in an organizational framework in which the R & D activity is structured by process phases (e.g., research development, engineering, and tooling) within each of which idea generation and submission occur. The *project-dominant* results in an organizational structure such that the R & D activity is primarily structured by projects within which all process phases are carried out. Our "phase-dominant" and "project-dominant" models of organized R & D evolved from our synthesis of the literature; they are not contained in the previous literature, although several of the case studies imply them. More empirical confirmation is needed, because these models have the potential for providing the analytical ability needed to relate structure and function in the study of organized innovation.

The preceding paragraph suggests a one-way implication--orientation toward R & D determines how the R & D activity will be structured. Our analysis suggests tha the converse implication also holds--how the R & D activity is structured determines which orientation toward R & D will be predominant within the organization, and hence the kind and number of innovations which will emerge.

During the 1945-1955 growth period for R & D, the structures and the control and coordination mechanisms that were being used with apparent success to manage other organizational functions were simply extended to this emerging activity. As a result, R & D was blended into the existing organizational hierarchy as a new division (a "corporate laboratory") or a new function in an existing division (a "divisional laboratory") and was subjected to the organization's usual hierarchical control and coordination mechanisms. In the absence of conceptual models, the implications for R & D of the hierarchical structure and the associated control and coordination mechanisms were not well understood at that time. Specifically, it was not understood that they would most likely lead to a phase-dominant orientation toward organized R & D.

In order to explain the significance of this implication, it is necessary to consider distinctions at the level of the specific innovation. The literature distinguishes between innovations that constitute incremental improvement or minor extensions, i.e., "incremental" innovations; and innovations which represent major breakthroughs or discontinuities, i.e., "discontinuous" innovations. Thus, "incremental" and "discontinuous" refer to the degree to which the innovation is an extension of, or a departure from, current practice or existing knowledge. Other distinctions made in the literature include: size, in terms of cost and/or personnel; time to completion; level of uncertainty, both technical and economic; number of technologies; number of organizational units; number of R & D phases; urgency with which the results are needed; and extent of contribution to organizational goals and objectives.

The phase-dominant orientation, and the resulting organizational structure, are consistent with the attributes of relatively small, short-time-horizon, single-phase innovations that are carried on within one organizational unit and that involve one or at most a few fields of technology (for brevity, "unimodular innovations). The phase-dominant orientation is not consistent with the attributes of relatively large, long-time-horizon, multiple-phase innovations that cut across organizational boundaries and that involve several fields of technology ("multimodular" innovations). Conversely, the project-dominant orientation is consistent with the attributes of multimodular innovations, but inconsistent with unimodular ones.

We have accordingly concluded that the hierarchical structures and mechanisms adopted for R & D during the 1945-1955 period led most U.S. organizations to a phase-dominant orientation of R & D. This orientation, in turn, slanted subsequent R & D activity toward unimodular innovations and away from multimodular innovations. When multimodular innovations were attempted they could be carried out only at high cost and with major disruption to existing organizational norms and structures.

Consider, for example, the Rubenstein and Radnor (1963) finding that, even though both corporate and divisional management agreed that the emphasis of R & D should be on "multimodular" R & D programs, "unimodular" programs were much more likely to occur in the decentralized organizations which they studied. We suggest the explanation that organizations with a phase-dominant orientation are more likely to identify and undertake multimodular ones. The current popularity of topics such as "matrix organizations" and "project management" in the management literature would suggest an attempt to move toward a project-dominant orientation in order to facilitate the research and development of multimodular innovations. However, difficulties are being experienced with these approaches also. The difficulties arise at two points: when the matrix or project organization must interact with the parent hierarchical organization and when the project is finished and the team members must reenter the parent organization.

The interrelations among organizational structure, function, and orientation toward R & D are complex and not well understood. Even less well understood is how these factors interrelate with such considerations as focusing mechanisms and project performance at the level of the individual scientist and engineer. It is highly unlikely that understanding of organized innovation will be accomplished by the myopic, piecemeal, noncumulative empiricism that now characterizes the literature. There is a critical need for higher-level, integrative theories and models that can guide the empirical studies and lead to cumulative results, both explanative and normative. We see an immediate need for an iterative relationship between theoretical and empirical research.

Several of the necessary theoretical constructs appear in the organization development literature--one of the few literatures to have recognized the relationship between structure and function. The literature defines such constructs as level of *complexity* in the organization's incentive system (a highly complex system can reward for a large number of different behaviors): level of *diversity* in the task structure (a highly diverse structure permits an individual to perform a number of different tasks): degree of *formalization* (a highly formalized organization has highly codified tasks); and degree of *centralization* (the extent to which power and decision-making authority is in the hands of a small proportion of the organizational members). These constructs are directly related to control and coordination mechanisms and focusing mechanisms: perhaps they are surrogate measures of how highly controlled, coordinated, and focused the innovative activity is within the organization.

The organization development literature views innovation in terms of an initiation phase and an implementation phase. One of the more interesting hypotheses in this literature is that the organizational conditions most supportive to innovation at the initiation phase (high complexity, high diversity, low formalization, low centralization) are just the opposite of those most supportive at the implementation phase (low complexity, low diversity, high formalization, high centralization). It appears that this literature contains a number of theoretical constructs and hypotheses that can serve as a start point for developing the higher-level integrative theories and models of the process of organized innovation.

D. THE UTILIZATION OF R & D RESULTS

One additional problem area was discussed in Chap. 3: the interface of R & D with the operating units of the organization and the effective utilization of R & D project results. With some exceptions, notably the Quinn-Mueller survey (1963) and the writings of Morton describing the Western Electric-Bell Laboratories interfaces (1968), the literature on this topic is nearly nonexistent. Our assessment of this problem area is that it is one of utmost importance but the one for which there is the least information in the available literature, owing perhaps to problems with proprietary information. We believe that studies of this area deserve high priority.

In summary, a number of functions have been identified at both the individual and organizational level: idea generation, idea submission, performance characteristics, and transfer from R & D. In addition, the influence of structural conditions has been examined, including characteristics of the organization, the structure of R & D within the firm, and characteristics of the specific innovation. We have also attempted to assess the influence of the orientation toward R & D which is held by the firm and its members.

In the course of drawing together the results of these several lines of investigation, we have developed a preliminary conceptual structure--a framework for integrating and cumulating across the many diverse and fine-grained studies of R & D. This integrative conceptual structure or some superior alternative to it--though we know of none at the present--needs to be refined to the point of enjoying broad acceptance by the research community. In the absence of such a unifying framework, research results will continue to be largely noncumulative and our understanding of organized innovation correspondingly piecemeal.

The need for such a conceptual structure, adequate for integrating and guiding research on organized innovation, is matched by the equally pressing need for an even larger framework which displays the linkages of this segment of the total process to the societal context in which it is embedded. The "input" side of this larger whole was sketched in Chap. 2. The "output" side involves the diffusion of innovations, which was the focus of Chap. 4.

IV. THE DIFFUSION PHASE

The field of diffusion research is not clearly defined either sociologically, in terms of a single, interactive community of scholars; or conceptually, in terms of a theoretical framework that yields a set of relevant variables and criteria of methodological rigor. On the contrary, diffusion research is conducted by several communities or traditions, each with its own distinctive intellectual orientation and commitment. Each of these traditions seeks, and finds operative, certain variables in the diffusion process, and in terms of these variables accounts for a portion of the phenomenon. Each tradition also leaves a large, unexplained residue.

We surveyed the three basic diffusion research traditions (the geographical, the economic, and the social-psychological) noting both their differences and their implicitly complementary findings. We also examined the so-called "classical model," which attempts to identify the basic elements of the diffusion process. Finally, we described what we take to be the dimensions of a general theory of diffusion. In the absence of a theoretical structure embracing at least these dimensions, we argued that understanding of the diffusion process can advance little beyond the present fragmented, myopically empirical, and noncumulative state of the art.

The so-called "classical model" of diffusion identifies the following elements of the process; "(1) *an innovation* (2) communicated via certain *channels*, (3) to members of a *social system*, (4) who adopt it over a period of *time*" (Rogers with Shoemaker 1971). We agree with the centrality of the first element (the characteristics of *innovations*) but regard its conceptual structure as inadequate to support needed research on the influence of such characteristics on adoption behavior. We also agree with the emphasis on *channels of communications*, but again do not feel that the conceptual structure supporting this element of the model is sufficiently strong to permit an integration of the several research traditions. In addition, it does not reflect adequately the complexity inherent in the various levels of aggregation involved in diffusion and the communications patterns within and between these levels.

With regard to the *social systems* concept, we again agree with its centrality, but suggest distinctions must be made among relevant social systems in terms of levels of aggregation and their roles in the diffusion process. That is, social systems may (1) function as adopters, (2) be composed of adopter units, (3) be part of a larger whole which is an adopter, or (4) function as propagators.

The final element identified by the classical model, *time*, is introduced as defining a measure of innovativeness (time to adoption) and a measure of diffusion, namely, the rate of adoption. Although a model or general theory of diffusion must account for these differences, the fact is that they are dependent rather than independent variables. Thus we question the emphasis on this methodological, as opposed to substantive, consideration.

Despite these deficiencies, the classical model does represent a first effort to move beyond the fragmentation of effort and results of the several research traditions. In addition to this fragmentation, which works against needed theoretical development, another weakness in the diffusion literature derives from what we have called the other-things-being-equal syndrome. As Rogers and Shoemaker have noted, 95% of the 6811 empirical generalizations they identified in diffusion research publications were two-variable generalizations--"whereas the real nature of diffusion is certainly a cobweb of interrelationships among numerous variables" (1971, pp. 93-94). Thus, the dominant procedure has been to investigate the relationship of some x and y either on some dubious assumption that all other variables are unimportant, not operative, or always function in the same manner, or on the scholarly dead-end that the generalizability of results is unimportant.

The myopic empiricism of the other-things-being-equal syndrome yields meager understanding of the diffusion process. It can be overcome only by an increase of multivariate research designs and development of an expanded and refined conceptual framework to guide empirical research. As a first step in this direction, we have offered a preliminary sketch of the dimensions of a general theory, derived from the existing, disparate literature and the classical model.

Four major dimensions of influence were identified as impinging upon the process by which innovations come to be adopted. The first involves the nature and *characteristics of sectors*. Sectors are social systems, the constituents of which are adopter units plus certain other units with whom the adopters regularly interact. Sectors may be public (e.g., state or local governments) or private (e.g., firms); they may be composed of either individuals or organizations; and they function as mechanisms of influence (social hierarchies) as well as communication.

The second major dimension of influence on adopter behavior is the *characteristics of adopters* themselves. Certain parallels were suggested between the influence of organizational structure on innovative behavior and on adoption behavior. We also examined the influence of organizations as social systems and as embracing smaller social systems (work and social groupings). Finally, we looked at the role of individuals, both as influencing the adoption behavior of organizations and as adopters in their own right.

The third dimension of influence concerns the *characteristics of particular innovations* themselves. Here we introduced a distinction between adopter-independent and adopter-dependent characteristics. As these terms indicate, the influence of the former arises directly from the nature or form of the innovation itself, whereas for the latter the nature and circumstance of the potential adopter is an intervening variable in the determination of their influence. As it turns out, quite similar lists of characteristics have been developed by various research traditions. These characteristics have not been sorted according to the adopter-independent/adopter-dependent distinction, so that there has been little success in attempts to assess their relative influence and how such influence operated.

The final dimension of the general theory which we have suggested involves the various *propagation mechanisms* active in the innovation process. These mechanisms as they couple with the potential adopter may be of three types; (1) material artifact-to-adopter linkages, (2) person/organization-to-adopter linkages, and (3) media-to-adopter linkages.

They may also function at various levels of aggregation; international, intersectoral, intrasectoral, and for organizational adopters, intra-adopter. And finally, this typology of propagation mechanisms must distinguish between those who are active advocates, pro or con a particular innovation, and those who perform an adoption-neutral function of disseminating information.

Such identification of the dimensions of a general theory of diffusion is, of course, but a first step towards overcoming the fragmentation of the several research traditions and the simplicity of the "other-things-being-equal" syndrome. What is missing, and which cannot be supplied *a priori*, is an understanding of how these several dimensions of influence are linked. In fact, the long-term potential of the conceptual structure we have offered lies in its heuristic value in eliciting a new level of inquiry concerning the linkages, interdependence, and relative influence of the process dimensions. It is meant to serve as a starting point which may be revised in the course of further research.

Lest this assessment of the state of the art of understanding the process of diffusion seem unduly pessimistic, let us repeat an observation made earlier. The perspectives and emphasis of the several diffusion research traditions are substantially different, but they are by and large not incompatible or in conflict. They are, in fact, implicitly--though rarely explicitly--complementary. The unexplained residue of one is often a major preoccupation of another. Thus, we would venture the judgment that the development of a comprehensive theoretical framework, and the emergence of a single, interactive, research community could, in fact, take place rather quickly. We would urge the assignment of the highest priority to the conceptual work necessary to make this a reality.

V. A FINAL WORD

We have tried to gather together what is known about technological innovation and assess it with an eye toward the future. In so doing we have:

pictured the process as an ecological one, not to be "trendy" but because simpler analogies failed;

conducted our investigation at several levels of aggregation, both indigenous and exogenous to the process itself, because understanding required it;

dealt with an unusually wide range of variables, because the research community has found them to be important;

identified a number of process phases and dimensions which synthesize and at times extend the conceptualization to be found in the literature;

considered both empirical and theoretical works, because the quest for understanding requires both; and

reported, assessed, hypothesized, and speculated, because there is a place--and a need--for all of the above.

Having done these things, we are left with only a final observation.

We have earlier noted an assumption that the innovation process--and the complex multilevel environment in which it takes place--can be *deliberately* altered with the predictable result of becoming more conducive to the creation, development, and diffusion of innovations. This assumption is not only implicit in the literature, it also underlies the increasing institutionalization of the process itself. Although trial and error dominates the present state of the art, a growing capacity to control the process is envisioned.

Such an act of faith is not foolish; on the contrary it is probably a necessary condition of a professional engagement in the process at all, or in its study. One probably cannot endure the early learning frustrations of any complex process (e.g., a game of chess) without some assumption that thoughtful diligence and practice will make us better. But it is worse than foolishness to assume that our understanding and success at chess will grow if we concentrate solely on our bishops and ignore the rest of the pieces, their relative position, their functions, and their dynamic interplay. They all have a distinctive contribution to make, and success comes from mastering their interdependent and complementary roles.

Likewise, our assumption that the innovation process will be increasingly subject to rational control is folly given the present fragmentation of emphasis and effort. Purely technical considerations constitute "a piece of the action," but there are many other pieces as well: economic, social-psychological, organizational, informational, etc. We have reported the often impressive progress of these specializations in demonstrating the influence of the variables with which they deal. But the process and its study will continue to be characterized by too many unexpected turns, frustrations, defeats, and unyielding puzzles as long as pieces and their linkages are ignored or discounted as unexplained residue. We have probably realized all of the easy gains from gross correlations of action and outcome, or from muscling things through.

The future of technological innovation--if the assumption of increased understanding and control is to be realized--lies in learning how to integrate the presently fragmented and ill-coordinated specialized concerns. Neither the university nor the firm is particularly good at such interdisciplinary or interorganizational cooperation. Yet both will have to learn, for the alternative is simply to wait on those unpredictable--and rare--"happy accidents" that might come our way; and however we may delight in the serendipitous we aspire to more, for noise is rarely music or the dropcloth a painting.

PART II: ASPECTS OF TECHNOLOGICAL INNOVATION

Part I of this report examines the process of technological innovation from a holistic or ecological perspective. By juxtaposing variables that are usually treated separately, by integrating the results achieved at different levels of analysis, and by meshing the several disciplinary perspectives, we sought to determine the state-of-the-art understanding--that is, what we know, do not know, and should know about the innovation process.

At the inception of this integrative study, however, we recognized that no state-of-the-art review and assessment could do justice to every important area of concern; the subject matter is too complex and the body of scholarly literature too large and too diverse. Thus we need to balance and complement our integrative thrust with more specialized treatments of selected aspects.

Part II provides this balance. It contains commissioned papers by nine distinguished scholars dealing with special aspects of the innovation process. These scholars, each of whom has made major contributions to the understanding of technological innovation, were asked to assess the state of the art in their research areas. The resulting essays, in addition to being an integral part of the Georgia Tech Innovation Project, stand on their own as unique contributions by major scholars in the field. We are deeply grateful for their contributions to this study and to our knowledge of technological innovation.

All project participants whose papers appear in Part II represent academia. But the innovation process itself is a world of action as well as of thought; it takes place not in the "ivory laboratory" of academia but in the "real-world" setting of corporations and government laboratories. Recognizing that, we also sought--and received--the assistance of consultants from this "real world." Although they did not prepare special papers for this study, their advice has been invaluable in keeping our efforts in contact with innovation itself, not just those who study and write about innovation.

The names of all participants--the scholarly contributors and the industrial advisors --are listed in the Preface that appears before Part I. We thank them again for their significant contributions to our work.

The Editors

6. INVENTORS: THE PROBLEMS THEY CHOOSE, THE IDEAS THEY HAVE, AND THE INVENTIONS THEY MAKE
Thomas P. Hughes

The innovation process may be defined as including a number of activities, among them problem identification; idea response; invention; research and development; introduction into use; and diffusion. However, innovation has been so thoroughly investigated by now that a definition as broad as the foregoing is distressingly inadequate. Nevertheless, the simple listing of five or six stages in the general innovation process is an opening reminder that innovation is a complex of interactive events. The listing also provides a way of setting the limited scope of this essay in perspective.

For the purpose of this study, *problem identification* means the inventor's perceiving a problem or problem complex and deciding to devote himself to seeking a solution to it. *Idea response* is the inventor's effort--active and passive (subconscious, perhaps)--to formulate concepts that will initiate the process that will solve the problem. Usually the inventor gathers information as he pursues--or even awaits--ideas. The idea response is a mental one and will become an *invention* only after the ideas has been given form, The initial form is often a drawing, a mathematical model, or equation; subsequently the idea is given form as a mechanical and electrical device or as chemical process. This invention is then brought to the stage at which it can be introduced to the market by *research and development*. The research is an information-gathering exercise and can be done by a literature search or by scientific experimentation. Development, an important part of the innovation process, often involves the redefinition of the problem, new ideas and research as the invention is tried in environments increasingly like the real use environment within which the innovation must function. It is important to add that the innovation process is not a simple linear one but a process involving backtracking to identify new subproblems, to elicit additional ideas, and to make new subinventions.[1]

The problem-identification, idea-response, and invention stages of the innovation process are the subject of this essay, with concentration on the inventors themselves as problem identifiers, men of ideas, and the makers of inventions. Moreover, we shall limit attention to American inventors during the period 1870-1930, an era of strong support for independent inventors and inventors in research laboratories. Directing attention to American inventors of this era is practical because the literature available on them is in relatively good supply.[2] By contrast, it would be extremely difficult to write about inventors since 1930 because few detailed biographies have been written about them. Not only are biographies about the earlier inventors available, but these books and essays are generally grounded on a broad base of source information, for the inventors of the heroic age of the independent inventor--men like Bell, Edison, and the Wrights--were public figures and were often asked to talk about themselves, their work, and even their "Eureka moments." And the inventors of the research laboratories that mushroomed in the United States after 1900 and before 1930 were publicized by corporations impressed by their achievements and not adverse to impressing public and shareholders as well.

Considering the biographies of individual inventors and the histories of individual inventions should prove fruitful, not only because they and their works are known, but because the historian is able to focus sharply on problem identification, idea response, and invention synthesized in a single mind. This assertion is not meant to deny collegiate inspiration and constructive activity, but to take advantage of the feasibility of depending on an individual's ideas and activities as a synthesis of the thoughts and activities in his environment. For instance, Thomas Edison, Elmer Sperry, and Thomas Midgley, who will be considered in this paper, drew upon and organized the intellectual currents and the technological talents known and available to them in order to identify problems and originate inventive responses.

Dr. Hughes is professor of the history of technology in the Department of the History and Sociology of Science at the University of Pennsylvania.

Often the historian lacks the articulated methodology and fails to provide the highly structured analysis associated with social scientists. However, the historian has a unique contribution to make through his detailed knowledge of the persons and events whose biographies and histories provide the information and the situations by which theory can be formulated through methodology and analysis. Some studies of innovation made by social scientists lack the richness and complexity that well-chosen case histories provide. On occasion, history, methodology, and analysis will be found combined in the work of a single scholar, as is the case with those whose works on innovations are considered in the following pages. In its history and analysis this essay will emulate their approach. Its essential methodology will be to focus on three major inventors, single out a major invention of each, and recreate the surrounding circumstances and the events of problem identification, idea response, and reduction to practice, or invention.

Edison (1847-1931), Sperry (1860-1930), and Midgley (1889-1944) not only span the period but they are regarded as representative types. Edison is generally considered an heroic inventor who, early in his career, worked alone and who, even after having established his laboratories in Menlo Park (and later West Orange), New Jersey, invented without becoming an adjunct of a large corporation. Furthermore, Edison is widely believed to have been a self-educated empirical inventor, a genius who worked without depending on science. (However, this image, as will be shown, is somewhat wide of the mark.)

Sperry, is a transitional figure--from heroic independent to industrial research scientist. He began a career--highlighted by more than 350 patents--as an independent who greatly admired Edison and probably imitated his style. At middle age, however, he established the Sperry Gyroscope Co. (later Sperry-Rand Corp.) and directed an electromechanical research laboratory populated by very bright, enthusiastic, and energetic young engineers and inventors. The symbol of Sperry's transition was his posing for a photograph wearing a white lab coat and peering through a microscope, a dress and pose virtually unknown to him.

Thomas Midgley represented for his generation the industrial research-scientist role fulfilled. He was not only generally accorded the status of scientist, but his method has been characterized by a noted historian of chemistry as one of exemplary scientific research. "His synthesis of the antiknock agent, tetraethyl lead, and the refrigerant, trichlorofluromethane...were," according to Eduard Farber, "beautiful pieces of pure, or at least deliberately planned chemical research." His biographer for the *Memoirs* of the National Academy of Sciences wrote, "in the matter of recognition for his scientific endeavor he was particularly fortunate." He received all four of the most important medals for chemical achievement: the Nicols Medal of the New York section of the American Chemical Society, 1922; the Perkin Medal of the Society of Chemical Industry, 1937; the Priestley Medal of the American Chemical Society, 1941; and the Willard Gibbs Medal of the Chicago section of the American Chemical Society, 1942.

Edison, Sperry, and Midgley were not simply inventors; they were inventor-entrepreneurs. Inventor-entrepreneurs are inventors who preside over the innovative process from its origins as a problem to at least the introduction of the invention into use. (They may also be involved in the diffusion process, but this stage will not be noted here.) The usual reason that the inventor-entrepreneurs studied here were not simply inventors was that they were determined to have their invention used, and they realized that to achieve such use they would have to take the initiative not only in the early phases of innovation but in research and development and marketing (introducing into use). In essence, as will be seen, they were inventors; in effect, they had to be entrepreneurs. Yet the evidence is that they found their work most satisfying when identifying problems and inventing solutions, not when presiding over and promoting the other phases. The evidence also tends to support the generalization that in America before 1930 most successful inventors were in fact inventor-entrepreneurs, for any list of the outstanding inventors--Edison, Thomson, De Forest, Hall, Wright, Sperry, Sprague, Acheson, and Midgley--reminds us that these men brought their inventions into use. Most of them seem to have gone so far as to organize companies or enterprises for their inventions.

Edison, then, was the heroic inventor, Sperry the transitional figure, and Midgley, the the putative research scientist. By choosing this variety of types flourishing in the period 1870-1930, the validity and general applicability of several interpretations of the the innovation process can be considered more effectively than if only one kind of inventor--the heroic, for example--were the subject of attention. We shall focus on tho those parts of the interpretations that are relevant to problem identification, idea reasponse, and the act of invention. Case histories of an Edison, a Sperry, and a Midgley invention will provide the information for critically evaluating the interpretations.

The interpretations are drawn from three essays on the innovation process: Abbot Payson Usher, *A History of Mechanical Invention* (1954); S. Colum Gilfillan, *The Sociology of Invention* (1935); and Thomas Parke Hughes, *Elmer Sperry: Inventor and Engineer* (1971).[3] There are other studies of note on the general character of innovation, but these three studies are suitable because they are based on extensive and intensive case histories of invention and because they are particularly concerned with the formulation of a theory about the identification of the problems, idea reaction, and invention. Other studies on innovation, such as those by Schmookler, Kuznets, Bright, Rosenbloom, Rogers, Rubenstein, Strassman, and Mansfield, direct attention to economic, institutional, and social factors and changes affecting the innovation process in general.[4] Jewkes, Sawers, and Stillerman (1958) present a large number of case histories and an essay based on them, but their principal concern is to demonstrate the importance and survival of the independent inventor in the 20th Century, not to develop a theory of innovation.

Usher's analysis (Table 1) highlights "the act of insight." He illustrates the act of insight by an example drawn from Kohler's experiments with chimpanzees at Tenerife. A chimpanzee, Tschego, had bananas placed outside her barred cage just beyond her reach. Within the cage, "somewhat to one side," were several sticks. Tschego was exasperated by her inability to reach the fruit and after a number of tries reluctantly abandoned the effort, discouraged. However, when several young animals outside the cage approached the bananas, she seized the stick, which had apparently gone unnoticed, and deftly maneuvered the fruit within reach--an act of insight. Judging by the prominence Usher gave the experiment in his chapter on "novelty in thought and action," his learning of it was his act of insight into the nature of invention. Furthermore, Usher constructed his theory of the nature of insight as a generalization of the experiment or incident. Usher (1954, p. 65) wrote that

> for purposes of generalized exposition, this analysis of the individual act of insight can be formalized as a generic sequence of four steps. The first step is the perception of a problem which is conceived as an incomplete or unsatisfactory pattern. Typically, the problem is an unfulfilled want. Gratification is made effectively possible by some fortuitous configuration in event or in thought, which present to the individual all the data essential to a solution. This step can be called the setting of the stage. For the great apes at Tenerife, this vital step was really the work of the experimenter; for the general process of invention this step is dependent upon pure chance, or upon the mediated contingency of a systematic effort to find a solution by trial and error. At low levels of empiricism, trial and error is commonly presented as aimless fumbling; at substantial levels of scientific activity, the trial-and-error approach is described as systematic experimentation. The setting of the stage lead directly to the act of insight by which the essential solution of the problem is found. But this does not bring the process to an end. Newly perceived relations must be thoroughly mastered, and effectively worked into the entire context of which they are a part. The solution must, therefore, be studied critically, understood in its fullness, and learned as a technique of thought or action. This final stage can be described as critical revision.

Confusion could arise from Usher's statement because he regards the "act of insight" as one step in a generic sequence that he calls "the individual act of insight." Judging by the context of the chapter in which he developed his act-of-insight thesis, one assumes Usher intended that the generic sequence be called "the emergence of novelty."

Usher believed than an individual act of insight--or the emergence of novelty--in human society, as contrasted with the act of the apes of Tenerife, became significant (that is, emerged as a major invention) only through cumulation. His explanation is complex and deserves detailed analysis, but in essence he seems to be saying that a major invention, like the incandescent light, the airplane, or the automatic ship pilot, result at least in part, from the synthesis by an inventor of his insight and a number

TABLE 1.—Concepts of innovation process.

General	Usher Emergence of Novelty	Gilfillan Principles of Invention	Hughes Inventor-Entrepreneur
Problem identification	Incomplete pattern Perception of problem	Growth and chance evokes system need or imbalance	Reverse salient in expanding technological system Congenial critical problem identified by inventor
	Setting of state	Evolutionary accretion of response components	Consideration of prior unsuccessful solutions Experimentation
		Nontechnological factors stimulate response	
Idea response Invention	Act of insight	Institution or person organizes congenial synthesis of responses	Discern weakest point Invention
Research and development			Development toward complexity of the use environment
Introduction into use			Introduction into use
	Critical revision		Postinnovational development

of earlier insights that were known to him because knowledge of them was preserved and available to him. Some of the earlier insights may actually be the inventor's own, but Usher wanted to stress that the organized social communication system of a social structure makes possible the complex syntheses of man as compared to the relatively simple novelties of the apes.

In order to explain novelty in thought and action, Usher stressed the emergence of an unsatisfactory pattern, which he equates with "perception of a problem" in his emergence-of-novelty sequence. In this instance, he was drawing upon Gestalt psychology, for he assumed that the inventor was able to comprehend the inadequacy of an existing technological device or process by the mental act of comparing it to a mental pattern encompassing it, related to it, but extending beyond it in effectiveness, as judged by function, efficiency, or some other external factors. This sense of inadequacy is the unsatisfactory pattern and the inventor feels tension until by an act of insight the implications of the pattern are fulfilled.[5] We shall return to Usher's generalized exposition on the act of insight and the emergence of novelty but Gilfillan's theory of invention must also be considered so that the two can be compared. Gilfillan explicitly labelled his theory of invention, in contrast to Usher who seemed reluctant to encompass his act of insight and emergence of novelty within such simple conceptual confines. In *The Sociology of Invention*, Gilfillan defined thirty-eight "social principles of invention." These principles, he wrote, were based primarily on his detailed study of invention in his companion volume, *Inventing the Ship*. He characterized his principles as social because of his conviction that invention is evolutionary, an accretion of "details, modifications, perfectings, minute additions," and is not solely the brainchild of an heroic individual inventor. His belief that invention is a social process is further revealed in his thirty-eight principles.

Not all the Gilfillan principles are relevant to the problem-definition, idea-reaction, and invention stages, as these concepts are used in the present essay. The relevant ones fall primarily within five of the seven categories within which he grouped the thirty-eight principles: (1) the nature of invention; (2) changes evoking invention; (3) factors fostering, retarding, and locating invention; (4) principles of change; and (5) inventors and other classes, and tendencies in the craft.[6] Within the category *nature of invention*, Gilfillan advanced the proposition that an important invention is an accretion of detail and an evolution resembling a biologic process. He also defined an invention as involving a complex of elements, including a process for using it; a method for building it; installations such as shops and factories for housing, supplying, and maintaining it; and diverse other elements. Under *changes evoking invention* he subsumed inventions themselves, for they introduce changes in complexes that alter the whole and necessitate other inventions, so that the whole can fulfill its functional implication. Other changes invoking invention are "increase of wealth, growth of population, industrialization, and institutional change." Among the *factors fostering, retarding, and locating invention* are leading firms that use inventions to foster production. Because their products have a set character such as automobiles, electric light and power equipment, or heavy chemicals, these firms tend to attract or "locate" inventions of relevant characteristics. Similarly, regional specialization of labor and production tends to attract a certain type of inventive activity or invention to particular regions. On the other hand, Gilfillan found that the vested interest of heavily capitalized firms leads them to oppose or retard inventions that would antiquate the means of production in which the firms have heavily invested. Under *principles of chance*, Gilfillan argued that inventions are seldom accidental but usually inevitable "when the time is ripe" because of evolution in general and changes--such as those mentioned above--in particular (inventions disturbing the stability of an existing complex or system; increase in population and industrialization). For Gilfillan, such inevitability helps explain frequent simultaneity of invention. As for *inventors and other classes, and tendencies in the craft*, he believed that perception of need depends primarily on the inventor class but secondarily on the informed and articulate in any society. The craft of invention revealed to Gilfillan a trend from the "accidental toward deliberate...and...from the individual source to the organized inventing group." Further, on inventors, Gilfillan found that inventions of a revolutionary character come from outsiders, while the perfecting devices come from those within the industry that the invention affects. "The inventors," Gilfillan stated, "are in partnership usually with enterprisers...."

Not all Gilfillan's principles have been summarized, but only those that pertain to problem identification, idea response, and invention. The principles have been summarized in an order and in a form much like Gilfillan's original statement so as to suggest the cast of his thought and the general character of his analysis before reorganizing and restating the principles so that they can be used in an analysis of the early stages of the innovation process as presided over by inventor-entrepreneurs. To achieve the objective at hand it seems appropriate to recast his principles in the following way: invention is the cumulation through an evolutionary process of a complex system; such as system is affected by new inventions in a way that necessitates still further invention so that the system can fulfill its function; evolving inventions and existing systems are stimulated and modified by nontechnological factors such as population increase and economic growth or recession; industrial firms and regions tend to preside over and attract inventive activity that maintains the momentum of their existing means of production and products, whereas inventors beyond the influence of the region or the industry tend to invent devices and processes that would antiquate or alter the status quo; the general trend toward specialization in industrial activity has been accompanied by a trend toward specialization in invention, which has manifested itself in the professional inventor. He responds, Gilfillan believes, to specific needs revealed more clearly by the specialization of labor and in industry.

The biography of Elmer Sperry is, among other things, a sequence of analyses of at least twenty of Sperry's major inventions and a less exhaustive discussion of many more of the 350 patented inventions that he made during his lifetime. From the analyses of the creative work of this professional inventor emerged a pattern of problem identification, idea response, and invention. Until more biographies of professional inventors are written in an analytical vein, one cannot state with authority that Sperry's style was typical or atypical, but comparing the analytical case histories that follow of other inventions and inventors with the Sperry style is a step in the direction of this goal.

Sperry's problem identification was complex and subtle, but it should be added that he never analyzed or articulated his method. Evidence of his activity supports the conclusion that he chose fields of endeavor in which his particular characteristics as an invention were needed. His particular characteristics began to emerge early in his career as he found himself adept in solving problems amenable to the electromechanical and the autmoatic-control approach. For example, patents that he acquired in his early twenties claimed improved design and controls for electric arc lamps and electric generators. Because other problems (such as those involving streetcars and electric automobiles) responded to the electromechanical and automatic-control solution, he acquired a generalized characteristic as an inventor. Because Sperry was a professional and not a dilettante he sought problems in various fields, but he instinctively sought problems amenable to his style because he sought environments, or fields, in which capital was available for inventors, in which a market for his kind of invention was probable, and in which critical problems were emerging. Furthermore, he gravitated to fields not yet populated by teams of corporation-supported inventors, engineers, and scientists. If he found such a field, or environment, in the nascent automobile industry, he would concentrate there; if he found it in the electrochemical, he would work in that. Later in his career, he discovered that the armaments race preceding World War I provided the funding, the market or need, the opportunity for the unsalaried professional, and the critical problems which attracted him as a professional (Hughes 1971, pp. 54-70).

Between 1880 and 1910, the period before he established the Sperry Gyroscope Co. and worked closely with the military, Sperry shifted his efforts from electric light and power to electrical mining machinery, electric traction, automobiles, batteries, and industrial chemistry. Superficially these shifts may seem irrational, but a deeper probe reveals, besides the electromechanical factor common to all and the automatic control shared by half, a sharp growth of investment in a rapidly emerging field common to all. Judging by the nature of his patent applications, Sperry seems to have lost interest in a field after about five years, which suggests that an inrush of inventors, engineers, managers, and corporations brought by capitalizaton, growing markets, and size convinced him that his special characteristics and circumstances could best be employed elsewhere. The research and development, or industrial, laboratory loomed on the horizon, and Sperry was canny enough not to try to compete.

Identification of critical problems was a professional capability of Sperry's and one crucially important for his survival as an independent inventor. When Sperry's numerous patents are examined and their claims considered, it becomes clear that he did not invent dynamos, arc lights, streetcars, or automobiles, though the title of his patents might lead to such superficial conclusions. In the case of his dynamo patent, he claimed automatic controls; in connection with the arc light, he invented a regulator for the feed of the carbons; and when inventing for streetcars, he contributed an operational control. His patent claims in these instances and in numerous others show that he solved very specific problems--which can aptly be labelled "critical problems."[7]

Critical problems can be defined as problems retarding technological and/or industrial change and as problems likely, in the opinion of the inventor, to be solved by invention. Sperry's letters, memoranda, notebooks, and other records reveal that he identified critical problems by close study of technical journals, patterns of patent applications, the patents of others, attendance at meetings of professional engineering societies, conversations, and his intimate knowledge of expanding technological systems. Articles in technical journals often told Sperry of the interests of other inventors and therefore of problems on which they were working; in weekly reports of patents granted, like those published in the *Scientific American*, he could discern a pattern of concentration on certain problems (for instance, a bevy of patents on arc-light regulation). By a close reading of the claims of these patents he could delineate the problem of focused attention more precisely; and the *Official Patent Office Gazette* regularly provided summaries of all patents. Sperry regularly attended the sessions of conferences of engineering societies, for there he might gather from fresh reports and papers more intelligence pertaining to critical problems on which other inventors were working. In his later years, Sperry was a part of the engineering establishment, and his conversations with his peers of the "invisible college" also helped him direct his inventive talents. However, Sperry's close attention to the work of others was probably matched by a careful check by them of what he was "up to."

Moreover, he identified problems through deep involvement in evolving technological systems. For example, he learned of the need for a whole family of shipboard inventions by realizing the implications of the centralized feedback guidance and control device (the gyrocompass) that he had invented and installed on American dreadnoughts. A complex and elegant shipboard system for gunfire control emerged as he and his small staff of engineers identified critical problems with the help of naval line officers and invented electromechanical solutions that depended on the reference system provided by the gyrocompass.[8]

So far in our discussion of the Sperry model we have focused on his identification of critical problems. Identification implies the prior existence of the problem that the inventor perceives and formulates. In the Sperry biography, these complexes were named "reverse salients." The concept originated in the military idea of an advancing line of battle punctuated by segments in which the advance has been significantly retarded. These segments of the line, reverse salients, are often seen as situations that must be improved if the advance is to continue. Similarly, advancing technological, economic, and social systems have reverse salients, bottlenecks, or imbalances calling for response. If an inventor perceives in the reverse salient a critical problem that he is inclined to attempt to solve by invention, then he has related the reverse salient with the critical problem--a step in the innovation process. Possibly the identification of a critical problem should be placed not only in the "problem identification" category but also in "idea generation," for the identification itself is an act of insight, probably not unlike idea response. Defining the problem is a big step along the route to solution.

Having identified critical problems--in reverse-salient areas--susceptible to his inventive style, Sperry had yet to experience the "Eureka moment," related to idea response. After a long and distinguished career, Sperry ruminated about this experience, which can assume mystical proportions in popular and ill-informed accounts of invention, or acts of insight. He said:

> Think as I may, I cannot discover any time in which I have felt in the course of my work that I was performing any of the acts usually attributed to the inventor. So far as I can see, I have come up against situations that seemed to me to call for assistance. I was not usually at all sure that I could aid in improving the state of affairs in any way, but was fascinated by the challenge. So I would study the matter over; I would have my assistants bring before me everything that had been published about it, including the patent literature dealing with attempts to better the situation. When I had the facts before me I simply did the obvious thing. I tried to discern the weakest point and strengthen it; often this involved alterations with many ramifications which immediately revealed the scope of the entire project. Almost never have I hit upon the right solution at first. I have brought up in my imagination one remedy after another and must confess that I have many times rejected them all not yet perceiving the one that looked simple, practical, and hard-headed. Sometimes

it is days and even months later that I am brought face to face with something that suggests the simple solution that I am looking for (Hughes 1971, pp. 293-294).

Having thus commented on problem identification, Sperry suggested the nature of idea response, at least for him. A close reading discloses that he set the stage by reading all he could about other inventors' efforts to solve the critical problem to which he was currently directing his attention. A clue to his movement toward fresh ideas is contained in his statement that he "tried to discern the weakest point and strengthen it," a remark that can readily be associated with Usher's Gestalt theory with its emphasis on the incomplete pattern. Continuing, Sperry brought up into his imagination "one remedy after another" until perhaps many days--or even months--later he realized the perfectly "simple, practical, and hardheaded" one. The act of imagination or insight defies precise logical analysis, but once Sperry had familiarized himself with prior efforts to solve the problem, his seeming ability to find the "weakest" point and his emphasis on practical and hardheaded "remedies" defined in part the matrix from which emerged that perfectly "simple" solution.

Having thus had the idea of an invention, Sperry then embarked on a process that can be categorized as invention, but invention merged imperceptibly into development. Analysis of Sperry case histories leads to the conclusion that he transformed the idea into an invention and then into a developed device ready for the market by continually modifying the original idea to adopt it to increasingly more complex environments until it functioned as a device or process in the real environment in which it was intended to be used. The invention-development process often involved initial notebook sketches with mathematical equations predicting performance.[9] Next followed the construction of a model, perhaps by a professional model builder, which was used in laboratory tests. (The laboratory environment was usually far simpler than the real environment in which the device or process ultimately would be used.) Afterwards the device was scaled up to full size and given experimental runs in a controlled environment. Finally, the complex full-scale developed invention was tried in real circumstances. The successive environments usually involved more variables, altered parameters, and new factors, which revealed the need for successive modifications of the first idea by revision and new invention. What had been an elegantly simple idea expressed as a mathematical equation at the start became, at the end, an almost incomprehensible complex of adaptations to increasingly complex environments. It should be added that, after introductions onto the market or into use, unanticipated environments were often encountered and necessitated postinnovation invention and development so that the device or process could survive.[10]

Having considered three theories of innovation as they pertain to problem identification, idea response, and invention, we now turn to three case histories in order to illustrate the theories and to test their validity. We cannot expect that each case will confirm each item in the theoretical frameworks that have been erected, but we may hope that the cases will suggest the items in each theory that are most helpful as explanations and will suggest a selection and a single theoretical synthesis. The resulting synthesis cannot be presented as having general validity beyond the case histories considered in this essay, but it can be taken as a hypothesis to be tested by broad application and might provide a basis for further research and analysis. It might also prove helpful as a hypothesis in formulation of policy to stimulate problem identification, idea idea response, and invention.

The first case to be investigated is that of Elmer Sperry's marine gyrostabilizer. It is suitable because detailed information relevant to problem identification, idea response, and invention for this case is available. Sperry's notebooks suggest that he identified gyrostabilization as a likely problem area of interest after reading in 1907 of experiments abroad using gyros to stabilize ships and monorail trains. His notebooks carry references to gyro articles in *Scientific American*, *Electrical Engineering*, and *Engineering*, technical journals he often read.[11] He seems to have been inclined to investigate this area initially because he had been fascinated by gyros ever since he had seen them demonstrated in a science class and because he had been interested enough some years earlier--for reasons the sources do not disclose--to have tested the stabilizing effects upon a rowboat of a small electrically driven gyro.

Initially, in 1907, Sperry's commitment to the gyro field was not substantial; he carried on other inventive activities simultaneously with more enthusiasm and resource expenditure. His first application of the gyro in 1907 was a circus trick: a wheelbarrow was to enclose a gyro, and a clown was to use the stabilized wheelbarrow to steady himself as he walked a tightrope. Nothing came of that despite approaches to Barnum and Bailey Circus. Next, he thought he had found a critical problem which, if solved, would bring funding by a rapidly expanding industry for which he had invented earlier. He proposed a gyrostabilizer for automobiles. Knowing that the automobiles of 1907 had a high center of gravity so that they would clear the bumps of the poor roads of the day and recalling the overturning of an auto that had threatened to injure his son seriously, he patented an automobile stabilizer similar in action to ship and monorail stabilizers that were being tested abroad.[12] Again, the response was negative because the automobile manufacturers considered it an expensive solution to a marginal problem.

Early in 1908, judging by the frequency of his notebook entries about the subject, he was substantially committed to the gyro applications, especially ship stabilization, and he found the "weakness point" in "earlier attempts to better the situation." This weak point assumed the proportions of a critical problem for Sperry after he decided that if he could correct it, he would have a significant invention. So convinced, he spent weeks in "idea generation."

The weak point, or critical problem, holding back acceptance of ship stabilization was (Sperry decided) the sluggish reaction of the well-publicized gyrostabilizers being tested by the inventor Ernst Otto Schlick in Germany.[13] The sluggishness allowed the ship to develop the momentum of a roll before the gyro reacted. Friction and inertia were well known to mechanical inventor Sperry; also well known to him, because of his earlier work on automatic controls for electrical dynamos, were techniques of controlling large masses with small sensors and servomotors (although Sperry did not call them so). In short, the idea generation came out of his imaginatively coupling his ability to identify the friction and inertia component of malfunctioning machinery (he was an artful diagnostician) and his automatic-control experience and inclination. Also to be appraised in any endeavor to understand Sperry's idea generation is the probability that he was aware of the technique used by the monorail-stabilizer inventor Louis Brennan "to accelerate the precession of the gyro." Simply to have read the infinitive phrase might have been enough to stimulate Sperry to imagine by analogy a stabilizer able to anticipate the roll of a ship and to head it off.

In essence, Sperry's patented "active" ship stabilizer involved an electric motor to activate the stabilizing response of the gyro before an unaided "passive" gyro could respond, with a delay caused by inertia, to the roll of the ship. He controlled the activating motor with what he called an actuator, or what would later be called a sensor. The sensor was a pendulum, sensitive to extremely small, incipient rolls. The pendulum operated switches, valves, or other means for controlling the activating motor causing the early gyro response. By "activating" the gyro, Sperry solved a critical problem that was obstructing the development of the gyrostabilizer. Comparative U.S. Navy tests of this "active" gyrostabilizer and the "passive" gyro proved the superior roll-quenching capacity of the former.

The ultimate intention here is not, however, to explain Sperry's technology but to analyze his inventive style. After conceiving of the active gyro and using sketches and mathematical formula to visualize and predict its behavior, he simulated a ship and an active gyro. He used a pendulum to simulate a ship and a small hand-controlled gyro to simulate the automatic device, which would be installed aboard ship. He then interested U.S. Navy captain, David W. Taylor, who was a naval engineer and in charge of the Navy's experimental model basin, in the gyrostabilizer as a means of providing a stable ship, or gun platform, that would increase accuracy of gunfire. However, Taylor and Sperry agreed that the early simulation did not introuduce sufficiently realistic conditions. Therefore, in order to develop Sperry's invention, Taylor provided a far more sophisticated ship simulation at the model basin, and Sperry had a more complex embodiment of his invention constructed.[14] With the data derived from these experiments, Taylor and Sperry felt confident that they had a close enough approximation of reality to design a gyrostabilizer for installation in 1911 aboard the torpedo-boat destroyer *Worden*. As was anticipated, sea trials led to postinnovation modifications.

Easily discernible in the Sperry marine-gyrostabilizer case are Usher's perception of of the problem, or incomplete pattern (the passive gyrostabilizer), his "setting of the stage" (Sperry's experimentation directed by his experience in automatic controls), the act of insight (active principle), and critical revision (development). Sperry's inventive style as described earlier is also manifest in the single case history given: concentration on new and rapidly expanding industrial or technological fields likely to have problems amenable to solution by his acquired characteristics as an inventor; the appraisal of earlier attempts to solve problems in the field (gyrostabilization) with particular attention to their weaknesses; identification of a critical problem that could be solved by invention; and the generation of ideas by synthesis and analogy from prior experience and need-parameters of the critical problem. The case history also illustrates Sperry's style of development.

Some of Gilfillan's principles are also fulfilled by the case history. The naval armaments race generated the funds that Taylor and the Navy used to develop Sperry's extremely expensive invention (an example of nontechnological factors such as industrialization that Gilfillan believed stimulated innovation). Also, the ship stabilizer was a complex of electric motors, feedback control devices, mechanical components, and materials that obviously support Gilfillan's contention that an invention is a complex system of components that have evolved over a long period of time.[15] Also, Gilfillan's statement that industrial firms or regions tend to attract inventions that maintain the momentum of existing systems is supported by the U.S. Navy's attraction to and support of Sperry and his invention, for it was another step in the improvement of naval gunfire, a technique undergoing dramatic change in the prior decades.

Another case history about which sufficient information is available to allow detailed analysis of problem identification, idea response, and invention, is Thomas Alva Edison's invention of a system of incandescent electric lighting.[16] Within the framework provided by Usher, one would agree that Edison's invention of the system of incandescent lighting was the culmination of a societal process. Before Edison turned to the field, electric generators of improved characteristics had been introduced in the 1860s and 1870s by Werner Siemens and Z. T. Gramme, to mention only two of many significant contributions.[17] The transmission of electric currents had been studied and achieved by numerous inventors, engineers, and scientists working with the telegraph and the underwater cable. Arc-light inventors like Charles Brush and Paul Jablochkoff had installed systems of arc lighting involving generators, transmission lines, and lamps. The experience obtained by these earlier acts of insight and by the emergence of novelties was recorded and available to Edison through technical articles, books, patents, and the devices themselves. Furthermore, the culmination of the societal process was part of the setting of the stage.

Usher's setting of the stage is an embracing concept. The societal cumulation can be thought of as an underlying cause. His simple illustration--the sticks and the bananas in close proximity--suggests an effort is also needed to find the immediate causes of Edison's act of insight into the essential characteristics of the system he intended to invent. (Besides the setting of the stage for this act of insight, there were many other stage settings for the acts of insight resulting in the emergence of the various components of the systems, such as the incandescent bulb.) The environmental factor that seems to have set the stage most immediately was Edison's Menlo Park laboratory. He had established it in 1876, equipping it with a library, office, chemistry laboratory, scientific instruments, machine shop, carpenter shop, and other special facilitires. He also staffed it with craftsmen, engineers, and scientists.[18] This array of facilities and talents probably set the stage upon which Edison's thoughts turned to inventions that were complex systems and that would utilize the varied human and physical resources at his disposal. A less heavily capitalized inventor would have probably attempted to invent a single component of the system.

Because Edison intended to invent a system involving many components, he perceived many incomplete patterns. The interaction of the various components he was inventing and developing revealed incomplete patterns in one another. For example, the emerging characteristics of the distribution system affected the evolving concept of the incandescent lamp. Edison, like Sperry, also contemplated the earlier efforts of other inventors to invent incandescent lamps, generators, and distribution systems; in so doing he doubtless focused upon the "weak points" in these earlier endeavors and, in finding them, perceived Usher's incomplete pattern. As indicated earlier, an analysis of the Edison history carried on at length would reveal incomplete pattern following upon incomplete pattern.

Usher's act of insight occurred many times for Edison as he was inventing the system.[19] Attention here will be directed to the act of insight that defined the characteristics of the system as a whole. Also like Sperry, Edison set up need parameters and searched for analogies. In the Edison case, his parameters were economic as well as technological. In 1878 he wanted a system of lighting that would compete with gas lighting prices when he embarked on a sustained effort to introduce a comparable incandescent lighting system. After having evaluated the various costs anticipated, he found that the copper costs of his distribution system was a major consideration. A prime technological objective was the invention of an incandescent lamp filament that would last for hours--not for the mere minutes of the earlier filaments of other inventors.

Edison's act of insight in response to these parameters was the concept of an incandescent filament that would minimize the amount of current that had to be carried in the conductors and so reduce the size of the copper conductors and the energy loss in them. He reached this concept by a complex route involving the imaginative application of science. In essence the idea reaponse was structured by Edison's familiarity with Ohm's law and his ingenious manipulation of the characteristics of his system within the structure provided by Ohm's law. He was also guided by his resourceful application of Joule's equation relating the energy in a system and the voltage and current relationship of the system. One of Edison's laboratory associates, Francis Jehl, in recounting the Menlo Park days, repeatedly recalls with admiration Edison's adept use of Ohm's law, but Jehl does not explain in sufficient detail how Edison gained insights with it.[20]

Edison's development of his system involved many critical revisions of the various components to increase their efficiency, practicality, and durability. One example of critical revision must suffice, because such revision falls within the category of development, which is beyond the scope of this essay. After trying his system of incandescent lighting experimentally at Menlo Park in 1879, Edison commercially introduced his central station on Pearl Street in New York City in 1882. For this station, he used 110 volts and two-wire distribution. Critical analysis of the system revealed that high transmission cost reamined a deterrent to the adoption of the Edison system for lighting in other than densely populated markets. His critical revisions included an important patent on a three-wire 220-volt system, which was tried out in Sunbury, Pennsylvania.[21] Despite this revision, Edison's low-voltage transmission system could not lower transmission costs for lightly populated areas to the level of the high-voltage transformer system introduced in America by George Westinghouse and his engineers in the late 1880s. (The Westinghouse innovation is a case of a response to an incomplete pattern or critical problem, for it remedied the high transmission costs of the Edison system.[22]).

Having considered some events of the Edison innovation within the Usher framework, we shall also analyze them according to the Sperry and the Gilfillan categories. When Usher saw the perception of a problem (incomplete pattern), the Sperry analysis would posit the identification of a critical problem and also the emergence of a reverse salient in an expanding front as an inseparable antecedent phenomenon. (Usher was not explicit in his step-by-step analysis on the emergence of the incomplete pattern or problem perceived.) The front would involve a cumulation of many components conserved and related by collective action over a long period. Where Usher's interpretation suggests that the Edison Menlo Park laboratory was an immediate cause (stage setting), the Sperry analysis has Menlo Park as a set of circumstances that inclines the inventor to perceive a particular critical problem within a reverse salient.

Edison's perception of incomplete patterns, such as the lack of durability in other inventor's earlier incandescent lamps and the inefficiency of earlier high-internal-resistance generators, should be named critical problems within a Sperry frame of reference.[23] Continuing the Sperry analysis, we should label the acts of insight conditioned in part by scientific notions and economic parameters simply solutions to the critical problem.

Similarities between the Gilfillan and Usher analysis categories can also be shown by the Edison case. Gilfillan believed invention to be a complex accumulation. The societal cumulation of improvements in generators, lamps, and transmission systems mentioned in conjunction with Usher is a relevant example. Edison's finding that advances in the design and functioning of one component in his sytem necessitated invention elsewhere in the system supports Gilfillan's invention-mothering-invention, as well as Usher's incomplete patterns. Gilfillan's conclusion that inventive activity tends to localize according to the special characteristics of the invention and of the environment in which it is invented relates to Usher's stage setting (taken to be Menlo Park, for the setting's character conditions the inventive drama enacted in it).

In the third case history, that of Thomas Midgley and the invention of a gasoline antiknock additive, we have a notably clear example of the emergence of a reverse salient in an expanding technological front. The reverse salient, it should be recalled from the Sperry analysis, is a subject area in which a perceptive inventor identifies a critical problem. The reverse salient on a technological front emerging in the Midgley case resulted in part from the rapid development of an industry, a relationship that Gilfillan anticipates by predicting that rapid industrialization and economic development stimulate invention.

Specifically, the reverse salient in the Midgley case occurred in the automobile production and use system, one that includes, among other industries, the petroleum and the automobile.[24] The technological problem emerged as the automobile-engine manufacturers sought to increase the compression ratio in the cylinders to raise efficiency but found that destructive knocking of the fuel frustrated the attempt. The reverse salient, then, emerged at the interface of two technologies, petroleum and automotive, an interface embraced by the concept of the automobile production and use system.

Dayton Engineering Laboratories Co., an independent R & D organization, responded to the interface problem. Charles F. Kettering, who presided over the laboratories, had invented a motor-generator set for electric lighting (Delco) in rural areas, a cash register, and a self-starter for automobiles, to mention only a few of his inventions.[25] The attention of Kettering and his staff was drawn to the knocking problem because of the tendency of the Delco unit to knock. Furthermore, the critics of battery ignition for automobiles attributed knocking to battery ignition. Since Kettering's self-starter depended on this form of ignition, the starter could not have been introduced if the critics had been able to discredit battery ignition by associating it with engine knock.[26] Kettering's need to turn to the problem of knocking because of the nature of two of his inventions illustrates Gilfillan's proposition that invention is the mother of necessity.

Kettering turned the antiknock project over to Thomas Midgley, who joined the Dayton Laboratory staff in 1916 after receiving a degree in mechanical engineering at Ohio State University. Midgley's search for the critical problem in the reverse salient led him to attempt to define precisely and to clarify the concept of knock. Trained in science as well as engineering, he did a literature search on the subject and concluded from the not altogether conclusive evidence that knock was not a pre-ignition problem as many believed, but a detonation resulting from a rapid and intense build-up of heat in the ignited fuel and a consequent explosion of the fuel. The shock wave from the explosion striking the cylinder walls caused the knock, Midgley believed.[27]

By engaging with specifics, Midgley moved nearer critical problem identification. Having decided that the general problem was detonation, Midgley, advised by Kettering, sought a method of increasing the volatility of the fuel by means of an additive; a more volatile fuel would absorb and disperse the heat of ignition rapidly, thus avoiding explosion or detonation. His quest for an inventive solution resulted in a short-lived Eureka moment (act of insight).

Reasoning by analogy, this ingenious and resourceful mechanical engineer drew on his limited knowledge of flora to try to solve a chemical problem. Knowing that red trailing arbutus bloomed under the snow, Midgley hypothesized that the early flowering resulted from the superior heat-absorbing ability of the trailing arbutus and that this quality was a function of redness. So Midgley asked for a red aniline dye to pour into the fuel, hoping thereby to distribute the heat of ignition. Because his laboratory had no aniline dye, he was given iodine as a substitute. The knocking decreased.[28]

Further experimentation disclosed that it was not the color but the chemistry and structure of iodine that made it an antiknock additive. As Usher argued in his analysis of innovation, an accident (in this case, the availability of iodine in the laboratory) resulted in an invention. However, the Eureka moment was a short one, because subsequent developments showed that iodine not only cut knocking but also turned the inside of the cylinder into a salt factory, with the cylinder walls as the raw material.

Nevertheless, Midgley had proved that there was a chemical that reduced knock and he assumed--and was encouraged by his assumption--that there were others with less corrosive side effects. Midgley's idea response was now what he mislabelled--in view of Edison's system and method--an Edisonian approach: he tried gasoline-soluble compounds of many elements. This approach proved to be a time-consuming and expensive wild-goose chase, because the making of compounds that were soluble proved difficult.[29]

However, accident again played a role in setting the stage. Reading casually in a newspaper, Kettering noted the discovery of a so-called universal solvent, selenium oxychloride. Long sensitized to the subject matter by their search for soluble compounds, Midgley and Kettering procured some of the compound. To their delight, the venture brought another antiknock agent to their attention. To explain this act of insight is not simple, but an incomplete pattern (untried soluble compounds), accident (casual encounter with a newspaper article), and inventor inclination (try out all elements possible in a so-called Edisonian style) should be noted.

Knowing that oxygen and chlorine had negative antiknock qualities in prior experimentation, Midgley used diethyl selenide instead of the oxychloride compound and found selenium an excellent antiknock element. However, this agent also had an unacceptable side effect--a nauseous odor that would fill garages and streets. At this point Midgley's idea response was imaginative. Exploiting a scientific notion, he converted a wild-goose chase into a "fox hunt." He had a notion of the orderliness of the periodic table that he had acquired in secondary school. If we accept his own account, it was this vague notion that led him to use the orderliness of the periodic table to bring a system or a method to his research (Boyd 1950, pp. 184-185).

Midgley used a variation of the periodic table prepared by R. E. Wilson, following the valence and atomic structure theory of Langmuir, as a three-dimensional plotting board. He focused on testing soluble compounds of the elements that fell near selenium in the orderly structure of the periodic table, and he inserted in a wooden model of the **periodic table** wooden pegs of varied length according to the antiknock property of the element. The high rise area led him to tetraethyl lead. Again, it is difficult to analyze the act--or acts--of insight involved in the use of the periodic table notion, but the Midgley case history has provided details closer to the heart of the matter than most stories of invention (Boyd 1950, pp. 185-186).

Many other problems were encountered in development before tetraethyl could be marketed by the Ethyl Corp., a company of which Midgley was an officer.[30] However, we shall not follow that story, for the focus of this essay is on problem identification, idea response, and invention (act of insight).

What emerges from the three case histories is a verification of the three interpretations of innovation defined initially. No one proved capable of fully ordering or explaining the sequence of events involved in problem identification, idea response, and invention, but together they bring clarification and understanding. The categories of the interpretations, such as "critical problem" and "incomplete pattern," embraced the varied events in each of the three case histories despite the three cases being accounts of invention by an heroic type inventor, a research industrialist type, and a figure who bridged the gap between the two. Furthermore, the abstractions or categories as ordered in a sequence by Usher and in the Sperry biography corresponded with the real chronological order of the case histories. In conclusion, we shall attempt to combine the concepts and interweave the sequences of the interpretations, in the hope of achieving a more complex and more subtly nuanced interpretation of greater general validity. Despite the increased complexity, the synthesis interpretation should bring more clarity to the process of problem identification, idea response, and invention, for the process really is a complex one and simplification probably brings distortion rather than clarification.

Before a problem can be identified, it must emerge. Generally, inventors do not create problems; they solve them. To understand the innovation process, one should thus consider the ways in which problems emerge. Too often this step in the process has been ignored. Judging by the Sperry case, a successful inventor has the ability to focus his attention where problems are emerging. This ability can be perhaps explained by an awareness, conscious or unconscious, that problems tend to emerge where change is dramatic. Dramatic change has been variously labelled industrialization, population increase, technological revolution, and economic development. In all cases, the change generates reverse salients in an expanding front, imbalances, and incomplete patterns--to name a few essentially interchangeable concepts. Incomplete pattern, reverse salient, and imbalance all imply a complex system or Gestalt configuration, otherwise there would be no incompleteness, no balance to achieve (it takes more than one to balance), nor advanced segments of a front in relation to which the reverse can be seen. Because an invention is often a complex of components, it too can be perceived as a dynamic assemblage from which emerge imbalances, reverse salients, and incomplete patterns necessitating subinvention so that the total invention will function effectively as a system.

How does an inventor identify a problem? In the first place, a professional inventor like Edison or Sperry discerns a critical problem within the general area of the reverse salient. A critical problem can be distinguished from the general problem, or reverse salient that has emerged, by a critical problem's specificity and amenability to solution. To illustrate, the steam engine, wireless telegraphy, the incandescent lamp, and the marine stabilizer were not invented; what was invented was a separate condenser, a superheterodyne receiver circuit, a high-resistance filament, and an active feedback control for a stabilizer--or some other component. An inventor identified a weakness in the steam engine or another device and decided that the correction of that weakness would greatly improve the performance of the device. In so doing, he identified a critical problem.

General problem areas, such as imperfectly functioning steam engines or ship stabilizers, may involve a number of weaknesses or critical problems. How does the inventor decide on which to focus? A number of influences are at work, but his prior education, experience, and demonstrated talent determine his expertise and the wise professional knows it and takes advantage of it. Otherwise, he might find himself working with a handicap relative to other inventors working in the same general problem area. Other influences determining the inventor's selection of a critical problem are his anticipation of funding, foreseeing a market, and availability of R and D facilities of an appropriate kind, such as laboratory ones.

Another influence working on the professional searching for the critical problem is the prior activity of other inventors who have worked in the same general problem area or reverse salient. The inventor can inform himself about the direction and focus of this prior work by an analysis of the literature and patents in the field, and by conversation with his peers. If he finds, for example, that a majority of the earlier patents or technical reports in the area were about incandescent filaments, he can assume that there is probably a critical problem at that point. Prior failures not only alert him about focus, but also inform him about dead ends and open avenues.

Having identified a critical problem, how does the inventor generate the idea response or come upon the solution? The Edison, Sperry, and Midgley cases are instructive, but do not answer the question satisfactorily. That is not surprising, because idea response is essentially creative, a human attribute that has long defied analysis. However, the cases considered suggest, among other things, that the creative inventor seeks one small step forward; he does not pretend to great leaps. Furthermore, he looks for a component in earlier manifestations of the invention that can be labelled a "weakness." The simple act of structuring his perceptions by the search for weakness predisposes him towards a means of improving upon the situation, for the act of identifying weakness presumes an imcomplete pattern or the existence of a remedy. Sperry saw sluggishness as a weakness because he knew from experience that inertia and friction could be overcome in machines.

Another means by which inventors find solutions, or generate idea responses, is by analogy. In the cases discussed, Sperry and Midgley used analogy. They found an analogy to the problem and then recalled the solution to the analogous situation and tried to modify the early solution to suit the unique particulars of the new problem. To do that the inventors had to have the mental ability to abstract the similar essences of superficially different problems and to particularize an abstraction of the earlier solution.

Many intellectual and emotional qualities besides the power of abstraction doubtless entered into the idea response in the cases discussed. One emotional quality valued highly by Sperry when appraising his qualities as an inventor, which may have contributed to the solution of problems by other inventors, was courage.[31] Obviously, courage is of many kinds, but it seems likely that the kind of courage Sperry related to invention was an ability to live with tension and unresolved problems. In his analysis of invention, Usher emphasizes an ability to contain tension. Psychological capability to withstand the tension arising from a well-identified but unsolved problem would sustain the long-term empiricism and painstaking routine and the carefully worked out and fulfilled methodology involved in the case histories of invention discussed. To put it differently, the inventive type must have the psychology to accept, even seek, problems.

The early stages of innovation are more complex than any one or any combination of the interpretations that have been examined. The reading of the surviving sources on many inventions and the inventions themselves sustain this argument. On the other hand, the exploration of case histories considered in this essay demonstrates that analytical interpretations do bring credible--and not simplistic--order out of the chaos of facts. What is obviously needed, in view of the paucity of studies now available, is an intense and sustained effort to research and write analytical histories of inventions and inventors, especially inventor-entrepreneurs. It is hackneyed to say that we know much more about our politicians, diplomats, and generals than our inventor-entrepreneurs; it is less obvious to say that what we do know about our inventor-entrepreneur, generally, is simplistic and adulatory. We need biographies and histories in innovation that are based on the same close reading of the sources and the same critical analysis that has come to be taken as commonplace in political and diplomatic history. With the 20th Century structured by technology, it is remarkable that more support and encouragement has not been given to the study of the history of man making his world with technology through innovation.

When such studies are done, the historian must shift to the appropriate sources. In the preceding essay, the analysis drawn from the Sperry biography was based on a close reading of the inventor's notebooks, his patents, and when possible, an inspection of the inventions themselves. Inventors' notebooks provide a day-by-day, even hour-by-hour account of the problem identification, idea response, and reduction to model. A political historian of high standards would not think of analyzing a political campaign without the hard, tiny facts of the process; yet theories of innovation are formulated without reference to the basic source material. It is surprising and shocking that Edison's notebooks have not yet been analyzed in a systematic manner nor has the pattern of his patents been discovered. As for patents, an analysis of Sperry's patents was a key to understanding his inventive style. Until these types of sources are used, it is fatuous to say that invention is a creative process that defies analysis. It should be added that the facts and the interpretations considered in this essay strongly recommend that any analysis be one that is shaped, or influenced, by a dynamic systems model. Each interpretation and each case reinforce the conclusion that complex change within even more complex change is the process under way when innovation is occurring. Furthermore, the events of innovation take on meaning within a systems context that has been labelled variously as expanding front; complex and dynamic societal cumulation; and economic, demographic, or industrial environment. Further study would probably prove that exploration of other systems concepts, such as output-input analysis, would prove extremely helpful to the historian and biographer in the study of the early phases of innovation.

NOTES AND REFERENCES

1. For a definition of innovation as used in the Georgia Institute of Technology/National Science Foundation study, of which this essay is a part, see Chap. 1 above.

2. The discussion in this essay is limited to three inventors so that the analysis can be in depth. There is a pressing need for analytical biographies and studies of inventors and inventions. Generally, biographies have been narrative and descriptive. Needed are detailed analyses of inventors' patents with a view to finding a pattern, and a sustained effort to discern the inventors' various styles of innovation. Inventors of the period 1870-1930 about whom biographies have been written include: Matthew Josephson, *Edison: A Biography* (New York, 1959); F. E. Leupp, *George Westinghouse: His Life and Achievements* (Boston, 1918); Robert Bruce, *Bell: Alexander Graham Bell and the Conquest of Solitude* (Boston, 1973); David Woodbury, *Beloved Scientist: Elihu Thomson* (New York,

1944): Junius Edwards, *The Immortal Woodshed: The Story of the Inventor Who Brought Aluminum to America* (New York, 1955); Fred C. Kelly, *The Wright Brothers* (New York, 1943); *The Papers of Wilbur and Orville Wright*, ed. by Marvin W. McFarland, 2 vols. (New York, 1953); Lee de Forest, *Father of Radio: Autobiography* (Chicago, 1950); Michael I. Pupin, *From Immigrant to Inventor* (New York, 1923); John W. Hammond, *Charles Proteus Steinmetz: A Biography* (New York, 1924); Thomas C. Martin, *The Inventions, Researches, and Writings of Nikola Tesla...*(New York, 1894); Lawrence Lessing, *Man of High Fidelity: Edwin Howard Armstrong* (Philadelphia, 1969); T. A. Boyd, *Professional Amateur: The Biography of Charles Franklin Kettering* (New York, 1957); and Thomas Parke Hughes, *Elmer Sperry: Inventor and Engineer* (Baltimore, 1971). An extremely helpful biliographical series is: Thomas James Higgins, "A Biographical Bibliography of Electrical Engineers and Electrophysicists," *Technology and Culture*, 2 (1961): 28-32, 146-165; "Biographies of Engineers, Metallurgists and Industrialists," *Bulletin of Bibliography*, 18 (1946): 207-210, 235-239 (1947): 10-12, 32; "Book Length Biographies of Chemists," *School Science and Mathematics* (1944): 650-665; and "Book-length Biographies of Physicists and Astronomers," *American Journal of Physics*, 12 (1944): 31-39, 234-236.

3. See especially pp. 56-83, in Usher (1954); pp. 3-13 in Gilfillan (1935); and pp. 64-70 in Hughes (1971).

4. Edwin Mansfield, Richard Rosenbloom, Nathan Rosenberg, Everett M. Rogers, James R. Bright, Albert H. Rubenstein, W. Paul Strassmann, and Simon Kuznets have also contributed essays to the present study. Their works and other studies of innovation are in the massive bibliography compiled for this project. Several books that deserve special attention because of their relevance for the case-history approach to the understanding of innovation are: Morison (1966); Hunter (1949); Hewlett and Anderson (1962); Passer (1953); Enos (1962a); (1966); and, Maclaurin (1949).

5. Usher (1954) uses Watt's conception of the essence of the separate condenser while strolling on a Sunday afternoon as an example of tension, incomplete pattern, and act of insight (p. 71).

6. Gilfillan (1935) pp. 5-12, provides an outline discussion of each category; expanded in his *Supplement* (Gilfillan 1971).

7. Sperry's patents are listed by title, date filed, date issued, and number in Hughes (1971), pp. 325-339.

8. Sperry Gyroscope Co., "The Sperry Fire Control System," *Bulletin 301* (1916).

9. Sperry Notebook No. 50 (10 March 1909 to 31 December 1909) exemplifies the excellence of the raw data on problem identification, idea response, and invention in the inventor's notes. Notebook 50 has information on the gyroscopes.

10. Sperry chose to learn of minor design problems by introducing a device into use when it was substantially but not completely ready. The user would identify for Sperry the minor malfunctions arising from environmental factors unanticipated in trials and tests. Users were informed of his policy. "Addendum to Reports on the Sperry Gyro-Compass..." (1916?), typescript.

11. Sperry Notebook No. 47 (6 May 1906 to 5 December 1907).

12. Filed 2 December 1907, "Steadying Device for Vehicles," issued 29 December 1908 (Patent No. 907 907).

13. Sperry identifies his inventive idea of countering "sluggish behavior" in his patent filed 21 May 1908, "Ship's Gyroscope," issued 17 August 1915 (Patent No. 1 150 311).

14. Report from David W. Taylor to Chief Constructor, Bureau Construction and Repair, 15 July 1910.

15. The complexity of the Sperry stabilizer system and similarly complex technological systems often demand diagrams for clarification; see Fig. 4.11 in Hughes (1971), p. 115.

16. This story of the incandescent lighting system has often been told. However, it has not been subjected to selection, emphasis, and analysis that follow. Although Usher analyzes the invention of the incandescent lamp as a case to support his theory of invention, the incandescent bulb was merely one component in the system that is analyzed here. In the most recent biography of Edison (Josephson 1959), the invention and development of the system is described in detail, but there is no attempt at systematic analysis.

Passer (1953) does not ignore Edison as inventor but the emphasis is on Edison and the process of innovation, involving development, marketing, and business entrepreneurship.

17. Werner von Siemens (1816-1892) is as well known a figure in the history of German technology as Edison. Siemens was not as prolific an inventor as Edison, but he is generally believed to have been more scientific. His style of invention and innovation merits the kind of analysis tried in this essay. Information about him is readily available. See, for example, Sigfrid von Weiher, *Werner von Siemens* (Göttingen, 1970) with a selected bibliography and reference to the Siemens Archive in Munich. In English, see, *Inventor and Entrepreneur: Recollections of Werner von Siemens* (Munich, 1966).

18. Intimate recollections of an Edison pioneer are found in the richly illustrated book by Francis Jehl, *Menlo Park Reminiscences*, 3 vols. (Dearborn, Michigan, 1937). Details about the laboratory setting are given.

19. Josephson (1959) describes Edison's invention of an incandescent light *system*; Frank L. Dyer and Thomas C. Martin, *Edison: His Life and Inventions*, 2 vols. (New York, 1930) knew him and stressed his determination to invent a system ("Inventing a Complete System of Lighting," Chap. 14, pp. 318-346).

20. Jehl (1937) I, 362-363 and II, 852-853. Edison's system concept, Ohm's law, and other aspects of Edison's invention of the incandescent-lamp system are discussed in more detail in my essay on Edison prepared for the Science Museum in London: T. P. Hughes, *Thomas Edison: Professional Inventor* (London, 1976).

21. A good collection of source material about the Edison innovation culminating in the Pearl Street Station is Payson Jones, *A Power History of the Consolidated Edison System, 1878-1900* (New York, 1940). On the development and critical revisions of the Pearl Street System, see Thomas C. Martin, *Forty Years of Edison Service, 1882-1922* (New York, 1922).

22. Thomas Parke Hughes, "Harold P. Brown and the Executioner's Current: An Incident in the AC-DC Controversy," *Business History Review*, 32 (1958); 143-165. On George Westinghouse, see Francis Leupp 1918 and Henry Prout, *A Life of George Westinghouse* (1922). No biography of William Stanley, who played a leading role in the Westinghouse ac innovation, has been published.

23. Edison's invention of an incandescent lamp has become a popular legend; other system components of importance--for example, the generator or dynamo--have been comparatively neglected. On the generator, see, *"Edisonia: A Brief History of the Early Edison Electric Lighting System"* (New York, 1904), pp. 27-60.

24. For an account of a technological problem emerging and an inventive response occurring at this interface (the automobile production and use system) see Thomas Parke Hughes, "Technological Momentum: Hydrogenation in Germany, 1898-1933," *Past and Present*, No. 44 (August 1969): 106-132.

25. Charles F. Kettering's inventive style is also well worth analysis. See the biography by his associate T. A. Boyd mentioned in Note 2 above.

26. Charles F. Kettering, "Biographical Memoir of Thomas Midgley, Jr., 1889-1944," *National Academy of Sciences: Biographical Memoirs*, 24 (1947): 365.

27. Thomas Midgley, Jr., and T. A. Boyd, "The Chemical Control of Gaseous Detonation with Particular Reference to the Internal-Combustion Engine," *The Journal of Industrial and Engineering Chemistry*, 14 (1922): 894-896

28. T. A. Boyd, "Pathfinding in Fuels and Engines," *Society of Automotive Engineers Quarterly Transactions*, April 1950, pp. 182-183.

29. Thomas Midgley, Jr., "From Periodic Table to Production," *Industrial and Engineering Chemistry*, 29 (1937): 241-242.

30. William Haynes, *American Chemical Industry*, 4 (New York, 1948), pp. 400ff.

31. Sperry wrote of the inventor's courage in an essay on the "Spirit of Invention..." in *Toward Civilization*, ed. Charles A. Beard (New York, 1930), pp. 47-68.

7. TECHNOLOGICAL INNOVATION AND NATURAL RESOURCES: THE NIGGARDLINESS OF NATURE RECONSIDERED

Nathan Rosenberg

I

Concern over the adequacy of natural resources has always been a central preoccupation of economists. The management of human enterprise at both the household and large societal levels has always been perceived to be dependent on the intelligent management of the limited resources made available by the natural environment. Indeed, this concern is evident in the very etymology of the word "economy," which derives from the Greek words *oikos* (house) and *nomos* (manage), or the management of household affairs. When economics as a discipline is conceived as dealing with the problems of the household writ large, the obvious and central question is how the resource endowment constrains the production of goods and services.

Some of the basic insights of classical economics emerged out of this preoccupation with natural-resource constraints. Malthus and Ricardo were both pessimistic about the long-term prospects for economic growth in an economy experiencing substantial population growth and with only a fixed supply of land available for food production. Indeed, as Malthus argued, in the "old" country population growth was always pushing society against the limits imposed by the fixed supply of land; subsistence incomes were, as a consequence, inevitable--at least for the working class. Indeed, it was in speculating about the consequences of Great Britain attempting to grow its own food supply as population continued to grow that classical economics formulated one of its central relationships-- the law of diminishing returns. It is worth noting that the formulation of this law clearly distinguished between a quantitative and a qualitative dimension.[1] For not only did continual population growth lead necessarily to a decline in the land/man ratio; in its Ricardian variant, it also created the necessity for resorting to qualitatively inferior soils--that is, soils in which an equal dosage of capital and labor led to a smaller increment to output than on land already under cultivation. In other words, there was both an extensive and an intensive margin of cultivation. The classical economists' concern with the benefits of free trade and the theory of comparative advantage followed directly from this vision of the deteriorating resource position of a growing economy in possession of a fixed supply of land.

The impact of the Malthusian conception is difficult to exaggerate. It not only forcefully focused on the implications of limited resources for the prospects for economic growth, but it specifically linked the problem, as it must be linked, to the rate of population growth. Malthus, it turned out, was not much of a demographer, but his formulation nevertheless has had an overwhelming influence upon the framework within which these matters have been discussed down to the present day.[2] If the Malthusian view seemed to play a subordinate role in the writings of professional economists after the "Marginal Revolution" of the 1870s, it was not so much because the Malthusian specter had been exorcised but because economists began to address themselves to a very different range of problems--problems dealing with the optimal allocation of a fixed amount of resources. John Bates Clark summarized this new focus very well in his *Distribution of Wealth*:

> In any given society, five generic changes are going on, every one of which reacts on the structure of society by changing the arrangements of that group system which it is the work of catallactics to study.
>
> 1. Population is increasing.
> 2. Capital is increasing.

Dr. Rosenberg, editor of the *Journal of Economic History*, has been professor of economics at the University of Wisconsin and (since 1974) at Stanford University.

3. Methods of production are improving.
4. The forms of industrial establishments are changing.
5. The wants of consumers are multiplying.[3]

Clark regards all these forces as belonging to economic dynamics. Economic statics, on the other hand, is the study of an economic system where none of these forces is present. Therefore (Clark argues) in the study of economic statics we must rigorously abstract from the operation of these forces.

> One function of economic society is that of growth. It is becoming larger and richer, and its structure is changing. As times passes, it uses more and better appliances for production. The individual members of it develop new wants, and the society uses its enlarging process to gratify them. The organism is perpetually gaining in efficiency, and this is promoting the individual members of it to higher planes of life (pp. 55-56).

But all these "generic" changes connected with economic growth, Clark argues, must be firmly placed aside in order to permit the study of economic statics. Clearly this is a level of abstraction within which there is no room, by assumption, for the operation of Malthusian forces.

Although Malthus and Ricardo had been primarily concerned with the adequacy of the supply of arable land to provide food for a growing population, the voracious appetite for natural resources of an industrializing economy shifted the locus of concern to other resources in the second half of the 19th Century. As early as 1865 the distinguished English economist W. S. Jevons published his book, *The Coal Question*, in which he warned that the inevitably rising costs of coal extraction posed an ominous and urgent threat to Britain's industrial establishment. Having demonstrated the British economy's extreme dependence upon coal--especially in the form of its reliance on steam power--Jevons went on to argue explicitly that there were no prospective reasonable substitutes for coals as a fuel.[4] After examining recent trends in coal consumption he demonstrated the inevitability of rising coal costs associated with the necessity of conducting mining operations at progressively greater depths. As a result,

> I draw the conclusion that I think any one would draw, that *we cannot long maintain our present rate of increase of consumption; that we can never advance to the higher amounts of consumption supposed. But this only means that the check to our progress must become perceptible considerably within a century from the present time;* that the cost of fuel must rise, perhaps within a lifetime, to a rate threatening our commercial and manufacturing supremacy; and the conclusion is inevitable, that our present happy progressive condition is a thing of limited duration.[5]

II

Even in the USA, a country of continental proportions and possessing a resource/man ratio far more favorable than that of Great Britain, expressions of concern over the adequacy of natural resources became widespread before the end of the 19th Century. The conservation movement emerged into the consciousness of American political life at just about the time that Frederick Jackson Turner was announcing *The End of the Frontier* (1893). The movement was a significant force in Americal political life from 1890 to 1920.[6] Although the spokesmen of the movement did not attack their concerns in an analytical or rigorous way, they did forcefully present a conceptualization of the problem which was to exercise a profound influence upon later thinking. In the words of Gifford Pinchot, Chief Forester of the United States, and an articulate leader of the movement,

> We have a limited supply of coal, and only a limited supply. Whether it is to last for a hundred or a hundred and fifty or a thousand years, the coal is limited in amount; unless through geological changes which we shall not live to see, there will never be any more of it than there is now. But coal is in a sense the vital essence of our civilization. If it can be preserved, if the life of the mines can be extended, if by preventing waste there can be more coal left in this country after we of this generation have made every needed use of this source of power, then we shall have deserved well of our descendants.

And,

> *The five indispensably essential materials in our civilization are wood, water, coal, iron, and agricultural products....*We have timber for less than thirty years at the present rate of cutting. The figures indicate that our demands upon the forest have increased twice as fast as our population. We have anthracite coal, for but fifty years, and bituminous coal for less than two hundred. Our supplies of iron ore, mineral oil, and natural gas are being rapidly depleted, and many of the great fields are already exhausted. Mineral resources such as these when once gone are gone forever.[7]

The conservationist view, therefore, was that nature contained fixed stocks of useful inputs for man's productive activities, which could be readily identified and measured in terms of physical units. These resources need to be husbanded carefully. Preference should be given to exploitation of renewable resources--agriculture crops, forests, fisheries--which should be operated on a sustained yield basis. Waste should be avoided in the exploitation of nonrenewable resources so as to pass on the largest possible inventory to future generations.

When, in the years immediately after the Second World War, economists turned their attention once again to the problems and prospects for long-term economic development, the Malthusian pressures again figured prominently. Underdeveloped countries all appeared to confront some serious demographic obstacle--already high population densities, rates of population growth that had sharply accelerated in the 20th Century, or both. Not surprisingly, Malthusian-type models made their appearance. In these models, that is to say, there existed a "quasi-equilibrium" such that temporary improvements which raised per capita income triggered off demographic changes which eventually restored the low level of per capita income at higher absolute population size.[8] Concern over the implications of population growth subsided in the USA and most high-income countries by the late 1950s when the postwar "baby boom" appeared to have exhausted itself. However, it reappeared forcefully in the 1960s as a consequence of the growing concern over pollution and environmental deterioration; and its possible consequences have been more recently dramatized in the apocalyptic Meadows Report, which attempts to focus attention on the long-term implications of continued growth of population and industrial production.[9] In this model, exponential growth of population and industrial output confront a world of finite resources--mineral deposits, arable land, etc.--and limited capacity of the natural environment to absorb the growing pollution "fallout." The finite limits appear to guarantee, not merely a ceiling to future growth, but a precipitous and disastrous decline in approximately a century.

III

In assessing the Malthusian view and its implications over the past century or so, a convenient starting point is the recognition that Malthus and Malthus-like models have led to predictions which have been demonstrably and emphatically wrong. Indeed, it is difficult to understand the persistence of widespread attachment to a hypothesis which has been so decisively refuted by the facts of history. As Stigler has said:

> What evidence could have been used to test the theory? If the subsistence level has any stability, and hence any significance, Malthus' theory was wrong if the standard of living of the masses rose for any considerable period of time. He did not investigate this possibility--and ignored the opinions of such authorities as Sir Frederick Eden that it had been rising for a century. His theory was also contradicted if population grew at a constant geometrical rate in an "old" country, for then the means of subsistence were also growing at this rate, since population never precedes food. Despite the rapid increase of population in almost all western European nations at the time, which he duly noted, he persisted in considering this as only a confirmation of his fecundity hypothesis....
>
> The "principle of population" had the dubious honor of receiving from history one of the most emphatic refutations any prominent economic theory has ever received. It is now fashionable to defend Malthus by saying that his theory applies to other places and times than those to which he and his readers applied it. This may be true, but it is tantamount to scientific nihilism to deduce from it any defense of Malthus.

It is an odd theory that may not some day and somewhere find a role; for every answer one can find a correct question.[10]

Not only have per capita incomes in the western world experienced a sustained increase over long periods of time, while population has grown at exponential rates (Kuznets 1966, p. 2), but the long-term historical record also fails to reveal any convincing evidence that the limited supply of natural resources and associated diminishing returns have significantly hampered long-term growth of these countries in the past. Clearly, no amount of evidence from past history will enable us to predict future relationships, for there is certainly no necessary reason to believe that the future will resemble the past. However, it is not our purpose to attempt anything so foolhardy as prediction, but rather to grasp more firmly and precisely the nature of the relationships between shifting patterns of resource scarcity and the innovative process.

Classical models have failed to make relevant predictions primarily because they adopted an excessively static notion of the economic meaning of natural resources and because they drastically underestimated the extent to which technological change could offset, bypass, or provide substitutes for increasingly scarce natural resources. A large part of what economists have had to say in recent years about the innovative process in relation to natural resources needs to be seen as a prolonged effort to break out of the restrictive conceptualizations inherited from Malthus and Ricardo.

The need to break out of this framework has been made increasingly apparent by numerous studies, each of which, in their different ways, has shown that natural resources have played a role of declining importance, at least within the favored circle of industrializing countries. Whereas classical economic models based upon fixed resources, population growth, and diminishing returns lead us to expect rising relative prices for extractive products and an increasing share of GNP consisting of the output of resources, such features have obviously not dominated the growth experiences of industrial economies. To begin with, the agricultural sectors have declined in their relative importance. A declining agricultural sector, as Kuznets has shown, has characterized all economies that have experienced long-term economic growth (Kuznets 1966, Chap. 3). Indeed, perhaps the most distinctive characteristic of growing economies has been the complex of structural changes associated with the declining importance of the agricultural sector. Moreoever, within agriculture itself, the implicit assumption that there were no good substitutes for land in food production has been belied by a broad range of innovations which have sharply raised the productivity of agricultural resources and have at the same time made possible the widespread substitution of industrial inputs for the more traditional agricultural labor and land--machinery, commercial fertilizer, insecticides, irrigation water, etc. As Schultz observed in 1951,

> It is my belief that the following two propositions are historically valid in representing the economic development that has characterized Western communities:
>
> (1) a declining proportion of the aggregate inputs of the community is required to produce (or to acquire) farm products; and
> (2) of the inputs employed to produce farm products, the proportion represented by land is not an increasing one, despite the recombination of inputs in farming to use less human effort relative to other inputs, including land.[11]

IV

The growing appreciation of the importance of technological change in raising resource productivity has led to several notable studies of the working of this process in agriculture. Attempts have been made to identify and measure the contribution of separate variables to the growth in agricultural productivity (including the impact of individual innovations), as well as to studying the factors accounting for the rate of diffusion of new innovations. The most notable attempt to measure the contribution of a single innovation was Griliches's pathbreaking study of hybrid corn, in which he estimated that the social rate of return over the period 1910-1955 to the private and public resources committed to research on this innovation was at least of the order of 700% (Griliches 1958). Parker, in his study of cereal production (wheat, corn, and oats), found that output per worker more than tripled in the U.S. between 1840 and 1911.

Parker estimated that 60% of the increase was attributable to mechanization, which raised the acreage/worker ratio, and that practically all the growth in productivity can be explained by the combination of mechanization with the westward expansion of agriculture. The most important improvements came in activities that had previously been highly labor intensive--especially the harvesting and post-harvesting operations. In fact, two innovations alone--the reaper and the thresher--accounted for 70% of the total gain from mechanization (Parker and Klein 1966).

To be sure, major improvements in productivity in agriculture were the result of innovations in other sectors of the economy. For example, the growth in agricultural productivity due to westward expansion and an increasing degree of regional product specialization in turn depended on improvements in transport facilities. Indeed, by the end of the 19th Century the railroad, the iron steamship, and refrigeration had created, for the first time in human history, a high degree of regional agricultural specialization on a worldwide scale.[12]

The impact of an innovation on productivity growth in agriculture, as in other sectors, is not only a function of its potential for reducing costs, but of the speed with which it is adopted. As a result, increasing attention has been devoted to the diffusion process in agriculture. Paul David has employed a threshold model of farm size to explain the very slow adoption of the reaper for the 20-year period before 1893 and the sudden rapid rate of adoption beginning in that year.[13] Griliches has shown how the spatial and chronological diffusion of hybrid corn can be explained in terms of economic factors shaping the profit expectations of farmers and seed suppliers. His model closely accounts for the early and rapid adoption in Iowa as compared to the later and slower adoption on the western fringes of the Corn Belt (Griliches 1957). A significant element of Griliches's diffusion model is its explicit recognition of the dependence of the adoption process on the need to undertake local adaptations.

An important attempt to provide a synthesis of a wide body of literature on the interaction between resource endowment and agricultural technology is represented in Hayami and Ruttan's *Agricultural Development* (1971). In this book the authors present a theoretical framework for examining the patterns of agricultural development in individual countries within which endogenous technological change plays a critical role. A crucial feature of their approach is a theory of induced innovation which incorporates a unique, dynamic response of each country to its agricultural resource endowment and relative input prices. Hayami and Ruttan argue that

> there are multiple paths of technological change in agriculture available to a society. The constraints imposed on agricultural development by an inelastic supply of land may be offset by advances in mechanical technology. The ability of a country to achieve rapid growth in agricultural productivity and output seems to hinge on its ability to make an efficient choice among the alternative paths. Failure to choose a path which effectively loosens the constraints imposed by resource endowments can depress the whole process of agricultural and economic development (pp. 53-54).

In developing their induced innovation model, Hayami and Ruttan postulate the existence of a "metaproduction function," an envelope curve that goes beyond the production possibilities attainable with existing knowledge and described in a neoclassical long-run envelope curve. It describes instead a locus of production possibility points which it is possible to discover within the existing state of scientific knowledge. Points on this surface are attainable, but only at a cost in time and resources. They are not at present available in blueprint form.

Within this framework Hayami and Ruttan study the growth of agricultural productivity as an adaptive response to altering factor and product prices. The adaptation process is conceived as the ability to move to more efficient points on the metaproduction function, especially in response to the opportunities being generated by the industrial sector, which offers a potential flow of new inputs.

> From the intercountry cross-section and time-series observations, the relative endowments of land and labor apparently have a significant influence upon the optimum path to be followed in moving along a metaproduction function. Where labor is the limiting factor, the optimum for new opportunities in the form of lower prices of

modern inputs is likely to be along a path characterized by a higher land-labor ratio. Movement toward an optimum position on the metaproduction function would involve development and adoption of new mechanical inputs. On the other hand, where land is the limiting factor, the new optimum is likely to be the point at which the yield per hectare is higher for the higher level of fertilizer input. Movement to this point would involve development and adoption of new biological and chemical inputs.

The partial productivity and factor input ratios presented earlier in this chapter suggest that those nations which have achieved relatively high levels of either land or labor productivity have been relatively successful in substituting industrial inputs for the constraints imposed by a relatively scarce factor, either land or labor. It seems possible to explain many of the ratios in agriculture among countries by hypothesizing that technical advance in agriculture occurs primarily as a result of new economic opportunities created by developments in the nonagricultural sector. The advances in mechanical and biological technology do not, however, occur without cost. Development of a more fertilizer-responsive crop variety, in response to declining prices of fertilizer, typically requires substantial expenditures on research, development, and dissemination, before it actually becomes available to farmers. Public investment in improvements in water control, land development, and other environmental modifications may also be required before it becomes profitable for farmers to adopt the newly developed varieties (Hayami and Ruttan 1971, p. 84).

The Hayami and Ruttan approach constitutes a very significant development because it is the most detailed attempt to date to specify the nature of the mechanisms through which technical change and adaption take place in response to shifting patterns of relative resource scarcities.[14]

V

Barnett and Morse (1963) have attempted a quantitative test of the implications of the classical model that increasing natural resource scarcity should make itself apparent in the secularly rising unit cost of extractive products generally--agriculture, minerals, forestry and fishing. To do so they examined data for the American economy over the period from about 1870 to the late 1950s. Their findings are that unit costs of extractive products, as measured by labor plus capital inputs required to produce a unit of net extractive output, declined through the period 1870-1957. For agriculture, minerals, and fishing, the trends are persistently downward. Indeed, the time sequence of the trends is particularly damaging to the increasing-scarcity hypothesis because in the cases of agriculture and minerals (which together account for about 90% of the value of extractive output) the rate of decline in unit costs was *greater* in the later portion of the period than the earlier. Forestry, in fact, is the only extractive industry in which the long-term trend in unit costs and relative prices has been upward since 1870. Since forestry has accounted for less than 10%, by value, of extractive output in the 20th Century, the influence of this sector has been swamped by the downward trends elsewhere. Moreover, the rising costs of forest products seems to have induced a large-scale substitution of more abundant nonforest-based products for forest-based ones--metals, masonry, and plastics--which may account for the rough constancy since 1920 in the unit cost of forest products.[15]

The declining importance of raw materials is further confirmed in the findings of input-output analysis. In her book *Structural Change in the American Economy* Anne Carter (1970) provides important insights into the process of technological change between 1939 and 1961 through a study of changing intermediate input requirements. In particular, she neatly documents, for this period, the decline in the quantities of materials required in the delivery of a given quantity of goods. The great viture of input-output analysis is that it enables us to understand the structural interdependence of the economic system by providing quantitative measures (input-output coefficients) of the interindustry flow of goods and services. Changes in such coefficients need to be interpreted with care, since they may alter not only as a result of technological change, but also as a result of changes in product mix over time, and because of substitution resulting from changes in the relative prices of inputs. However, it can also be counter-argued that much of what is called "substitution" is, itself, made possible by *earlier* technological change.[16] Indeed, one of the most significant findings of Carter's book is precisely her careful demonstration that "there now seems to be a technological basis for greater substitutability as relative price conditions change."[17] Carter shows that economic change has been associated with an increasing reliance on general sectors--producers of services, communications, energy, transportation, and trade. What is most interesting for our present purposes is that these increases have been offset by decreases in other sets of coefficients, most conspicuously in the general, across-the-board declines in the contributions of producers and materials. Technological change has been forcefully associated with a significant expansion in the kinds and qualities of materials and in improvements of design generally. Moreover, Carter demonstrates how technological change has been expanding the range of substitutability among materials.[18] For example, the traditional dominance of steel in many uses, has been successfully challenged by improved aluminum, plywood, and prestressed concrete. The growing importance of plastics and chemicals, and the changes in product design associated with such new and versatile materials, is thoroughly documented.

An attempt to provide an overall estimate of the role of resources in the growth of the American economy finds quite conclusively that "Since about the beginning of this century, the resource base has been playing a noticeably smaller, and in many respects different, role in growth than it did before then. General economic growth has been less closely and clearly tied to abundant resources than it was previously."[19] In quantitative terms Fisher and Boorstein find that the output of resources, expressed as a percentage of United States GNP, has declined from 36% in 1870 to 27% in 1900 and to 12% in 1954 (p. 43). Furthremore, it is apparent that a decline of this magnitude can in no way be accounted for by an increasing importation of primary products and resource-intensive goods,[20] although it is of course true that our reliance on imports of certain resources--e.g., metallic ores--has vastly increased in recent decades. In fact, after a comprehensive study of a wide range of prices and other data, Potter and Christy found that the available evidence did not support the apprehensions of the Paley Commission of a decade earlier, that the long-term decline in natural resource prices was finally and permanently coming to an end for the United States.[21]

In a suggestive article, "The Development of the Extractive Industries," Anthony Scott constructs a three-stage model for dealing with the sequence through which man has gone in the exploitation of his natural environment (Scott 1962). In the third stage, where the application of science makes a sophisticated degree of control and manipulation over the environment possible, older scarcities are deliberately bypassed in the systematic search for substitutes from abundant and convenient sources; the classic early 20th Century example is the Haber process of extracting nitrogen from the atmosphere. Scott makes the important point that there is no economic demand for specific minerals as such, but rather for certain *properties*, and that an advanced technology makes it possible to attain these properties from materials available in great abundance.

> Demand for minerals is *derived* from demand for certain final goods and services. Therefore, certain properties must be obtainable from the raw materials from which such services and types of final goods are produced. Man's hunt for minerals must properly be viewed as a hunt for economical sources of these properties (strength, colour, porosity, conductivity, magnetism, texture, size, durability, elasticity, flavour, and so on). For example, there is no demand for "tin," but for something to make copper harder or iron corrosion-free. No one substitute for tin has been found, but each of the functions performed by tin can now be performed in other ways. Tin's hardening of copper (as in bronze) has been supplanted by the use of other metals. Food need no longer be packed in tin cans. Hence the immense capital investment that society might have been forced to undertake to satisfy its former needs for tin from the minute, low-grade quantities to be found in many parts of the world have been replaced by simpler investments in obtaining other materials. Chief of the replacements for tin is glass, made from apparently unlimited quantities of sand and with little more energy than is needed to bring metallic tin to the user. Lead and mercury are also being bypassed in similar fashion; zinc and copper may be next[22] (p. 81).

Thus, although there may be no close substitutes for tin in nature, there may be excellent substitutes for each of the properties for which tin is valued. The implications of this position are, of course, far reaching.

VI

These findings have had a great deal to do with what may be called a steady downgrading of the importance of natural resources. In his 1951 article Schultz called attention to the fact that R. F. Harrod, in his notable book *Toward a Dynamic Economics* published in 1948 "...saw fit to leave land out altogether."[23] Although it is doubtful that many economists concerned with economic growth (as against Harrod's primary concern with cyclical phenomena) would wish to go that far, there has certainly been increasing agreement on some weaker propositions. This trend was made strikingly apparent in 1961 with the publication of a collection of conference papers in a volume edited by J. J. Spengler, *Natural Resources and Economic Growth*. Although the participants approach the subject from a wide variety of perspectives, there are two propositions which recur so frequently as to amount almost to a consensus: (1) the pervasive influence of classical economics in its Malthusian-Ricardian variant had resulted in a vast exaggeration of the importance of natural resources and an overstatement of the constraints which they imposed on an economy's possibilities of development; and (2) the relative importance of natural resources is a declining function of development itself. Kindleberger, for example, adopted an almost patronizing tone in conceding the economic significance of natural resources: "It may be taken for granted that some minimum of resources is necessary for economic growth, that, other things being equal, more resources are better than fewer, and that the more a country grows the less it needs resources, since it gains capacity to substitute labor and especially capital for them."[24]

In a similar spirit, Barnett and Morse conclude that

> the Conservationists' premise that the economic heritage will shrink in value unless natural resources are "conserved," is wrong for a progressive world. The opposite is true. In the United States, for example, the economic magnitude of the estate each generation passes on--the income per capita the next generation enjoys--has been approximately double that which it received, over the period for which data exist. Resource reservation to protect the interest of future generations is therefore unnecessary. There is no need for a future-oriented ethical principle to replace or supplement the economic calculations that lead modern man to accumulate primarily for the benefit of those now living. The reason, of course, is that the legacy of economically valuable assets which each generation passes on consists only in part of the natural environment. The more important components of the inheritance are knowledge, technology, capital instruments, economic institutions. These, far more than natural resources are the determinants of real income per capita (1963, pp. 247-248).

VII

It should by now be obvious that economics has in the past been burdened with an excessively restrictive definition of natural resources. More precisely, economists (and conservationists) worked with a conception of natural resources that encompassed the prevailing state of technology but failed to recognize how profoundly technological changes required a redefinition of the *economic meaning* of the natural environment. Purely physical or geological definitions of resources, even if they are exhaustive,[25] are not very interesting.

> The Plains Indian did not cultivate the soil; neither coal, oil nor bauxite constituted a resource to the Indian population or, for that matter, to the earliest European settlers in North America. It was only when technological knowledge had advanced to a certain point that such mineral deposits became potentially usable for human purposes. Even then the further economic question turns, in part, upon accessibility and cost of extraction. Improvements in oil drilling technology (as well as changing demand conditions) make it feasible to extract oil today from depths which would have been technically impossible fifty years ago and prohibitively expensive twenty years ago. Similarly, low grade taconite iron ores are being routinely exploited today

which would have been ignored earlier in the century when the higher-quality ores of the Mesabi range were available in abundance. Oil shale, known to exist in vast quantities--for example, in the Green River formation in Colorado, Utah, and Wyoming--is not yet worth exploiting but might well be brought into production if petroleum product prices rise very much above their present levels. The rich and abundant agricultural resources of the Midwest were of limited economic importance until the development of a canal network beginning in the 1820s with the completion of Erie Canal, and later a railroad system which made possible the transportation of bulky farm products to eastern urban centers at low cost. Natural resources, in other words, cannot be catalogued in geographic or geological terms alone. The economic usefulness of such resources is subject to continual redefinition as a result of both economic changes and alterations in the stock of technological knowledge. Whether a particular mineral deposit is worth exploiting will depend upon all of the forces influencing the demand for the mineral, on the one hand, and the cost of extracting it, on the other (Rosenberg 1972, pp. 19-20).

The extreme sensitivity of any definition of natural resources to changing economic conditions as well as to changing technology needs to be emphasized. The President's Materials Policy Committee reported an estimated 30 billion tons of coal were recoverable at 1951 costs levels. However, as Dewhursts has pointed out, "A rise of 50 per cent in prices and costs...would mean a 20-fold increase in estimated recoverable reserves."[26] Not only is there a widening range of substitution among resources with respect to end uses--new materials, such as plastics and aluminum, and older ones, such as glass, have been replacing wood as a packaging material--but there is often also a range of *sources* from which a given input may be extracted. Oil, for example, is found in nature not only in its crude form. Enormous reserves of shale oil are known to exist and "the technical feasibility of recovering oil from shale has been clearly demonstrated."[27] Furthermore, oil can be produced from coal, a much more abundant resource. Finally, the extraction of oil from tar sands is a clear possibility. A significant increase in crude oil costs may be expected to activate such possibilities.[28]

The manner in which sharply rising prices of timber products (which quadrupled between 1870 and 1954) triggered off technological and other adaptations is illuminating. Conventional materials, such as iron and steel, were substituted in many uses. In construction, which is by far the largest single user of lumber, traditional masonry and other building materials as well as aluminum were substituted. More recently, technologically new products such as plastics and fiberglass have served as substitutes. Moreover, new techniques for economizing on wood requirements have been developed that do not involve the substitution of competitive materials, or which, in effect, substitute cheaper woods for more expensive woods--as in plywoods and wood veneers. Other technological changes, such as the self-powered chain saw, the tractor, and the truck, have made previously inaccessible forest stands available for exploitation by reducing the cost of extraction and transport. Yet other technological changes have significantly increased the size of our forest resource based by making possible the exploitation of low-grade materials which had previously gone unused.

Although wood pulp has been manufactured for about 100 years in this country, it was the rapid advance in sulphate pulping technology in the 1920s that released the industry from its dependence on spruce and fir of the Northeast and made it possible to utilize southern pine for pulp. This led to the "sulphate revolution" of the South and was largely instrumental in increasing national consumption from 6.8 million cords in 1926 to 33.4 million in 1955. The South's share of production rose from 1.1 million to 19.2 million cords in this period while pulpwood consumption in the West was increasing from 0.5 to 6.4 million cords. By the mid-1950s the South's share of woodpulping capacity had risen to over 55 percent of the national total.[29]

In sum, the main tradition of economic thinking has been dominated until comparatively recently by a view of the economy's natural resource base which was excessively restrictive because it ignored the dynamic interactions between technological change and natural resources and because it ignored a whole range of additional adaptations which are a mixture of pure technological change, redesigning, and substitution. Thus, rising fossil fuel costs may lead to the more rapid development of techniques for the exploitation of atomic or solar energy, but they will doubtless also lead to the utilization of lower-quality oil sources, to the redesign and modification of automobile engines, to the substitution of smaller for larger cars, to the use of more blankets and thermal underwear, and perhaps to greater immigration to the southern tier of states. The point is not that the rising fossil fuel costs are therefore of no consequence. That would be patently absurd. Rather, the point is that an advanced industrial society has available a wide range of mechanisms by which this rising costs and its consequences may be offset, and the historical evidence suggests overwhelmingly that such rising costs have not, *in the past*, constituted an insurmountable obstacle to sustained increases in both total population and per capita incomes. One of the main thrusts of past technological developments has been to reduce the economy's dependence on any specific resource input and to widen progressively the possibilities for materials substitution.

> As a result, although particular resources of specified quality do inevitably become increasingly scarce, the threat of a *generalized* natural resource constraint upon economic growth by no means follows from this. It now seems clear that the discussion of the role of natural resource constraints upon economic growth cannot be usefully pursued within the framework of asking how long it will be before we "run out" of specific resource inputs defined and estimated in physical units. That is simply not an interesting question, partly because there are seldom sharp discontinuities in nature, and partly because, by making it possible to exploit resources which could not be exploited before, technological change is--in economic terms if not in geological terms--making continuous *additions* to the resource base of the economy. What we are more normally confronted with are limited deposits of high quality resources and then a gradually declining slope toward lower-grade resources, which typically exist in abundance. The much greater profusion in the earth's crust of lower-grade resources than high-grade resources is one of those geological facts of life which we can--and, if the past is any guide at all, will--learn to live with. Our technological adjustment to this fact has been apparent in recent decades. As the high quality Mesabi iron ores approached exhaustion, the major steel companies not only turned increasingly to foreign sources, but directed their research toward the exploitation of the enormous deposits of hard, low-grade taconites. Techniques such as beneficiation have proven to be highly successful in resisting the pressures toward higher cost, even though the iron ore which we exploit today is inferior in quality to the iron ore which we exploited fifty or sixty years ago. Similarly, although we import large quantities of bauxite because our domestic reserves of high-grade bauxite are now inadequate, alternative sources of aluminum such as clays are, quite simply, immense. As a matter of fact, the earth's crust contains far more aluminum than it does iron. The economic exploitation of these and many other low-grade resources turns primarily upon the question of fuel and power costs, since the technologies of low-grade resource utilization are highly fuel-intensive in nature (Rosenberg 1973, pp. 116-117).

VIII

The current round of intense concern over natural resources, which began in the late 1960s, contains an important new element. Not only does it assert the inadequacy of the resource base to sustain continued growth of population (because of the limited supply of arable land) and industrial production (because of the limited supply of mineral resources). Insofar as models such as The *Limits to Growth* and its companion, Jay Forrester's *World Dynamics* (1971), do that they are essentially, for all their elaborate systems dynamics and computer methodology, Malthus in modern dress. That is, they deduce the conclusion, from certain restrictive behavioral assumptions, that continued exponential growth is impossible in a world of finite resources. The new element in the present debate is the prominence, not to be found in Malthus, of concern over the pollution problem and the assertion of additional limits to growth imposed by the increasing incidence of environmental pollution.[30]

Although an extended appraisal of the unique aspects of the new concerns is impossible here, they are subject to the same criticisms that have been made of earlier static approaches to the resource problem. That is to say, all such extrapolations of recent trends that point to some collapse of the social system years hence fail fundamentally to take account of the human capacity to adapt and to modify technology in response to changing social and economic needs. Indeed, a central feature of such models (like the dog that did *not* bark in the night) is the total *absence* of any social mechanism for signaling shifting patterns of resource scarcity and for reallocating human skill and facilities in response to rising costs, either the rising direct costs of raw materials or the rising indirect costs associated with environmental pollution.[31] One need not be a capitalist apologist to argue that the price mechanism does perform such a function, albeit imperfectly, or to believe that it might perform these functions even better under altered arrangements which no longer allowed private firms to treat our water courses and atmosphere as if they were free goods.[32]

These models constitute a reversion to Malthusian Fundamentalism in another, closely related way. They allow no possibility that mankind's taste for offspring and therefore his reproductive practices will be altered *as a result of the growth experience itself*. In both the demographic and technological realms, therefore, such purely deductive models ignore the mass of dramatic evidence of the past 150 years or so that human behavior undergoes continuous modification and adjustment as a response to the changing patterns of opportunities and constraints thrown up by an industrializing society.[33]

IX

The thrust of these critical remarks is *not* that population growth, pollution, and increasing scarcity of key natural resources are unimportant problems, or that technology may be confidently relied on to provide cheap and painless solutions whenever these key variables, and their interactions, begin to behave in problematic ways; quite the contrary. The main complaints against such apocalyptic, neo-Malthusian models, built upon a global scale and incorporating naive behavioral assumptions, are, first, that they define the problems incorrectly; and second, that they deflect attention from more "modest" but genuine questions that ought to figure more conspicuously among the agenda of social science research. Consider the following sample.

What are the determinants of human fertility? How is it likely to respond to future changes in mortality, income, urban densities, pollution, education levels, female employment opportunities, new goods (including new contraceptive techniques)? The very recent reductions in American fertility levels, which may actually turn out to be a short-run phenomenon, indicate forcefully that demography is a subject about which our present state of ignorance is truly momentous. Malthusian-type models that simply accept certain population growth rates as somehow exogenously determined and then go on to demonstrate the awesome power of exponential growth over time periods of a century or more, should surely be losing their capacity to fascinate all but the intellectually most immature. What sort of alterations in our system of property rights and our tax structure hold the greatest promise for the control of pollution? What are the technological possibilities that an altered incentive system might lead to the development of "cleaner" technologies, and at what cost?

What kinds of technological and social adaptations can we visualize if the cost of key resources, such as fuels, should increase substantially in the future?

What improvements can be made in our present mix of private and public institutions devoted to the "production" of useful knowledge?

Instead of the almost total preoccupation with supply considerations, is it possible to anticipate future shifts in income and social organization that may reduce the need for resources by shifting demand away from resource-intensive goods?

What can we find by opening up the "black box" of technological change? *How* responsive are human agents with the requisite talents to the market forces that continually shift the prospective payoff to inventive and innovative activities? And--a very different question--what can be said in a systematic way about the responsiveness of nature? That is, given the incentives, what factors shape the prospects for success in overcoming various kinds of scarcities? Is there any meaningful sense in which it may be harder or easier to make resource-saving innovations than capital-saving innovations, or either of these innovations than labor-saving innovations? Is a dollar spent on agricultural invention likely to yield the same eventual social return as a dollar spent on manufacturing or on service? What of petrochemicals vs textiles, or building materials vs transportation, or pharmaceuticals vs machine tools? Instead of computing the number of years before we "run out" of specific mineral deposits, can we project reasonable estimates for the changing resource costs of alternative technologies?

Finally, of what relevance is the historical experience of the currently rich countries? For one thing, the industrializing countries of the 19th and 20th Centuries were often able to overcome their own resource deficiencies by imports of primary products from the less developed world. When--or if--these regions are able to enter successfully into the stream of industrialization, with what alternatives will they be confronted in overcoming their own resource deficiencies? But, perhaps most fundamental, what factors have determined the differential effectiveness with which societies have responded to the problems and constraints posed by their unique resource endowments? For, as Simon Kuznets has observed:

> It need not be denied that in the distribution of natural resources some small nations may be the lucky winners at a given time (and others at other times). But I would still argue that the capacity to take advantage of these hazards of fortune and to make them a basis for sustained economic development is not often given. In the 19th century, Brazil was commonly regarded as an Eldorado--and indeed enjoyed several times the position of a supplier of a natural resource in world-wide demand; yet the record of this country's economic growth has not been impressive, and it is not as yet among the economically advanced nations. The existence of a valuable natural resource represents a permissive condition, facilitates--if properly exploited--the transition from the pre-industrial to industrial phases of growth. But unless the nation shows a capacity for modifying its soical institutions in time to take advantage of the opportunity, it will have only a transient effect. Advantages in natural resources never last for too long--given continuous changes in technology and its extension to other parts of the world.
>
> To put it differently: *every* small nation has some advantage in natural resources --whether it be location, coastline, minerals, forests, etc. But some show a capacity to build on it, if only as a starting point, toward a process of sustained growth and others do not. The crucial variables are elsewhere, and they must be sought in the nation's social and economic institutions.[34]

NOTES AND REFERENCES

1. Furthermore, the simplifying assumption was usually employed that capital and labor were applied in doses of fixed proportion, thus collapsing a three-variables model to one of two variables. Note that this procedure also has the effect of ignoring the possibilities for offsetting by capital formation the decline in worker productivity due to the deteriorating resources/man ratio.

2. For an incisive analysis of the Malthusian theory, see K. David, "Malthus and the Theory of Population," in Paul Lazarsfeld, Ed., *The Language of Social Research*, Glencoe, Ill.: Free Press, 1955. See also Mark Blaug, *Economic Theory in Retrospect*, Homewood, Ill.: Irwin, 1968; revised ed., Chap. 3.

3. J. B. Clark, *The Distribution of Wealth*, New York: Kelley and Millman, 1899; p. 56.

4. W. S. Jevons, *The Coal Question*, London: Macmillan, esp. Chap. 7, "Of Supposed Substitutes for Coal."

5. Ref. 4, p. 215; Jevon's italics.

6. Samuel P. Hays, *Conservation and the Gospel of Efficiency: The Progressive Conservation Movement, 1890-1920*, Boston: Harvard University Press, 1959.

7. Gifford Pinchot, *The Fight for Conservation*, 1910; pp. 43, 123-124; quoted in Barnett and Morse 1963, p. 76. (Emphasis supplied by Barnett and Morse.)

8. See, for example, R. Nelson, "A Theory of the Low-level Equilibrium Trap in Underdeveloped Countries," *American Economic Review* 46:894-908, 1956; Harvey Leibenstein, *Economic Backwardness and Economic Growth*, New York: Wiley, 1957; Chap. 10.

9. D. H. Meadows et al., 1972.

10. George Stigler, "The Ricardian Theory of Value and Distribution," *Journal of Political Economy* 60:187-207, 1952. Mark Blaug has made the same point: "If Malthus' theory were indeed a theory, we would want to ask: What would happen if the theory were not true? The answer is, or ought to be, that income per head would rise, not fall, with increasing population. The history of Western countries, therefore, disproves Malthus' theory. The defenders of Malthus reply: But what of India today? No one denies that India is overpopulated because the death rate was lowered by the introduction of Western medicine, thus divorcing population growth from the current level of income. It follows that India would be better off if she could also "Westernize" her birth rate. But what has this piece of advice to do with Malthusian theory of population?" Mark Blaug, Ref. 2, pp. 79-80.

11. T. W. Schultz, "The Declining Economic Importance of Agricultural Land," *Economic Journal* 61:727, 1951. Schultz presents quantitative evidence to support his generalization, which he holds "...to be empirically valid for such technically advanced communities as France, the United Kingdom, and the United States" (p. 739).

12. See A. J. Youngson, "The Opening of New Territories," Chap. 3 in M. M. Postan and H. J. Habakkuk, Eds., *The Cambridge Economic History of Europe*, Vol. VI: *The Industrial Revolutions and After*, Cambridge: Cambridge University Press, 1965.

13. Paul David, "The Mechanization of Reaping in the Ante-bellum Midwest," in H. Rosovsky, Ed., *Industrialization in Two Systems*, New York, Wiley, 1966.

14. For further treatments of technological change in agriculture with particular emphasis on their impact on agricultural productivity, see: V. Ruttan, "Research on the Economics of Technological Change in American Agriculture," *Journal of Farm Economics*, vol. 42, November 1960, and the numerous citations contained therein; R. Loomis and G. Barton, *Productivity in Agriculture: United States, 1870-1958*, U.S. Dept. of Agriculture Technical Bulletin No. 1238, Washington, D. C., 1961; Z. Griliches, "Productivity and Technology," appearing under "Agriculture" in *International Encyclopedia of the Social Sciences*; Z. Griliches, "The Sources of Measured Productivity Growth: U.S. Agriculture, 1940-60," *Journal of Political Economy*, vol. 71; August 1963; Irwin Feller, "Inventive Activity in Agriculture, 1837-1900," *Journal of Economic History*, December 1962; W. Rasmussen, "The Impact of Technological Change in American Agriculture, 1862-1962," *Journal of Economic History*, vol. 22, December 1962; W. Parker, "Agriculture," Chapter 11 in L. David, R. Easterlin, W. Parker et al., *American Economic Growth*, New York: Harper & Row, 1972. For an informative, wide-ranging survey of worldwide food prospects and the possibilities for their augmentation by use of scientific knowledge, see *The World Food Problem*, A Report of the President's Science Advisory Committee, Washington, D. C.: Government Printing Office, 1967; vol. 2.

15. The material in this paragraph is drawn from Chap. 8 of Barnett and Morse (1963), "The Unit Cost of Extractive Products."

16. As Carter points out: "There is also research directed specifically toward the use of low-priced materials, which tends to blur even more the operational distinction between technological development and price substitution" (p. 85).

17. *Ibid*. For empirical evidence to support this proposition, see Chap. 6 above.

18. The degree of substitutability, the number of such substitutes, and the quantity in which they are available are, of course, critical to long-run arguments over the possibility for avoiding rising costs imposed by the increasing scarcity of particular mineral deposits. On this issue, Barnett and Morse point out that "...near-perfect substitutes are better scarcity mitigators than imperfect ones. If consumers are virtually indifferent to whether they obtain Btu for heating, cooling, and lighting from gas, coal, or electricity, a change in the proportions in which these commodities are used will not, in itself, affect welfare. The same is true of rayon or silk, cinder-block or wood-frame construction, margarine or butter, and so on. The point is relevant to the present discussion because, by and large, scarcity-induced substitutions often involve commodities that can be employed or consumed in widely varying proportions without making much difference in welfare terms. Preferences for different forms of heating fuel, construction, clothing, or even foods are largely matters of socially conditioned taste, and are therefore quite flexible over time. Long-run elasticities of substitution are thus greater than short-run elasticities." Barnett and Morse, *op. cit.*, p. 130.

19. Joseph Fisher and Edward Boorstein, *The Adequacy of Resources for Economic Growth in the United States*, Study Paper No. 13, materials prepared in connection with the Study of Employment, Growth, and Price Levels, Washington, D. C.: Government Printing Office, 1959; p. 42.

20. The basic empirical justification for this statement is the very low level of American imports expressed as a percentage of GNP. See U.S. Dept. of Commerce, *Historical Statistics of the U.S.: Colonial Times to 1957*, Washington, D. C., 1960, p. 542. For a careful study of the changing natural resource composition of U.S. foreign trade, see J. Vanek, *The Natural Resource Content of U.S. Foreign Trade, 1870-1955*, Cambridge, Mass.: MIT Press, 1963. See also the detailed assessments for individual materials in President's Materials Policy Commission, *Resources for Freedom*, Washington, D. C.: Government Printing Office, 1952.

21. N. Potter and F. T. Christy, Jr., *Trends in Natural Resources Commodities: Statistics of Prices, Output, Consumption, Foreign Trade, and Employment in the U.S., 1870-1957*, Baltimore: Johns Hopkins University Press, 1962.

22. For an interpretation of the British industrial revolution as revolving around the substitution of abundant, inorganic materials for increasingly scarce organic ones, see E. A. Wrigley, "The Supply of Raw Materials in the Industrial Revolution," *Economic History Review*, August 1962.

23. "The Declining Economic Importance of Agricultural Land," p. 725. Schultz cites the following statement from page 20 of Harrod's book: "...I propose to discard the law of diminishing returns from the land as a primary determinant in a progressive economy... I discard it only because in our particular context it appears that its influence may be quantitatively unimportant."

24. *Natural Resources and Economic Growth*, p. 172.

25. Improvements in the technology of resource *discovery*, for example, have been an important factor in continually falsifying the pessimistic predictions of imminent exhaustion of oil reserves periodically issued by the U.S. Geological Survey. See Hans Landsberg and Sam Schurr, *Energy in the United States*, New York: Random House, 1968; p. 98.

26. J. F. Dewhurst and Associates, *America's Needs and Resources*, New York, 1955, pp. 765-766. Chapter 31 of this book contains much useful data on individual materials.

27. Fisher and Boorstein 1959 (p. 59).

28. Similarly, a sufficient rise in the price of fresh water would activate techniques for the desalination of ocean water which are already established but are, at present, prohibitively costly. Even our *present* state of technological knowledge provides us with many such options which, in effect, place an upper limit to the prospective rise in raw-material prices.

29. V. Ruttan and J. Callahan, "Resource Inputs and Output Growth: Comparisons between Agriculture and Forestry," *Forestry Science* 8:78, 1962.

30. *Why* this environmental concern should have emerged when it did is, itself, an interesting sociological question. One interpretation is that it is primarily a demand phenomenon associated with rising incomes. "[I]n relatively high-income economies the income elasticity of demand for commodities and services related to sustenance is low and declines as income continues to rise, while the income elasticity of demand for more effective disposal of residuals and for environmental amenities is high and continues to rise. This is in sharp contrast to the situation in poor countries where the income elasticity of demand is high for sustenance and low for environmental amenities. The sense of environmental crisis in the relatively affluent countries at this time stems primarily from the dramatic growth in demnad for environmental amenities." V. Ruttan, "Technology and Environment," *American Journal of Agricultural Economics* 53:707-708, 1971.

31. "Forrester's model is unsound in basic structure. He fails to build in feedbacks whereby societies utilize knowledge to overcome resource depletion and pollution. When specific mineral resource deposits begin to run out in modern societies, the prospect of rising costs feeds back signals. Man then undertakes to find new ones or substitutes, and he develops technology to utilize formerly uneconomic deposits. In response to

feedback signals, he counters pressures upon agricultural land by devising hybrid plants, high yield varieties, fertilizers, pesticides, transportation, and refrigeration. Responding to feedback, he overcomes anticipated energy shortages by exploiting deeper or new resources of great volume. In modern nations he has been successful in devising new technologies not merely to maintain real output per capita but to increase it. The real costs per unit or output have declined greatly in the face of population increases. In the U.S. the unit cost of agricultural products, in terms of labor and capital, has fallen by more than half and of mineral products by three quarters since the Civil War, and the rate of decline has accelerated since World War I." Harold Barnett, review of Jay Forrester's *World Dynamics* (1971) in *Journal of Economic Literature*, September 1972, p. 853.

32. A well-known feature of the price mechanism as a signaling device is that it fails to provide socially correct signals where externalities are present, as is notoriously the case with pollution phenomena. In large measure, this failure is due to the peculiarities of our property-rights system, which allows private individuals to make unrestricted use of common-property resources in the discharge of their pollutants. For an intelligent discussion of a range of policy alternatives for dealing with the problem of the pollution of common property resources, see J. H. Dales, *Pollution, Property and Prices*, Toronto: University of Toronto Press, 1968.

33. For a searching and suggestive analysis of the demographic response over the course of American history, see R. A. Easterlin, "Does Human Fertility Adjust to the Environment?" *American Economic Review Papers and Proceeding* 61:399-407, 1971. Easterlin's conclusion is "...that both theory and the empirical research done so far on historical American fertility suggest that human fertility responds voluntarily to environmental conditions. If this is so--and it seems hard to ignore the evidence--then the nature of what is called 'the population problem' takes on a radically different guise. The question is not one of human beings breeding themselves into growing misery. Rather, the problem is whether the voluntary response of fertility to environmental pressures results in a socially optimal adjustment. In thinking about this, it seems useful to distinguish between the potential for population adjustment and the actual degree of adjustment. The staggering change in American reproductive behavior over the past century and a half clearly demonstrates the immense potential for adjustment. Whether, currently, the degree of adjustment is socially optimal remains a matter for research" (p. 407). See also the same author's book, *Population, Labor Force, and Long Swings in Economic Growth: The American Experience*, New York: National Bureau of Economic Research, 1968.

34. S. Kuznets, "The Economic Growth of Small Nations," in E. A. G. Robinson, Ed., *Economic Consequences of the Size of Nations*, London, St. Martin, pp. 27-28. Elsewhere Kuznets stated: "[A]ny emphasis on relative scarcity of irreproducible resources, as a factor in determining low levels of economic performance extending over a *long* period, must be countered with the question why no successful effort has been made by the victim of such scarcity to overcome it by changes in technology. To be retained, the hypothesis must, therefore, be rephrased: the have-not societies are poor because they have not succeeded in overcoming scarcity of natural resources by appropriate changes in technology, not because the scarcity of resources is an inexorable factor for which there is no remedy. And obviously human societies with low levels of economic performance are least able to overcome any scarcities of irreproducible resources by changes in technology; but this is a matter of social organization and not of bountifulness or niggardliness of nature." S. Kuznets, *Economic Change*, New York: Norton, 1953; p. 230.

8. THE ECONOMICS OF INDUSTRIAL INNOVATION: MAJOR QUESTIONS, STATE OF THE ART, AND NEEDED RESEARCH
Edwin Mansfield

1. INTRODUCTION

This paper attempts to do three things. First, it describes briefly some of the major questions concerning industrial innovation that have interested economists in recent years. Second, it discusses and evaluates the kinds of research that have been carried out by economists and others to help resolve these questions. Third, it describes the sorts of research that, in my opinion, are needed to provide a more satisfactory understanding of these questions.

Needless to say, the economics of industrial innovation is a very broad area. To cut the subject down to a size that can be handled in a paper of the suggested length, I shall focus here on the innovation process in the individual business firm and devote particular attention to the effects of the organization and decision-making procedures and forecasting techniques of the firm on the innovation process. Also to be considered are the effects of industrial organization on the rate of innovation, and the extent to which industrial innovations can be forecasted.

I shall take "innovation" to mean the first application on a commercial scale of a new or improved process or product. Clearly, the process leading to a successful innovation is often a complex web of interrelated processes involving the formation of a new concept, the development of a new device based on this idea, and the probing and stimulation of a market for the device. Sometimes this entire process goes on within a single organization, but frequently that is not the case.

2. SOME LEADING QUESTIONS CONCERNING INDUSTRIAL INNOVATION

Economists have shown a very lively interest in a wide variety of questions concerning industrial innovation. Any attempt to pick out a relatively small number of these questions for special attention is bound to be influenced by one's interests and biases. In this section, I describe five broad problem areas that seem to me to be of fundamental importance, but it should be recognized that others might pick out a different set of questions. Also, note once more that I am concerned here solely with the innovation process within the individual firm.

The first set of questions concerns the nature of the innovation process. How does innovation occur? What inputs does it require? How profitable is it to the innovator? How risky is it? What types of information are used? Where does this information come from? What is the role of formal research and development (R & D)? This set of questions has stimulated a considerable amount of work, both here and abroad.

Second, economists are concerned with the very difficult question of whether or not the existing rate of innovation is optimal, or at least satisfactory. Assuming that one can estimate the rate of innovation, to what extent is the rate of technological innovation in a particular firm smaller or greater than it should be? Clearly, the answer may differ, depending on one's values. For example, an individual firm, weighing the benefits and costs of particular innovations, may come to a conclusion different from that of society as a whole. What, if anything, can be said on that score? And to what extent is our knowledge potentially useful to public and private policy makers?

Third, economists are interested in the determinants of the rate of innovation. For example, to what extent can organizational and managerial changes influence the rate of innovation? To what extent can changes in the characteristics of markets influence the

---

Dr. Mansfield is professor of economics at the Wharton School of the University of Pennsylvania in Philadelphia.

rate of innovation? More broadly, if the rate of innovation in particular firms or industries should be increased (or decreased), how can that be done? Obviously, this set of questions complements the second set. The second set is concerned with the proper direction and rate of change of the rate of innovation, whereas this set is concerned with how such a change can be effected.[1]

Fourth, economists are concerned with the relationship between industrial organization and the rate of innovation. To what extent does successful innovation in particular industries require very large firms? To what extent must industries be composed of a few large firms in order to achieve a high rate of innovation? How much conflict is there between the preservation or creation of a competitive industrial structure and the achievement of a high rate of innovation? In recent years, economists and policymakers have shown a considerable interest in this set of questions.

Fifth, the past decade has seen a spectacular increase in the amount of attention devoted by social and management scientists to technological forecasting. Books, articles, journals, and firms devoted to this subject have appeared in large numbers. Exploratory technological forecasting[2] attempts to predict the technological state of the art that will be attained and used at future times if certain conditions--such as levels of research support--are met. How adequate are the techniques of technological forecasting? To what extent can they forecast the rate of innovation? This too is a very important set of questions.

3. THE INNOVATION PROCESS: NATURE, INPUTS, RISKS, AND RETURNS

In this section, we describe briefly the existing state of the art with respect to the first set of questions. There are many models or descriptions of the phases involved in the process leading to an innovation. For example, Myers and Marquis (1969) say that the process begins with a new idea that involves the recognition of both technical feasibility and potential usefulness or demand. But this new idea, or design concept, is only the formulation of a problem that seems worth pursuing. The next step, according to Myers and Marquis, is the problem-solving activity--the search processes and research and development--that, if successful, results in a solution to the technical and commercial problems related to the idea. Finally, there is the stage in which the new device is introduced commercially. Of course, there is no guarantee that a given project will go through all of these stages. As we shall see, most projects do not clear all the technical and commercial hurdles required for completion and implementation. Moreover, as Langrish et al. (1972) and others point out, the sequence of activities leading to an innovation often fails to conform to any simple linear model.

Another way that the innovation process has been broken down is shown in Table 1. The first stage is applied research. The second stage is preparation of project requirements and basic specifications, which consists largely of routine planning and scheduling but also often involves coordination with marketing to try to insure that the proposed innovation will be a success commercially. The third stage is prototype or pilot plant design, construction, and testing. The fourth stage includes preparation of detailed manufacturing drawings, tooling, and the design and construction of manufacturing facilities. The fifth stage is manufacturing start-up, and the sixth stage is marketing start-up. (Of course, not all these stages need occur in every case.) This breakdown of the innovation process has been used by the Commerce Department's Panel on Invention and Innovation (1967) and by Mansfield, Rapoport, Schnee, Wagner, and Hamburger (1971).[3]

Table 1 shows that the third and fourth stages account, on the average, for about two-thirds of the total costs of a commercialized product innovation, based on data from Mansfield et al., (1971). Table 1 also shows that there is an enormous amount of variation among innovations in the percentage of total costs attributable to each of these phases. Based on a sample of chemical innovations, much of this variation can be explained by the importance of the innovation (as measured by sales volume), the extent of the innovator's experience in the relevant technological area, the size of the innovating firm, and other such variables. Specifically, a simple econometric model (Mansfield et al. 1976) can explain about one-half of this variation.

TABLE 1.--Percentage of total cost of innovation arising in each stage of innovative activity. (Source: Mansfield et al. 1971.)

Stage	Percentage			
	Chemicals	Machinery	Electronics	Whole sample
Applied research: Mean	17	3	4	10
Standard deviation	17	5	6	
Specifications: Mean	13	4	3	8
Standard deviation	17	6	3	
Prototype, pilot plant: Mean	13	41	44	29
Standard deviation	8	17	17	
Tooling and manufacturing facilities: Mean	41	37	30	37
Standard deviation	29	10	18	
Manufacturing startup: Mean	8	4	14	9
Standard deviation	8	6	9	
Marketing startup: Mean	7	11	6	8
Standard deviation	12	12	5	

Table 2 shows the average percentage of total elapsed time that occurred in each stage of the innovative process. Clearly, the stage that generally goes on longest is the stage during which the prototype or pilot plant work is done. Also, considerable

TABLE 2.--Percentage of total elapsed time in each stage of innovative activity. (Source: Mansfield et al. 1971.)

Stage	Percentage			
	Chemicals	Machinery	Electronics	Whole sample
Applied research: Mean	62.0	10.6	13.4	28.7
Standard deviation	28	18	19	
Specifications: Mean	34.6	9.2	12.7	18.8
Standard deviation	28	14	9	
Prototype, pilot plant: Mean	35.0	57.0	58.9	50.3
Standard deviation	22	21		
Tooling and manufacturing facilities: Mean	21.9	41.3	29.8	31.0
Standard deviation	18	17	16	
Manufacturing startup: Mean	7.8	7.8	19.4	11.7
Standard deviation	6	13	7	
Time overlap*: Mean	59.1	26.0	27.5	37.5
Standard deviation	49	31	26	

*The overlap measure here is $\left(\sum_{j=1}^{4} L_j\right)/T \times 100$, where L_j is the overlap in months between stages j and $j+1$ and T is total elapsed time of the project.

overlap exists among these phases, depending on how rapidly the innovation is carried out and on the size of the firm. Finally, there is a time-cost tradeoff; the costliness of reducing time is a function of the extent of the attempted advance in the state of the art and the size of the innovating firm (Mansfield et al. 1971). According to Frank Lynn (1966) the entire process from basic establishment of technical feasibility to commercial introduction tends to take much less time than it did prior to World War II.

Many studies have attempted to shed light on the extent to which firms are induced to undertake innovations by technical opportunities or by market factors. In other words, to what extent are innovations triggered by market needs, and to what extent are they triggered by technological opportunities? Obviously, this is not an easy distinction to make, and data based on it are not easy to interpret. But based on the studies conducted by Carter and Williams (1957), Goldhar (1970), Langrish et al. (1972), Myers and Marquis (1969), the Materials Advisory Board of the National Academy of Sciences (1966), Baker, Siegman, and Rubenstein (1967), and others, many more innovations are initially stimulated by the recognition of market and/or production need than by recognition of a technological opportunity. Specifically, about three-quarters of the innovations included in these studies seemed to be stimulated initially by a recognition of such need. Schmookler's findings (1966) based on patent statistics, point in essentially the same direction.[4]

Studies have also been carried out to determine the source of the major information inputs that evoked the design concept for the innovation, as well as the information used in the subsequent problem-solving activity. According to Myers and Marquis (1969), in about 25 per cent of the cases they studied, an information input initiated active work on the problem. With regard to the solution of problems already being worked on by the firm, the major information inputs that contributed to the solution of the problem were predominantly general in nature, widely diffused, and generally accessible. (Active search, analysis, experimentation, and outside help were involved in getting the major information input in only about 24 per cent of the cases.) Personal experience and personal contacts were the principal sources of information for the cases they studied.[5]

Because these studies indicate that only a minority of the innovations in many industries are based to any significant degree on formal R & D, some people have been led to conclude that R & D is not a very important source of innovations.[6] This conclusion may be quite wrong, partly because the innovations in these studies were not weighted by their relative importance. Based on a sample of chemical innovations, findings by Mansfield et al. (1976) indicate that R & D plays a more important role in bigger, more important innovations (as measured by sales volume of new products) than in smaller, less important innovations. Also, the finding in Langrish et al. (1972) that larger technological changes tend to be of the "discovery push" type seems to point in the same direction. Much more information is needed on this score.

One of the most obvious characteristics of the innovation process is its riskiness. For example, an early study by Booz, Allen and Hamilton (1960) concluded that 7 out of 8 hours devoted by scientists and engineers in major firms to technical product development are spent on unsuccessful projects, and that 5 out of every 10 products that emerge from R & D fail in product and market tests, and only 2 become commercial successes. A detailed study of the R & D portfolio of several chemical and drug firms indicated that, on the average, about 43 per cent of the projects that were begun were not technically completed, about 45 per cent of those that were technically completed were not commercialized, and about 62 per cent of those that were commercialized were not economic successes (Mansfield et al. 1971).

An extremely important characteristic of any economic activity of the firm is its profitability. Far too little is known about the profitability of innovative activities of various sorts carried out in various industries. As I have pointed out elsewhere (Mansfield 1972b), what information we have (based largely on the studies of Griliches (1958, 1964), Mansfield (1968a), Minasian (1969), and Fellner (1970) seems to indicate that the rate of return from the investment in technological innovation has been quite high. But for a variety of reasons, which I have tried to spell out in detail (1972b), this information is too weak to support very confident conclusions concerning the proper direction for private or public policy.

4. NEEDED RESEARCH ON THE INNOVATION PROCESS

From the discussion in the previous section, it is evident that our understanding of the innovation process has improved during the past decade, owing to the efforts of researchers who have approached the subject from a variety of points of view. Yet it is also evident that we know far too little about many key aspects of this process. Specifically, in my opinion, at least three types of studies are badly needed. First, we need much better estimates of the profitability of the investments of firms in innovations. Confining our attention to the private returns (since the social returns will be discussed in the next two sections), we can visualize at least three types of studies that could be carried out. (1) Studies could be made of the profitability of the R & D investments of several firms, based on their internal records and past evaluations made by the firms themselves. (2) Holding other factors constant (statistically), a comparison could be made of the profitability of innovators and noninnovators (or unsuccessful innovators) in particular industries. (3) Estimates could be made of the rate of return from R & D, based on the estimation of production functions with R & D included as an input. All three types of studies are being carried out on a small scale,[7] but much more work is needed.

Second, more studies are needed of the sources of innovation and the characteristics of innovators. A great many studies have taken place in recent years, including Jewkes's study (1969) of about 60 inventions; Hamberg's study (1966) of about 25 inventions; Enos's study of petroleum refining (1962); Mansfield's study (1968a) of the iron and steel, petroleum, and bituminous-coal industries; OECD's studies (1969) of electronic computers, electronic components, plastics, nonferrous metals, and a number of other industries; the study of the pharmaceutical industry in Mansfield, Rapoport, Schnee, Wagner, and Hamburger (1957); the study of the chemical industry in Mansfield, Husic, Rapoport, Romeo, Villani, and Wagner (1976); and the study of 84 innovations in Great Britain by Langrish, Gibbons, Evans, and Jevons (1972). But more needs to be done. For example, more attention should be paid to failures and to organizations with mediocre or poor performance. Also, we still lack reliable and reasonably comprehensive data for a great many important industries. But perhaps the most important need is for a more sophisticated level of analysis. In particular, more attention needs to be paid to research design, quantification, and analysis, and less should be paid to description. Indeed, a good deal might be accomplished at this point by bringing together the existing case studies and descriptive material, and by subjecting this body of material to more intensive analysis based on uniform concepts and measurement techniques, as well as on more incisive hypotheses.[8]

Third, data have been gathered recently concerning the probability of technical completion of R & D projects in various firms. Data have been also gathered concerning the probability of commercialization (given technical completion), and the probability of economic success (given commercialization) for each firm. These probabilities reflect how efficient and productive a firm's innovative activities are, as well as the inherent riskiness of its R & D portfolio. Attempts have been made by Mansfield and Wagner (1975) to construct econometric models to explain the large differences among firms in the size of these probabilities. Although the results (some of which are discussed in Sec. 8 below) seem quite encouraging, existing work does little more than scratch the surface. Much more work is needed.[9]

5. THE ADEQUACY OF THE RATE OF INNOVATION

We turn now to the second set of questions in Sec. 2: to what extent is the rate of innovation in particular firms or larger sectors of the economy what it should be? This is a very difficult question, and one that has received relatively little attention. First, consider this question from the viewpoint of a single firm. Various groups within the firm view the desirability of a rapid rate of innovation differently. Top management views it according to its preference for, or aversion toward, risk. To the production people, innovation may represent a threat that disrupts the established way of doing things and is to be resisted (Burns 1961). The marketing people may, as Schon (1967) points out, try to make safe statements within their own professional territory, and push off as much uncertainty as they can onto the R & D people. They have a vested interest in change, but even they may balk at taking big risks. They may prefer to work on relatively minor projects where there is a reasonably high chance of technical success. The firm's workers may, of course, view innovations, particularly labor-saving ones, with distrust. Since each of these groups within the firm may have quite a different idea of the optimal rate of innovation, it is clear that the answer to this question will vary, depending on how one weights various parts of the firm.

If we regard the views of top management as being of particular significance (since top management is in charge of the firm's total operations), we find that such managers often feel that their firm's rate of innovation is less than it should be, given the amount they invest in innovative activities and given their objectives. A variety of reasons is given for this finding, but one of the most important is that the marketing and production departments of the firm are unable or unwilling to utilize the output of the R & D department effectively. For example, according to the major executives of 18 large firms, the probability of technical completion, the probability of commercialization (given technical completion), and the probability of economic success (given commercialization) would be 10-25 per cent higher, on the average, if the marketing and production people did a better job of exploiting R & D output (Mansfield and Wagner 1975).

Although this may be one important barrier to a more adequate rate of innovation, there are many others as well. For example, marketing and production people often retort that the R & D department carries out projects that are too far removed from the realities of the marketplace. Since many of these other factors are discussed in Sec. 7, we shall not take them up here. For present purposes, the principal point is that, judging from small surveys and impressions gained from conversations with firms, many managers seem to feel that the rate of innovation of their firms is not what it should be; and judging from Langrish et al. (1972), the same is true in the United Kingdom. But there is very little hard evidence concerning the extent of this dissatisfaction, how widespread it is, or how justified it may be.

Next, let us consider this same question from the point of view of society as a whole. Because the results of research are often of little direct value to the sponsoring firm but of great value to other firms, there is good reason to believe that, left to its own devices, a market economy would allocate too few resources to R & D. Also, because research and development is risky for the individual firm, there is good reason to believe that a market economy would underinvest in innovative activities. These defects of the market mechanism have been recognized for many years. For example, Pigou (1928) described them in the 1920s. In recent years, they have been discussed at length by Arrow (1962a) and Nelson (1959).

However, since our nation already supplements the market's provision of resources for innovative activity in a variety of ways, the relevant question is not whether the market mechanism requires supplementing, but whether the type and extent of supplementary support provided at present is too large or too small, and whether it is allocated properly. Most of the economists that have studied this question seem to be of the opinion that we may be underinvesting in particular types of R & D in the civilian sector of the economy, and the estimated marginal rates of return from certain types of civilian R & D seem very high. The major studies in this area are by Griliches (1958, 1964), the President's Commission (1966), Mansfield (1968b), Minasian (1969), and Nelson, Peck, and Kalacheck (1967). Although the evidence from practically all the available studies seems to point in the same direction, it should be recognized that this evidence is very weak. A detailed description of the nature and weaknesses of this evidence is given in a paper devoted largely to this subject (Mansfield 1972b).

In the previous two paragraphs, I have assumed implicitly that more output is a good thing. In recent years, some economists have argued that we are rich enough, that economic growth is likely to be accompanied by higher levels of environmental pollution and other disamenities, and that it is foolish to grasp for more output. Although the anti-growth economists, led by E. J. Mishan (1967), have had an undeniable impact, economic growth remains a goal of practically all governments of the world. Whether or not economic growth is worth its "costs" is a political question that must be settled by the peoples of the world. Thus far, they seem to have opted on balance for growth.[10] At the same time, however, many observers feel that fuller account should be taken of the divergences between the private and social costs of introducing various kinds of new technology. This feeling is, of course, at the heart of the current concern with technology assessment.

Another factor that has loomed large in recent discussions of the socially optimal rate of innovation is the deficit in our balance of payments. According to many observers, our technological lead over many foreign countries has shrunk considerably in recent years, due in part to the fact that Western Europe and Japan have recovered completely from World War II. Some observers, such as Boretsky (1971), regard this reduction in our technological lead as having been an important factor leading to the deterioration in our balance of payments. That may be true, although many other factors certainly were important as well. However, despite some arguments to the contrary, the fact that an industry is finding it increasingly difficult to compete with foreign producers is not an adequate social justification for more R & D--or other forms of investment in innovation --in that industry. Additional R & D may not have much of a pay-off there; and even if it does, the additional resources may have a bigger pay-off somewhere else in the economy.[11]

6. NEEDED RESEARCH ON THE ADEQUACY OF THE RATE OF INNOVATION

From the discussion in the previous section, it is clear that far too little is known about the optimal rate of innovation for particular firms or larger segments of the economy. Much of the existing literature consists of rather superficial attacks on technological change or one-sided accounts of the benefits of innovation. In part, the lack of better information in this area is due to the inherent difficulty of the problems, as well as the fact that, in the final analysis, the socially optimal rate of innovation is as much a political as an economic question. Nonetheless, it seems to me that at least four types of studies would be extremely useful in promoting a better understanding of the questions in this area.

First, attempts should be made to obtain better estimates of the social rates of return from innovative activities of various kinds. Without such estimates it is extremely difficult to formulate public policy on technological change. The estimation of social rates of return is a very difficult task, because innovations generally result in benefits and costs of many sorts that do not accrue to the innovator. For example, a new product is likely to result in uncompensated benefits to firms and consumers that use the new product; it may result in uncompensated benefits and costs to workers and suppliers of inputs used by the industry producing the new product; it may trigger or suggest improvements in other industries or it may have adverse effects on them; and it may have adverse or positive effects on other public groups through its environmental or other impacts.

Although the problems involved in estimating the social returns from various kinds of innovations are formidable, they do not seem insurmountable, particularly when one recognizes that even rough estimates would be useful for many purposes. Although it is doubtful that we shall be able to make very accurate estimates, there is no reason that intensive research should not enable us to make estimates of social rates of return that are far better than those currently available. After all, in many other areas of economics, worthwhile attempts have been made to make such estimates. And in this area, Griliches (1958) and others have already achieved some success.[12] Much more work needs to be done. Moreover, as stressed in Sec. 4, we need better estimates of the private rate of return from innovative activity of various sorts. Such estimates would go a long way toward indicating what rate of innovation is optimal from the point of view of the individual firm. For some recent results in this area, see Mansfield et al. (1976).

Second, studies are needed of the ways in which firms trade off various kinds of risk: (1) the chance that a project will not be technically completed; (2) the chance that a project, if technically completed, will not be commercialized; and (3) the chance that a project, if commercialized, will not be economically successful. On the basis of detailed studies we have made of particular firms (and the results of internal audits made by other major firms), I suspect that some firms tend to make an improper (from their own point of view) trade off between these types of risk. The people who choose projects sometimes focus too much attention on the first type of risk; and to reduce this type of risk, they pick many relatively unambitious projects. Unfortunately, however, since such projects tend to have high levels of the second and third types of risk, this choice of projects may not be optimal from the overall point of view of the firm. Very little work has been carried out to test, modify, and extend hypotheses of this sort. Yet such work obviously is needed if firms are to get the most out of their investments in innovation.

Third, studies are badly needed of the optimal extent and pattern of international specialization with respect to technology. Most countries cannot hope to have strong capabilities for innovation in most areas of technology, because of their limited R & D (and other) resources. To what extent should a country specialize in particular areas of science and technology? What determines the particular areas that a country should specialize in? As Pavitt (1971) and others have pointed out, a certain amount of specialization seems essential, even for a big country like the United States. But the optimal strategy--for a single country or for broader groups--has not been studied in any real depth, even under highly simplified conditions. It is, of course, a much more difficult problem than the relatively simple, static problem regarding specialization and trade with respect to goods. This question is of fundamental importance, and deserves much more attention than it has received. Of course, related to this question are many questions concerning international technology transfer, the multinational firm, and so on, some of which are discussed in Mansfield (1972).

Fourth, more work is needed to develop more satisfactory measures of the rate of innovation. At the firm level, we are beginning to get data based on counts of innovations (weighted or unweighted) carried out by particular firms. At the firm and industry levels, attempts have been made to estimate the rate of technical change by shifts over time in the production function. But these methods clearly are crude at best. Inherently, the construction of measures of the rate of innovation is a very difficult task, and I doubt that we shall obtain really satisfactory measures in the foreseeable future. But I think that considerable improvements can be made. What is needed is a stronger melding of the contributions of the economic theorist and the econometrician with those of R & D managers and scientists and technologists; economists should work more closely with the latter.[13]

7. FACTORS INFLUENCING THE RATE OF INNOVATION

We turn now to the third set of questions put forth in Sec. 2: what factors influence the rate of innovation? Obviously, many factors are influential, including the nature and extent of a nation's scientific capability,[14] the size and quality of its educational system, the nature of government policy regarding technological change (including regulatory policies, tax policies, patent policies, procurement policies), the attitudes and strength of labor unions, and the quality and attitudes of management. Also, reasonably full employment, coupled with a high rate of investment, tends to promote innovation (and the rapid and relatively painless adaptation of the labor force to technological change), and programs that spread more equitably the social costs of technological change--for example, retraining programs that help displaced workers--reduce resistance to innovation.

Basically, the rate and direction of industrial innovation depend on the extent to which various kinds of new technology are stimulated by changes in product demand, changes in input proportions and availability, and other factors influencing the profitability of technical change. The effects of these factors have been discussed by Habakkuk (1962), Hicks (1963), and others. In this paper, my task is to focus on factors operating at the level of the firm (or the industry), not on the broad factors listed in this and the previous paragraph. Even when the topic is cut down in this way, the task is not an easy one, since there are so many factors that could be included. Without making any pretense that the following discussion is any sense exhaustive, I shall attempt to describe--as briefly as I can--what we know about the effects of six important factors on the rate of innovation.

The first factor is the way in which the firm couples R & D on the one hand, with marketing and production, on the other. Many studies, including Freeman (1973),

National Academy of Sciences (1967), Mansfield, Rapoport, Schnee, Wagner, and Hamburger (1971), and Lawrence and Lorsch (1967), indicate that effective innovation requires a close coupling of these functions. Although it would be an exaggeration to say that we understand this factor in detail, there seems to be general agreement among managers concerning its importance, and the available studies seem to bear out that view. Moreover, a fair amount is known about the techniques that firms use to promote a closer coupling of these functions. However, very little is known about the cost effectiveness of particular techniques of this sort.[15]

A second factor that seems to influence a firm's rate of innovation is its organization. According to Burns and Stalker (1966), innovation is less likely to occur in firms where problems and tasks are broken down into specialities and where there is a strict vertical chain of command, than in firms where there is no strictly defined hierarchy, where communication resembles consultation rather than command, and where individuals have to perform tasks in the light of their knowledge of the tasks of the whole firm. Many observers, like Schon (1967), are impressed by the difficulty of carrying out radical innovations in the large, established firm; they feel that such innovations tend to be spearheaded by new firms and "invaders" from other industries.[16] A number of large firms--Du Pont, Exxon, ICI, and others--have tried to obtain some of the advantages of the small firm by creating a number of teams in their development department that operate somewhat like small firms (Peterson 1967). Like most other areas of organization theory, the results in this area are often qualitative and descriptive, and lines of causation are difficult to establish. Existing knowledge is far from adequate.

A third factor that influences a firm's rate of innovation is the way it manages its R & D and innovative resources. For example, some firms allow projects to run on too long before evaluating their economic (as distinct from their technological) potential. Judging from detailed data for 16 major firms, a firm's probabilities of success (both technical and economic) seem to be directly related to how quickly it evaluates the economic potential of projects (Mansfield and Wagner 1975). Also, some firms and government agencies, according to Klein (1965) and others, tend to suppress and underestimate uncertainty, and they do not run parallel approaches in many cases where they should do so. According to still other observers, there is frequently a tendency for firms to devote too large a part of their resources to developments arising from their own laboratories and to neglect or resist developments that are "not invented here." This may be a particularly troublesome tendency because, according to many studies (Mueller 1962, Mansfield et al. 1971), external sources of technology are frequently of great importance in the innovation process. Our knowledge of the importance of these and other such problems--and the best ways to overcome them--is extremely limited. All that we really have at present are some clues, many of which are based on little more than visceral reaction by managers, engineers, and management scientists. Similarly, despite the pioneering work of Pelz and Andrews (1966), Gordon (1972), and others, our understanding of the determinants of scientific and engineering creativity and productivity is limited.[17]

A fourth factor that may influence a firm's rate of innovation is the extent to which it uses various analytical techniques for the management of innovation that have come into prominence in recent years. Among such techniques are PERT and related scheduling methods, project selection methods based on sophisticated quantitative comparisons of costs and benefits, and technological forecasting methods (discussed in Sec. 11 below). Although there is a large volume of literature on such analytical techniques, it is not clear that they really have found very extensive use. Moreover, when they have been used, it is not at all clear what their effects have been. In view of the large errors in the estimates used in the project selection techniques, and the shortcomings of the models themselves, it would seem advisable for firms to use these techniques with considerable caution (Mansfield et al. 1971).

A fifth factor that influences a firm's rate of innovation is the size and sophistication of the relevant market. The scale of the market influences the extent to which the firm can spread the fixed costs of developing and introducing an innovation. Also, the sophistication of market demand obviously influences the rate of innovation. For example, the large government demand for advanced electronic components, scientific instruments, and electronic computers certainly promoted innovations in those areas in the USA. In recent years, many European observers attributed the "technology gap" partly to differences in the size and sophistication of markets (Pavitt 1971). Of course, the relevant market is not necessarily the national market. Tariffs and other barriers can be overcome, although that often is not easy. Pavitt (1971) has done some work in this area, but his findings are not easy to interpret because his sample is so small and other factors are not held constant.

A sixth factor that may influence the rate of innovation is the availability of risk capital. Innovation is often a risky business; and as noted above, radical innovations frequently are the product of small businesses run by people with little knowledge of the market for venture capital. On the other hand, the financial institutions, wealthy individuals, and others who are in a position to put up the money typically have relatively little appreciation of technical matters, and there are enormous problems in making a proper appraisal of such new ventures. It seems likely that the rate of innovation depends on the size of the pool of venture capital, and on the efficiency and imagination with which it is lent out. But little is known beyond that. According to the Charpie report (1967), "the alleged absence of potentially available venture capital is not really the problem," but as the authors of the report would be the first to admit their evidence is not very strong.[18]

8. NEEDED RESEARCH ON FACTORS INFLUENCING THE RATE OF INNOVATION

From the discussion of the previous section, it seems clear that existing knowledge concerning the factors influencing the rate of innovation, although more extensive and reliable than what was available a decade ago, is still too weak to provide the information needed by public and private policy makers. There are many studies that are well worth carrying out. Three kinds of studies that I regard as being of particular interest are as follows. First, very detailed studies might be made of the innovative performance of a variety of firms in the same industry. Data might be obtained from the firms themselves concerning the innovations they introduced, the percentage of their R & D projects that were successful, the profitability of their new products, and many other such variables. Based on these performance measures, an attempt might be made to relate a firm's innovative performance to the characteristics of its organization, its decision-making process, its personnel, and so forth. Although correlation does not prove causation, it should be possible to test many relevant hypotheses, at least partially, in this way. Some attempts (Mansfield and Wagner 1975) have been made to carry out studies of this sort, but they have done little more than scratch the surface.

Second, studies might be made of the effects of changes in organization, corporate strategy, personnel, and a variety of other factors on a firm's innovative performance. It is fairly common for a firm to undergo a major change in organization or tactics. Yet there have been few attempts to get detailed information from such firms concerning various measures of their innovative performance before and after the change, and to compare the difference in performance with other reasonably comparable firms that did not experience the change in organization, tactics, etc. Mansfield and Wagner (1975) attempted one such study, but it is based on only a few firms. Much more work of this sort is needed.

Third, it is high time that we obtain reliable and comprehensive information regarding the extent of the venture capital that is available, the criteria used by investors to determine whether or not they will invest in a particular project, and the experience of various innovators in obtaining such capital. Some people claim that there is insufficient capital of this sort. The Charpie report (1967), on the other hand, concludes that more than $3 billion of potentially available venture capital exists in this country. It seems to me that this is a researchable subject, and that more work on it is needed. For one thing, it would be very interesting to review the records of various investors to see the nature of the projects that were turned down. To what extent does it appear that good bets were rejected? Of course, no one can predict with accuracy what would have happened if a project had been approved rather than rejected; but with the advantages of hindsight, it might be possible in many cases to reach a reasonable consensus. On this basis, it might be possible to estimate more precisely than is now possible the probability that a good project will be turned down, as well as the probability that an unsuccessful project will be accepted. This is an area that has been discussed ad nauseam, but with few facts. It should be possible at this point to obtain much better information on this topic than currently exists.

9. SIZE OF FIRM, MARKET STRUCTURE, AND THE RATE OF INNOVATION

Economists have long tended to argue that economic efficiency is promoted by competition. For example, Adam Smith wrote about two centuries ago that "monopoly...is a great enemy to good management." In recent years, however, some economists, led by Joseph Schumpeter (1950) and John Kenneth Galbraith (1952), have argued that a competitive economy is likely to be inferior in a dynamic sense to an economy including many imperfectly competitive industries (i.e., monopolies, oligopolies, etc.). It is easy to grant that a certain amount of market imperfection may promote the rate of innovation. But some of these economists go beyond this to say that the "modern industry of a few large firms [is] an almost perfect instrument for inducing technical change" (Galbraith 1952).

If true, this is an extremely important point. But is it true? Does the evidence indicate that an industry dominated by a few giant firms is generally more progressive than one composed of a larger number of smaller firms? Based on the studies of Jewkes, Sawers, and Stillerman (1969), Hamberg (1966), Mueller (1962), and others, it appears that the role of the small firm or the independent inventor may be very important at the stage of concept formation and the initial, relatively inexpensive stages of R & D. Moreover, although the development of these ideas often requires more resources and different kinds of management skills than some of these small firms have, the investment required for development and innovation is seldom so great or so risky that only the largest firms in the industry can do the developing or the innovating. For example, our studies of the drug, steel, petroleum, and coal industries indicate that the firms that carried out the largest number of innovations--relative to their size--were not the very biggest firms. Indeed, in the steel and drug industries, they tended to be surprisingly small firms (Mansfield 1968a, Mansfield et al. 1971). Table 3 summarizes some of our results.

TABLE 3.--Relationship between size of firm and innovative performance in some major U.S. industry groups. (Sources: Mansfield 1968a, 1976; Mansfield et al. 1971.)

Industry	Time period	Did the largest four firms account for a bigger share of innovations than their market share*?	Size of firm that carried out the most innovations (relative to its size)
Iron and steel	1939-1958	No	Relatively small
Petroleum	1939-1958	Yes	About 6th largest
Bituminous coal	1939-1958	Yes	About 6th largest
Ethical drugs	1950-1962	No	About 12th largest

*Innovations are weighted by economic importance (as measured by sales of products and cost reductions of processes). Sales or capacity were used as measure of firm size.

According to studies made by Scherer (1965c), Mansfield (1968a), Grabowski (1968), and others, it also appears that the largest firms generally do not spend more on R & D, relative to their size, than somewhat smaller firms. There seems to be a threshold effect. A firm has to be a certain size to spend much on R & D, as defined by the National Science Foundation. But beyond a certain point, increases in size no longer bring any disproportionate rise in R & D expenditures. This relation seems to hold in all industries studied, with the exception of chemicals. Of course, the threshold varies from industry to industry, but, according to Scherer (1965c), increases in size beyond an employment level of about 5000 employees are generally not accompanied by a more than proportional increase in innovation inputs or outputs. Thus, the evidence does not seem to indicate that giant firms devote more resources, relative to their size, to inventive and innovative activities than their somewhat smaller competitors. Moreover, there is some evidence that, per dollar of R & D, the biggest firms get less inventive and innovative output than smaller firms. Scherer (1970) and Mansfield (1968b) review this evidence in some detail.

From the point of view of antitrust policy, it is particularly important to explore the relationship between the extent of concentration in an industry and the industry's rate of technological change. Far too little work has been carried out in this area to support firm or confident conclusions. But certain patterns are beginning to emerge. For example, our studies seem to indicate that innovations are not utilized and accepted more rapidly in more concentrated industries than in less concentrated industries. On the contrary, it often appears that innovations spread more rapidly in less concentrated industries (Mansfield et al. 1976). In general, most studies of the relationship between industrial concentration and the rate of technological change seem to conclude that a slight amount of concentration may promote more rapid invention and innovation. Very fragmented, splintered industries like construction do not seem to be able to promote a rapid rate of technological advance. But beyond a moderate amount of concentration, further increases in concentration do not seem to be associated with more rapid rates of technological advance. In part, that is so because of the accompanying decrease in the number of independent sources of innovation. Thus, the available evidence does not seem to indicate that we must permit considerable concentration in order to promote rapid innovation.[19]

In addition, three other points should be noted. First, there is evidence that new firms and firms entering new markets play a very important role in the promotion of technological change. Existing firms can be surprisingly blind to the potentialities of new ideas. They can become bureaucratized and tradition bound. An important way that their mistakes and inertia are overcome in our economy is through the entry of new firms. Moreover, there have been many cases where the threat of entry by new firms or firms in other industries has stimulated existing firms to carry out significant technological advances.

Second, it is generally agreed by economists that the ideal market structure from the point of view of stimulating technological change is one where there is a mixture of firm sizes. Complementarities and interdependencies exist among larger and smaller firms. There is often a division of labor, with smaller firms focusing on areas requiring sophistication and flexibility and catering to specialized needs and bigger firms focusing on areas requiring larger production, marketing, and technological resources. There seem to be considerable advantages in a diversity of firm sizes; no single firm size is optimal in that respect. Moreover, the optimal average size is likely to be directly related to the costliness and scope of the inventions that arise. However, in general these factors do not mean that the very biggest firms in an industry are required to promote rapid technological change.

Third, some industries composed of exceedingly small firms are so fragmented that technological change is hampered. Firms cannot afford to do much R & D and they may find it difficult to use new techniques. The risks involved in innovation are too great for any single firm. In such cases, it may be good policy to supplement the R & D supported by the private sector with publicly supported R & D. An obvious example is in agriculture, where the government has supported a great deal of R & D for many years. However, we know very little about the frequency with which such cases occur. Much of the available evidence is qualitative and by no means unambigous.

10. NEEDED RESEARCH CONCERNING SIZE OF FIRM, MARKET STRUCTURE, AND INNOVATION

Based on the discussion in the previous section, it is evident that, although we are beginning to assemble the data and techniques needed to answer the important questions in this area, we still have a long way to go before we can be confident of the answers. Studies of many kinds are needed. Three types of studies that, in my opinion would be very useful are as follows. First, we need more information concerning economies of scale in particular types of R & D. There are many reasons for thinking that there are economies of scale in R & D up to some point--"lumpiness" of capital equipment used in R & D, advantages from specialization of labor, reduction of risk due to the law of large numbers. However, we know very little, industry by industry, concerning the extent of these economies of scale for particular types of work, or about the size of R & D establishment beyond which further increases in size bring little or nothing in the way of further efficiencies for the type of work in question.

Second, we need to know much more about the costs of technological innovation in various industries. A number of recent studies have focused attention on "technological thresholds"--that is, minimum sizes of firm--that must be achieved before a firm can be a successful innovator. For example, in the electrical equipment industry, Freeman (1965) has presented evidence indicating that these thresholds are very high.[20] It seems obvious that they are lower in many other industries, but we know little about their size or rates of change. As Pavitt (1971) remarks, "few generalizations can be made about the thresholds necessary for effective industrial innovation." In considerable part, that is because we have so little data concerning the costs of innovation in various industries. Clearly, such costs are greatly in excess of R & D costs alone, but beyond that, little is really known.

Third, we need detailed data identifying the major innovations that have occurred in recent years, as well as the innovator in each case, for many more industries. As matters stand, we have such data for only about a half dozen industries. Clearly, much more such data are needed. Further, it would be useful if more attempts were made to estimate the relationship between size of firm and innovative performance among particular types of innovations. It seems likely that this relationship will vary in a particular industry, depending on the characteristics of the innovations. Although the overall relationship between size of firm and innovation has been studied (in the small number of industries for which data are available), no attempt has been made to classify innovations in various ways, and estimate the relationship in each class. Such studies might shed considerable new light on the subject.

11. TECHNOLOGICAL FORECASTING[21]

We turn now to the fifth set of questions put in Sec. 2: what is the current state of the art with respect to technological forecasting? It is generally agreed that technology is a difficult variable to forecast, because there is so much uncertainty concerning what will be produced by R & D efforts, what breakthroughs will occur, and when. How do the people engaged in technological forecasting go about making their forecasts? According to various surveys, as well as the leading early texts on technological forecasting (Bright 1968, Cetron and Ralph 1971, and OECD 1967), simple intuitive projections seem to play a very important role in exploratory technological forecasting. For example, suppose that a firm or government agency wants to forecast the maximum speed of commercial aircraft a decade hence. One way of obtaining such a forecast is simply to ask an expert, or group of experts, to guess as best they can what the maximum speed will be at that time. Certainly this approach is straightforward enough and relatively cheap. But it runs into a number of difficulties. First, technologists are no more in agreement about the future than economists are, so that the answer is likely to vary depending on the choice of expert. Second, even when based on the opinion of distinguished experts, such forecasts can contain large errors. For example, Vannevar Bush predicted in 1945 that a 3000-mile rocket would "be impossible for many years" (Mansfield 1968b).

To cope with some of the problems involved in simply asking a group of experts for a consensus guess, Helmer and Gordon, while at RAND Corp., formulated the technique known as the Delphi method, which attempts to utilize expert opinion more effectively (1964). For example, to forecast the maximum speed of commercial aircraft, users of the Delphi method would ask a number of experts to formulate separate and independent estimates. Then the median and interquartile range of the estimates would be communicated to each of the experts, and they would be asked to reconsider their previous answers and revise them if they wanted to. Then those people whose answers lie outside the interquartile range would be asked to state why they disagree to this extent from the other members of the group. Then their replies would be circulated among the group, and the members would be asked once again to make a forecast. This iterative process would continue until there was a reasonable convergence of the individual forecasts.

The Delphi method has been used in fields as diverse as defense, pharmaceuticals, political science, and educational technology. According to its developers, it is a useful tool for technological forecasting. However, it is obvious that the results of the Delphi method can be no better than the foresight of the individual experts. And as noted above, this foresight can be very imperfect. Moreover, by relying so heavily on a consensus, the Delphi method assumes that collective judgment is better than individual judgment. This is a dangerous assumption, as evidenced by the many important technological advances that have been made by individuals and groups that acted contrary to prevailing majority--and elite--opinion.[22] Another technique that plays an important role in exploratory technological forecasting is simple trend extrapolation (Lenz 1970, Cetron and Ralph 1971). For example, to forecast the maximum speed of commercial aircraft for a decade ahead, one could obtain a time series of the maximum speed of such aircraft at various points in history, and project the historical trend into the future. In fact, this simple sort of extrapolation technique has been used in the U.S. Department of Defense where much of the work on technological forecasting originated. It has also been used in commercial work of various kinds. Of course, economists themselves have long used such techniques. For example, about 20 years ago, John Kendrick discussed the use of such crude techniques to forecast productivity change in the American economy (1954).

The problem with naive extrapolation techniques of this sort is that, unless the fundamental factors determining the technological parameter in question operate much as they have in the past, previous trends will not necessarily be a good guide to the future. For example, a host of factors, including the allocation of R & D resources and the pressure of environmental concerns, may see to it that the maximum speed of commercial aircraft increases at quite a different rate than it has in the immediate past. Or consider productivity increase: there is considerable evidence that it has not occurred at a constant rate in the United States. The moral, well known to economists, is that a naive projection of historical trends is a dangerous business, particularly when long-term forecasts are being made.

Besides trend extrapolation, the technological forecasters have adopted another old favorite of the economic forecasters--lead-lag relationships. For example, to forecast the maximum speed of commercial aircraft, one could plot the maximum speed of commercial aircraft against the maximum speed of military aircraft. Finding that commercial speeds have lagged military speeds, one might be able to use this relationship to make the desired forecast. Of course, here too the problem is that the historical relationship may not continue into the future. For a discussion of these and other technological forecasting techniques, see Roberts (1968).

Finally, there has been some experimentation with somewhat more sophisticated models. Input-output analysis has been used by Carter (1970) and others to forecast future input requirements. As present, its use is still in its infancy, and all that can be said is that it represents a promising area for research. Also, econometric models of aggregate output, which include R & D expenditures, have been used for certain kinds of technological forecasting, as we saw in Sec. 5. In addition, econometric models of the rate of diffusion of innovations are beginning to be used for certain kinds of technological forecasting by government agencies and business firms. For an account of the accuracy of some forecasts of the rate of diffusion of numerically controlled machine tools, which were made on the basis of my own model (Mansfield 1973a).

12. NEEDED RESEARCH CONCERNING TECHNOLOGICAL FORECASTING

Based on the discussion in the previous section, it seems fair to say that most of the techniques commonly used for technological forecasting seem crude, even by the standards of the social and management sciences. In view of this crudeness, it seems unlikely that the results can be at all accurate. But as matters stand, one cannot even be sure of that, since there have been no studies measuring the track record of various kinds of technological forecasting techniques. Such studies seem to be called for. It would be useful to have some idea of how well these techniques have performed under various circumstances, and of which sorts of techniques seem to do better under particular kinds of circumstances. Without such information, it is hard for anyone to make decisions concerning the types of exploratory forecasting activities that are worth carrying out.

To prevent misunderstanding, I want to add that there is no doubt that technological forecasting is a necessary part of the decision-making process in firms and government agencies. Just as one cannot avoid forecasting the economic future--explicitly or implicitly--one cannot avoid forecasting the technological future. But that does not mean that it is necessarily worthwhile for a firm or government agency to support any formal work in technological forecasting. Whether or not it is worthwhile to support such work depends on whether--under the particular set of circumstances facing the firm or agency--the potential gains seem to outweigh the costs.[23] And given the lack of reliable data regarding the likely gains from various kinds of technological forecasting, this is not an easy comparison to make.

In my opinion, the emphasis at present ought to be on studies leading toward a better basic understanding of the process of technological change. Until the fundamental processes are somewhat better understood, it seems unlikely that much improvement will occur in exploratory forecasting techniques. The area that is perhaps best understood at present is the diffusion process--and this is the area where forecasting currently seems most effective. Needless to say, I am not suggesting that a moratorium be declared on technological forecasting until we understand the basic processes more thoroughly. What I am suggesting is that more emphasis be placed by researchers and practitioners on the accumulation of the basic knowledge that is required if this field is to become more of a science and less of an art.

13. CONCLUSIONS

In conclusion, recent studies of industrial innovation have provided us with a much fuller understanding of some of the leading questions concerning the innovation process. But despite the progress made in the last 10 or 15 years, we still know very little about many of these questions. This is no criticism of the people working in this area. Given the inherent difficulty of the problems and the small number of people working on them, a great deal has been accomplished.

In view of the importance of the questions in this area, more research is badly needed. There is no indication that these questions are impervious to research. On the contrary, the research carried out in the past 10 or 15 years has resulted in considerable progress. But much more work is needed. In this paper, I have tried to indicate specific types of studies that might be carried out. This is merely a sample of the kinds of work that are needed--and a sample that obviously reflects my own interests and biases. But I hope it may be of use as an indication of where we stand in this area and what needs to be done.

NOTES AND REFERENCES

1. Needless to say, the answer to the second set of questions depends on the answer to the third set. In other words, the proper rate of innovation depends, of course, on how changes can be effected in the rate of innovation, and on how costly they may be.

2. The literature on technological forecasting generally distinguishes between exploratory and normative technological forecasting. Normative technological forecasting attempts to specify a rational allocation of resources to influence the rate and direction of technological change. The quantitative project selection methods discussed in Sec. 7 are examples of normative technological forecasting techniques.

3. This breakdown is designed for product innovations; for process innovations, changes (of a rather obvious sort) would have to be made.

4. See Usher (1954) and Ruttan (1959) for discussions of the processes of problem perception and solution involved in the acts leading to successful innovation.

5. See Allen (1966) for a discussion of the relative performance of various channels in transferring technical information. Many studies refer to the importance of

various kinds of "gatekeepers." Also see Rosenbloom and Wolek (1970) for an interesting study of how engineers and scientists acquire information. Further, see National Science Foundation (1967).

6. For example, see some of the discussion in "Bringing the laboratory down to earth," *Fortune*, January 1971.

7. Some studies along these lines are being carried out under my direction; for example, see Mansfield et al. (1976).

8. This statement should not be interpreted as a criticism of existing studies. In an area that is as poorly understood as this one, a sensible first step is to gather some reliable descriptive material. What I am suggesting is that, at this point, further replications of such descriptive studies are likely to encounter sharply diminishing returns. More attention should be focused now on the formulation of more sophisticated analytical techniques and hypotheses.

9. Another area that needs attention is the study of the small improvements that Arrow (1962b), Hollander (1965), and others have shown to be so important. Researchers tend to concentrate their attention on the big innovations, and pay too little attention to these small improvements that, over a period of time, frequently are as important as the big innovations.

10. In addition, of course, Jay Forrester, Dennis Meadows, and their colleagues have made some well-publicized studies of "the limits of growth." These studies have been criticized severely by economists. For a discussion of both the views of Forrester and Meadows and those of Mishan, see Solow (1973).

11. For some discussions of the relationship between innovation and foreign trade, see Cooper (1971), Gruber, Mehta, and Vernon (1967), Hufbauer (1966), Keesing (1967), Mansfield (1972a), National Academy of Engineering (1971), National Bureau of Economic Research (1970), National Bureau of Standards (1966), OECD (1969), and Vernon (1966).

12. As part of the project in Note 7, we have done some studies of social and private rates of return from innovations; see Mansfield et al. (1976).

13. For a discussion of some of the problems, see Kuznets (1962).

14. For some data bearing on the role of science in technological innovation, see Sherwin and Isenson (1966), Illinois Institute of Technology Research Institute (1969), and Price and Bass (1969).

15. Much of the literature in this area suffers from the fact that it does not even attempt to address the sort of question that, to an economist, would seem basic: What are the costs and benefits of particular actions? However, we recognize that such questions are easier to pose than to answer.

16. Also, studies by Schon (1967) and others conclude that the existence of a champion, a missionary, a zealot who fights strongly--and sometimes not in strict accord with the "rules"--for an innovation is an important contributor to the application of new technology. Thus, it is sometimes argued that organizations that can tolerate or nurture such individuals may tend to be most innovative. For a somewhat more cautious view, see National Academy of Sciences (1967). Also, see Morrison (1966).

17. For a view of the findings of the behavioral science literature concerning the factors influencing the performance of scientists and engineers, see Martino (1973).

18. Also, see Charpie (1972).

19. For relevant discussions, see Comanor (1965), Fellner (1951), Hennipman (1954), Maclaurin (1954), Pavitt (1971), Peck (1962), and Schmookler (1966), as well as those cited in text.

20. Freeman defines the threshold as the minimum R & D expenditure per year, rather than the minimum size of firm. Clearly both measures are useful.

21. In the next two sections, I have borrowed freely from parts of Mansfield (1973b).

22. Another well-known techniques based on subjective estimates is the "cross-impact method," also pioneered by Gordon and Helmer. According to this method, one estimates the effects of one event on the probability of occurrence of other events. Then one goes through a Monte Carlo process to estimate the unconditional probabilities of occurrence of various events (Gordon and Becker 1972).

23. Thus, Quinn (1967) is, of course, quite correct in stressing that technological forecasts need not be completely, or even very, accurate to be useful. The correct test is whether their value exceeds their cost.

9. TECHNOLOGICAL INNOVATION IN FIRMS AND INDUSTRIES: AN ASSESSMENT OF THE STATE OF THE ART
Richard S. Rosenbloom

This paper considers the base of knowledge that might illuminate the ways in which technological innovation is or could be influenced by public policies and the administrative practices of firms. It is directed toward readers interested in discovering which are the major forces stimulating and guiding the behavior of firms with respect to the creation and utilization of new technology.

A better understanding of the processes of innovation is necessary if society is to be able to influence them. The existing base of knowledge offers relatively little that is directly useful for policy makers, but it provides an extensive foundation for research that would do so. The "state of the art" invites a policy-oriented synthesis, one built on a conceptual framework that would interlink currently disparate traditions of inquiry.[1] We argue that the concept of corporate strategy provides a useful--and unexploited--framework for a policy-oriented research on technological innovation.

The forces influencing the development of technology have been analyzed at different levels of aggregation, ranging from studies of entire societies to examination of minute particulars of a single innovation. Table 1 suggest five levels of analysis, with examples of work at each level. At the upper end of this range, historians, economists, anthropologists, sociologists, and others have identified social and cultural forces interacting with science and technology and have examined the nature of some of those interactions. For example, the eminent sociologist Robert Merton tells us that "the substantial and persistent development of science occurs only in societies of a certain kind, which provide both cultural and material conditions for that development" (Merton 1973). Merton stresses that there are "*reciprocal* relations between science, as an ongoing intellectual activity, and the environing social and cultural structure," and that "the nature and extent of these interchanges differ in various societies, depending on the state of their science and of their institutional systems of economy, politics, religion, military, and so on" (p. 176; emphasis in original).

Nathan Rosenberg, writing from the perspective of an economic historian, explains the relationship of technology to American economic growth in strikingly similar terms (Rosenberg 1972). Technological change, in his framework, is both situational and environmentally (socially) influenced. That is, a given change is viewed "as a successful solution to a prticular problem thrown up in a particular resource context" (p. 21). Rosenberg's careful review of the American experience shows how an "economic matrix" of demand pulls and supply constraints (or their absence) shaped the main paths of historical change in technology.

An improved understanding of the paths of historical change at the level of entire societies is valuable in its own right and offers a helpful background for policy research and the deliberations of policymakers. But it can offer few affirmative prescriptions for policy and action. Policy-oriented research must be directed at smaller units of social organization, addressing such questions as the following. Do the various sectors of industrial societies differ systematically in their capacity to generate or accept technological innovation? What are the causal factors? In what terms can the apparent differences between local governments, farms, and manufacturing firms be

---

Dr. Rosenbloom is David Sarnoff Professor at the Harvard Graduate School of Business Administration. The author gratefully acknowledges the benefit of numerous discussions of this topic and of drafts of this paper with W. J. Abernathy, D. Bodde, M. Horwitch, H. Kolodny, I. Kusiatin, and C. K. Prahalad.

TABLE 1.--Alternative levels of aggregation in studies of technological development.

Level	Illustrations of Research at This Level	
	Authors	Typical Variables
1. Society as a whole	Merton (1973)	Values, norms
2. Sectors	Rosenberg (1972)	Factor costs
3. Industries	Schmookler (1966) Scherer (1970)	Demand Market structure
4. Firms	Mansfield (1968b) Woodward (1965)	Size of firm Organization structure
5. Innovations	Myers and Marquis (1969) Mansfield (1968b)	Sources of information Investment and profitability

usefully explained? What about differences within a sector, for example, between manufacturing industries such as furniture, petroleum, and electronics? Or, taking even narrower segments, within a given industry, how to explain the differing capacities for and receptivities to change among firms?

Important differences do exist. Scholars from several disciplines have sought explanation through the "lenses" of their respective fields. For example, Rosenberg calls attention to the variation in technological change across industries:

> Technological change is not something which has emerged in a random way from all sectors of the economy. It is, rather, the result of certain acquired problem-solving skills which, in our history, have been heavily concentrated in specific sectors of the economy. Throughout the early stages...these skills were heavily concentrated in metallurgy, machine tools, steam power, and engineering; later, in large part, as a result of advances in science, this focus shifted to chemistry-based, electrical- and electronic-based and, more recently, biology-based industries (1972, p. 53).

Rosenberg's analysis depicts the influence of economic factors, particualrly the role of market prices, relative factor costs, and limitations of supply. Another influential economic variable is growth in demand, which Schmookler has shown is directly related to the intensity of technological development (Schmookler 1966). But Rosenberg's observation, quoted above, also draws our attention to noneconomic factors. Such factors as the development of science, the education of engineers, and the availability of particular kinds of problem-solving skills are recognized as interacting with economic determinants of the pattern of change.

In the next two sections of this paper we discuss the traditions of research on innovation in firms and industries (levels 3-5 in Table 1). The scope of our inquiry is intended to be broad enough to include the full range of relevant factors but at the same time to maintain a focus on issues that are pertinent to policy makers and administrators. We shall begin at the most concrete level: studies of particular innovations.

INNOVATIONS

An examination of recent traditions of research at this level suggests the following generalizations. First, among these scholars, there is a useful and widely shared paradigm of the innovation process. Despite variations in terminology and the minor particulars of definitions, there is consensus on the distinctions to be made between stages in the process embracing the creation of ideas, the demonstration of their practicality, their development for and introduction to routine use, and their diffusion within a population of users. Second, there are useful taxonomies identifying and describing elements of this process, associated activities and outcomes, and the kinds of situations in which innovations arise. Third, empirical research using these conceptual tools has discovered certain characteristic patterns in the dynamics of innovation and has helped to delineate several of the embedded microprocesses, such as information flow.

A dominant theme in recent studies of technological innovation has been the idea that "innovation is not a single action but a total process of interrelated subprocesses":

> It is not just the conception of a new idea, nor the invention of a new device, nor the development of a new market. The process is all these things acting in an integrated fashion toward a common objective (Myers and Marquis 1969).

Systematic research on the innovative process is comparatively recent. A necessary step in any field of inquiry, of course, is the development of generally accepted categorizations of the phenomena of interest; a shared taxonomy is a requisite for the cumulation of research. There now is a widely shared taxonomy for the tasks that constitute the process of innovation. One statement of this scheme is offered by Utterback (1974), who groups the activities into three constituent phases:

> *Generation of an idea* involves synthesis of diverse (usually existing, as opposed to original) information, including information about a market or other need and possible technology to meet the need.
> *Problem solving* includes setting specific technical goals and designing alternative solutions to meet them.
> *Implementation* consists of the manufacturing engineering, tooling, and plant and market start-up required to bring an original solution or invention to its first use or market introduction. Diffusion takes place in the environment and begins after the innovation is first introduced (p. 612, emphasis added).

Given a useful classificatory scheme, the next logical step is to try to observe the characteristics of elements of that scheme as they are found in the real world. In this way one can build up a picture of the process in some detail. The problem, of course, is that innovation occurs in different ways under different circumstances; it is not a process that is capable of deterministic description. The response has been to seek descriptions identifying central tendencies, based on observation of the relative frequencies of occurrence of various characteristics associated with elements of the innovation process. For example, the taxonomy may be viewed as dividing innovation activities temporally, into sequential "stages" marked by critical events, such as the demonstration of feasibility or the first introduction to use on commercial scale. Within this framework, empirical research gives us estimates of the characteristic duration of each stage. Both Lynn, in a study of 20 innovations, and Enos, in a study of 35 outside the petroleum industry, found that an average of 14 years elapsed between invention and commercialization of innovations introduced in the years following World War II (Lynn 1966, p. 4; Enos 1962b, pp. 229-321). Other descriptive studies treat such characteristics as the sources of inventions (Jewkes et al., 1958) and innovations (Mueller 1962); adoption rates in the diffusion of innovations (Mansfield 1968b, pp. 114-119); and the linkage of scientific discovery to technological innovation (Battelle 1973). Studies of this sort provide the basis for a kind of "demography of innovation."

The base of knowledge useful for understanding the demography of innovation has been growing rapidly in recent years. Utterback's recent review article (Utterback 1974, Table 1) identifies 17 studies documenting the characteristics of more than 2000 innovations. Consistent patterns have emerged from these studies despite their differences in time period, scope, methodology, and definitions. The patterns disclose central tendencies (e.g., long time lags between invention and commercial use) but also large variations (e.g., a range of 3 to 79 years, with standard deviation of 16 years in the time lags in 35 innovations studied by Enos (1962b, pp. 305-309). One naturally asks: to what extent can the variations be given systematic explanation?

The simplest level of explanatory research observes an association between variation in the characteristic of interest (e.g., time lags) and variation in other observable characteristics. Lynn, for example, shows that the time lag is less for consumer products than for industrial goods (1966, p. 4), whereas Enos sorts the data a different way and concludes that "mechanical innovations appear to require the shortest time interval chemical and pharamaceutical innovations next. Electronic innovations took the most time" (p. 309). Although many findings are based on simple classifications of this sort some studies employ more complex and powerful multivariate statistical tools. Mansfield for example, uses a regression model to demonstrate the relationship between the rate of diffusion of an innovation and its profitability and required investment (Mansfield 1968a, Chap. 7).

Explanations of this sort treat the stages of the innovation as a sequence of "black boxes." The mechanisms that could account for observed relationships remain mysterious, or are inferred from general premises, such as the expectation that individuals and firms will seek to maximize profits. Another strategy of research is to delve into the black boxes, making explicit the microprocesses of innovation, in order to describe and perhaps explain innovative outcomes more adequately.

One of the most widely quoted descriptive studies of this sort is the study of 567 industrial innovations initiated by Sumner Myers and the National Planning Association, and subsequently reported by Myers and Marquis (1969). Myers collected data on innovations judged to be commercially successful and important to the innovating companies, but of a scale and complexity that classes them as "incremental" changes (as opposed to those new products and processes that may launch a new industry).

The analysis by Myers and Marquis centers around an information flow model of the innovative process. They identify the source of the technology, that is, whether or not it was original to the innovating firm; factors leading to the initiation of the innovation, including technological, market, production, and administrative forces; a number of characteristics of the acquisition and utilization of information used in innovation; the cost of development; and the innovation's impact on the production process. For the most part, the findings of Myers and Marquis are demographic rather than explanatory. Their results do suggest that the characteristics of information flow vary among kinds of innovations and industrial settings. But their work offers no higher-level explanation, or theory, of why these variations are observed.

One of the most ambitious efforts to explicate the most influential administrative factors within the innovation process is the study and failure in industiral innovation (SAPPHO) conducted at the University of Sussex (Achiadelis et al., 1971). In their introduction the SAPPHO investigators observe: "Typically, studies of innovation have highlighted 'single factors' in the process, in which the ultimate success is believed to have hinged. Accepting that innovation is a complex sequence of events, involving scientific research as well as technological development, management, production, and selling, it was felt that those single factor interpretations were less than satisfactory..." (p. 5).

Project SAPPHO identified and evaluated 29 pairs of innovations in two industries, chemicals and scientific instruments. Each pair was selected to provide one successful and one unsuccessful attempt to introduce comparable innovations. Interviewers documented 201 characteristics of each case and then compared the members of each pair to establish whether and in what way they differed on each characteristic.

The SAPPHO study offered five principal conclusions concerning factors contributing to successful innovation. Three pertain to the innovating organization's relations with its external environments. Successful innovators had significantly better understanding of user needs, paid more attention to marketing tasks, and made more effective use of external technical resources. Two other organizational characteristics seemed consistently and strongly associated with success: more efficient performance of development tasks and the assignment of responsibility to a more senior executive.

Despite its large database and imaginative use of multivariate statistical techniques, the SAPPHO study has little explanatory power. The findings identify instrumentalities associated with success, but like those of Myers and Marquis, they are not supported by adequate theory. Are characteristics such as "the use of external information sources" of inherent significance, or are they merely artifacts of more fundamental factors? One cannot tell. Nor can one safely say how far these findings can be generalized. Significant differences were found between the two industries studied but they cannot be explained within the SAPPHO framework. Finally, these findings cannot easily be operationalized, e.g., it helps little to advise managers to be "more efficient" in development tasks.

Although the descriptive paradigm of the innovative process is relatively well developed, even at the level of some constituent microprocesses such as information flow, no framework has yet emerged to tie together the factors which are offered in explanation of variations in innovative outcomes. As the Sussex group has argued, single factors are inadequate to provide explanations. The multiple factors recognized so far in explanatory research have often been defined eclectically and adventitiously, with interdependencies assumed away or simply left unexplained. But common-sense observation tells us that many of these factors almost certainly are linked systemically. It is unlikely, for example, that an organization's awareness of user needs and its concern for marketing tasks are independent factors.

Some of these shortcomings could, in principle, be alleviated by a larger scale of inquiry. The methodology of SAPPHO could be employed on a larger sample structured to yield data about the pertinent characteristics of innovations, firms, and industries. As a practical matter, if one were to account for differences between product and process innovations of several levels of complexity within enough industries to permit examination of more than a single characteristic of the industries, the task would become unmanageable large. More fundamentally, larger databases are a poor response to the inherent lack of explanatory theory.

A more attractive approach is to seek an integrative theory rather than to try to embrace, empirically, all the plausibly significant descriptive variations between cases. The principle of parsimony suggests that we should seek a framework that capture those systemic relationships in terms that would focus empirical examination on a few fundamental variables. A concern for utility of findings implies further that the framework should treat variables that can be influenced by decision makers in the real world; that is, characteristics that are "malleable."

In summary, research at this level of aggregation, in which the primary unit of analysis is a particular innovation, has contributed a useful body of descriptive findings, but only limited explanatory findings. Extensions of this tradition of research, whether by enlarged scope or more powerful technique, are unlikely to alter that conclusion. For more adequate explanation, we shall have to move the focus of inquiry to higher levels of aggregation, where we can build on available theoretical frameworks pertinent to the behavior of firms and industries.

FIRMS AND INDUSTRIES

As we shift the focus from the characteristics of particular innovations to the "innovativeness" of firms or industries, the metrics of research become more complex. There is no widely accepted measure of the innovativeness of an organization or industry. It is widely believed, however, that there are substantial differences among firms and industries in this respect. Strong support for this view can be found in the data on industrial R & D expenditures, a commonly used proxy for innovative effort.

In the United States, the performance of R & D is heavily concentrated in six industries, and within them is concentrated among the largest firms. Thus, according to a recent survey by the National Science Foundation, 20 companies spend more than half the funds expended for R & D within industry and 100 firms account for more than three-fourths of industrial R & D performed (NSF 1971).[2] Six industrial categories stand out as "R & D intensive." Table 2 lists these industries and gives 1971 data for expenditure for company funds on R & D by the largest firms (by sales dollars) in each industry. The 49 corporations representing the eight largest in each industry accounted for slightly more than half of all company funding for R & D reported in 1971.[3]

There are two well-developed streams of inquiry into the differing innovative behavior of firms and industries. One, by economists, examines the technological implications of firm size and industry market structure. The other, by behavior scientists, examines the technological implications of firm size and industry market structure. The other, by behavior scientists, examines the relationship between technology and the design of organizational systems. In some cases these are alternative explanations of the same phenomena. In the next section we shall suggest an effort to explore the linkages between these modes of explanation. But first, let us look at each separately.

The most pertinent lines of economic inquiry deal with the technological implications of differing firm sizes and industry market structures. There is disagreement

TABLE 2.--R & D funds and sales volume (millions of dollars) in key U.S. industries.

Industry	SIC	Sales of Top 8* as Percent of top 20 Firms	Company R & D Funds			
			First 8 Firms		9-20th Firms*	
			$	Percent Sales	$	Percent Sales
Chemicals and allied products	28	60.6	542	3.6	305	3.1
Industrial chemicals	281-282	67.5	598	4.0	169	2.4
Drugs and medicines	283	64.4	-	-	-	-
Other chemicals	284-289	68.0	-	-	-	-
Machinery	35	66.0	886	5.7	208	2.6
Electrical equipment and communications	36, 48	76.0	1496	4.4	205	1.9
Radio & TV receiving equipment	365	94.0	-	-	-	-
Communications equipment and electronic components	366-367, 48	79.0	984	4.1	179	2.9
Other electrical equipment	361-364, 369	84.5	-	-	-	-
Motor vehicles and other trasportation equipment	371, 373-375, 379	89.5	1382	2.9	34	0.6
Aircraft and missiles	327, 19	69.0	759	4.4	172	2.2
Professional and scientific instruments	38	75.0	386	5.1	71	2.8
Scientific and mechanical measuring instruments	381-382	65.0	28	1.5	22	2.2
Optical, surgical, photographic, and other instruments	383-387	81.0	394	5.4	60	3.4

*Ranked by sales volume.
Source: NSF 1971, Table B-44.

concerning both the magnitude and character of the effects of these variables.[4] Most of the empirical research centers on the quantitative aspects of the presumed relationship, that is, on rates of change, rather than on the character of the technology or on the mechanisms by which the rate or character of change might be influenced. Most of the quantitative research, moreover, necessarily measures R & D spending or productivity change as surrogates for innovation.

Up to a point, increasing firm size seems to be associated with greater R & D intensity. Scherer notes that companies with 5000 or more employees are more likely to engage in R & D and to do it more intensively than their smaller counterparts. He suggests that a firm size of roughly $75-200 million seems to be "good for invention and innovation" (p. 361). But the most pertinent comment on the quantitative studies remains Markham's apt conclusion: "Clearly, any answer to how inventive and innovative efforts are affected by firm size hangs on an extraordinarily slender reed that may alternatively bend upward or droop downward, depending on the species of statistical zephyrs blowing at the time." (Markham 1965).

The quantitative studies of the relationship of market concentration to inventive or innovative effort appear to exhibit greater statistical consistency--confirming the conjecture that R & D efforts, at least, are more intense in highly concentrated industries. Other interindustry differences confound the analysis, however. Thus, when an index of technological opportunity is taken into account, the correlation with concentration is greatly weakened (Scherer 1970, p. 374). Scherer suggests, nevertheless, that "a modest degree of oligopoly is beneficial in fields of limited technological opportunity" (1976,

p. 376). As Markham observes, however, the difficulty with these analyses is more conceptual than statistical (p. 331). No theory explains why some oligopolies are innovative in technology and others not; for example, the synthetic fiber and cigaret industries have similar high concentration ratios and quite different behavior. Correlation alone fails to "explain" the differing behavior of firms and industries.

The empirical analyses of the relationship of firm size and market structure to technological innovation have recently been reviewed once again by Markham. After a careful appraisal of a baker's dozen of such studies, he reaches conclusions similar to those of his 1965 appraisal, summing up as follows:

> If, as economists assume, business outlays are at least in some sense governed by their profitability, data are beginning to suggest that large vertically integrated companies with large relative market shares have a greater incentive to invest in R & D than other firms. However, the really important questions of *why* R & D spending appears to be more profitable for such firms is as yet unanswered, except for perhaps suggesting the hypothesis that firms having those characteristics have a strategic advantage when it comes to exploiting the results of industrial R & D.[5]

Technical and economic forces, though doubtless influential, are by themselves unlikely to provide an adequate explanation of the differences between firms and industries. Administrative factors also can impact on innovative effort and effectiveness. A progression of studies in the field of organizational behavior indicates that technology is closely interdependent with the administrative characteristics of organizations.

The central ideas of the organizational studies are these. First, that there is no one "best way" to structure an organization; rather, what is best in a particular case is contingent upon the task of the particular organization. This contingency theory obtained its fullest statement in the work of Lawrence and Lorsch,[6] but it was foreshadowed by others, including Woodward and Burns, to be discussed below. Second, the differing characteristics of technologies employed in the operations of firms may condition the effectiveness of alternative organizational forms. A third idea, implicit in the first two, is that firms with certain organization structures may have more difficulty than others with the adoption of technological change.

A seminal work in this line of inquiry was Joan Woodward's study of 110 British manufacturing firms (1965). Woodward found that a series of measures of organizational structure bore no general relationship to a firm's size or its success. Patterns emerged only when firms were classified by types of production (e.g., unit and small batch; mass production), revealing a direct relationship between these measures of technology and the organizational variables. Within each technological category the more successful firms exhibited the same organizational characteristics, leading Woodward to conclude that "one particular form of organization was most appropriate to each system of production." As this conclusion implies, technical change that altered the basic method of production led to fundamental organizational change, although such change, in Woodward's experience, was generally responsive to emergent problems rather than a planned concomitant of the technical change.

Woodward's findings have been tested by a series of other studies in the United States and the United Kingdom, with mixed results.[7] There is no consensus on an appropriate metric for "technology," nor on which characteristics of "structure" are likely to be related to it. Size of firm is a confounding factor in these studies and the lack of an accepted larger theoretical framework has added to the difficulty of interpreting empirical results.

Another influential early study, by Burns and Stalker, investigated the interaction of technology and organization in variety of manufacturing firms in the United Kingdom (1961). Burns argues that choice of the appropriate organizational system depends on characteristics of the task (especially rates of change in technology and markets), the informal system (particularly the individual commitments to personal, political, or status goals); and the quality of leadership. Idealized systems of management, dubbed "mechanistic" and "organic," are defined as polar types. In stable task environments, firms employ mechanistic systems, in which there is a steep hierarchy, formal rules and procedures, jobs are defined from above, and spans of authority are narrow and sharply defined. Organic systems, with contrasting characteristics, prevail in environments of rapid change. When firms with mechanistic organizations tried to exploit more rapidly changing technologies and entered new, more dynamic markets, they were generally unable to make the necessary adjustments in organization.

Lawrence and Lorsch developed a more comprehesive framework for analysis fo the relationship of organization to environment.[8] At the core of their "contingency theory" of organization design is the notion that a successful organization achieves a "good fit" within each subunit (e.g., sales department, factory, etc.) between (a) characteristics of the tasks implicit in the segments of the external environment with which that unit interacts; (b) characteristics of the formal structure of the units; and (c) characteristics of the personal orientation of the managers in the subunit. In empirical studies of firms in the plastics, food, and container industries, they tested this idea using measures of environmental diversity and uncertainty, degree of formality of structure, and three measures of managerial orientations,[9] with results that tend to confirm the basic contingency hypothesis.

High-performing organizations were found by Lawrence and Lorsch to have a higher degree of differentiation between units than others in the same environment and also to achieve a higher degree of integration across units. By integration they mean "the quality of the state of collaboration...required to achieve unity of effort"--a result that obviously grows more difficult to achieve with greater differentiation. Although high integration seems required for success whatever the characteristics of the environment, the particular patterns of collaboration required and the mechanisms (through formal structure or modes of conflict resolution) that will achieve them appear to vary with environment.

Emerging from these studies is the notion that the individual firm, defined by its purposes, tasks, and administrative design, can most usefully be viewed as an open system in continuing interaction with its environment, defined in terms of economics and technology. Richard Normann focuses this idea specifically on technological innovation in products (Normann 1971). He suggests that a product defines the type of relationship that exists between a firm and a segment of its environment. He argues that a product ties the firm to a specific portion of the environment, namely "the domain, which is that part of the environment...with which the organization is in more or less constant interaction." The characteristics of successful products must be consonant with the needs and values of the demain and, according to Normann, can also be considered "a mapping of the value and power structure of the organization."[10] Product variations may not alter this relationship, but "reorientations" (i.e., the addition of new dimensions, or radical changes among them) or the introduction of new kinds of products often lead the firm into new domains, with generally profound consequences for the value and power structures, as well as the formal design, of the organization.

THE NEED FOR SYNTHESIS

We have mentioned only a small part of the research in microeconomics and in organization theory that bears on the interaction of social, economic, and behavioral characteristics of particular organizations with their creation or use of technology. Two central tendencies can be discerned in recent research of this kind. In one mode, the tools and outlook of economics prevail; the problems addressed are predominantly those of the public policy maker; and the substantive focus is on the external environment of the organization, with simplifying assumptions about the behavioral processes inside the "black box" of the firm. In the alternative mode, the tools and outlook are those of organizational behavior (sociology, psychology, cultural anthropology, social psychology); the problems are those of the managers of organizations; and the substantive focus is on the "inner environment" of life in organizations, with simplifying assumptions about the processes generating events in the external environment.

The two modes of inquiry have been linked by concern for interrelated problems, but by little else. For most of these scholars, the problematics of technological innovation have been pursued as a secondary concern in work aimed at contributing to the empirical or theoretical basis of the economics of firms and industries or the design of organizations. Research in which understanding the problem of innovation has been central to the investigator's purposes has tended to draw unevenly from these discipline-oriented modes. Even when there is comprehensive attention to findings of both sorts--as in the report of the SAPPHO study--intellectual synthesis has been frustrated by the lack of an adequate integrative theoretical structure.

An intellectual synthesis is needed if policy research on technological innovation is to develop findings whose normative implications could confidently be adopted by policy-makers and administrators. That is not to say that it is unhelpful to have research conducted in the modes identified above; on the contrary, with such work there would be little foundation for the needed synthesis. Nor does it mean that economists should try to become organizational theorists, or vice versa. What we do suggest is that policy oriented research must find a framework that is capable of embracing several kinds of variables: economic and social factors both within the firm and in its external environment.

Consider the implications of the view of the firm as an open system in continuing interaction with its environment. Clearly, this view suggests that external forces, e.g., the behavior of a competitor or the needs of customers, might explain much of the observable behavior of firms in the tasks of innovation. Public policies, of course, are an important influence shaping the environments for most firms. But internal characteristics are also important; for example, organization structure or the kind of technology employed in existing operations.

The top managers of a firm can influence both its internal characteristics and the environments affecting it. For example, managers, by shaping strategy to lead the firm into new environments, alter the firm's domains (in Normann's sense).[11] Product policy is not the only sphere in which this power resides, although it is probably the most potent. Financial policy can expose the firm to the fluctuations of uncertain short-term sources or insulate it by reliance on retained earnings. Manufacturing policy can be used to integrate backward to stablize access to materials or technology. Procurement can choose between spot (uncertain) or contractual (more stable) sources. Changes which lead the firm into an environment with new (to the firm) characteristics evoke pressure for changes *within* the firm.

To a limited but still significant degree, managers also can alter the characteristics of a given environmental domain. For example, as an alternative to forward integration into final markets, materials producers can educate and otherwise influence the behavior of fabricators (their customers) (Corey 1956). Automotive manufacturers stabilize the environment of their assembly plants by the franchise system for dealers, which permits them to "push" cars onto dealers' floors, thus stimulating sales effort and smoothing short-term fluctuations in consumer demand (White 1971, Chap. 9). Contractual agreements for procurement, and leasing policies for certain producers' goods, may have analogous effects.

The "open-system" view of the firm consequently implies that explanations of innovative behavior may be seriously misleading if they take into account only characteristics of its internal organization or only aspects of its external environment. Both must be considered. For example, the research discussed in the preceding section shows that the structural characteristics within organizations in the real world will be correlated with the characteristics of their particular environments. Hence an observable relationship between an environmental characteristic and innovative behavior may be heavily conditioned by the organizational characteristics associated with that environment. The converse is also true. To develop a sound understanding we must treat both organization and environment explicitly and carefully.

How are we to bring both organizational and environmental contexts simultaneously into the perspective of applied research on the innovation process? I do not see great promise in the kind of eclectic empiricism by which long lists of economic, social, behavioral, and other "factors" in the environment of innovations are identified and then sorted out by multivariate statistics. The needed synthesis should be sought at a higher level of abstraction. In my opinion, the concept of corporate strategy framework might make it possible to reduce the multitude of apparently relevant "factors" to a parsimonious few.

STRATEGY

"Corporate strategy is the pattern of major objectives, purposes, or goals, and essential policies and plans for achieving those goals, stated in such a way as to define what business the company is in or is to be in and the kind of company it is or is to be." In these words Kenneth Andrews (1971, p. 28) summarizes the main points of the concept of strategy that has been built up over more than a decade by the contributions of a group of scholars located primarily at the Harvard Business School.

The framework for study of corporate strategy developed by this group embraces both the formulation and implementation of strategy, which are treated as interacting administrative processes. Emphasis on this interaction has been a distinctive characteristic of the Harvard group, whose research and writings define and explore the implications of the subtle interplay between the processes of formulation and implementation.

Strategy formulation results in the top management's commitment to a set of goals, policies, and programs that match the organization's distinctive resources to perceived opportunities in a changing environment. Profit objectives are made explicit and the means of financing matched to the implied balance of risks and returns. Commercial goals are set, defining market segments and channels, and the products or services (functionally defined) to be offered. The core idea in the establishment of an economic strategy for the firm is the careful assessment and subsequent matching of (environmental) opportunity and (organizational) capability.

The process of implementation creates the organizational conditions necessary for realization of a chosen strategy. If defines the structure of the formal organization, its information and control systems, and the "style" of management direction. Implementation changes the capabilities of the organization; environmental factors change autonomously and in response to corporate initiatives; both kinds of change create a need for continuing reassessment and reformulation of strategy.

An example cited by Bower illustrates the concept of strategy at a concrete level:

> In 1964 Crown Cork and Seal was a major producer of metal cans and crowns. Broadly speaking, its strategy was "to be a world-wide manufacturer of cans sold as packages (rather than metal containers) to meet the marketing needs of the container customer using strength in specialty markets of the U.S. to achieve a stable domestic base for international growth." Given resources far more limited than those of its competition, these objectives were to be achieved by a lower level strategy: "To be a manufacturer of a narrow line of containers for technologically 'hard to hold' products achieving growth by bringing existing products to new markets and by rapid adaptation, development, and marketing of competitors' inventions. In turn, these objectives were to be achieved by a series of subsidiary functionally oriented policies which capitalized on the agility small size gave Crown, the near homogeneous quality of *most* container products, the high cost of transporting empty cans, and the high cost of changing over fabricating machines. Crown's policies were 1) to modernize plants, 2) installing multiple lines of fabricating machinery requiring infrequent changes in set-up, 3) which permitted rapid customer service, which 4) was reinforced through geographic dispersion of plants to sites near customers, which 5) together with plant flexibility provided a basic selling point of service and low transport costs for exploitation by 6) a customer-oriented sales force responsible for accounts rather than products. These in turn were supported by 7) a limited number of industry specialists and 8) applications-oriented R & D group which allocated 60% of its funds to test packaging for new products (Bower 1967, pp. 17-18).

One should expect the interaction of Crown Cork's strategy and its innovative behavior to be substantial. Since the generation of ideas is influenced by contact with areas of need, it is significant that the character and intensity of relevant contacts are determined largely by the commitment to decentralized plants and a customer-oriented sales force. The allocation of resources to development was shaped directly by corporate goals and the decision to emphasize new uses for existing technology and rapid adoption of competitors' inventions. An aspect of implementation, the structuring of R & D efforts in relation to other organizational units such as plants and sales offices, creates linkages and barriers that are significant in the transfer of technology and the introduction of innovation.

This brief illustration is suggestive of the ways in which the concrete implications of a particular strategy can influence the allocation of resources to and the effectiveness of mechanisms for various kinds of technological development within the firm. The other cases, the logic of a strategy, by itself, leads a firm to engage in, or disengage from a business field in a way that has important consequences for the development of technology in the society at large.[12]

The advent of the diesel railroad locomotive is a case in point. It is a classic example of the overthrow of a tight oligopoly by an outsider backing a superior technology. The three American manufacturers of locomotives followed identical strategies which embraced a single technology--steam--and emphasized the marketing of individualized designs for each customer and thus custom manufacture of the product. All three producers remained wedded to that strategy to its conclusion, which saw them withdrawing from locomotive production.

The outsider, as is well known, was General Motors. GM's entry into this field arose from the fortuitous (for GM shareholders and for the U.S. economy) coincidence of several factors. One was the interest in diesel technology of GM's technical genius, Charles Kettering. The GM technical organization experimented with diesels throughout the 1920s, but Kettering's intense interest seems to have stemmed from his purchase of a diesel-powered yacht in 1928. Kettering developed no new principles, but he perceived that the two-cycle diesel could be developed into a practical large engine meeting the power-weight requirements necessary for railroad operation. A second factor was the willingness of Kettering's backers in GM to stick with the project until the practicality of the concept could be demonstrated (Sloan puts this in 1930) and solutions found for the many detailed engineering problems encountered prior to production of a working full-scale engine (produced for the Chicago Fair in 1933). Third was the ability of GM to acquire, in 1930, the Winton Engine Company, a leading manufacturer of large diesel engines, and the Electro-Motive Engineering Company, an engineering, design, and sales organization that had a close business relationship with Winton. These organizations, acquired at a time when Kettering had demonstrated the practicality of his engine, but had built no commercial-scale engine of the new design, provided the muscle and bone of GM's future Electro-Motive Division.

But more than good fortune was involved, for chance favors the prepared organization, as it does the prepared mind. The strategy articulated for GM by Sloan and his associates in the 1920s, and the new structure they created to implement it, prepared GM for the diesel opportunity and thus had a powerful influence on the course of this important technological innovation. First, the new structure of decentralized, largely autonomous operating divisions meant that the GM central-office staff had the freedom from operating pressures to evaluate ventures that were as different from automobiles as refrigeration, aviation, and, of course, diesel locomotion. The same structure meant also that a segment of GM's technical staff, the Research Laboratory under Kettering, had a similar freedom. It is pure speculation, but one wonders whether the idea would have arisen at all in GM if the corporation had retained the then more conventional functional organization form.[13]

GM's structure may have created the opportunity for a nonautomotive venture to gain serious attention, but GM's strategy certainly shaped the new venture in two significant ways. First, it sanctioned the patient support of Kettering's development work and the investments in Winton and Electro-Motive. GM, over the years, has made many different products, but its sales have always been overwhelmingly from automotive goods. When it has diversified, the scope has been limited; as Sloan puts it: "We have never made anything except 'durable products,' and they have always, with minor exceptions, been connected with motors."[14] Even within these narrow boundaries, the diesel venture, in 1930, appeared as opportunity to be pursued. As Sloan recalls, "We were not at this point certain about the future of the U.S. automobile market....We had a natural interest in any enterprise *within our scope* that offered us a reasonable opportunity to diversify" (p. 346, emphasis added).

GM applied to this business the design, manufacturing, and service policies of its automotive business, rather than following the then established practices of the trade.

This was the second way in which GM's strategy influenced the venture. GM built standard locomotives, priced them lower than the traditional custom designs, produced them in volume to realize economies that would make the lower price profitable, and established a service organization and stocks of standardized repair parts to back up its sales. As a consequence of this strategy, GM had both the motive and the staff to conduct R & D aimed at continuing improvement, unlike the steam manufacturers, which were organized to build to customer specifications.

After some experimentation with passenger locomotives, the strategy was put to the test in 1935, when GM offered a standard diesel switching engine, priced 10% below its steam counterpart. GM began production on 50 units ($3.6 million worth) before receiving a single order. The engine sold easily. Deliveries began in May 1936; within seven years GM had delivered 768 switchers and had cut the price a further 20% to pass along operating economies to its customers. Soon after World War II the steam locomotive, and more important, the companies making steam locomotives, had passed from the industrial scene.

CONCLUSION

This paper began with the assertion that the existing base of knowledge concerning technological innovation "invites a policy-oriented synthesis." We are prepared now to identify a direction for inquiry that seems promising as a way of producing such a synthesis.

Implicit in our discussion has been a view of technological change that should be made explicit. Technological change is defined here as the introduction to practice of new and more useful ways of serving human purposes. Technological change is both specific and diffuse. Specific, identifiable changes in goods or techniques, such as the diesel locomotive, the tear-top beverage can, or magnetic encoding of checks, symbolize the power and pervasiveness of innovation. Also important are the myriad of minor changes, so diffuse as to be largely unidentifiable, that in the aggregate account for an appreciable fraction of measured productivity improvement.[15]

In principle, we can speak of "an innovation" as the basic unit of technological change (Marquis 1969). This basic unit of analysis, however, embraces a wide range of magnitudes of scale and complexity. The use of the Concorde in place of the Boeing 707 and the use of the aluminum bolt in place of a steel bolt are both "innovation." Every innovation implies a prior sequence of tasks concerned with idea generation, problem solving (development), and implementation. This conception of the process of innovation, too, embraces situations varying greatly in scale and complexity. One innovation may be conceived, developed, and put into practice by an individual in one place within a single day; another, like the Supersonic transport, may be the result of activities spanning decades in time, continents in space, and involving tens of thousands of individuals. Yet we may say that the characteristics of all innovations are influenced not only by factors inherent in the innovation process but also by what Merton has called "the environing social and cultural structure."

Policy makers affect the outcomes of the innovation process largely by influencing its environing social and cultural structure. Research that identifies the influence on innovation of malleable elements of that structure is, therefore, most relevant for policy purposes.

The collection of forces outside the process of innovation that influence its behavior or outcomes we shall call its "context." The *structural context* is the enduring pattern of social, economic, and political forces that impinge on the process in every situation. In the case of the GM diesel development, the structural context included such factors as GM's decentralized organization, the existence of a central research laboratory, and the oligopolistic structure of the locomotive industry.

Structural context can be defined at several levels of aggregation; the immediate work group, a larger organizational unit or department, the corporation, and industry, or a whole society. A division into two categories--organizational and environmental-- will be sufficient for our purposes. Since the activities in the process of innovation commonly take place within multimember organizations,[16] the characteristics of the organizational milieu, its "internal environment," constitute a coherent set of forces. We shall refer to them, collectively, as the *organizational context*. They include, among other factors, the organization's goals, formal structure, leadership and resources. The formal boundary of the organization defines the distinction between organizational context and the remaining contextual forces, including external technical and economic factors, as well as social, political, and other structural forces.[17] Collectively, these forces constitute the *environmental context*. They include, among other factors, technological trends, government policies, competitive behavior, and demand trends.

The core of the idea just advance arose in Joseph Bower's study of planning and investment in a large multiproduct manufacturing corporation. His observation of the persistent influence of organizational or structural forces on the technical, economic, and political acts of managers led Bower to the hypothesis that general managers could manipulate those forces to evoke desired behavior (Bower 1970, p. 318). There is a great deal to support the hypothesis, both with respect to resource allocation within firms and more generally, with respect to innovative behavior. Bower's analysis required attention only to what we have called the organizational context, but students of technological innovation will need to enlarge the notion, recognizing the impact on innovation of contextual forces outside the firm, as well as those within it.

The viewpoint toward the innovation process stated above is illustrated schematically in Fig. 1. Technological change, the "dependent variable" of primary interest, is viewed as the result of a social process, which is customarily described as comprising three phases of tasks.[18] The tasks in the process of innovation in turn interact with their context, which may be roughly differentiated into forces within the organizational system performing the tasks of innovation and those external to that system.

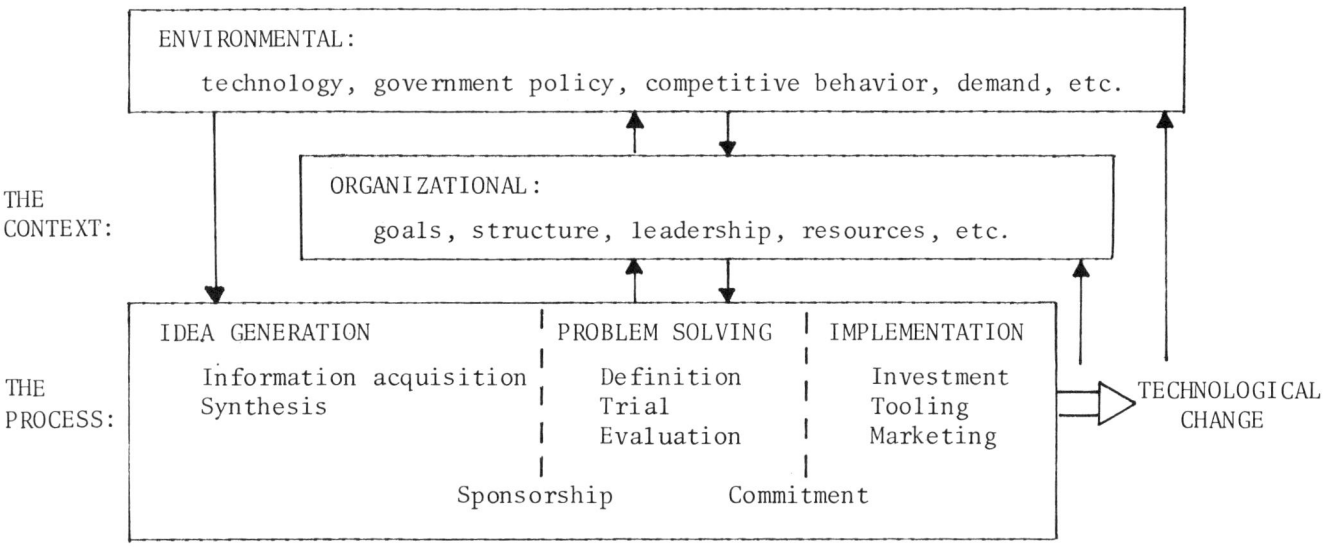

FIG. 1.--Conceptual scheme.

How are the rate and character of technological change influenced by factors in the organizational and environmental contexts of business? That is a more precise restatement of the question posed in the first paragraph of the paper and directs our attention to the sets of factors shown on the top half of Fig. 1. They can meet the test of malleability, since the elements of the organizational context are within the control of corporate management, whereas external factors are influenced both by corporate behavior and public policy. To clarify the relationship among these factors, however, and to meet the test of parsimony, we need a higher-level integrative framework. We suggest that that purpose can be served by the concept of corporate strategy, as defined in the previous section.

The strategy framework is particularly appealing because it integrates in two relevant dimensions. First, the concept of strategy formulation calls for a perspective that cuts across the boundary of the organization, matching capability (an aspect of the organizational context) with opportunity (an aspect of the environmental context). Unlike the modes of inquiry discussed in our section on firms and industries, the strategy framework demands explicit attention to technical, economic, social, political, and behavioral considerations simultaneously, as it embraces factors within the firm and external to it. In the second dimension, i.e., *within* the organizational context, the concept of strategy implementation requires the translation of higher-level strategic abstractions into more concrete and implementable terms. It is this aspect of strategy that gives shape to the organizational context. The twin concepts of formulation and implementation might be the basis for development of a theory of the linkage between organizational context and environmental context.

Research within the broad framework of corporate strategy could perhaps begin to clarify the apparent interactions among organizational goals, structure, and leadership as influences on technological innovation. All three factors have appeared in past research as influential variables. Any combination of the three might explain, for example, why a large firm might counter a competitor's large R & D investment by increased advertising, acquisition of a new product line, or other methods not imitative of the competitor.[19]

Commitment to different goals clearly is one factor explaining different technological strategies by firms in similar environments. From Bower's description of Crown Cork's strategy one can readily see why this firm did not imitate American Can's initiative in establishing a central corporate R & D laboratory. Leadership also is a significant variable affecting innovative performance. In defining the determinants of organizational design (and, by implication, of effectiveness in adaptation to innovation), Burns and Stalker list "relative leadership capacity of top people" as one of three main factors (1961, Chap. 6). Argyris (1965b) argues that "openness" or "authenticity" of behavior facilitates risk-taking and problem-solving, characteristics that are essential for innovation. More generally, the behavior of key individuals appears significant in most histories of innovation. But individual behavior is greatly conditioned by other elements of organizational context, including structure.

The implication of the preceding is that one should attempt to keep purpose, structure, and leadership in view simultaneously when analyzing the organizational context of innovation. As already noted, much of the research on innovation acknowledges the significance of these elements of strategy, but none makes explicit use of a framework incorporating several of these elements in concert. Systematic work on corporate stategy, on the other side, has paid uneven (and generally little) attention to technology and virtually none to the concepts or findings of the several traditions of research on technological innovation. Ansoff's discussion of corporate strategy does not integrate technology as an element, although he subsequently devotes many pages to planning for R & D.[20] An article by Cohen and Cyert (1973) setting forth a conceptual framework for strategy formulation makes no mention whatever of technology. Andrews, summarizing the framework developed by the Harvard group, discusses technology, but in far too limited a scope. He recognizes technological change as an important determinant of opportunity: "From the point of view of the Corporate Strategist, technological developments are not only the fastest unfolding but the most far reaching in extending or contracting opportunity for an establish company" (Andrews 1971, p. 60). But technology is treated as something outside the firm, apparently autonomous in its unfolding.

Technology, however, is neither autonomous nor wholly "outside" the control of managers. It is more useful to view it as interacting with the firm, part of its capability, shaped by as well as a shaper of corporate action. At times, to be sure, its principal significance is as part of the environment, when it emerges as a threat or an opportunity to which the firm must respond strategically. In other cases as, for example, in GM's development of the diesel locomotive, the technological capability of the firm conditions its strategic commitments. In still other cases, as in the Crown Cork example, a certain posture toward technology is one of the strategic tasks implied by the implementation of a defined strategy.

One is tempted to conceptualize strategy as a unitary variable to be introduced into empirical studies. James Utterback, in an unpublished paper, has suggested precisely that, in a proposal that a firm's "competitive strategy" (e.g., cost minimizing or sales maximizing) to be treated as an explicit variable.[21] Bruce Scott and his associates have developed an appealing scheme to classify firms according to the character and extent of their integration and diversification ("strategy") and their formal structure (Scott 1973). Although concerned with issues other than technological innovation, Scott does offer the conjecture that firms in one category--diversified strategy with divisional structure--are better equipped than others to exploit the results of R & D. Andrews's definition of corporate strategy, cited above, is admittedly diffuse in comparison with the crisp classificatory schemes suggesting by Utterback or Scott. Strategy, however, is a multidimensional concept that still requires the careful, clinical appraisal of specific situations (as in the Crown Cork example); much rich detail can be lost in premature efforts to construct a single simple scale. It is appealing to think that the introduction of a powerful and overlooked variable will shed important new light on our subject, but I doubt that any single variable would have such an impact.

The argument of this paper is that the concept of strategy offers a *framework* which can help us to design empirical studies which might generate "explanations" of the process of technological change in firms and industries using only a few highly salient variables. Whether this argument is correct or not is an empirical question that deserves to be answered.

NOTES AND REFERENCES

1. Although various segments of the literature on innovation are discussed below, no effort is made to offer a complete survey of the foundations of the field. Utterback (1974) reviews the results of numerous studies of innovations. Zaltman et al. (1973) is a recent and useful summary of social and behavioral aspects of innovation within formal organizations. Mansfield (1969) is a standard work on its subject. Rogers with Shoemaker (1971) surveys and synthesizes a wide range of research emphasizing socio-cultural aspects of the adoption and diffusion of innovations.

2. The 100 companies with the largest R & D programs accounted for 39% of net sales and 38% of total employment of all R & D performing companies. Expenditures for industrial R & D totalled $18.3 billion in 1971, of which $10.6 billion (58%) was company funds.

3. If one included the 20 largest firms in each industry, the total R & D spending would increase by $1 billion. R & D expenditure for the 9th-20th companies represent a smaller percentage of sales than for the largest firms in each industry, but still more than for firms in most other industries. In only two other 2-digit industries do the largest firms report company R & D expenditures of close to 2% of sales (Rubber Products, SIC 30; and Stone Clay and Glass, SIC 32). The aggregate dollar expenditures by the top 8 firms in those industries are much smaller ($171 million and $109 million) than the amounts spent in the industries in Table 1.

A large aggregate expenditure ($344 million) is reported for the eight largest petroleum firms, but this sum accounts for less than 1% of their sales.

4. A useful and comprehensive summary is given by Scherer (1970, Chap. 15).

5. Jesse W. Markham, "Market concnetration and innovation," unpublished draft, March 1973.

6. Paul R. Lawrence and Jay W. Lorsch, *Organization and Environment: Managing Differentiation and Integration*, Boston, Harvard Business School, 1967.

7. An excellent review of this literature is given by Khandwalla (1973, pp. 1-8).

8. Note 6.

9. They measured the "time orientation" of managers in terms of the degree of immediacy of feedback expected as a consequence of action and further classified managers as "task oriented" (primarily concerned with technical content of job) or "socially oriented" (primarily interested in relations with other people). A third measure, which they termed "goal orientation," was not found to be related to environmental uncertainty.

10. Pp. 205-206. For example, the power and size characteristics of automobiles made in Detroit can be viewed either as consequences of a certain market segment's preferences or of the ideology of managers of the Big Three.

11. This point is made most clearly by Charles Christenson, "The contingency theory of organization: A methodological analysis," Harvard Business School, Working Paper 73-76. Using rigorous analysis of the "open system" premises of the Lawrence and Lorsch work, he shows that the apparent normative implications of the idea of "good fit" may be misleading.

12. Bower calls attention to the more general implications of a global strategic choices and uses the withdrawal of General Motors Co. from the airframe industry as an example (p. 17).

13. Chandler (1962, Chap. 3), describes Sloan's reorganization of GM in 1925-27.

14. Alfred Sloan, *My Years With General Motors*, p. 341. This description of the origins of GM's diesel locomotive business is taken from Sloan's account.

15. See Hollander (1965) and Enos (1962a). Some diffuse changes benefit society but are overlooked in the conventional measures of productivity, since they occur in government organizations or within households.

16. Individuals may continue to account for inventions of significance and many adopters of innovation are individuals or small proprietorships (farms, retail stores, etc.), but the diverse skills and greater financial strength of larger organizations are needed, almost without exception, for development and commercialization of technology of any great significance.

17. This is not an unambigous boundary (e.g., as when several organizations collaborate on an innovation) but it is of heuristic value, at least, to distinguish characteristics intrinsic to organizations from those in their general environment.

18. Progress in the problem-solving and implementation tasks of innovation requires the establishment and maintenance of organizational commitment; these behavioral factors are listed along with the strictly technical-economic tasks in Fig. 1.

19. Markham (1965, p. 326) raises this possibility as an illustration of the limitations of using market structure as a primary explanation of innovative behavior.

20. H. Igor Ansoff (1965). It is interesting that this book has no index entry for "technology."

21. "The process of innovation in firms with differing competitive strategies," September 1973.

10. REVIEW AND ASSESSMENT OF THE METHODOLOGY USED TO STUDY THE BEHAVIORAL ASPECTS OF THE INNOVATION PROCESS

Charles F. Douds and Albert H. Rubenstein

The purpose of this paper is to assist in the overall task of improving the Research and Development/Innovation (RD/I) process, with which the Office of National R & D Assessment and many other parts of the National Science Foundation and the whole U.S. government are concerned.

To help improve the knowledge base for such improvements in the RD/I process, we must also improve the research process that generates knowledge about the factors influencing its outcomes, and the consequences of its outcomes. If we are increasingly to substitute credible research findings for ad hoc opinions as the basis for this improvement (through policy making, organizational design, or management procedures), we must improve the ways in which innovation research is structured, data are collected, and inferences are made. It is not obvious that descriptive studies alone, no matter how rigorous and credible, can provide a sufficient basis for improvement of innovation. We certainly need careful analytical and normative studies that examine the nature of the process and its connections with the rest of society. However, our best chances for improvement of the process lie in a healthy interaction between such normative/analytical studies and empirical studies of how the process actually works in the "real world."

Given this premise--the need for both kinds of studies in effective interaction--our task in this paper to to examine the methods used in conducting the empirical-research component of the total research effort on the innovation process.

If we are to substitute results from the research process for rules of thumb and private experience as a basis for policy formation and innovation process design, we want assurance that these results are produced in a manner that provides high credibility, replicability, and communicability--important characteristics of "good" scientific results.

Unfortunately, we are far from this situation at present. The field of innovation "research" is currently diverse, loose, and full of gaps and contradictions. It is not cause for discouragement, however, for many science-based fields such as engineering and medicine have been or still are in this condition. Much of the disorder in "our" field --studies of the innovative process--is due to lack of standard or even fully disclosed methodology for conducting studies, collecting data, and making inferences from the findings (despite the complex trappings of statistical tests and analytical methods).

In addition, there is the persistent problem of "amateurism" or lack of adequate training among many of the people who conduct or attempt to conduct innovation studies. This is a situation which is not tolerated in other fields, where adequate training is a *sine qua non* before an experimental or field investigation is undertaken.

The deceptive thing about our kind of research is that, for some people, all that appears to be involved in data gathering is "writing down a list of interesting questions for a questionnaire (or interview) and mailing it out (or interviewing some people)." A lot of bad and useless data are collected in this way, and a lot of time and patience are expended by respondents with little promise of a useful payoff to either themselves or society.

In this paper we describe the range of various methodologies used in the study of the innovation process: in particular, the behavioral aspects of that process--the decisions

---

Dr. Rubenstein is Professor and Director, Program of Research on the Management of Research and Development, Department of Industrial Engineering and Management Science of the Technological Institute of Northwestern University in Evanston, Ill. Dr. Douds is Associate Professor at DePaul University in Chicago.

made, the information exchanged, the attitudes expressed, and the nontechnical as well as technical results achieved. For example, a major consequence of a particular innovation project started in the R & D laboratory of a company may be the realization that the company's product lines or production processes are technically or economically obsolete and that an intensive program of improvements is needed. The consequences of an innovation project either initiated by or involving the marketing department may be indications of the way in which selling or servicing of the company's products needs to be improved. Either or both of these consequences may lead to changes in personnel, skills, decision processes, attitudes, power relations, etc.

Our procedure in this paper is to present a set of factors for classifying existing and future studies of the innovation process and then to present a set of criteria for evaluating a sample of extant studies in the context of "scientific studies of the innovation process."

Our purpose in this evaluation is less to evaluate the specific studies per se, in terms of their intrinsic scientific merit, than to examine them as candidates for contributing to the body of knowledge needed for improvement of the innovation process. That is, the emphasis is on the factors involved in assessing their credibility and usefulness as a basis for policy making, organizational design, or management in the innovation process. This paper is not, and cannot be, a text on field research methods, but we hope it will contribute to improving the research situation in the field of innovation studies.

THE SITUATION AND THE NEED

The innovation process is central to the advancement of an economy. Through innovation, change is brought about. Discoveries in science and technology through a variety of processes are singled out, combined, revised, transformed, commercialized, etc., with a variety of impacts on the society. For a variety of motives, individuals, organizations, and governments desire to exert some degree of influence over the process of innovation in its many forms. One major focus is on *technological* innovation--devices, machines, chemicals, drugs, etc.--that have demonstrable economic and/or social consequences. But on what do decision makers base their actions when they seek to affect the process?

This question, easily asked, is not so easily answered. Philosophers and historians have made observations on various aspects of the process for many years. Economists have sought to examine the effects of technological innovation on primarily the macro aspects of economies. "Science policy" has become a specific focus of concern in government circles and among associated specialists. This upsurge of interest and concern has focused increased attention on our understanding--or lack thereof--of the innovation process.

Three distinct but overlapping areas of innovation-related activity by students of the innovation process can be distinguished (Rubenstein 1973):

1. *Research*, of a descriptive, theory-building nature. Its main task is to explain and provide a basis for predicting the behavior of individuals, organizations, industries, etc., involved in innovation, and the outcomes of the process.

2. *Policy formulation*, of a prescriptive nature. Its main task is to set standards and criteria; to indicate what courses of action are desirable and "correct"; and to provide guidelines for legislation and other kinds of social decision making.

3. *Design*, of actual, ongoing organization strategies, structures, and processes. It involves the implementation of implicit or explicit policy recommendations based on implicit or explicit theories, models, or hypotheses about the relationships among the variables believed to be operating and manipulable in the process.

These distinctions indicate the potential importance of research on the innovation process. Design actions are being taken by government and private organizations continually. Decisions by executives related to the innovation process are based on direct personal experience, beliefs about what ought to be done, observation of the behavior of their peer groups, and predictions of the consequences of such actions. Such decisions are a part of "political" (in the broadest sense) processes. If the information input taken into account in such decisions is based primarily on the experiences of specific individuals or speculative theorizing, then the quality of the resulting programs rests solely on the idiosyncratic characteristics of the decision makers involved. Except fortuitously, the results can be no better than the insights and "gut feelings" of the decision maker in the midst of the pressures of political processes of which he is part. To the extent that he draws on policy recommendations formulated by others, he may attain a degree of objectivity apart from the political processes of firms and governments. But clearly policy-formulation decisions must themselves draw on external information sources if they are not also to be speculative or idiosyncratic.

Design of policies and systems intended to affect innovation is going on continuously. But much of it is unsystematic, cut-and-try design which involves little systematic analysis of the likely effects of various design configurations on the outcomes of the innovation process itself, or likely side effects on the participating organizations, other sectors of the economy, and society as a whole.

There is a large amount of literature on policy for this area, but very little of it is based on such theory and research results as exist. In addition, many of the policy prescriptions are in direct conflict with each other, with equally distinguished observers and practitioners in disagreement over basic issues.

It is not too surprising that this should be the situation. There has been little systematic basic research on the actual behavior of individuals, groups, or organizations, and on the consequences of that behavior for effective innovation. There is an abundance of material that relates to this problem area but, as is characteristic of a newly developing field, there is relatively little "proven knowledge" as yet. Much of the lore is contained in the experience- and intuition-based "wisdom" literature.

FORMS OF KNOWLEDGE (Adapted from Douds 1971)

The approaches reflected in the literature of the field can be seen to take three forms: the discursive approach; the case-study approach; and the field-study research approach. Each contributes to the search for knowledge--the "truth-seeking" process. Each has sets of procedures or methodologies that guide it.

The Discursive Approach

There is an abundance of writing of a discursive nature about the R & D/innovation process. It contains a wide variety of material serving many purposes--speeches, conference papers, etc. Much widsom garnered from the practical experience of managers and administrators is contained in it. Many propositions or theories are implied. But as the implicit propositions are collected from various sources, one finds many inconsistencies or contradictions. Some of them represent viable differences in explanations of the way the phenomena work, and as such they are of great interest for more structured research. Unfortunately, these interesting propositions derived from practical experience are immersed in a mass of other propositions stemming from nonexperientially based sources--observations of "experience" that do not accord with what actually took place, wishful thinking, etc. Part of the literature is also persuasive in nature, intended to advocate a particular cause, budget item, organizational approach, or specific decision.

The methodology of this literature is very diffuse. If a consistent basis exists at all, it is a methodology of "exposition and rhetoric." This literature, particularly the "wisdom" literature, provides one of several starting points for the truth-seeking process. It is not amenable to a structured analysis for the verifiable "truths" it may contain, for it is not organized for verification, but it does serve many useful functions. It draws attention to the field, creates awareness of the importance of problems in the area, stimulates action, and can provide insights and propositions that are amenable to test and potential verification by more "scientific" methods.

The Case-Study Approach

Typically, case studies provide a narrative description of the circumstances involved in given situations or events. They generally deal with specific, limited sets of events. Historical and biographical literature is also of this character. Some valuable insights, especially with regard to the individual, are provided by historical and biographical literature and case studies, but little of this material lends itself to generalization to a larger population. It can provide the source of some propositions capable of being tested, but the outcomes reported in case studies are often those of unique and sometimes unusual cases. One should hesitate to base actions of any magnitude on generalizations drawn only from one or a limited number of such reports where the structure of the methodology does not involve *a priori* propositions to be tested or a reasonable *a priori* framework for guiding data collection and testing speculations or hypotheses.

Cases of innovation often appear in a form that can be characterized as "technologic chronologies," in which the emphasis is on chains of concepts, discoveries, successes, failures, and other events described in terms of the science and technology involved, more so than the people, management, and organizational factors involved. This type is well illustrated by the TRACES report (IIT 1968). It traces primarily the chronology of the scientific concepts and technological accomplishments leading up to the six innovations selected for the study. The chain of events is traced back into the last century to show the relative contribution to the innovations of scientific to technological inputs. It appears that this project was stimulated in part by one of the findings of Project Hindsight (Sherwin and Isenson 1966), namely, that the ratio of scientific to technical inputs in a number of key events for weapons system innovations was low over the timespan considered. The TRACES report notes the difficulties in determining what events are to be included in the chronology. It is not surprising that they find a much higher ratio of relevant events than in Project Hindsight. The case studies and charts do illustrate well the diversity of inputs and some of the missed opportunities that could possibly have led to the innovation at an earlier time.

Lipetz (1965) has compiled about 320 case histories of scientific and technological activity. He notes a number of difficulties in assembling the studies and hence with interpreting them, originating with the varied intentions of their authors. This difficulty with inference is illustrated by the two sets of case studies above: Project Hindsight and the subsequent TRACES report, each funded by different organizations.

Whereas in the discursive approach there is little evidence of a paradigmatic truth-seeking methodology, in case studies there does often tend to be evidence of structured approaches. However, they are many and varied, depending in part on the purposes for which the case was prepared as well as the background and training of the case writer.

In this early stage of the innovation-research field, it is appropriate that case studies be of a chronological, narrative character, since there is little theory on which to draw and a great deal of uncertainty about the nature of the phenomena involved. Such studies are more useful if they also treat the behavioral, organizational, and environmental aspects. However, the insights obtained depend strongly on the training, skill, perceptiveness, and other such personal characteristics of the case *writer*, as well as the reader. In the discursive literature the writer often seeks to persuade the reader to some attitude or action. The writer's role is intentionally an active one. In this "midway" literature--midway to providing the verifiable predictive capability associated with scientific theory and engineering practice--the case writer's role is not intended to be active. But it must necessarily be so, for he can report only a finite amount of information from the greater possibilities of what could have been reported. Such cases can help to structure a field. One can discover factors common to several cases, and relationships among these factors may be suggested. A single case may suggest an important element or relationship not reported in other cases. Case-study information may be of use in generating alternatives for action as well as propositions for testing. Case studies do not provide tests of propositions and, in aggregate, form a somewhat hazardous basis for determining the frequency of occurrence of various factors. One usually has little information to determine how the sample of cases examined compares with the population of potential cases, and infrequently knows the methods and criteria used to select the material reported. A bias may also be present in many case studies: it is hardly worth the considerable effort required to collect the data unless the case is notable in some regard.

Price and Bass (1969) reviewed several key studies of technology transfer. Most include case-study descriptions (e.g., MAB 1966; Sherwin and Isenson 1966; IIT 1968) as well as analysis, to a varying extent, of variables quantified to some degree. This latter aspect is an important first step in studies oriented toward prediction.

However, the case study can serve other functions. It can serve as a device around which to center a learning experience. For the reader, that usually happens in a structured, classroom situation. But it can also serve in the same manner for the authors, particularly when several work together, as is clearly evident in the report of the Materials Advisory Board and specifically noted by Tannenbaum:

> ...the Committee is convinced that we, as individuals, have been by far the greatest beneficiaries of the studies which have led to this report. The generation and discussion of these cases was an intensive learning experience (MAB 1966, p. 20).

There is another use of the case-study approach. One may have a theoretical model that structures or explains the relationship between certain conditions or variables in a range of situations. This model is more than a listing of factors to note existing prior to, or developed during, a case study. The case is prepared with particular attention paid to the factors contained in the model. The input filter is made explicit, and the reader as well as the author can judge if the model does *not* fit the situation. If it does not fit, then the model must be reworked. If it fits, then further studies testing the model may be worthwhile. If it appears to fit the situation, at best it can only be said that it fits this *one* situation.

This is the approach adopted by Nelson (1962), in which he articulates not only the technological concepts involved in the invention of the transistor but also some aspects of the manner in which the individuals involved worked together, the coupling of the groups, and the role of the managers and managerial policy involved. Marschak (1962) takes a similar approach in presenting an extensive case history of a complex systems development program (also at the Bell Laboratories) to test the viability of a single hypothesis. A number of other propositions concerning the management and behavior of systems development are derived from the case history. Several other case studies concerned with the development of complex weapons systems are presented in Marschak, Glennan, and Summers (1967, Chap. 3). These reports include less successful, as well as more successful, cases.

Extreme cases are particularly useful in generating insights or "concept-testing" propositions and models. *Parallel cases* with different outcomes provide similar opportunities if they have been developed comparably. However, it is extremely difficult to determine when cases are truly "parallel" and to assure that their *preparation* is comparable.

Parallel cases have particular value in that they can provide an explicit basis for comparison. It is through such comparisons that meaningful "information" is extracted from "data." A single case, or a set of single cases, provides a great deal of data to the reader. The basis for inference is necessarily left implicit. Information is extracted by unstructured comparisons the reader makes between the data of the case and other data available to him from his background, other cases, etc. Parallel cases if selected, for instance, on the basis of differing outcomes with analogous initial conditions, can point the way to variables that may be associated with the differing outcomes. They can make a significant contribution to theory building where propositions are formulated in such a manner that they can be tested and generalized beyond the immediate set of cases.

The Field-Study Approach

The field-study approach, as the term is used here, refers to those research activities developed with a structure that is often intended to permit generalization beyond the immediate sources of information. Ideally, data are obtained from a representative sample of a population in such a manner that inferences drawn from the sample observed can be attributed to the population as a whole. In practice, there are considerable difficulties relating both to theory and feasibility that limit the researcher's approach to this ideal of representative sampling.

Data may be collected from institutional records, publications, questionnaires, interviews, observation by a trained observer, and other methods limited only by the ingenuity and resources of the researcher. They are "hard" data in that they are collected in a systematic manner, supposedly do not depend on the idiosyncratic characteristics of the data collector, and (theoretically) are reproducible within an acceptable error tolerance.

In the early stages of the development of a field of research it is appropriate to seek to identify the key variables involved in the phenomena. Such studies typically are concerned with "state" variables. They may reveal factors that are important in the process, but they reveal little about how the process works. Variables of potential interest are identified by a variety of means, including prior studies in related fields (Folger and Gordon 1966 is an illustration); case studies (the MAB 1966 report is of itself a brief case study of this process); the "wisdom literature;" and personal observation or experience. Myers (1966) contains many propositions derived from a combination of all three sources. Means are then found to collect data on the variables and to analyze them. A whole repertoire of specialized techniques for "instrumentation," "measurement," and analysis to provide meaningful findings is available in the behavioral sciences, just as there are instruments, measurements, and techniques appropriate to the investigation of other scientific disciplines. The results may be presented as straightforward numerical tabulations, which sometimes are quite revealing of themselves, or they may be analyzed for statistical correlations or causality. Correlations based on variables that are typically included in such studies may indicate important relations between the inputs to the process and the process and the outputs of the process, but more often than not they fail to explain the process. For some purposes, the information provided by studies that stop at this point is sufficient. However, if the process dynamics or cause and effect are to be determined, more elaborate research designs are required.

A great number of disciplines can contribute to our understanding of the innovation process. Each discipline and area of specialization has developed its own variations of the truth-seeking process. Some of them address a very macro level (e.g., economic theory) and others a very micro level (e.g., personality theory). Useful propositions and research methodologies are found in such disciplines and specialities as:

- Anthropology
- Bureaucratic theory
- Business and public administration
- Economic development
- Economics of innovation
- Engineering management
- Entrepreneurship theories
- Group dynamics
- History of science and technology
- Individual learning theory
- Organizational climate
- Psychology
- Role theory
- Sociology of innovation

In these and related disciplines a few major strategies are represented in the literature:

- abstract speculative theorizing without recourse to empirical data
- manipulation of large amounts of data abstracted from the "real world"--e.g., data on exports, license fees, patents, etc.
- mathematical model building
- simulation in laboratories or by computer
- survey research by mass questionnaire or interview administration
- field studies of single cases

- comparative field studies
- longitudinal, or "real time," behavioral studies
- field experiments

A major difference among these general strategies centers around the extent to which they directly deal with the dynamics of the behavioral processes--attitudes, decisions, and actions--*in situ* where the actual technological innovations, transfer, and utilization are taking place, in contrast to operating at one or more levels of abstraction from the actual phenomena as they are occurring.

Even if we limit our concern, as we do here, to the general strategies that utilize studies of *behavior in the innovation process*, we may still employ a wide range of methodologies. Some are traditional to a particular discipline; others are adapted or invented to fit the needs of a particular problem. The net result is that many studies make a contribution to the theory building and verification of a particular discipline, yet an overall understanding of the dynamics of the social and organizational process of innovation does not clearly emerge. Such understanding could contribute directly to policy formulation and provide a more informed basis for the organizational or political decision maker.

In the remainder of this paper we explore the possible effects of the methodologies used in empirical, behavioral studies on their results: their communicability, quantifiability, reliability, and validity or generalizability.

FIELD RESEARCH STRATEGIES

Many approaches are used in designing field studies; there are no firm rules as to which method is the most appropriate. Many factors affect the selection of the method. We can distinguish at least three types of factors, concerned with

- the nature of the phenomena being investigated and the purpose of the study
- the decisions made in the field by other people
- the resources available to the researcher

The resources of the researcher include money, time, and talent. Even with the talent to design the best possible study, it is of little value if the researcher is not in a position to follow through with it because of inadequate funding, other commitments for his time, unavailability of appropriately trained staff, etc.

When data are to be collected in the field, decisions are involved that are not made by the researcher. He can only seek to influence organizations to allow him access, and once having gained access, to obtain the cooperation of the many people that may be involved. Another aspect in some types of studies is that in order to carry them out certain types of events have to occur to yield the requisite data. These events may occur at infrequent, unpredictable intervals. More likely, events may occur during the course of a study, such as an economic crisis or company reorganization, that introduce dominating changes in the variables being observed. The practicalities of field research almost inevitably compromise the research design and implementation to some degree.

Clearly, the nature of the phenomena to be investigated affects the design of the research. In many fields, major breakthroughs have required the development of new methodologies, techniques, and instruments. Many specific methodologies, designs, and techniques are used in the study of phenomena related to organizations. However, it is not a reasonably well-organized and codified assemblage such as the chemist has at his disposal. The techniques come from a variety of academic disciplines, each with its own preferences and styles of research. As Vroom notes:

> One cannot help being struck by the fact that there is substantial disagreement among contributors [to organizational research] on such matters as the boundaries of the field of inquiry, the most pressing problems for study, and the relative usefulness of different research methods or theoretical systems....[I]t can be argued that different contributors are working from vastly different sets of architectural plans and that many are ignorant of or unsympathetic towards the activities of large segments of other contributors (1967, p. ix).

The situation with regard to substantiated knowledge about the R & D/Innovation process seems to be similar. However, some general comments can be made about the applicability of the several broad types of research strategies. We shall consider:

- Field surveys
- Comparative case studies
- Field studies
- Longitudinal studies
- Field experiments

As a field research strategy, a *survey* is particularly appropriate when relatively little is known about a situation. Simply obtaining data about that situation may enable some order to be given to it. Surveys can be relatively cheap and can sometimes be performed relatively quickly.

A straightforward survey[1] is designed to produce data. It does not produce information. "Information" in the sense of "meaning" requires that a basis for comparison, contrast, or evidence of change exists. In a survey, as we are using the term, this basis for comparison is left up to the individual--the author and the reader. When there is little knowledge in a field and a survey is conducted, "information" is obtained in the sense that there is a change from having no data to having data. But inferences about relationships contained within the data are obtained on the basis of the user's background, training, and insight. Another user will not necessarily arrive at the same conclusions. The remainder of the techniques listed above involve features to overcome this limitation.

A mass of data, numbers, or prose contains a great deal of variation. When we are looking for a particular message--that a relationship noted elsewhere or hypothesized to be operating exists in the current data--much of this variation constitutes "noise" or *error* for our purpose. Diamond (1959) illuminates this point well:

> Error embraces every sort of variation in the phenomena which we are observing, except what results from the one source which is the object of our study at the moment, and which we try to bring into prominence by the design of our investigation. It includes all the accumulated effects of other influences, whether suspected, understood or not understood (p. 6).
>
> Error does not carry any recognizable badge, for when we change our point of view, to focus on a different problem, what had been error may become information, and what had been information may become error (p. 7).

The five research strategies provide alternative ways of handling the error that will be present in the data. We commonly attempt to sort out the information from the error in a given set of data by applying statistical techniques. This approach presumes that we have preconceived ideas of what we are looking for, no matter what the source of the ideas--experience and intuition or carefully developed theoretical reasoning. But a great deal is involved in research methodology besides analytical techniques for data. Diamond (1959) notes:

> Statistical method is only one of several ways by which we strive to isolate information--the perceived effect of the experimental (or hypothesized) variable-- from error--the contamination which enters our observations because of the uncontrolled effects of other influences. The first, and no doubt the most important way, is by the *clear formulation of our problem*, which tells what kinds of data most deserve our attention. Then come the efforts at experimental control, by which we hope to eliminate some extraneous factors and to limit the scope of others (p. 7).
>
> If we succeed uncommonly well in our efforts to eliminate error, the outcome of the investigation may be so clear that there will be no need for statistical analysis....[T]he correct appraisal of information must be preceeded by an appraisal of potential error. *Experimental technique and research design strive to reduce error; statistical technique measures the error which remains* (p. 8).

The single-case-study approach is appropriate to use in early tests of some theories, especially in the preliminary exploration of theories explaining processes. It can provide an indication of possible substantiation for the theory, as we described earlier. When a *comparative case study* strategy is chosen, an explicit basis for comparison is built into the research design. Information about the theoretical model is extracted from the differences in the events occurring in what one hopes are parallel situations. However, our confidence that the differences cannot be explained in other ways depends heavily on personal evaluation. When there is a relatively well-developed theory prior to the design of the study, the research problem can be clearly formulated and procedures for obtaining the data most deserving of attention can be worked out. Often there is little to be gained by using more advanced designs if the theory is not adquately developed.

Field Studies are frequently used to investigate one or a number of specific propositions that can be expressed in a relatively simple form--one or several variables are hypothesized to be associated with one or a few other variables. If these relationships are essentially concerned with the state of a system, rather than its processes, a cross-sectional field study is usually appropriate. Much of the field research in the literature is of this form. This strategy choice also has the advantage to the researcher of making his resource requirements much more predictable than the strategies discussed below. Field studies can vary greatly in complexity, depending on the number and nature of the variables involved, as well as on the complexity of the hypothesized relationships.

Field studies may range from tentative explorations of hypotheses to rigorous hypothesis-testing efforts. That is, their intellectual underpinnings may range from weak and diffuse, as in the early stages of development of theory and understanding, to strong and well developed at later stages. In either case, the intensity of the initial design effort is important, because of the uncertainties about the phenomena in the early stages of developing understanding, and, as understanding matures, the necessity rigorously to eliminate other explanations--rival hypotheses.

The methods used for extracting information in field studies as they are commonly reported in the literature primarily emphasize the statistical-data-analysis aspect. A wide range of these techniques may be applied, but nevertheless the initial design of the study can have a major influence on the amount of error in the results--error not necessarily measured by statistical tests. The general lack of replication of field studies limits the generalizability of the findings about the propositions and theories being tested.

Longitudinal studies are needed when a process explanation is to be investigated thoroughly. Since a longitudinal study extends through time, and many phenomena having to do with people and their institutions involve time spans of many months or years, the resource investment in a longitudinal study can be considerable. With adequate design, process hypotheses can be investigated with cross-sectional field-study strategies. This strategy compresses the duration of the study and thus inherently limits the overall confidence that can be placed in the results, since different people, organizations, products, etc., are used to obtain information about different sequences in the process. The longitudinal study, by following the focal unit in time, overcomes this limitation. However, the resources required to support it imply that it should be used primarily in the later stages of theory testing. (There may be some phenomena that can only be brought to light by being studied over long actual time spans; that is of course the particular skill of the historian.)

A major weakness of cross-sectional field studies is that, in general, cause-and-effect relationships cannot be ascertained. Although it may be possible to determine that two variables are associated, it cannot be determined that a change in one (which may be subject to change through managerial action or policy implementation) leads to a change in another (a sought-for outcome). For instance, it is fairly well established that job satisfaction and enhanced productivity tend to be associated, but it is uncertain to what extent improving satisfaction improves productivity, or how improved productivity causes increased satisfaction. (There is evidence for both.) Longitudinal studies and field experiments often allow cause and effect to be determined, since the sequence of events is observed.

Field experiments are appropriately used when cause-and-effect relationships are to be determined and when there are well-identified alternative explanations of phenomena.

Two general types of field experiments can be identified:
- Natural experiments
- Administrative experiments

Natural experiments capitalize on "spontaneous" events. By administrative experiments we mean an experiment that is conceptualized and deliberately planned out beforehand so that one or more variables are caused to change with others being controlled in some fashion. An organizational manager (administrator) is ordinarily involved in the planning, perhaps as the experimenter.

As an aside, we note that the word "experiment" is used in a variety of ways. It may be used in the sense of personal learning, as when an individual experiments with a new pocket calculator. To a more knowledgeable observer, the results of his actions are fully predictable, but the user is uncertain of the forthcoming results. Another sense refers to making changes where the concern is with the utility of the intended outcome in socio-economic settings. It is the action testing of the manager as he institutes a new order procedure, for instance. The third sense, the one that is used here, is concerned with propositions testing to establish cause-and-effect relationships or to enhance the confidence in theory (Douds 1972).

By *natural experiment* we refer to a situation where a change occurs in an organization (or relationship between organizations) without deliberate intervention by an "experimenter." More particularly, the change occurs in a situation where current data are already being collected, so that measurements of the consequences of the change and other impinging factors can be made as they occur, or soon after. The observed change, the outcomes of interest, and other possible reasons for the actual outcomes can be related to each other at least in time sequence, if not in terms of direct causality. The natural experiment is in contrast to the typical managerial "experiment" where, even if the outcomes are measured well, the factors that might be significantly influencing or "causing" the outcomes are not known. The deliberately made change may be only one of the possible explanations for the observed outcome. In some cases, the change in observed outcome might have occurred even if the deliberately made change had never taken place; or the change might have occurred sooner, later, or in different form.

The essence of conducting a natural experiment is to be prepared to observe and capitalize on changes and variations that occur in the actual situation being observed. Various types of data must be collected often enough so that a basis for comparison exists to permit extraction of information about the influence of factors operating in the situation (Douds 1973).

Administrative experiment refers to the situation where the experimenter has sufficient control deliberately to introduce changes. As Campbell and Stanley (1963) point out, in experimental design in the sense of Fisher (1925, 1935), the experimenter has complete control of the situation. He can schedule treatments and measurements for optimal statistical efficiency, with complexity of design emerging only from that goal of statistical efficiency. However, the extent of control that can be achieved in organizational experiments is vastly different from that in, say, agriculture. Discussing the problems of experimentation in the field, Campbell and Stanley point out a variety of necessarily more complex designs needed to control for various threats to the internal and external validity, and hence the generalizability, of the findings. They term the type of experiments that can be performed in the field "quasi-experimental designs."

One of the more significant reasons for the paucity of true experiments, or even quasi-experiments, in organizations is the difficulty the external researcher has in securing and maintaining the cooperation of the organization. The chances of carrying through the experiment are far greater when a manager in the organization is the experimenter. Managers are constantly experimenting (action testing), but without deliberate (quasi-) experimental designs (hypothesis testing). Thompson (1969, 1974) has initiated a program to train research and engineering managers in transferring their technical methodology skills to administrative experimentation.

Specific examples of these several strategies as they are followed, and elaborated upon in practice, are discussed in the following sections. Both the methodological strengths and weaknesses of particular studies will be given. The reader should note that we are not praising or criticizing the studies *per se*. We are using them as examples of the strengths and faults of the many studies listed in the bibliography, rather than commenting individually on all the studies. Certainly we must be able to assess reported work adequately; but it is also important that work not yet started be done well.

FIELD SURVEY

Litvak and Maule (1973) conducted a survey of Some Characteristics of Successful Technical Entrepreneurs in Canada because no data were available about the founders of small technologically based Canadian firms. The Canadian government was developing policy and legislation intended to stimulate innovation in Canada, and there were questions about how appropriate it was, whether intended beneficiaries were aware of it, etc. Some of the data obtained directly answered such questions. Other data obtained would enable the reader to infer answers or to infer certain relationships between sets of variables.

There are at least two basic issues in a survey, whether it contains factual data (e.g., age of respondent, annual sales of the firm), opinion questions, or both: how reliable and how representative it is. *Reliability* involves such problems as:

- Will the same person answer the questions the same way on a second administration?
- If the wording of a question is varied somewhat, will the same person still answer much the same as before?
- Will different people interpret the same set of questions the same way as each other?

The first problem involves the stability of the responses over time. The second and third problems involve whether or not the survey is getting at the meaning the researcher intends. Many techniques are used to enhance and measure the reliability of data-collection instruments. Standard research methodology textbooks often discuss the measurement of reliability at length. However, perhaps the most fundamental and best method of initially enhancing reliability of self-administered questionnaires, the form typically used in surveys, is for the researcher to administer the questionnaire personally to a limited number of people in pilot tests. Discussion with the respondents and examination of responses often reveals needed improvements.

Survey reports seldom mention any aspect of reliability or that the instrument was pilot tested. (This is the case for the Litvak and Maule report.) It could be that such preliminary work is taken as a *sine qua non* of a professional conducting a survey. However, not all professionals reporting surveys are professionals in survey technique.

The second basic issue in surveys is how *representative* the findings are of the sample of respondents and of the whole population. The first aspect is simply handled by reporting the numbers or percentages of respondents giving various answers. As in most survey reports, Litvak and Maule include a variety of tables and give quantitative data in the text up to the point where the environment for entrepreneurship is discussed. Evidently this section is based on a number of open-ended questions concerning opinions about availability of venture capital, accessibility to special government innovation-stimulating programs, etc., and measures to be taken to promote entrepreneurship. Lack of quantitative data may make the section more readable, but it provides no basis for the reader to evaluate the statements presented.

A second aspect of representativeness, generalizability to the whole population, involves procedures used to obtain the sample of respondents. Unfortunately, in field studies it is often not clear what the population is. Hence the many available well-documented techniques cannot be readily applied. There may be other reasons the sampling technique may be less than "ideal." When a sample cannot be taken by statistical techniques, the reader can still judge for himself how to evaluate the data in terms of the intended population if the basis for selection is revealed. The Litvak and Maule report simply states that questionnaires were mailed to 76 "selected entrepreneurs throughout Canada" without indicating the basis for selection or what "throughout" meant.

One other problem of generalizability involves people who do not respond to mailed questionnaires. Often, a mail response rate of 20 to 40% is obtained, depending upon subject matter, length, etc. After a follow-up letter, Litvak and Maule obtained a response of 56%, perhaps indicating the importance the respondents gave to the subject matter in a 10-page questionnaire. Even with such a relatively high response rate, we still do not know whether the nonrespondents differ from the respondents in a characteristic way that introduces biases into the data. The usual ways of answering this question are either to continue the follow-up efforts, or to contact a sample of the nonrespondents by phone or in person and ask a few key questions that should reveal their lack of representativeness if it is present.

COMPARATIVE CASE STUDIES

Here we shall consider the use of case histories in pairs and in parallel, by bringing together diverse case studies after the histories have been compiled or by collecting data about parallel cases within a prestructured framework. Work by Marschak, which includes most of the necessary comments on strengths and limitations, illustrates the former strategy; an article by Kruytbosch illustrates the latter.

Again, we note that the strengths and weaknesses of the examples discussed here are illustrative of what is found in the literature; similar criticism and praise could be given for any other published study. The ones discussed here were chosen only because they illustrate various points we wish to make.

In an earlier section we discussed the general characteristics of case studies. We noted that in terms of the testing of hypotheses, individual case studies have serious limitations. When case studies are used in groups, these limitations may be alleviated to some degree. In *Strategy for R & D* (Marschak, Glennan, and Summers 1967) the authors present a number of similar cases about the development of radar, jet engines, and aircraft to illuminate two hypotheses. The first is that "Major uncertainty, at least in the earlier phases of development, is a natural and inevitable property of a program that seeks a real technical advance" (p. 50). The second hypothesis is a lengthy one concerning the effect of uncertainty and its consequences on making heavy or light commitments in the early phases of development.[2]

Since major uncertainty occurs in all the case histories, it appears reasonable to conclude that the first hypothesis would be supported in a field study designed for statistical hypothesis testing. "Predictions about the effort required to make a given design attain a specified performance are subject to large errors, especially at the start of development" (p. 136). Indeed, a statistical test of the data would probably turn out to be unnecessary if such a field study were carried out, especially since the hypothesis borders on being tautological.

Marschak noted that the second hypothesis is harder to formulate and test for the following reasons:

> In the first place, few of the programs fall close to one or the other of the two extremes. Those that do not are not relevant to the hypothesis. It requires some subjective interpretation, moreover, to place even the less ambiguous programs in one category or the other.
>
> A second difficulty in testing the proposed hypothesis is that of grouping the projects studied into significant pairs. In such a pair one project is of the inflexible type and the other of the flexible type while both are similar with respect to the circumstances that might explain major differences in the projects' total experiences...[and] both achieve the same order of technical advance.... We can hope to approximate this ideal test only very crudely.
>
> Finally, if the histories presented are to support the hypothesis they either should be roughly a random sample from each of the areas of technology or they ought to cover these areas completely. Since histories often take a great deal of effort to compile, one has to strive for the former alternative. The preceding histories, however, were not chosen at random--they were largely chosen because the programs seemed interesting and their study promised to be feasible (pp. 138-139).
>
> Marschak also notes that the studies differ from one another in scope and style for several reasons: they were prepared by several different authors; the availability of historical material varied considerably; and military security or company confidentiality

prevented inclusion of all the available material (p. 50). He concludes with respect to the second hypothesis: "Some of the histories, then, fall into pairs that support the proposed hypothesis and others do not. None appears to fall into pairs that directly conflict with the proposed hypothesis..." (p. 139).

It is clear that a great deal of uncertainty remains about the hypothesis. In fact, from an earlier comment of Marschak, it appears that examination of at least some of these case histories may have been the origin of the hypothesis:

> Chronologically these cases are the earliest research efforts in the book. The case histories were done at a stage in the total RAND program on R & D management when hypotheses were more crudely stated than today. Indeed, the case studies were perhaps the crucial step in the formulation of conjectures as to what constitutes good or bad development. Because of their place in history of RAND thought on development, the questions posed during the collection of data and the format for the presentation of the data are not the same as if the studies were started afresh today (p. 3).

It is not surprising, then, that the cases do lend support to the hypothesis, since they evidently provide its origin. The narrative is not clear on the point of how many of the case histories were involved at the time the hypothesis was formulated. But this is characteristic of the approach of the historian. He cannot replay history. But when it comes to being able to make odds on future outcomes, tests of a priori propositions are to be preferred to a posteriori deductions. Nevertheless, such hypotheses have their utility:

> We note, finally, that even if the project-history approach falls far short of confirming a hypothesis, it may still be useful to the policy-maker. For the policy-maker, unlike the student of R & D, must make decisions here and now.... Although he cannot be certain his choice is the correct one, such studies can make his uncertainty a great deal smaller (p. 139).

We have noted that the comparative case study is useful for probing theoretical formulations. Kruytbosch (1972) examines the theory of bureaucracy favored by many sociologists and raises questions about its ability to explain the effectiveness of groups. In an innovation-producing firm--a large aerospace company--he compared the history of two groups doing similar work through two reorganizations. The first brought the groups into being. The second moved one group in its entirety into a position at least as strong as it had before; the second group was disbanded and the individuals were relocated in two different divisions.

The study is not detailed enough to compare the groups on multiple dimensions of productivity and effectiveness. Data were collected by interview on two occasions separated by a number of months. However, on the single criterion of survival as a measure of leadership and group effectiveness, Kruytbosch notes that clearly one group was effective and the other was not.

The groups are compared on the basis of a number of factors: preexisting social ties, internal social system including cohesiveness and morale, character of the leader, technique of administration, etc. As the author notes: "A hardly startling implication would be that a cohesive work team organized around a 'hot' technical specialty and spearheaded by astute leadership is an effective anchoring device for R & D personnel in the industrial environment. Obviously this is subject to a number of conditions such as the state of the art, the state of the relevant sector of the economy, and so on." The comparisons that can be made on specific points can help to clarify how this happens, although such inferences are left to the reader. Kruytbosch does explicitly point to one aspect evidenced in his study that has been neglected in the literature on bureaucracy and leadership style, although it was noted many years ago:

> D. Pelz's two-decade-old study (1955) opened up a perspective regarding supervisory practices as taking place within a larger organizational system context. He found that worker attitudes and performance were more strongly related to perception of their supervisor as having influence on departmental decisions up the line than to any variations in conventional management practices. The study

suggested that the way a work unit is tied into the larger organizational structure may be a crucial variable in group stability and performance.

The case serves to support the proposition arising from Pelz's data and the limitations of bureaucratic theory. It gives rise to broad questions for theory and research: "The material developed in this study suggests that a question facing students of organization is how to develop models of large-scale organizations depicting fluid, loosely coordinated, open-ended arenas within which groups...range rather freely and develop their productivity, or fail, and regroup into new constellations." Though raising such questions, the case material is not used in the paper to develop new specific hypotheses (which is certainly the author's prerogative).

We have already noted the difficulty in using material from a variety of cases to test hypotheses and theory, especially when the separate cases all deal with "successes" or events out of the ordinary. Comparative cases such as this one provide a rich source for obtaining insights. It is unfortunate that so few truly parallel cases are available, developed on comparable bases with reasonable documentation of a number of variables and parameters in the situation.

FIELD STUDIES

The number of field studies of the RD/I process is quite a bit greater than other types of studies--survey, longitudinal, or field experiment. The *IEEE Transactions on Engineering Management* has made a consistent effort to publish such work since 1960. From that year through 1973 about 42 articles containing field data more or less closely related to the RD/I process appeared. Thirty-seven of them were field studies developing or testing propositions or instruments. In this section we consider some of the characteristics and criteria relevant to evaluating field studies. Rather than review particular studies, as we are doing in other sections where there are very few of each type, we provide a few tabulations of particular aspects of the field studies appearing in the *Transactions* to give some appreciation of the state of the art in field research. We limit this survey to the *Transactions* because it is the professional refereed journal with the longest history of specialization in studies of the RD/I process. The Bibliography at the end of this chapter lists all the articles surveyed from the *Transactions* plus a number of others published in other refereed journals. Our intention was to include all papers relevant to behavioral aspects of RD/I that we could find within the resources available to us for preparation of this paper.

We have noted that a field study may serve a number of purposes. Most studies in one fashion or another seek to add to our knowledge of factors involved in the RD/I process and relationships existing that interrelate these factors. Other studies occasionally report a pilot test of a design or the development of a measure for a single variable of particular concern. In any of these cases, the reader implicitly has two general questions that he evaluates: utility and credibility.

Utility (to him) of the subject matter depends solely on his own interests and concerns; only he can determine the criteria to apply. The question is central to the researcher, of course, for if he can do research that will be perceived as being of high utility to increased understanding or solution of a relevant, important problem, he will be making a significant contribution.

A more complex question is the believability or credibility of the results to the reader. There is no one right way. In general, one asks that he not depend solely on intuition. One looks for data, in one or more of many possible forms, from which information about the factors, variables, and interrelationships can be extracted. One looks for internal consistency with prior related work (or explicit reformulation). It is not enough that the results of the study--the "information"--be interesting. A major factor determining credibility is the design and execution of the research--the methodology and the procedures employed. The following questions provide one framework in which to evaluate the confidence, the believability one can place in the results of a field research project. Because there are so many ways that field research can be done well, the questions are intended to provide a broad perspective. The results of a study with a sample size of three are not necessarily to be discounted. Results with a statistical confidence level of 0.001 are not necessarily meaningful (Winch and Campbell 1969). No meaningful, broadly applicable guidelines can be given that deal with specifics of this sort. A great many broad, as well as specific, issues are involved. The questions suggested here, and a few specific points to be raised, may assist in the assessment of sets of, or particular examples of, field research studies.

1. *What Was It That Structured the Design of the Study?*

The initial aspect of this question involves the connection of the study with prior work. It may be rooted in one or more specialized disciplines--psychology, sociology, economics, history of science, etc. The context of RD/I may have appeared to the investigator to provide characteristics suitable for testing some particular theoretical aspect in one discipline, or the disciplines may have appeared to the investigator to provide aspects of theory applicable to the phenomena. The difference between these two approaches can be considerable. On the one hand, there may be the test of a theory perhaps largely irrelevant to RD/I. On the other hand, the theories or hypotheses may be central to RD/I. At another level of consideration, one researcher, through lack of familiarity with the phenomena, may overlook important aspects of the theory.

The connection of the study with prior work may be less close than implied above. It may involve the exploration of an area in which there has been little prior work, either from the standpoint of the phenomena concerned or the application of a particular type of theory; the researcher is exploring new fields. As a practical matter we should also note that a study labeled "exploratory" may mean, in some cases, that it is the researcher himself who is new to the field. Both human propensities and the diffuse nature of the literature in this field make it difficult to avoid "reinventing the wheel."

It is also possible that no connection to prior work is expressed, for instance when the project involved research on a practical problem of concern to an engineer, scientist, or manager in an organization involved in the RD/I process. He may be short on theory, but he is likely to be very familiar with many practical aspects of the phenomena.

The other aspect of this question deals with the concrete manifestation of what structured the study, as expressed in its report. Papers in refereed publications usually contain a reasonably clear statement of purpose. Yet even they often lack an explicit statement of the *researchable* questions addressed. It is here that the researcher has the opportunity to sort out the parts of the problem that can be based on measurable, empirical data and those that can be based only on judgment, experience, wisdom, or intuition.

Kerlinger (1965, pp. 19-20) provides three criteria for a good research question: (a) It should express the relation between two or more variables. (b) The research problem should be stated clearly and unambiguously in question form. A statement of purpose alone is not adequate. "Questions have the virtue of posing problems directly. The purpose of a study is not necessarily the same as the problem of a study." (c) The question should at least imply the possibility of empirical testing. "A problem that does not contain implications for testing its stated relation or relations is not a scientific problem. This means not only that an actual relation is stated, but also that the variables of the relation can somehow be measured. Many interesting and important questions are not scientific questions simply because they are not amenable to testing."

On a much less stringent set of criteria--whether the researchable questions could reasonably be inferred (and setting aside marginal cases or studies that involve the evaluation of a technique)--we find that of 37 field studies relevant to the RD/I process appearing in the *Transactions* from 1960 through 1973, 57% contained explicit or reasonably implicit researchable questions or a clear statement of purpose. (We set 16% aside.) Only 27% contain explicit question statements, and not all of those would satisfy Kerlinger's other criteria. (Of course, we have no way of knowing how many of the studies developed unreported explicit researchable questions during the early course of their work and how many of the reported questions were developed after the study was complete.) Although the absence of explicit researchable questions does not necessarily mean poor research, lack of encouragement to use this approach can lead to ill-defined or poorly formulated research. The record implies that we have considerable room for improvement.[3]

2. Are the Nature of the Information to Be Derived and the Characteristics of the Study Design Evident?

At this point we are simply asking whether the author, going beyond the statement of the problem, makes clear specifically what information is sought and the nature of the method for attaining it.

One method of making clear exactly what information is sought is to state one or more *a priori* hypotheses or propositions. However, if little is known about the phenemonon, one may be seeking initially to discover what factors are present and then what relationships may exist among them. One may also be reporting work on the measurement of a particular concept or variable. Various types of propositions can still be used to structure the research design more clearly in such cases, but the necessity for doing so is less.

Again, taking a not too stringent view as to what constitutes a clear proposition statement, in the *Transactions* from 1960 through 1973, we find that only 24% of the articles reporting field studies of the RD/I process had reasonably clear statements of propositions or hypotheses. (Again, 16% were set aside because they involved evaluation of a technique or instrument, or because they were considered marginal.) Thus, in 54% of the cases it becomes much more difficult to determine what information was sought, and one is led to wonder how one can judge whether the overall methodological design was appropriate to the problem. The information discovered and reported may be of value, but that depends on the purpose to which it is put. There is a difference between information suggestive for further research and information that clearly describes the relationship among variables.

Given the premises of a study, the design basis for extracting the information may or may not be evident. To evaluate the findings adequately one needs to know:

- What were the variables?
- What was actually measured (the indicators of the variables)?
- How were the indicators measured?

Measurement can take many forms, since research may appropriately involve thoughts, feelings, decisions, etc., as well as more concrete evidence and results of behavior such as forms of contracts signed or pieces of equipment delivered on particular dates for given costs. Measures can be relatively "hard"--such as the dates when equipment was due and when it was actually delivered--or judgmental--such as classifying the feelings described in interviews about how the designers and customers felt when the delivery was due but the equipment was not ready. Of course, just because the data are in numeric form does not mean that they are more desirable than if they are of another sort, as we shall consider below. The question being asked at this point is: Was the design basis for selecting the desired data made explicit so that the overall structure of the research project can be determined? The answer is important in the assessment of the meaning of whatever was discovered.

3. How Does One Know What Was Actually Measured?

The variables of a study involving social and organizational behavior are often convenient constructs rather than physical entities. In natural phenomena there are no such physical entities as "gross national product," "leadership style," etc., though our training and discipline may have brought us to the state where such concepts take on a reality and life of their own. But through various operations we can invest such concepts with meaning and usefulness. In general, it is necessary to measure something other than the named variable itself. Some other indicator, or set of indicators, must be used as a surrogate for the given concept.

Perhaps the simplest way to show what was measured is to present the instrument itself. Unfortunately, the expense of publication or sometimes proprietary concerns make this practice an uncommon one. For other types of measures--for example, Apstein's (1965) measure of the "effectiveness of military laboratories" where judgments are made in a particular manner by a number of qualified judges--the procedure to arrive at the measurement may be described. As important as the measures are to the credibility of the results of a study, it is surprising that the behavioral science community, as scientists, accepts any reports that do not make clear the content of the basic operations involved in reported measures.

Because this question is so vitally important, most articles do provide at least some indication of the substantive content of the measures, at least for key variables, and show at least what the indicator was for that variable. But many questions remain. The central question is: Are we measuring what we think we are measuring? The question leads to the nature and meaning of one's variables, the concept of *validity*, which also involves the related concept of *reliability*.[4]

When only the indicator or procedure for measuring a variable is presented, with no other data, one must deal with "face" or content validity. Face validity implies that the relevance of the measuring instrument to what one is trying to measure is apparent "on the face of it." The issue can only be resolved as a matter of judgment. In the making of this judgment, two major questions must be answered: "(1) Whether the instrument is really measuring the kind of behavior that the investigator assumes it is, and (2) whether it provides an adequate sample of that kind of behavior" (Selltiz et al. 1965). The assumption that the behavior or property being measured in a test is actually that behavior is most likely to be met in proficiency and achievement measures such as those of mechanical skills. When other types of constructs are involved, Peack (1953, p. 285) notes that "More often than not, additional meanings are smuggled in and the assumption is made that the observations are, in fact, interpretable as a sample of a known universe...."

Content validity is determined in terms of the question: Is the content of this scale or test representative of the universe of the content of the property being measured? Such universes of content items exist only theoretically, so content validation consists largely of judgment (Kerlinger 1965). Content validity and face validity are essentially the same, raising the same questions and utilizing the same methods of judgmental processes.

Regardless of whether or not an indicator adequately measures the concept it is supposed to, there is always some degree of uncertainty in any measurement. Before one can adequately interpret the results obtained from a series of measurements, an estimate of this uncertainty is needed. Reliability coefficients for an instrument provide this measure. They indicate the stability, consistency, or predictability of the response to an instrument or method of measurement. This consistency may involve the responses from one item to the next within a given instrument (inter-item reliability), or from one measurement form to another (equivalence). For the first two types, two forms of the same scale, or multiple forms--depending on the particular reliability measure used--are created and the consistency of response determined by comparison of one set with another. This procedure provides a coefficient of *inter-item* reliability. The correlation between the scores of a test given at one time and the same test given after an interval indicates how stable the scores are, and so provides a coefficient of (time) *stability*. Reliability measured on the basis of one administration indicates inter-item stability. Both inter-item and stability measures provide an indication of the error of measurement. Reliability measured on the basis of equivalence between different forms of measurement merges into the concept of validity, as we shall see.

In the 37 articles from the *IEEE Transactions of Engineering Management*, not one provides a reliability coefficient and only two provide some form of evidence concerning the reliability of their measures. (In three of the articles a reliability measure could be considered unnecessary.) It is notoriously difficult, but not impossible, to establish a reliability measure for certain types of data collection procedures such as interviewing and participant observation. However, for many forms of questionnaires it is quite easy to calculate a reliability coefficient such as Cronbach's *alpha* (1951). Computer programs are readily available and employed, as done by one of the authors and colleagues in a study of factors affecting communication between R & D groups (Douds 1970). Since many field studies do utilize questionnaires, it is to be hoped that reliability coefficients will come to be more frequently reported so that better informed judgments can be made about the quality of the data.

If the reliability of a measure is low, it is not at all certain what is being measured. The "noise" (error) is obscuring the "signal" (the desired measure). At this point two questions arise: How can we get a better measure of the signal? Are we measuring what is intended? But on the other hand, if we succeed in getting a measure with a high reliability coefficient, we can only say that we are getting a strong signal--we still cannot be sure that we are measuring what we intend. It is difficult to get measures of high reliability (and not always desirable to do so because of other attendant phenomena), especially when organization processess are under study.

Both these difficulties may be alleviated by the use of *multiple measures*. If one has a number of quite different indicators for a given variable, any one of which has a relatively low reliability, but *all* of which "point in the same direction"--that is, they are all moderately correlated--we can then feel more comfortable with and certain of a variable than if we have but a *single* measure, even if it is of high reliability. Speaking of interviews and questionnaires, Webb, Campbell, Schwartz, and Sechrest (1966) note:

> *But the principal objection is that they are used alone.* No research method is without bias. Interviews and questionnaires must be supplemented by methods testing the same social science variables but having *different* methodological weakness (p. 1).

In the remainder of their book they primarily address the issue of finding a variety of measures for given phenomena, measures that are unobtrusive so that the process of making measurements does not disrupt or change the phenomena observed as the conventional techniques of interviewing, observation, and questionnaires can do.

The final aspect we shall consider, related to the question, "How does one know what was measured?" is the notion of *construct validity*. Here we are specifically asking, Have we measured the concept of the variable that we intended to measure? Does our mental construct hold up when we measure it? Any given concept can be defined as being related to certain concepts and distinctly different from others. This insight, in conjunction with the notion of capitalizing on methodological weaknesses by utilizing different methods with different types of weaknesses, has led to a method for establishing construct validity.

Construct validity is of central concern to the development of theory. Cronbach and Meehl (1955) first made the concept of construct validity explicit, although it has been implicit in the history of science. As long as the concept of phlogiston was assumed to be a necessary constituent of all combustible bodies, progress in the fields we now know as chemistry and thermodynamics was limited. The introduction of new, more stable constructs integrated into a theoretical structure that allows a variety of new, "substantiable" predictions to be formed, is a major goal of scientific activity. The definitions of such constructs imply sets of propositions about their relations to other variables. "Thus, in examining construct validity, it is appropriate to ask such questions as: What predictions would one make, on the basis of these sets of propositions, about the relationships to other variables of scores based on a measure of this construct? Are the measurements obtained by using this instrument consistent with these predictions?" (Selltiz et al. 1965, p. 159.) Consequently, construct validation involves not only the measuring instrument but the underlying theory, and the operations become involved and extremely time consuming. It becomes difficult to disentangle the validation of a construct, the instrumentation of its measures, and the theory of which it is a part.

These difficulties were mitigated and the notion of construct validation made more operational within the life span of a researcher by Campbell and Fiske (1959). They have provided a specific technique, the multitrait-multimethod matrix, that provides a quantitative focus on the adequacy of the measure of the construct in question *before* hypotheses involving other variables are considered. This approach partially disentangles the validation of constructs from the total research process in attempts to validate a theory. At least from a theoretical point of view, it is reasonable to ask just what a proposition supported by field data means if the concepts in the variables have not been validated? One can only proceed on little more than faith that the concepts are meaningful in the real-world phenomena, that the measures actually used have some tenable relationship to the variable, and that they have some reasonable degree of stability over time. All that is in addition to a number of assumptions about the parameters of the situation, most of which were not measured (and in practice most of which could not be measured).

Campbell and Fiske propose two major requirements to be met in validating a construct: demonstrate that (1) the various independent measurement methods for the same construct are convergent; and (2) the measures of a construct ("trait") discriminate it from other constructs ("traits") from which it is intended to differ. Instruments can be invalidated by too high correlations with scales for other variables. The first shows *convergent validity* of multiple methods; the second shows *discriminant validity* among various constructs and methods. More than one construct and more than one method have to be employed in this validation process. The relation among the constructs can be shown in what Campbell and Fiske call a "multitrait-multimethod matrix," presenting all the intercorrelations for each of several traits measured by each of several methods. Several criteria are provided to guide estimation of the construct validity.

Campbell and Fiske observe that

> reliability and validity can be seen as regions on a continuum. Reliability is the agreement between two efforts to measure the same trait through maximally similar methods. Validity is represented in the agreement between two attempts to measure the same trait through maximally different methods. A split-half reliability is a little more like a validity coefficient than is an immediate test-retest reliability, for the items are not quite identical. A correlation between dissimilar subtests is probably a reliability measure, but is still closer to the region called validity.
>
> Some evaluation of validity can take place even if the two methods are not entirely independent (1959, p. 83).

Such is the case in the example of the use of the multitrait-multimethod matrix provided in Douds (1970). In his example, the "multitraits" involved are four task interdependence factors, and the "multimethods" are responses to two separate instruments. The questionnaires have some items in common, so that they were not completely independent. Nevertheless, their use illustrates the application of the technique and provides some indication of convergent and discriminant validation of the constructs involved.

If we are to develop theories of the RD/I process, theories that are meaningful and ultimately useful in practice, we must pay attention to validating our constructs and our instruments. From the current state of the literature in this regard, any improvement would be welcome.

4. *What Is the Basis for Confidence in the Results?*

In the quest for information about the processes operating in the RD/I process and verification of our beliefs, we find two distinct approaches being taken. The previous section has emphasized an empirically based challenging of concepts and findings through measurement procedures. We do not wish to emphasize it unduly, for there is a long tradition of producing useful results that utilizes intellectual reasoning based on observation, compilations of data, or historical documents. Even those who would take a strictly statistical approach are advised by a statistician:

> We must use all available weapons of attack, face our problems realistically and not retreat to the land of fashionable sterility, learn to sweat over our data with an admixture of judgement and intuitive rumination, and accept the usefulness of particular data even when the level of analysis available for them is markedly below that available for other data in the empirical area (Binder 1964, quoted in Webb et al. 1966).

At one extreme, as a method of analysis for elucidating information from the error contained in our data, is the purely statistical approach; at the other is pure reasoning. Neither extreme is likely to be found in field studies in practice; most published studies employing empirical data tend in one direction or the other. When "reasoning" is the basis for extracting information from the data, the methodological design basis tends to be left implicit in published reports. When some type of statistical analysis is used, the design basis tends to be more explicit. However, the methodology can be explicitly designed in either case.

In the 37 field studies from the *Transactions* 27% used reasoning as a basis for establishing their findings. An analytical design was present in 68% of the cases. The remaining two cases (5%) presented data with an implicit design, but the relationship between the data and what was said was sufficiently unclear so that they could not be classified.

The set of studies from which these tabulations have been derived is presented in Table 1. Each study has been coded for the characteristics we have considered here: presentation of explicit (or implicit) researchable questions; reasonably clear statement of propositions; and the design basis for extracting information from the data collected. The citations for these studies are given in the bibliography at the end of this chapter.

The questions we have suggested as underlying the assessment of field research are:

1. What was it that structured the design of the study?

   - Were explicit researchable questions formulated?

2. Is the nature of the information to be derived and the characteristics of the study design evident?

   - Are explicit hypotheses or propositions given?
   - What were the variables?
   - What indicators of the variables were measured?

3. How does one know what was actually measured and how well?

   - Reliability of the measures?
   - Validity of the measures and constructs?

4. What is the basis for confidence in the results?

   - How was information extracted from the data?

The same questions apply to the evaluation of longitudinal studies and to field experiments.

LONGITUDINAL STUDIES

Longitudinal studies extend through time with some selected focal units--persons, ideas, projects, groups, etc.--as the subjects of measurement. Observations are made at repeated times of inputs to the focal unit or states of the unit, conditions existing in and around the unit, and outputs of the unit. With sufficient frequency of appropriate measurements, the dynamics of the process or processes giving rise to the observed states may be determined. "Observed states" may include outputs of the focal unit (e.g., patents, reports, product items), or they may be states of the focal unit itself (e.g., attitude toward computer terminal use, status of work on an idea, level of productivity).

A longitudinal study could appear to be similar to a case study, but a case study may be developed either retrospectively after all events have transpired, in real time as the events occur, or overlapping both periods. The data for a longitudinal study are primarily collected in "real time" so that the measures taken are not unduly changed by the distortions that are likely with the passage of time.

A case study, roughly speaking, is primarily descriptive and retrospective; it is often a narrative account with supporting data of a series of related events presented as one body, the case itself. A longitudinal study may be less concerned with "the case" (if one is evident) or multiple cases; it focuses on the phenomena. A longitudinal study, as we are using the term, is more like an exploratory or proposition-testing field study in that it seeks to discover or explain a phenomenon by establishing support for relationships among two or more variables. It is subject to the same research design problems--including measurement, reliability, meaningful extraction of information from data, and validity of inferences--which we have previously considered. Its special property is that it makes it possible with proper design to determine that one type of event (or a particular combination) precedes an effect. Often this property allows inferences about cause-and-effect relationships to be made. But when they are made, the nature of the process by which it is done is crucial. The inferences must be evaluated

TABLE 1.--Some methodological characteristics of selected RD/I field studies.

	Research Question	Proposition Statement	Design Basis	
Kaplan 1960	?	N	R	
Marcson 1960a	Y	N	R	
West 1960	N	I	E	
Walton 1960	Y	N	R	
Marcson 1960b	Y	N	R	
Glosky 1960	Y	N	R	
Burns 1961	Y	N	R	
Krohn 1961	Y	Y	E	(1)
Taylor, Smith, Ghiselin 1961	→	→	E	(2)
Storer 1962	Y	Y	E	
Dunlop 1963	→	→	E	(3)
Glaser 1963	N	N	E	
Gordon 1963	Y	→	E	(3)
Pelz 1964	Y	I	E	
Knight 1965	Y	Y	E	(4)
Apstein 1965	Y	N	E	
Rubin, Steadry, Willits 1965	N	Y	E	
Davis 1965	I	N	E	
Eyring 1966	Y	N	R	
Andrews 1967	N	N	E	
Rubin, Seelig 1967	Y	N	E	
Mordka 1967	N	N	R	
Baker, Siegman, Rubenstein 1967	Y	N	E	
Erickson, Gantz, Stephenson 1968	Y	Y	E	
Goodman 1968	N	Y	E	
Harrold 1969	?	N	?	
Rudelius 1969	Y	N	R	
Frischmuth, Allen 1969	Y	Y	E	(4)
Cooper 1971	Y	N	R	
Stephenson, Gantz, Erickson 1971	I	N	E	
Roberts, Wainer 1971	Y	Y	E	
Baker, Siegman, Larson 1971	N	N	E	
Utterback 1971	Y	Y	E	
Holland 1972	N	Y	E	
Wolek 1972	Y	→	?	(5)
Gemmill, Thamhain 1973	N	N	E	
Aram, Javian 1973	N	Y	E	

Y = Yes, present	(1) Generalizations discussed
N = Not present	(2) Develops measure of productivity
I = Implied	(3) Evaluation of a technique
E = Explicit design	(4) Notable/unique design
R = Reasoning/implicit	(5) Method for describing work
? = Uncertain	

with particular care.

In this sense of being able to provide support for inferences about relationships among variables, the design considerations for a longitudinal study are like those of a field study. The key feature distinguishing the longitudinal study from the field study, of course, is that in the field study the state of a number of focal units is determined at one point or one limited period in time, whereas in the longitudinal study the states of the focal units are measured repeatedly. In the field study, the focal units are often selected so that they will be at different stages of maturation, development, etc., because there must be variance in at least one of the variables if one hopes to show an association with another, dependent, variable. The focal units may have arrived at the states measured in the field study in many different ways. One can only hope that the statistical design for selecting the sample (which often cannot be actually realized in practice) is adequate to cancel out this potential source of error (differences in state of the focal unit) in the inferences made from the data. The longitudinal study, on the other hand, contains data that can potentially reduce this source of error, if they are properly utilized.

The data for a longitudinal study are primarily collected in real time--that is, more or less as the events are occurring. This "more or less" involves a difficult, basic methodological choice in the design of longitudinal studies. In general, it is not possible to take measurements continuously in a practical situation involving the behavior of individuals. The researcher would like to obtain measurements whenever "significant" events occur in most types of studies--"significant" events in the sense whatever is involved makes a difference to the outcomes, the dependent variables, being observed. But the time of occurrence of these events may not be at all predictable, and the cost of making repeated measurements in terms of annoyance to the participants (and of the reactive effects of measurements, etc.) may be appreciable. The costs may even be high enough to threaten the continued cooperation of the participants. The costs to the researcher in time and funds may also be appreciable. The trade-off is between the costs of collecting unnecessary data and not getting data that may be important to the results, plus the costs of upsetting or straining the field site relationships. Procedures can be devised to make the trade-offs acceptable. Many techniques have been tried and many are yet to be devised. Their application in any given situation has to be examined with care. As yet there are no established rules to determine when the application of any particular one is appropriate.

An alternative approach is to collect data at regular intervals. If the nature of the phenomena involved is such that changes cannot take place rapidly, then this choice is appropriate, but in studies of behavior this assurance is rarely present. However, in field situations, regularly scheduled data collection by the researcher may be the practical choice. In evaluating the results of a study one must consider the inaccuracies of data with varying times between the occurence of events or changes in the situation and the time when the data are collected. Even if the instrumentation is reliable, the quantity measured may have changed from the value it had at the time it would have been desirable to measure it. Again, there are no specific rules as yet to evaluate the choices in designing the methodology or evaluating the results. Many factors are involved--the nature of the variables, the nature of the phenomena, characteristics of the instruments, characteristics of the participants, etc., and the overall, combinatorial effect of these in the context of the methodology and procedures chosen as well as their individual features taken one at a time.

The time span covered by a longitudinal study may vary considerably. The defining characteristic is that the change in a phenomenon through time is measured. If appreciable change can occur in a few days and enough measurements are made to determine that the change has (or has not) occurred, such a study could be considered to be a longitudinal study. Common usage would seem to be that if a study in organizations is called "longitudinal," time spans of the order of months or years are involved. One of the authors (Rubenstein) is involved in a longitudinal study of the development of the information-seeking styles of potential medical researchers. It has involved collecting data from medical students at four- to six-month intervals over a three-year time span. A major problem in this type of study is the loss of subjects, not only from such factors as movement among schools or unwillingness to continue cooperation, but in this particular case the uncertainty as to how many and which students will enter the focal area of medical research (Rubenstein 1971).

On the other hand, for some phenomena a longitudinal study may involve only two widely separated data points. Again, loss of respondents is a major problem and a potential threat to the validity of the findings. Farris (1969a) attempted to determine to what extent certain factors preceded high technical performance by individual engineers and scientists, and to what extent the levels of these factors were the consequences of prior performance. Among the factors were high involvement in their work, high influence on their work goals, contact with a relatively large number of colleagues, a diversity of work activities, etc. The study involved a replication of Pelz and Andrew's original work (1966) in three of the original laboratories obtaining new information from 151 engineers. The original data were collected in 1959 and the replication was carried out in 1965. Farris tests six hypotheses and reports his surprise in finding that, "in every instance performance was found to be related significantly to subsequent levels of the organizational factor with which it was associated. In no instance was an organizational factor more strongly related to subsequent performance than to previous performance." Farris notes that in our performance-oriented society, we have tended to view performance as the end result of other conditions, not as a *cause* of the conditions, and correspondingly have done little research from this standpoint. However, he does not discuss the potential threats to the validity of his study. He does not mention, for instance, what proportion of the original respondents he had in his second sample. Conceivably, many of those people that left the organizations could have done so for reasons related to the factors under study. The period from 1959 to 1965 was a "sellers' market" for engineers.

Since the findings of this study are surprising and are potentially useful to management, it would appear to be one of the many areas in the RD/I process where it would be worth while to set up longitudinal studies. In this particular case, one including a wider range of data collection methods to help insure that one measures what he thinks he is measuring, somewhat more frequent data collections, and tracking of a sample of the individuals who leave the firms, would be helpful.

FIELD EXPERIMENTS

As of the beginning of 1974, the literature reflects very little interest in carrying out field *experiments* on the RD/I process, despite the great need for more credible evidence on how the process works. The process of experimentation, when well performed, enables one to probe propositions, hypotheses, and theories very deeply, for the essence of experimentation--in the scientific sense--is to minimize and (one would hope) eliminate other possible explanations for the findings. In some sciences it is possible to do so by control of all conditions (temperature, pressure, etc.) so that only the theory being tested can adequately explain the results; rival explanations are gradually eliminated by comparisons among quantities of data collected with well-understood instrumentation by trained observers, often over long periods of time. In the behavioral sciences the threats to validity of findings are so numerous that in general it is not possible for the researcher to control all the factors that may be influencing the results obtained. A similar situation prevails in agronomy and zoology. In such fields meaningful techniques for experimentation have been developed by use of the technique of randomization (with statistical analysis of data) to allow the uncontrolled variables to "average each other out." However, in the social setting it is often not possible to employ random selection of people, groups, or organizations. Even when it is possible to do so, the simple *knowledge* that one has been designated as a member of an experimental or control group may of itself influence the results should it become known.

Nevertheless, it is possible to carry out experiments in social settings. In one sense, managers are constantly "experimenting," in the sense that they are constantly making administrative changes in the design, configuration, and procedures of their organizations to obtain desired end results. They manipulate independent variables to obtain specific outcomes. The experimenter in the tradition of science manipulates independent variables in order to understand their relationship with the dependent (outcome) variables. (Of course, the researcher may desire a particular end result on occasion, and a manager may sometimes be concerned with just understanding.) There are many opportunities for gaining knowledge of how innovation processes, for example, operate by the incorporation of experimental-type designs in the day-to-day actions taken by organizational managers and government administrators.

For this type of situation, a "quasi-experimental" design approach has been described by Donald T. Campbell:

> The phrase "quasi-experimental design" refers to the application of an experimental mode of analysis and interpretation of bodies of data not meeting the full requirements of experimental control....Unplanned conditions and events may also be analyzed in this way where an exogenous variable has such discreteness and abruptness as to make appropriate its consideration as an experimental treatment applied at a specific point in time to a specific population. When properly done, when attention is given to the specific implications of the specific weaknesses of the design in question, quasi-experimental analysis can provide a valuable extension of the experimental method (1968, p. 259).

The crux of his approach lies in the analysis of threats to the validity of the findings:

> Too often a scientist trained in experimental methodology rejects any research in which complete control is lacking. Yet in practice no experiment is perfectly executed, and the practicing scientist overlooks those imperfections that seem to him to offer no plausible rival explanations of the results. In the light of modern philosophies of science, no experiment ever *proves* a theory, it merely *probes* it. Seeming proof results from that condition in which there is no available plausible rival hypothesis to explain the data. The general program of quasi-experimental analysis is to specify and examine those plausible rival explanations of the results that are provided by the uncontrolled variables. A failure to control that does not in fact lend plausibility to a rival interpretation is not regarded as invalidating (p. 259).

Campbell and Stanley (1966) present eight generic dimensions of threats to internal validity and four dimensions of threats to external validity. Internal validity deals with the nature of the findings themselves: Did in fact the experimental treatments make a difference in a specific case? "Internal validity is the basic minimum without which any experiment is uninterpretable" (p. 5). External validity is concerned with the question of the extent to which the findings can be extended to other settings--their generalizability and representativeness. The threats to validity are briefly described in Table 2. Campbell and Stanley provide 12 experimental designs applicable to different types of situations with different patterns of validity threats. Since the *specific* rival hypotheses differ in differing specific situations, one can select a design for a given experiment that has minimal weaknesses.

There are very few administrative or natural experiments in the literature on RD/I. A natural experiment was analyzed by Hill (1970). He shows a deleterious effect arising from the introduction of an autocratic/authoritarian leadership style into an applied science laboratory in Australia. The supporting agency measured the performance of the laboratory primarily by its output of publications. Hill used the annual reports of the laboratory for a 13-year period to obtain measures of publication output, authorship, and staff turnover. The laboratory had been under one director for six years when he then brought in an associate director. The scientists were administratively divided into two groups, one half responsible to the director, the other half to the associate director. As part of a study of group structures in R & D labs, Hill spent several weeks living in the laboratory. During this time he collected data by interview and observation on the leadership styles of the two men in addition to data on the scientists' groups. The natural experiment analysis was conducted following year 13.

The introduction of the associate director with his different style of leadership constituted the experimental treatment. The continuity of research-project areas and men assigned to them enabled Hill to divide the publication and turnover data into two groups for comparison over the entire 13 years. If the scientists had been assigned at random to the two directors, the procedure would have been closer to a classic experimental design. However, since the assignments were based on research areas, the differential factors operating in the two groups are not accounted for either by randomization or control. The design is then Campbell and Stanley's "No. 10: Nonequivalent control group" with time series data (1966).

TABLE 2.--Dimensions of threats to validity.

THREATS TO INTERNAL VALIDITY

History: Other specific events in addition to the experimental variable occuring between a first and second measurement.
Maturation: Processes within the respondents that are a function of the passage of time per se (not specific to the particular events), including growing older, more hungry, more tired, etc.
Testing: The effects of taking a test a first time upon subjects' scores in subsequent testing.
Instrumentation: The effects of changes in the calibration of a measuring instrument or changes in the observers or scorers upon changes in the obtained measurements.
Statistical Regression: Operating where groups of subjects have been selected on the basis of their extreme scores.
Selection: Biases resulting in differential recruitment of respondents for the comparison groups.
Experimental Mortality: The differential loss of respondents from comparison group.
Selection-maturation Interaction: In certain of the quasi-experimental designs, such as the nonequivalent control group design, an interaction of maturation and differential selection is confounded with (that is, might be mistaken for) the effect of the experimental variable.

THREATS TO EXTERNAL VALIDITY

Reactive or Interaction Effect of Testing, in which a pretest might increase or decrease the respondents' sensitivity or responsiveness to the experimental variable and thus make the results obtained for a pretested population unrepresentative of the experimental variable for the unpretested universe from which the experimental respondents were selected.
Reactive Effects of Experimental Arragements, which preclude generalizations about the effect of the experimental variable for persons being exposed to it in nonexperimental settings.
Interaction effects between *selection* bias and the *experimental* variable.
Multiple-treatment Interference, a problem wherever multiple treatments are applied to the same respondents, and a particular problem for one-group designs involving equivalent time samples or equivalent materials samples.

Source: Campbell and Stanley 1966; Campbell 1968.

The director was characterized "as generally democratic in the way he dealt with people, participating to an extent but allowing them freedom in their own decisions on the direction of their research" (Hill 1970, p. 14). The associate director, on the other hand, was characterized as autocratic. He "sought to control the fine detail of his subordinates' research. He chose the division's scientific projects and then dictated what each scientist was to do; he sought to control projects whose research discipline was outside his own professional knowledge and dictated courses of action his subordinates considered entirely incompetent; they would carry out the research he requested and bring him negative results only to be told to vary conditions in a minor way rather than abandon the project" (p. 14).

The time-series data show an almost immediate effect on publication rate and a less pronounced effect after some time delay on turnover. This latter variable tends to be a conservative measure because of the small number of research positions available in Australia; scientists there cannot change jobs as easily as their American colleagues.

Hill advances a number of rival hypotheses to explain these data. For instance, the associate director might not have been a highly competent scientist to begin with, but from the information available it appears he was. In keeping with the spirit of the quasi-experimental method Hill, in examining a variety of alternative explanations, is able to dismiss some but retain others. However, he does not explicitly analyze them in terms of the internal and external threats to validity. His discussion focuses mostly on the question: To what extent do these observed results come from the change in this one variable in this particular case? (i.e., internal validity). The questions relating to external validity--generalizability--are not considered in the same manner. (Of course the social system of Australia and the institutional system of science there are limiting factors of unknown magnitude.)

Proposal solicitations appear to be a useful context in which to establish experimental designs. A given proposal request provides a common problem or set of subproblems that are worked on in parallel by several organizations. The variance arising from different task characteristics is reduced and one significant area for rival explanations is thus minimized.

Allen and Marquis (1964) performed a study that is essentially a natural experiment in which they compared the effects of prior experiences or knowledge of given problem areas to success in solving a set of problems. The problems were six technical components required in proposals submitted by eight companies to two government proposals requests. A total of 26 possible solutions were involved.

The "experimental treatment" was the existence or absence of specific prior experience or knowledge in the technical area. This factor could potentially constitute a "biasing set" to the solutions the companies eventually came up with. The proposal evaluators judged whether or not each proposed solution was successful. Although the design is not such that rival hypotheses are removed, the data clearly indicate that having prior experience does not per se necessarily lead to successful solutions. "Prior experience only with a technique which is not appropriate for the present problem usually results in an unsuccessful solution (negative biasing set). However, if such laboratories consider more than one alternative approach, the probability of success is increased form zero to one half." On the other hand, a positive biasing set can be introduced when the laboratory has had only prior experience with a technique appropriate to the given problem. The effects of experience and knowledge are similar.

Administrative experiments can take a number of forms: the experimenter may in fact be the administrator; the experimenter may be an administrator and a researcher working together; the experimenter may be an outsider working with the permission of an administrator; etc. An experiment of the last form is reported by Rubenstein, Rath, et al. (1973) on behavioral factors influencing the adoption of an experimental information system. The research sites were 12 medical groups (matched by specialty). The treatment involved installing a new state-of-the-art scientific and technical information system, including facsimile and other special communication equipment in six research hospitals. The equipment was directly linked to a research library with a specially trained librarian to handle the physicians/researchers' information requests. The system was designed to be at the forefront of the art, and also to be economically feasible if the hospitals chose to continue its use after the experiment was completed.

A number of factors affecting internal and external validity were considered in the design of the experiment:

> First, because the research was to study group reactions to the introduction of a new information service, unrelated sets of researchers and clinicians were rejected in favor of well-defined work groups. Second, because specific organizational constraints often affect information-seeking behavior, groups were selected from several hospitals. Third, groups were chosen from more than one discipline because an individual's specialty may also influence his information needs and searching habits. Last, because it was a matter of speculation whether an individual's orientation (clinical, research, or supervisory) affects his information-seeking style and thus his reaction to new services, groups were selected in which individuals of all three orientations were included (p. 30).

Ninety-nine individuals in 12 groups, two in each hospital, participated in all phases of the experiment. The first phase of the study consisted of a questionnaire filled out by each participant giving data about his personal background, organizational environment, and present habits of information seeking. The second phase was the introduction of the advanced information retrieval system for seven weeks with all activity on it being monitored in detail. Forty people used the system, submitting a total of 141 requests. Many of these requests were for reprints of anywhere from ten to 30 articles. The last phase, six months after the system was discontinued, was an evaluation by each person of the system and his use of it. The results of the experiment provide useful information on a number of aspects of individuals' patterns of information seeking and some information about the effect of group structure (cohesiveness, presence or absence of information gatekeepers, and their use/nonuse of the system as contrasted to supervisors' use/nonuse of the system) on the groups' use of the system.

Complete randomization was not practical. However, the three orientations of the individuals provide a basis for establishing comparisons to extract information from the data and to reduce some of the threats to validity. All individuals had the opportunity to expose themselves to the experimental treatment (i.e., to make use of the system); some did not, others did to varying extents. Use/nonuse provides another basis for comparison. The behavior measured was "use of the system." Because of the study design, it is not possible to make comparisons between people who had the opportunity to decide whether or not to use the system--those included in the 12 experimental groups-- and control groups of those who continued their normal information-seeking style with no outside influences. To include such a "pure" control group would have required a quite different mode of data collection. However, in testing the hypothesis concerning group structure a basis of comparison is built into the design with those groups serving the function of a "control" that had a low degree of cohesiveness and neither a gatekeeper or a supervisor that were users of the system. In addition, the experimental treatment (the new system) was introduced on an alternating ON-OFF basis, in which one set of groups constituted "controls" while the other set was receiving the treatment. Beyond the first cycle, of course, such an arrangement is subject to a number of threats to validity identified in Table 2.

In terms of Campbell's threats to validity, the following points may be noted. *History* would appear to be fairly well accounted for by including groups from a number of hospitals. However, all were in the same city and so a common event could have occurred (e.g., a professional society meeting in one of the specialties). *Maturation* over seven weeks for mature professionals would not appear to be a significant factor. Since there was no repeated administration of the same questionnaires, *testing* would not appear relevant. *Instrumentation*, however, is more of a problem. Most of the questionnaires were devised for the study, although they were based upon prior work on information-seeking behavior (Trueswell, Rath, and Rubenstein 1968). *Selection* biases are more difficult to assess. However, since the experiment dealt with intact groups, the dominant bias would be the factors involved in the willingness of the individuals and groups to participate. *Experimental mortality* arose because 11 group members declined to participate beyond the first questionnaire. (The total membership of the groups was 110.) Also, high turnover and other factors led to 48% of the initial respondents not completing the final evaluation questionnaire.

Of the four threats to external validity, only the *multiple-treatment interference* does not apply since there was no multiple treatment. *Reaction or interaction effects of testing* arise because of initial phase of the research with its initial negotiations with the groups, questionnaires, and interviews. *Reactive effects of experimental arrangements* are clearly illustrated by the five respondents who said they felt they should use the system just to cooperate in the experiment. Other respondents gave this reason along with other reasons. *Interaction effects* between *selection* bias and the *experimental variable* may be present because of the groups' initial willingness to cooperate. The attitude and importance of information-seeking to these groups as compared to others--other kinds of scientists, engineers, etc.--have been investigated by the Northwestern group (Trueswell et al. 1968; Werner 1969; Kegan 1969; Thompson 1970; Moor 1972), but comparisons have not yet been made. From this cursory evaluation it would appear that the internal validity of the design is reasonable, but that the external validity is limited. However, considerably more information than is presented in the article and a more thorough analysis than we have performed here are needed to make a more detailed assessment. Our purpose here has been only to illustrate a few specific instances of threats to validity.

Administrative experiments may also be carried out by managers within a firm. In an article written for managers, Meyer (1969), then General Electric's Manager of Personnel Research, reports a series of studies he performed with the cooperation of George Litwin of Harvard Business School. In an experiment conducted in a large engineering department they attempted to see if they could "...help change the climate of an organization, as perceived by its employees, from a negative to a positive one." They found that it was possible to create a change towards a more favorable one, although the mechanisms by which it was achieved were somewhat different from what they had anticipated. The study is not presented in enough detail so that one can analyze it fully from a "truth-seeking" viewpoint, but it does illustrate the following points to conclude our discussion of administrative experimentation:

1) Perhaps most important administrative experiments can be performed in operating organizations.

2) Useful results can be obtained. Meyer notes that the positive results were encouraging and that further experiments were planned.

3) Administrative experiments can be facilitated by use of an outside consultant--in this case a university-based researcher.

4) Communication of the results of the experiment to others so that society can benefit, either from the utility of the experimental treatment or the contribution to understanding of organizational functioning, can be a problem. Results may be reported inadequately or not at all because managers are not used to such writing; because the company considers part of the data embarrassing or the results to give them competitive advantage; because the consultant considers the techniques proprietary; and for other reasons.

If more emphasis is placed upon administrative experimentation and interest continues to develop, as it seems to be doing, we can look forward to the problem of communication of results diminishing and to more reasonably designed experimental tests subjecting our commonly accepted managerial beliefs and theories, as well as the controversial ones, to the relative rigor of verification in practice.

NOTES AND REFERENCES

1. There are other types of surveys, intended for other purposes, that do not necessarily suffer from the limitation described here.

2. "When the predictions available at the start of a development program are used as the basis of heavy commitments to a highly specified design, then in the fortunate cases when the predictions turn out to be correct, time and money may have been saved compared with the postponing of such commitments until later in development. If the predictions are seriously wrong, however, costly revisions will be required. The initial uncertainties of development are such that the gains due to heavy initial commitments in the fortunate cases are outweighed by the costly revisions of the unfortunate cases" (p. 50).

3. Examination of articles relevant to the RD/I process in other publications indicates that the pattern above exists elsewhere. Figures were not tabulated for other publications in part because of differences in editorial policies between publications, and of differences in policies through time in a given publication, and because different aspects of the same research project may be presented in more than one publication.

4. Portions of this section are adapted from Douds (1970).

BIBLIOGRAPHY OF FIELD RESEARCH ON RD/I

This bibliography contains much, but not all, of the published research on RD/I. No working papers, theses, or dissertations are included.

ABBREVIATIONS. The following abbreviations are used for the most frequently occurring journals.

ASQ = Administrative Science Quarterly
Trans. IRE EM- = Institute of Radio Engineers Transactions on Engineering Management
Trans. IEEE EM- = Institute of Electrical and Electronics Engineers Transactions on Engineering Management (IRE became IEEE with vol. 10 in 1963)

Allen, T.J. Studies of the problem-solving process in engineering design. *Trans. IEEE* EM-13: 72-83, 1966.

Allen, T.J., and D.G. Marquis. Positive and negative biasing sets: The effects of prior experience on research performance. *Trans. IEEE* EM-11: 158-162, 1964.

Andrews, F.M. Creative ability: The laboratory environment and scientific performance. *Trans. IEEE* EM-14: 76-83, 1967.

Apstein, Maurice. Effectivenes of military laboratories as a function of contract activity. *Trans. IEEE* EM-12: 44-50, 1965.

Aram, J.D., and Setrak Javian. Correlates of success on customer-initiated R & D projects. *Trans. IEEE* EM-20: 108-113, 1973.

Argyris, Chris. *Integrating the Individual and the Organization*. New York: Wiley, 1964.

Argyris, Chris. *Organization and Innovation*. Homewood, Ill.: Dorsey Press, 1965.

Argyris, Chris. On the effectiveness of research and development organizations. *American Scientist* 56: 344-355, 1968.

Arthur D. Little, Inc., and Industrial Research Institute, Inc. Getting Over the Barriers to Innovation: Public Policy Options. Cambridge, Mass., 1973.

Avery, R.W., Enculturation in industrial research, *Trans. IRE* EM-7: 20-24, 1960.

Baker, N.R., J. Siegman, and J. Larson. The relationship between certain characteristics of industrial research proposals and their subsequent disposition. *Trans. IEEE* EM-18: 118-123, 1971.

Baker, N.R., J. Siegman, and A.H. Rubenstein. The effects of perceived needs and means on the generation of ideas for industrial research and development projects. *Trans. IEEE* EM-14: 156-163, 1967.

Barnes, L.B. *Organizational Systems and Engineering Groups*. Boston: Harvard University (School of Business Administration), 1960.

Battelle-Columbus Laboratories. Science, Technology, and Innovation. Columbus, Ohio. Prepared for the National Science Foundation/NSF-c667, February 1973.

Bruno, A.V. New product decision making in high technology firms. *Research Management*, September 1973.

Burns, Tom. Research, development, and production: Problems of conflict and cooperation. *Trans. IRE* EM-8: 15-24, 1961.

Carter, C.F., and B.R. Williams. The characteristics of technically progressive firms. *J. Industrial Econ.* 7: 87-104, 1959.

Cooper, A.C. Spin-offs and technical entrepreneurship. *Trans. IEEE* EM-18: 2-7, 1971.

Davis, Keith. Mutuality in understanding of the program manager's management role. *Trans. IEEE* EM-12: 117-123, 1965.

Dunlop, R.A. Developing an experience register for engineers. *Trans. IEEE* EM-10: 6-15, 1963.

Erickson, C.E., B.S. Gantz, and R.W. Stephenson. Development of the job description form: Its use in the functional categorization of professional jobs. *Trans IEEE* EM-15: 119-129, 1968.

Evan, W.M., and Guy Black. Innovations in business organizations: Some factors associated with success or failure of staff proposals. *Journal of Business* 40: 519-530, 1967.

Eyring, H.B. Some sources of uncertainty and their consequences in engineering design projects. *Trans. IEEE* EM-13: 167-181, 1966.

Farris, G.F. Some antecedents and consequences of scientific performance. *Trans. IEEE* EM-16: 6-16, 1969.

Frischmuth, C.S., and T.J. Allen. A model for the description and evaluation of technical problem solving. *Trans. IEEE* EM-16: 58-64, 1969.

Gemmill, G.R., and H.J. Thamhain. The effectiveness of different power styles of project managers in gaining project support. *Trans. IEEE* EM-20: 38-44, 1973.

Glaser, B.G. The impact of diffeential promotion systems on careers. *Trans. IEEE* EM-10: 21-25, 1963.

Gloskey, C.R. Research on a research department: An analysis of economic decisions on projects. *Trans. IRE* EM-7: 166-173, 1960.

Goodman, R.A. Organization and manpower utilization in research and development. *Trans. IEEE* EM-15: 198-204, 1968.

Gordon, Gerald. The problem of assessing scientific accomplishment: A potential solution. *Trans. IEEE* EM-10: 192-197, 1963.

Haberstroh, C.J., J.A. Baring, and W.C. Mudgett. Organizing for product innovation. *Trans. IEEE* EM-15: 20-27, 1968.

Hall, D.T., and E.E. Lawler. Job characteristics and pressures and the organizational integration of professionals. *ASQ* 15: 271-281, 1970.

Harrold, R.W. An evaluation of measurable characteristics within Army laboratories. *Trans. IEEE* EM-16: 16-23, 1969.

Hertz, D.B., and A.H. Rubenstein. Personnel practices and communication in 41 industrial research laboratories. *Personnel* 28: 247-250, 1951.

Hill, S.C. A natural experiment on the influence of leadership behavior patterns on scientific productivity. *Trans. IEEE* EM-17: 10-20, 1970.

Hlavacek, J.D., and V.A. Thompson. Bureaucrac and new product innovation. *Academy of Management J.* 16: 361-372, 1973.

Holland, W.E. Characteristics of individuals with high information potential in government research and development organizations. *Trans. IEEE* EM-19: 38-44, 1972.

House, R.J., A.C. Filley, and Steven Kerr. Relation of leader consideration and initiating structure to R and D subordinates' satisfaction. *ASQ* 16: 19-30, 1971.

Hower, R.N., and C.D. Orth. *Managers and Scientists*. Boston: Harvard University (School of Business Administration), 1963.

Howton, F.W. Work assignment and interpersonal relations in a research organization: Some participant observations. *ASQ* 7: 502-521, 1962.

IIT. Technology in Retrospect and Critical Events in Science (TRACES). Chicago: Illinois Institute of Technology (Contract NSF-c535), December 1968.

Kaplan, Norman. Some organizational factors affecting creativity. *Trans. IRE* EM-7: 24-30, 1960.

Knight, K.E. Some general organizational factors that influence innovative behavior. *Trans. IEEE* EM-12: 2-9, 1965.

Kornhauser, William. *Scientists in Industry: Conflict and Accommodation*. Berkeley: University of California Press, 1962.

Krohn. R.G. The institutional location of the scientist and his scientific values. *Trans. IRE* EM-8: 133-139, 1961.

Kruytbosch, C.E. Management stules and social structure in 'identical' engineering groups. *Trans. IEEE* EM-19: 92-102, 1972.

Lawrence, P.R., and J.W. Lorsch. Differentiation and integration in complex organizations. *ASQ* 12: 1-47, 1967.

Litvak, I.A., and C.J. Maule. Some characteristics of successful technical entrepreneurs in Canada. *Trans. IEEE* EM-20: 62-68, 1973.

MAB (Materials Advisory Board). Report of the ad hoc committee on principles of research-engineering interaction. Washington, D.C.: NAS-NRC, July 1966. (AD 636529.)

Marcson, Simon. Role concept of engineering managers. *Trans. IRE* EM-7: 3-34, 1960.

Marcson, Simon. Role adaptation of scientists in industrial research. *Trans. IRE* EM-7: 159-166, 1960.

Marquis, D.G., and T.A. Allen. Communication patterns in applied technology. *Am. Psychologist* 21: 1052-1060, 1966.

Marschak, T.A. Strategy and organization in a system development project. In R.R. Nelson, Ed., *The Rate and Direction of Inventive Activity*. Princeton, N.J.: Princeton University Press, 1962.

Marschak, Thomas, T.K. Glennan Jr., and Robert Sumers. *Strategy for R & D*. New York: Springer-Verlag, 1967.

Maw, I.L., and A. Addison. Attitudes of research and development management toward management training. *Trans. IEEE* EM-12: 134-138, 1965.

Millen, R.N. An industrial dynamics simulation of the process control/business control interface of a large firm. *Trans. IEEE* EM-19: 118-124, 1972.

Mordka, I. A comparison of a research and development laboratory's organization structures. *Trans. IEEE* EM-14: 170-177, 1967.

Moskowitz, Herbert. R & D managers' choices of development policies in simulated R & D environments. *Trans. IEEE* EM-19: 22-30, 1972.

Myers, Sumner, and D.G. Marquis. *Successful Commercial Innovations*, NSF 69-71. Washington, D.C.: National Science Foundation, 1969.

Nelson, R.R. The link between science and invention: The case of the transistor. In R.R. Nelson, Ed., *The Rate and Direction of Inventive Activity*. Princeton, N.J.: Princeton University Press, 1962.

Normann, Richard. Organizational innovativeness: Product variation and reorientation. *ASQ*, June 1971.

Ozanne, U.B., and G.A. Churchill. Five dimensions of the industrial adoption process. *J. Marketing Research* 8: 322-328, 1971.

Pelz, D.C. Some social factors related to performance in a research organization. *ASQ* 1: 310-325, 1956.

Pelz, D.C. Interaction and attitudes between scientists and the auxiliary staff: I. Viewpoint of staff. *ASQ* 4: 321-356, 1959.

Pelz, D.C. Interaction and attitudes between scientists and the auxiliary staff: II. Viewpoint of scientists. *ASQ* 41: 410-425, 1960.

Pelz, D.C., and F.M. Andrews. *Scientists in Organizations*. New York: Wiley, 1966.

Rankin, A.C. The administrative processes of contract and grant research. *ASQ* 1: 257-294, 1956.

Roberts, E.B., and H.A. Wainer. Some characteristics of technical entrepreneurs. *Trans. IEEE* EM-18: 100-109, 1971.

Roberts, E.B., and J.B. Sloat. Effects of incentive contracts in research and development: A preliminary research report. *Trans. IEEE* EM-13: 181-187, 1966.

Rogers, E.M., and F.F. Shoemaker. *Communication of Innovations: A Cross-cultural approach*. New York: Free Press, 1971.

Rosenbloom, R.S., and F.W. Wolek. *Technology and Information Transfer: A Survey of Practice in Industrial Organizations*. Boston: Harvard University (School of Business Administration), 1970.

Rothman, R.A., and Robert Perruci. Organizational careers and professional expertise. *ASQ* 15: 282-294, 1970.

Rubenstein, A.H., G.J. Rath, R.D. O'Keefe, J.A. Kernaghan, E.A. Moore, W.C. Moor, and D.J. Werner. Behavioral factors influencing the adoption of an experimental information system. *Hospital Administration*, Fall 1973, pp. 27-43.

Rubin, I.M., and W. Seelig. Experience as a factor in the selection and performance of project managers. *Trans. IEEE* EM-14: 131-135, 1967.

Rubin, I.M., A.C. Stedry, and R.D. Willits. Influences related to time allocation of R & D supervisors. *Trans. IEEE* EM-12: 70-79, 1965.

Rudelius, W. Methods of analyzing the impact of program stretch-outs. *Trans. IEEE* EM-16: 23-25, 1969.

Science Policy Research Unit, University of Sussex. Success and Failure in Industrial Innovation: Report on Project SAPPHO. London: Center for the Study of Industrial Innovation, 1972.

Shepard, H.A. The value system of a university research group. *Am. Sociological Rev.* 19: 456-462, 1954.

Shepard, Clovis, and Paula Brown. Status, prestige, and esteem in a research organization. *ASQ* 1: 340-360, 1956.

Smith, C.G. Consultation and decision processes in a research and development laboratory. *ASQ* 15: 203-215, 1970.

Smith, C.G. Scientific performance and the composition of research teams. *ASQ* 16:

486-496, 1971.

Steade, R.D. A study of the interface between research and development. *Trans. IEEE* EM-13: 63-67, 1966.

Steiner, G.A., and J.O. Vance. Managing the commercial-technical enterprise. *Trans. IEEE* EM-18: 109-114, 1971.

Stephenson, R.W., B.S. Gantz, and C.E. Erickson. Development of organizational climate inventories for use in R & D organizations. *Trans. IEEE* EM-18: 38-50, 1971.

Storer, N.W. Research orientations and attitudes toward teamwork. *Trans. IRE* EM-9: 29-33, 1962.

Taylor, C.W., W.R. Smith, and B. Ghiselin. A study of the multiple contributions of scientists at one research organization. *Trans. IRE* EM-8: 194-201, 1961.

Utterback, J.M. The process of technological innovation within the firm. *Academy of Management J.* 14: 75-88, 1971.

Utterback, J.M. The process of innovation: A study of the origination and development of ideas for new scientific instruments. *Trans. IEEE* EM-18: 124-131, 1971.

Vincent, H.F., and Abbas Mirakhor. Relationship between productivity, satisfaction, ability, age, and salary in a military R & D organization. *Trans. IEEE* EM-19: 45-53, 1972.

Walton, Eugene. What is the role of the government laboratory? A questionnaire study in one government laboratory. *Trans. IRE* EM-7: 114-117, 1960.

Walton, Eugene. Gauging organizational health: A questionnaire study in a government laboratory. *Trans. IRE* EM-8: 201-205, 1961.

Weiss, Robert. *Processes of Organization*. Ann Arbor: University of Michigan (Institute for Social Research), 1956.

West, S.S. The ideology of academic scientists. *Trans. IRE* EM-7: 54-62, 1960.

Wolek, F.W. Engineering roles in development projects. *Trans. IEEE* EM-19: 53-60, 1972.

11. ASSESSING THE KNOWLEDGE OF INNOVATIONS IN NEGLECTED SECTORS: THE CASE OF RESIDENTIAL CONSTRUCTION
W. Paul Strassmann

The sectors of the economy that have been relatively neglected by analysts of technological change are not necessarily the sectors that have had the fewest technological advances. The neglected sectors may be those where change has been difficult to codify and to measure, and therefore to analyze, at an abstract level. They are sectors where advances are qualitative, indirect, and somewhat intangible, so that it is difficult to aggregate and even to compare one case with another. Indeed, studies of these sectors are likely to be case studies or collections of cases, often chronological, with few abstract and sophisticated hypotheses under test.

Analysis of technological change is most sophisticated where the individual discoveries or adaptations have common objective dimensions other than those of the innovators or adopters. One such dimension is the relation between physical output in pounds, gallons, calories, kilowatts, etc., and inputs in similar units. Another dimension puts the changing input-output ratio in terms of money, net costs, and revenues. These terms are vulnerable to extraneous economic trends, but still manageable. Manufacturing, agriculture, mining, transportation, and power generation are all sectors with technological changes that raise productivity in such comparable, measurable ways. Consequently, technological change in these sectors, especially manufacturing, has been studied with ever increasing sophistication, to such an extent that one must remind oneself that technological change is not synonymous with advances in manufacturing.

The true output from manufacturing should not, of course, be seen as a flow of objects with pricetags, but as the flow of services from the accumulating stock of such objects. The pragmatic turn of mind that rejects this proviso is the same that steers researchers away from spending time on almost untestable theories about innovations in sectors that directly produce services with these qualitative, somewhat intangible characteristics. However, the problems and opportunities for society and inventors may be in those sectors even if they do not pay off conveniently for general technological analysts.

Thus, in the last quarter of the 20th Century, we need better medical care, better systems for paying and collecting bills, better postal services, better garbage collection, better education, and other improved service activities that have little in common with the physical production of mining, agriculture, and manufacturing. The dominant share of our productive activities, whether measured in man-hours or as GNP, is now devoted to these sectors. There would be no harm in understanding the innovation process here a little better. A problem with studying technological change in these neglected sectors is that they are poorly understood in their nontechnological aspects as well. Their more qualitative, less physical output makes them difficult to analyze in <u>all</u> respects.

As an illustration, consider the sector to which I have devoted some years: residential construction. This sector is a kind of halfway house. It does produce physical objects, and these objects even have a crude common denominator: floorspace. At the same time, no one loses sight of the flow of services in the form of shelter and amenities that dwellings are supposed to yield over an extended period of time, and

---

Dr. Strassmann is a professor of economics at Michigan State University in East Lansing.

everyone knows that these services are interrelated with trends in the urban setting.

TECHNOLOGY AND THE GENERAL UNDERSTANDING OF THE HOUSING SECTOR

The role of housing expenditures in the consumer's budget and the role of construction in the economy as a whole are both poorly understood. We do not know exactly what determines the volume of expenditures, nor how expenditures are related to net social benefits. We do not know how far actual benefits diverge from a feasible optimum using the existing technology.

The sector is poorly understood compared with others, mainly because of its complexity. Consumers may not only be renters buying a flow of housing services; they may also be investors who regard the dwelling as a reserve of savings. Durability causes stock-flow analytical problems. Production must always be appraised in terms of its effect on existing dwellings and their changing uses. Some analogies with the economics of machinery and transportation equipment may apply, but the immobile attachment of a building to a site raises a host of unique problems. For one thing, if the commodity is immobile, then its producers and consumers must go to it, instead of the reverse. Moreover, the product cannot be appraised separately from questions of land use, public utilities, and urban design. In addition to complexity, both instability and shifts from site to site have rendered data collection difficult and discouraged research. When behavior in a sector as a whole is largely unexplained, a single facet like technological innovation is likely to remain undeciphered.

The results of both these analytical failures are pervasive housing problems throughout the world. In developing countries the urban population is projected to rise by 1 billion from 1960 to 2000, or four times the rise from 1920 to 1960. The rate of building minimally acceptable dwellings is less than one-third of what it should be. As a result shantytowns without water, sewer systems, etc., are doubling in size every 4-6 years and will soon shelter most of the population. Cities like Calcutta are said to be threatened with collapse--a state that some observers say would be hard to distinguish from the present state.

High rents, overcrowding, lengthy waiting periods, and numerous inconveniences keep housing close to the top of complaints in developed countries. In Tokyo in 1970 some 813 000 households lived in one-room apartments. Blocks and blocks of boarded-up housing, new and old, are typical of large American cities. But the average Russian family is said to have only one-fourth of the space of the American. Even some private construction and ownership is encouraged in the Soviet Union to help reduce shortages. Housing is the sector where private capital persists the longest and reappears first in planned economies, and it is the sector where planning is accepted first in market economies. In this sector, remedies lag and mistakes fester in the form of ugliness, pollution, disease, and psychoses. As Churchill said, "We shape our buildings; thereafter they shape us."

Not all these urban problems have a technological facet; and when they do not, they simply cannot be solved with an invention, a material, or a design. Where a material change can be part of the solution for a more general problem, the material innovation must be contrived in terms of prevailing tastes and laws and the process for changing them. Otherwise, we sink millions into Pruitt-Igoes that must be detonated. Or we attempt Operations Breakthrough that break down. Families in the 1970s detest "industrialized modular housing" as much as families did "prefabricated housing" in the 1940s. Following the trend, some universities discontinued their "industrialized housing" courses.

Instead of technological faith in specialized experts, we hear of rebellion against dependence "on contractor-oriented bureaucratic systems...a breaking wave of reaction against authoritarian solutions to technocratically posed problems."[1] People are encouraged to lay their own concrete blocks, row on row. Of course, it is easier to crowd another mobile home in the slot between two others. In a decade or so, salvaging abandoned mobile homes may be a regular industry. For a while some architects favored homes with disposable, modular parts or "cartridges" that could be plugged in and out. As Julius Stulman put it, in *Creative Systems in Housing* (World Institute Council, 1971), there will be a "flow of precision, factory-built houses and housing complexes whose cartridges can be replaced periodically with listed trade-in values, incorporating new developments of science and materials." Fortunately, environmentalism seems to have kept that idea from getting on the ground.

A popular diagnosis of inefficiency and failure in building is that the industry was bypassed by the Industrial Revolution. Although false, the diagnosis has better durability than some houses. Carpenters and bricklayers do not make houses in the same way as they did in the 18th Century. Building codes, restrictive trade unions, and dull, portfolio-guarding bankers to inhibit change in the form of a super cookiecutter that could plunk down a condominium every thirty seconds. There has been a steady series of innovations that let us build expensive but inadequate housing more efficiently. Ingenious fixtures and components can be assembled in ways that, although basically conventional, would have astonished us a generation ago. For example, in the Lansing, Mich., area, Hugh Spall found that the most significant innovations in single-family housing were roof trusses, thin-coat plaster, oversized bricks, power nailers, finger-jointed wood, prefinished mouldings, prehung doors, and critical-path scheduling.[2] In European countries that took to industrialized panel building in a major way, adding innovation after innovation, the finding seems to be that innovations in conventional building occurs at a rate that keeps the productivity of the two approaches within 10 per cent of one another.

There are better explanations for the inadequate amount and direction of technological change in dwelling construction. One is what Marion Bowley has called "the system." This "system" is not a way of building at all but a set of barriers among materials producers, designers, builders, owners, and, in the case of Britain, quantity surveyors. The Industrial Revolution had transformed building with innovations that too often caused fraud and catastrophe. Separation of powers with legal checks and balances restored predictability and safety but paralyzed many types of innovation. Architects no longer designed building in terms of marginal savings in construction methods or materials production but merely became experts in the art of arranging space and appearances.

A DIGRESSION ON THE INDUSTRIAL REVOLUTION

One problem, therefore, is defining technological change in building methods too narrowly, as a way of fabricating a product without considering the urban context and complexities of habitation. The other "knowledge gap" is a simplistic or false view of the historical sequence of events. A simplistic explanation of the way we arrived at the present can inspire simplistic and futile plans for the future. What follows is a summary sketch of the interrelation between technological and organizational development in the Industrial Revolution. Thorough research will be needed to substantiate, to modify, or to refute this suggested interpretation.

The British Industrial Revolution after 1750 inspired some technological changes based on advances in knowledge and others consisting of empirical factor substitutions. As usual, the factor substitutions depended on relative price changes of materials and labor, and among materials. Customers for 18th Century buildings were willing to save money with substitutes of cheap materials and cheap methods. Molded clay, or terracotta, especially in the form of Coade Stone after 1769, was a substitute for repetitive stone oranments carved at high cost. After 1815, with improved iron making and less military demand, cast-iron ornaments replaced terracotta. Stucco was a cheap substitute for dressed stone. Colored gypsum plaster imitated marble, and John Nash even painted cast-iron columns to make them look like stone.

If traditionalism could not withstand economic substitutions in ornamentation, it was unlikely to stop less visible changes, particularly in buildings used for production. English cotton mills began to use cast-iron columns for interior supports in the 1790s. In 1801, for the Salford Mill, Boulton and Watt made complete interior frames with a forerunner of the I-beam, the inverted T-beam. By 1830 such frames were common in both Britain and France. Cast glass was introduced to Britain from France in 1776, and in 1847 the firm of Hartley made the unremarkable but popular innovation of leaving glass unpolished for roofs of factories and railway stations. All these changes were strictly empirical.

New knowledge, especially a theory of structures, was mainly developed in France and often inspired by problems of domes and bridges. Builders of Roman baths and medieval cathedrals did not know much about stresses, strains, and the strength of materials. However, the Greek or Gothic appearance of mid-19th Century buildings does not mean that building techniques were unchanged. Embedded behind the cast-iron facade were great advances in knowledge.

Much of this knowledge was first developed in the effort to build better bridges. Urbanization and larger volumes of trade made the old bridges inadequate, extravagant, or dangerous. Gautier's early 18th Century *Traité des ponts* still cited tradition as sufficient justification for the proportion between arches and supporting piers, though he urged *les savants* to study the problem scientifically. By 1729 Bernard Forest de Bélidor's *La science des ingénieurs* supported traditional methods with a simple theory of statics. Robert Hooke in England had shown that the distortion in any material is proportional to the force producing it, but this principle did not acquire importance until much later. By the late 1700s the discipline of structural mechanics had been created out of these beginnings, complete with mathematical expressions for critical loads, the static equilibrium of shearing forces, and the elastic behavior of materials.[3] In 1770 Emiland-Marie Gauthey designed a materials testing machine to help apply these theories to the reconstruction of the dome of Sainte Genevieve in Paris, later the Panthéon Français. Thus knowledge, developed in bridge building and at the Ecole des Ponts et Chaussées, began to spread to other types of structures. As in many industries, major theoretical and experimental advances, together with economic changes, accelerated around the mid-18th Century.

However, these theories and tests improved and rationalized types of construction after they were already being built. Theories did not come first. This sequence of theory after practice was repeated for iron construction around 1830 and for reinforced concrete around 1875. New materials with unfamiliar properties came into use and led to structural theories that meant greater safety and economy. In 1826-1830 Eaton Hodgkinson and William Fairbairn carried on experiments and developed theories about inverted T-beams.[4] But by this time internal structural cast iron frames were already fairly common in Britain.

Science followed invention faster in the case of concrete reinforcement. Inventions really have no precise beginning, but if we take 1867, the date of Joseph Monier's main patent, as the beginning of reinforced concrete, then only ten years elapsed until the publication of Thaddeus Hyatt's report on the relation of bar reinforcement to location and loads of structural parts.

Changes in technique affected not only large public structures. The technology of ordinary housing was also transformed. The two-story wooden and brick houses of the past century may look like 18th Century houses, but they result from a production process thoroughly overhauled. Cheaper iron for tools, especially for saws, as well as for nails and bolts, made possible a large number of changes in carpentry. More efficient mechanical sawing greatly reduced the cost of thin boards. Sawn softwood took the place of axe-dressed hardwood. The major invention of balloon framing could follow as a result, and is generally attributed to Augustine D. Taylor of Chicago in 1833. This most common type of wooden American house therefore could not even exist in the 18th Century. Hammering was not transformed, but what was being hammered was different.

In a similar sense, bricklaying was not transformed, but brickmaking was. In the early 19th Century, clamp-burning of bricks took the place of more primitive kilns. In the 1830s came pugmills for grinding the clay and mechanical wire cutting of extruded clay into bricks. Experiments were conducted with pressed and molded bricks and various forms of drying. By 1856 more than 230 patents had been granted for brickmaking in Britain.[5] It may be suggested that brickmaking is not really a part of the building industry. It is not. But it was. In the 18th Century, bricklayers often made their own bricks.[6] Vertical disintegration of building from material making was one of the consequences of the Industrial Revolution. As we shall see in the next section, these changes in organization, associated with innovation, first led to progress, then stagnation.

CHANGES IN ORGANIZATION

Designers, constructors, and materials producers were not yet separate and insulated occupations rigorously prevented from much coordination around 1800. Nor had restrictive building codes and trade unions come into existence. As a result experimenting was easy, and major and minor catastrophes followed--usually at the expense of others, often insurance companies. Failures naturally led to further research, experiments, and theories. Resulting knowledge was often converted into standardized tables of correct practice for a kind of self-policing. Building codes are these formulations turned into laws, often written under pressure from insurance companies.

The dates of these steps vary with the sector of the industrial. For concrete in the United States, for example, the self-policing stage did not come until the foundation of the American Society for Testing Materials (1898), the American Concrete Institute (1905), and the Portland Cement Association (1916). Britain had no rules governing the use of reinforced concrete until 1913. American codes acquired their ultraconservative characteristics after the collapse of an eight-story hotel at Benton Harbor, Mich., in the course of construction in 1924.

Restrictive trade-union practices were unimportant as the Industrial Revolution gathered momentum. The Webbs observed that,

> during the eighteenth century there is, as we shall see, no lack of information as to combinations of workmen in nearly every other skilled trade....But of combinations in the building trades we have found scarcely a trace until the end of the century.[7]

Even the rule of seven years' apprenticeship was being abandoned. Seven years had been excessive, anyway. Speaking of bricklayers, Adam Smith gave up his normal caution and said flatly that he considered "no species of labor more easy to learn."[8] Since speculative builders could lower standards of all types, in the words of Postgate, "the trade was soon flooded with cheap and half-skilled labor which was content with rates prevailing in other trades."[9] In 1815 Thomas Cubitt started to employ his craftsmen on an ordinary-time basis to work on whatever jobs he had for them.[10] Labor was less involved in production decisions during this period than either earlier or later.

Inflexibility due ot lack of coordination among specialized building professions ("the system" according to Marion Bowley) was not yet a problem during the Industrial Revolution. Such specialization was only beginning in those years and was itself a product of the new methods and the greatly increased scale of building. In the 18th Century a building expert like Batty Langley could still label himself as "carpenter and architect," but the trend was against designs by craftsmen.[11] After Andrea Palladio's work was translated into English in 1715, the view spread that unschooled craftsmen could not attain the scholarly and esthetic level of true architects.[12] When four architects were included in the Royal Academy in 1768, one could say that the profession had at last been officially recognized in Britain. However, an architects' club was not formed until 1791. This club became the British Institute of Architects in 1834 and the Royal Institute of British Architects (RIBA) in 1866.

The architects' efforts to keep others out of designing did not mean simultaneous and voluntary withdrawal by architects from organizing sitework. This withdrawal came about with the rise of general contractors, a rise that has been described by N. H. Port.[13] In the late 18th Century, building by measure had replaced the ancient system of material purchases by the owner and paying workers or master tradesmen piece rates or daily wages. By 1784 such direct labor payments were unusual.[14] Building by measure, the new system, meant that master craftsmen worked for the owner or architect and added a percentage (usually 15 per cent) for profit--a cost-plus system. General contracting then took its place in the first third of the 19th Century. But not until 1887 did the RIBA prohibit architects to take profit-making positions in building. However, the trend had been in this direction for some decades, and the prohibition merely formalized accepted standards.

Speculative builders, of course, used their own designs. These builders were rejected as legitimate designers by architects and were also not allowed in the official organization of general contractors. It is possible, however, that theirs was the largest volume of business. In France the Corps des Ingénieurs des Ponts et Chaussées had developed a system of general contracting in the 18th Century.

Meanwhile a special profession of measurers also emerged in England. During the mid-18th Century the builder employed one measurer for drawing up his bills and the owner employed another to check on the first. These measurers were at first paid by the day and, beginning in the 1760s, by commission. Since there was no agreement on either costing or physical measuring, the owner's measurer could easily come to a lower figure than the builder's measurer. As a result, lawsuits were common, and the profession of measurer was despised as unproductive and dishonest. This system of paying by measure gave way to general contracting in the 1820s, but the hated measurers survived in England by becoming an independent profession of quantity surveyors. Before construction began, the surveyor would draw up lists of quantities on which builders could bid, each using his own price estimates. A Royal Institute of Chartered Surveyors was founded in 1868, and by 1907 its members were barred from employment by building firms. Architects had been barred from measuring quantities as early as 1834.

We had, therefore, in the Industrial Revolution, not only a process of technological advance and occupational specialization in building, but also one of erecting barriers between specialities. These barriers were not only monopolistic conspiracies but nominally guarantees of competence, impartiality, and honesty. Both the specialization and the barriers were needed because both ignorance and fraud were problems. With more complex buildings for unprecedented uses, one person's knowledge was not enough for efficiency at design, sitework, and materials production. The barriers among specialities were a system of frontiers that could be policed to reduce fraud in buying materials, in hiring and supervising workers, and in other aspects of designing and contracting.

But specialization with permanent barriers implies a static situation. It implies that coordination is no problem, that consultation is not needed as a continuing activity that the interrelations among the sectors of the building process are given, not changing. Only if these assumptions are true, can barriers be consistent with maximum efficiency. If the assumptions happen to be false, then acting on them can make a potentially dynamic sector technologically rigid.

By the mid-19th Century given types of buildings were no doubt safer and built at lower cost. But this achievement came at the price of a complex and rigid system of organization that had only began to loosen up. By 1900 architects no longer knew with accuracy what alternative designs would cost, and they had little stake in keeping costs as low as possible. Hence innovations in techniques tended to come less often, and if so, from engineers, speculative builders, or manufacturers not bound by "the system." That "system," however, was a response to the dramatic but uncertain effects on building of the Industrial Revolution.

RECENT TRENDS

Breakdown of the "system" is under way in many countries, unfortunately with a considerable revival of fraud. Performance standards in building codes and government technical institutes as monitors of the building process (e.g., the French *agrément* system) are progressive substitutes for the inhibiting approach of the system.

Government-sponsored building research in industrialized countries began in the first quarter of this century. One might therefore suppose that a study of technological change could center on the work of these public building research institutes, that their annual reports, documents, and in-house histories would almost be a ready-made account of achievements and failures, together with explanations. In the United States, however, that is not the case. Public research was allowed as a support for private enterprise and as a safeguard for the public, but not as a progenitor of innovations. To be safe from private and Congressional attack, the Building Research Division of the National Bureau Standards must insist that it "does not develop new products, materials, or systems for commercial use, nor establish the performance limits that determine their acceptability for particular applications."[15]

The National Bureau of Standards began testing building materials shortly after its establishment in 1901. These activities were expanded and broadened with the creation of a special Division of Building and Housing in 1921. The Division was supposed to encourage simplification and standardization, to promote better state and municipal building codes, and to coordinate scientific, technical, and economic building research. The aims were mostly wishful preambling. After being renamed "Building Technology Division" in 1947, its research did become more profound, and invention and development were allowed "to serve the special needs of the Government."

The Building Research Division, as it was recently named, is allowed to "study new technical processes and methods of fabrication of materials in which the Government has a special interest," as well as basic and applied research, "including demonstration of the results of the Bureau's work by exhibits or otherwise as may be deemed most effective." It "seeks to promote economy and effectiveness in the design, construction, and use of buildings and in the manufacture of building materials and systems." Nevertheless, the fiction has been maintained that new inventions and construction methods were not generally being developed.

One may reasonably ask what innovation involves besides determining the performance needs of occupants, the diagnosis of flaws in current practice, the analysis of alternatives, and the development of tests and standards for application. Perhaps it is a patent, a brand name, and a label. The problem with such government research is that if it is bad, it is a waste of taxpayer's money; and if it is good, it threatens private enterprise, or may seem to. The conflict should not be exaggerated, however. The American Concrete Institute, the Common Brick Manufacturers' Association of America, fire insurance companies, the Porcelain Enamel Institute, and other institutions have often cooperated with the Bureau of Standards with research associateships and other measures.

However, trade associations may approach their favorite Congressmen in order to forestall, paralyze, or terminate a program. All this happened to the Civilian Industrial Technology Program during 1961-1963. John Kenneth Galbraith, Jerome Wiesner, Richard R. Nelson, and others had taken one look at dwelling construction and concluded that failure to make radical building innovations was a curious oversight whose correction was long overdue. The program they launched was ended by a small group of Congressmen. Dorothy Nelkin's *The Politics of Housing Innovation* is not as broad as its title would suggest, but it does document this one case well. A series of similar studies would be useful (Nelkin 1971).

A more important reason for technological failure in overhauling dwelling construction methods is failure to consider "externalities," that is, a "four-walls-and-roof" approach. An array of dwellings succeeds or fails in part because of nearby anemities, transportation facilities, and public utilities of all types. Raw shelter is a temporary and fragmentary answer to the search for housing. The provision of all these utilities is not a once-and-for-all investment in "infrastructure," a durable foundation for whatever else one might like to build or do. If anything, the need for utilities will rise exponentially and undergo qualitative transformation, thus calling for both organizational and technological innovations. As an example, consider garbage--or euphemistically, solid waste. If it is not to be incinerated or hand carried, should it be ground and flushed, compacted and rolled, or sorted out and recycled? The banning of 2200 incinerators in Washington, D. C., left the city with 500 additional tons of waste per day to dump somewhere, somehow. Better dwellings and efficient utilities must be designed in terms of one another. A city is not a mere accumulation of disconnected pieces, but a system of interrelated cells.

However, the cells or households of the urban system are not homogeneous. Bachelors, aged couples, families with babies, etc., have different needs, and these life-cycle aspects are only the beginning of the differentiation that people seek in their habitats. Where there is freedom to move, any technology that ignores these differential wants is a recipe for vacancy. Successful design minimizes neither vacancy rates nor cost--except in terms of differential demands for quality. A zero vacancy rate would mean that every chain of moves would have to be totally prearranged and completed in a single day. For such precision and speed, minimal differentiation among houses would maximize their transferability or liquidity. Variety complicates choice and delays moves, but it may be worth its price, part of which society pays in the form of vacant dwellings. Among the variations that people seek are those of density; any dispersion immediately raises land, transport, and other costs. Technological change in housing must raise variety and quality, as sought by the occupants, and not merely lower cost, as defined by engineers.

Innovations that do nothing to lower costs of walls, windows, and roofs will therefore continue to appear and should indeed be welcomed even where there are housing shortages among the poor. Such innovations may raise the convenience of fixtures, improve air circulation, heighten safety, or merely alter appearance. Per se they do not imply anything one way or the other about correcting housing shortages.

MASS-PRODUCED HOUSING

A thorough study, which rather unexpectedly discovered the complexities of the housing innovation process, was prepared by the Committee on Industrialized Housing of the National Academy of Engineering during 1970-1972. This Committee was set up to assess the role that large American corporations might play in the provision of housing through the development of a sophisticated mass production technology. Its technical staff came from government, universities, and some of the largest corporations. Over 700 housing systems and subsystems were codified as analytical source material. The committee developed a process model focusing on constraints that have to be eroded to let large corporations or consortia enter the housing field.

The assumption was that, given housing needs, government support, and changing attitudes, the time had come to use corporate skills to develop an industrialized delivery system for housing. The large companies had the capital, sophisticated management, and technicians to turn housing into high-volume production components, both cheaper and more profitable for the expanding, restructured markets. Had not the Europeans already developed highly integrated, low-cost systems? Had the automobile not shown what mass production, quality control, less skilled labor, and sophisticated delivery scheduling could do? The committee was prepared to go on to an implementation phase after a preliminary testing of these and other assumptions.

With no deep literature on the nature of innovation in this neglected sector to draw on, the committee had to discover the lay of the land for itself. It concluded that for the 1970s in the United States, "an improved technology yielding only increased production can, at best, reduce costs to the consumer by less than ten per cent." Investment in novel productive capacity to restructure the industry could improve quality, but was not needed otherwise, and would in fact be so unprofitable as to be "extremely discouraging." The problem was that

> The Committee found institutional constraints inhibit high-impact technical response to the housing problem. Recent similar findings in such diverse areas as public transportation, shipbuilding, resource allocation, and health care reinforce emerging private and public awareness that exotic technical solutions cannot, by themselves, be expected to resolve socially-based problems.[16]

Unlike housing, the automobile was a highly autonomous subsystem free to move about on a vast public infrastructure. Housing can function only if external supporting systems are created, and mass-produced housing involves, not just the narrow strips that roads consist of, but land development on an enormous scale. European New Towns have government support for integrated finance and land development that is not matched by American policy. What progress has been made in more intelligent zoning, statewide building codes, and less restrictive trade-union practices, is no more than a small, initial step. Moreover, the European standards of density, quality, and amenities simply do not fit current American tastes and social problems. The committee does not mention it, but the most successful American application has been the Larsen Nielsen System in Puerto Rico, where people cannot yet afford to be as choosy as North Americans but where building wages are a multiple of those in Latin America. Even that attempt failed.

Like the companies that have been "shaken out" of the industrialized housing field (or dream), the committee had to learn that housing innovation takes more than tooling up for making objects of good design at a rate of thousands per week. Given the rate of progress in complementary activities, the volume of any one housing prefabricator is not likely to exceed a few thousand per *year* for a long time.

One company that was "shaken out" of mass-producing modular housing was Fruehauf Corp. Its president, Robert D. Rowan, knows the pitfalls of the innovation process firsthand:

There are savings to be made in mass producing housing. Unfortunately, these savings are dissipated by the cost of protecting modules against weather and protecting them for shipment. Then there is the cost of transportation to the site. By the time you are through, you have dissipated much of the savings derived through mass production. Also conventional construction today is far more efficient than it used to be, especially in the construction of townhouses and condominiums--the type of housing in which we were interested. Another problem we had was that our house was too good. We were determined not to build the slums of the future, and ...we were in a position to give a 40-year guarantee against faulty workmanship.

We discovered the wife really bought the house and she wasn't interested in quality, but rather in the cupboard space, closet space, color of drapes. In addition, the more we got into this, the more we realized that we were limited in the number of designs we could produce. Obviously, the term 'massproduce' indicates some consistency or conformity, and we found that people aren't willing to buy....And if we couldn't mass produce, we couldn't generate the cost reductions. I've concluded, therefore, that the future isn't modular housing, but modular components. For example a bathroom....

Some of our major customers, like Sea Land and Sea Train, were running complete ships [to Vietnam]. Their containers began disappearing and when they investigated, they found that people were living in them. This gave us the idea. In examining the construction techniques of modular housing, and of containers, we found they were very similar....In looking back on what we tried to do, our first mistake was limiting ourselves to townhouses. We could have produced individual homes or condominium apartments without a great deal of alteration, and certainly if we get back into housing it will be on the basis of covering the entire industry rather than the very small part that we were initially involved in. The lesson has been costly for us. We invested $7 million in this venture.[17]

DIFFUSION

Obviously, the characteristics of innovators, information transmitters, and influential opinion leaders can be studied and classified without considering the changes themselves and their repercussions. An example of such a study if *The Adoption and Diffusion of New Architectural Concepts Among Professional Architects* by Charles W. King and Thomas E. Ness of the Graduate School of Industrial Administration of Purdue University. A search of the Diffusion Documents Center at Michigan State University showed that virtually no comprehensive research had focused on the change orientation or the process by which architects adopted new concepts. Consequently, a questionnaire with over 600 items was developed and filled out with the cooperation of 120 practicing architects in the greater Chicago area. The answers were then sorted out with cross-classification, regression, and multiple-discriminant analysis. Case studies were also undertaken of (1) high strength, low alloy, weathering steel; (2) prestressed concrete; (3) foamed plastics; and (4) school construction systems.[18] In general, however, specific innovations were not part of the interviewing procedure because "it would have required an almost exhaustive survey of the characteristics of such innovations."

The conclusion of the study was that innovators, transmitters, and influentials, compared with their less active counterparts, had a stronger professional architectural orientation; were wealthier, more cosmopolitan, and more exposed to selected printed media; and had a less favorable attitude toward television, radio, and advertising. It asserted that this sort of information will be useful to salesmen of new building materials, which may very well be true; but more doubtful is whether it will, as is also suggested, help the building industry catch up with Henry Ford. This diffusion study should be compared with the one by Spall cited above.

Spall interviewed the managers of 20 firms building single-family houses in the Greater Lansing Area, or 27 per cent of the total number of firms. After determining the characteristics of the managers and the firms, Spall asked whether each of eight innovations were used or not, and why. The list, given above, was compiled from the literature and from consultations with experts. These eight were specified as the most important introduced in the previous 25 years. A similar approach was used by Strassmann in Peru and by James Spillane in Iran and Colombia. Using multiple regression analysis, Spall concluded that the entrepreneur's extent of accepting these eight innovations was *not* correlated with his years of formal education, his experience in the industry, his subjective receptiveness to innovations, the size of his firm, and usual volume of business, nor with the possibility of increased competition. Acceptance or rejection depended purely on estimates of profits from adoption. This estimate depended in part on whether the man was building custom-designed houses or speculative dwellings for the mass market. Differences in uncertainty did affect the chance of adoption, not in an absolute or binary sense, but through changes in profit estimates.

THE HOUSING STOCK AND RESEARCH PRIORITIES

Normally we associate research and development or technological change with the production of better, cheaper, and more goods. But dwellings are a stock, and the services that flow therefrom may be enhanced either by additions to the stock or by its preservation. A desirable form of R & D would therefore concern itself with improving the durability and maintainability of structures. However, maintenance is not only a question of design, but a human, social thing. It must be learned and desired, and destructive urges must be overcome; and, where appropriate preconditioned values are lacking, they must be instilled. A way of doing it is to give families a high stake in preservation through ownership. Another way is to give developer-managers a stake in better occupant selection by deferring and raising some payments for construction for half a dozen years.[19]

A dwelling will lose value if its supply rises relative to demand. But what causes this shift; is it good or bad; and should it be fostered or hindered by R & D? A house may look unfashionable and have obsolete but functioning fixtures. If it is filtered to the poor, they get a bargain that may be unbeatable by new bricks and mortar. But if deterioration does impair functioning and if its prevention is expensive, then the bargain will be temporary. The quality of the dwelling will rot to match the poverty of occupants, and the city will rot with its dwellings. Causes of obsolescence, deterioration, disposability, and abandonment must be analyzed with a partly technological, partly social scientific research program.

Not enough is known about the potential of dwellings and their occupants to specify Federal research priorities in detail. Many detailed innovations have been tried in recent decades without basic housing problems being overcome. They add up to a demonstration that partial solutions will not be good enough. Housing looms too big in the budget, on the landscape, and in the national economy to respond to marginal tinkering here and there. Fundamental technological innovations must be sought in terms of who are to be occupants, where, and who is to release resources for construction and maintenance, and how.

CONCLUSIONS

Dwelling construction has characteristics that do not apply to the other neglected, less physical, economic sectors. For one thing, the physical aspect of structures is so pronounced that some countries define industry as the combination of manufacturing, mining, *and* construction. The way structures are designed and set up is easily studied out of context with the superficial conclusion that building has never fully experienced an Industrial Revolution. To experience one, "institutional barriers" must be removed.

But the bull-fence-cow metaphor is inadequate. Certainly there are arbitrary, obsolete, and vested-interest-serving barriers to change. However, many were a response to complexities, indirect repercussions, and dangers that became painful in the course of the Industrial Revolution and that simply will not go away. On the contrary, everyone knows that they grow more serious as population and productivity grow.

In this type of neglected sector, technological change is not just a question of having or not having "barriers." It is a question of solving social organizational technical problems in terms of one another. Once we had to learn from Veblen, Redlich, and Passer that invention and innovation were not, as Schumpeter saw them, separate technical and business activities. Engineer-entrepreneurs brought both technical and commercial insight into the development of a new product from the outset.[20] Now social and political insights need to be brought in as well.

In housing, health, education, and other public services such technical-commercial-social-political problem solving has, in fact, been under way, though perhaps at an inadequate pace. The process is not being analyzed comprehensively because elegant, testable tools of analysis are not easily fashioned for what is clearly an interdisciplinary task. Practitioners from each discipline modestly obey the imprints they received in graduate school infancy. In economics the tendency is to divert attention from diffuse social repercussions and conflicts with models that treat inventions themselves as so many manufactured objects pouring out of innovation factories. This diversion of attention inhibits what is nominally promoted, better technological change.

REFERENCES

1. J. F. C. Turner and R. Fichter, *Freedom to Build*, New York: Macmillan, 1972; p. viii.
2. Hugh Spall, "Factors Influencing the Receptiveness of Homebuilders to Cost Reducing Innovations in Greater Lansing," Ph. D. Dissertation, Michigan State University, 1971. Another dissertation on housing innovations is that of Frances Ventre at MIT.
3. S. B. Hamilton, "Building and Civil Engineering Construction," in Charles Singer, et al., *A History of Technology*, Oxford: Oxford University Press; vol. 4, p. 487; also C. W. Condit, "Buildings and Construction," in Melvin Kranzberg and Carroll Pursell, Eds., *Technology in Western Civilization*, New York: Oxford University Press, 1967; vol 1, pp. 367-392, 602-619.
4. R. A. Jewett, "Structural Antecedents of the I-beam, 1800-1850," *Technology and Culture* 8:346-362, 1967; 415-429, 1968.
5. Humphrey Chamberlain, "The Manufacture of Bricks by Machinery," *J. Royal Soc. Arts*, 6 June 1856.
6. M. H. Port, "The Office of Works and Building Contracts in Early Nineteenth Century England," *Economic History Review*, April 1967, p. 96.
7. Sidney and Beatrice Webb, *History of Trade Unionism*, rev. ed., London: Longmans, Green, 1920; rev. ed., p. 9.
8. Adam Smith, *The Wealth of Nations*, New York: Modern Library, 1937; p. 103.
9. R. W. Postgate, *The Builders' History*, London: National Federation of Building Trade Operatives, 1923; p. 28.
10. Marion Bowley, *The British Building Industry: Four Studies in Response and Resistance to Change*, Cambridge: Cambridge University Press, 1966; p. 335.
11. Hamilton, Ref. 3, p. 442.
12. *Ibid.*, pp. 470-471.
13. Port, Ref. 6, pp. 94-119.
14. D. Knoop and G. P. Jones, "Rise of the Mason Contractor," *J. RIBA*, 1061-1071, 1935-36.
15. P. R. Achenbach, *Building Research at the National Bureau of Standards*, Building Science Series 0, Washington, D. C.: U. S. Government Printing Office, 1970; p. 2.
16. Committee on Industrialized Housing, National Academy of Engineering, *Industrialized Housing*, Washington, D. C., 1972; p. v.
17. "Corporate Thoughts from a Successful Alumnus," *Transacta* 6:11-12, 1973.
18. C. W. King and T. E. Ness, *The Adoption and Diffusion of New Architectural Concepts Among Professional Architects*, doctoral dissertation by Ness and Paper No. 235, Lafayette, Ind.: Purdue University (Institute for Research in the Behavioral, Economic, and Management Sciences), 1969; p. 17.
19. J. Y. Halperin, W. G. Rosenberg, and T. W. White, "State HDA Creates Plan That Can Solve Subsidy Problems," *Apartment Construction News*, April 1973.
20. Thorstein Veblen, *Absentee Ownership and Business Enterprise in Recent Times: The Case of America*, New York: B. W. Huebsch, 1923; p. 102. Fritz Redlich, *History of American Business Leaders*, Ann Arbor: Edwards Brothers, 1940; vol. 1, p. 101. H. C. Passer, *The Electrical Manufacturers, 1875-1900*, Cambridge, Mass.: Harvard University Press, 1953; pp. 1, 66-67, 180-181.

12. DIFFUSION OF INNOVATIONS PERSPECTIVES ON NATIONAL R & D ASSESSMENT: COMMUNICATION AND INNOVATION IN ORGANIZATIONS

Everett M. Rogers with John Dudley Eveland [1]

> In short, while Rogers' diffusion model may be useful in understanding the adoption of simple innovations among aggregates of individuals, it appears to be of little value for explaining the implementation of organizational innovations. *Neal Gross et al. 1971, p. 22.*
>
> Give a small boy a hammer, and he will find everything he encounters needs pounding. *Abraham Kaplan 1964, p. 3.*

INTRODUCTION

The purpose of this essay is (1) to describe briefly the so-called "classical diffusion model," and (2) to discuss its particular applicability to national R & D assessment, a current problem of interest to the National Science Foundation. We argue that there are severe limitations in the classical diffusion model, due to several of its implicit assumptions (which have only recently been recognized), and so major modifications must be effected before it can be fruitfully applied to broader fields than were originally intended. One of the most important modifications is necessitated by the fact that technological innovations often diffuse to, and within, organizations, so we must be centrally concerned with how organizational structure affects diffusion. This broadening of the new diffusion approach to incorporate fuller consideration of organizational variables is the general theme of the present paper.

So essentially we intend to bring about an intellectual merger between two "invisible colleges" of social science researchers: (1) those investigating the diffusion of innovations, and (2) those studying organizational behavior.[2] The former field is represented by some 2400 research publications dealing with how an *innovation* (defined as an idea, object, or practice perceived as new by the relevant unit of adoption) is communicated via certain *channels* over *time* to members of a *social system*.[3] Only some 373 of the 2400 diffusion publications (about 17 per cent) deal with how innovations are adopted by an organization. Unfortunately, "treatments of innovations vis-à-vis organizations tend to follow the individual-oriented approaches" (Zaltman et al. 1973, pp. 10-11). In almost all these 373 researches, sample survey data are gathered via mailed questionnaires from only the chief executive of the organization (such as a school superintendent, a factory manager, etc.). As a result of this methodological approach, we have gained considerable knowledge about how innovations spread *from* organization to organization,[4] such as from hospital to hospital, but we lack an understanding of how communication and decision-making processes are involved in innovation *within* organizations. Such intra-organizational innovative behavior cannot be understood properly if it is only viewed through the eyes at the top.

In this paper, we recommend a new approach to studying the innovation process in organizations, one that is less "macro" in scale and that takes organization structure more fully into account. Such perspective borrows directly from the considerable behavior-science literature on organizational behavior and change. These researches usually consist of intensive case studies of one (or at most several) organizations, with main focus on the organization's structure and how it affects human behavioral change.

In this essay we shall bring these two invisible colleges together. We begin with a review of the classical diffusion model, featuring some of the implicit biases/assumptions in diffusion research, and pointing out alternative methods for overcoming them in future studies.

---

[1] E. M. Rogers is Professor of Communication at Stanford University; John Dudley Eveland is Assistant Professor of Management at Eastern Michigan University.

THE CLASSICAL DIFFUSION MODEL[5]

The study of the diffusion of new ideas really began in the late 1930s when sociologists investigated the spread of hybrid seed corn from agricultural scientists to Iowa farmers. Today, 2400 research publications later, we understand a great deal more about the way in which new ideas diffuse among such varied audiences as physicians, Colombian peasants, suburban housewives, industrial plant managers, and Australian aborigines.

Central to the investigation of diffusion are four key elements: (1) an *innovation*, (2) *communicated* via certain *channels*, (3) to members of a *social system*, (4) who adopt it over a period of *time*.

1. THE *INNOVATION*

An *innovation* is an idea, practice, or object perceived as new by the relevant unit of adoption.[6] It matters little, as far as human behavior is concerned, whether or not an idea is "objectively" new as measured by the lapse of time since its first use or discovery. It is the *perceived* newness of the idea for the receiver that determines his reaction to it. If the idea seems new and different to the individual, it is an innovation.

"New" in an innovation idea need not just be new knowledge. An innovation might be known by an individual for some time; that is, he is aware of the idea, but he has not yet developed a favorable or unfavorable *attitude* toward it, nor has he *adopted* or *rejected* it. So the "newness" aspect of an innovation may be expressed in knowledge, persuasion, or regarding a decision to use it.

Characteristics of Innovations

It should not be assumed, as it often has been in the past, that all innovations are equivalent units of analysis. This is a gross oversimplification. Although it may take an educational innovation like "new math" only five or six years to reach complete adoption, another innovation like team teaching may require several decades to reach widespread use. The several characteristics of innovations, as perceived by the receivers, contribute to their differential rate of adoption.

1. *Relative advantage* is the degree to which an innovation is perceived as better than the idea it supersedes. The degree of relative advantage may be measured in economic terms, but often social prestige factors, convenience, and satisfaction are also important components of relative advantage. Again, it matters little whether or not the innovation has a great deal of "objective" advantage, or judged by experts in the field. What does matter is whether or not the individual *perceives* the innovation as being advantageous. The greater the perceived relative advantage of an innovation, the more rapid its rate of adoption.

2. *Compatibility* is the degree to which an innovation is perceived as consistent with the existing values and past experiences of the receivers. An idea that is not compatible with the prevalent values and norms of the social system will not be adopted as rapidly as an innovation that is compatible. The adoption of an incompatible innovation often requires the prior adoption of a new value system.

3. *Complexity* is the degree to which an innovation is perceived as difficult to understand and use. Some innovations are readily understood by most members of a social system; others are not, and will therefore be adopted more slowly.

4. *Trialability* is the degree to which an innovation may be experimented with on a limited basis. New ideas that can be tried on the installment plan will generally be adopted more quickly than innovations which are not divisible. Essentially, an innovation that is trialable represents less risk to the individual who is evaluating it.

5. *Observability* is the degree to which the results of an innovation are visible to others. The easier it is for an individual to "see" the results of an innovation, the more likely he is to adopt.

The first attributes of innovations just described do not constitute a complete list of those innovation qualities affecting rate of adoption, but they are the most important characteristics, past research indicates, in explaining the rate of an innovation's adoption.[7]

Both the "advantage" and "compatibility" dimensions contain a large component which may be called the degree to which the innovation is perceived as implying a commitment to further change, which in turn requires an understanding of the relative threat of change implied by the innovation. Our language for describing this phenomenon is much more developed for individual, than for organizational, psychology.

All these characteristics of innovations are *perceptual* quantities. The same concrete referent may be perceived quite differently on any or all of these dimensions by different individuals. Within the organization, there may be as many sets of perceptions of a proposed innovation as there are individuals in the organization. This phenomenon makes the assumption of the equivalence of innovations even more hazardous in the case of organizational studies than in the study of individuals.

Given that an innovation exists, and that it has certain attributes, communication between the source and the receivers must take place if the innovation is to spread beyond its inventor. Now we turn our attention to this second element in diffusion.

2. COMMUNICATION *CHANNELS*

Communication is the process by which messages are transferred from a source to a receiver with the intent to change his behavior. The communication *channel* is the means by which the message gets from the source to the receiver.

Diffusion is a subset of communication research that is concerned with *new ideas*. The essence of the diffusion process is the human interaction by which one person communicates a new idea to one or several other persons. At its most elementary form, the diffusion process consists of (1) a new idea, (2) an individual who has knowledge of the innovation, (3) another individual who is not yet aware of the new idea, and (4) some sort of communication channel connecting the two individuals. The nature of the social relationship between the source and the receiver determines the conditions under which the source will or will not tell the receiver about the innovation, and the effect that the telling has.

The communication channel by which the new idea reaches the receiver is important in determining his decision to adopt or reject the innovation. Usually the choice of communication channel lies with the source and should be made in light of the purpose of the communication act and the audience to whom the message is being sent. If the source wishes to simply inform his receiver about an innovation, mass media channels are often most rapid and efficient, especially if the audience is large in number. *Mass media channels* are all the means of transmitting messages that involve a mass medium, such as radio, television, film, newspapers, magazines, etc., which enables a source of one, or a few individuals, to reach an audience of many.

On the other hand, if the source's objective is to persuade the receiver to form a favorable attitude toward the innovation, an interpersonal channel is more effective. *Interpersonal channels* involve a face-to-face exchange between two or more individuals.

So the source should choose between mass media and interpersonal channels on the basis of the receiver's stage in the innovation-decision process. That brings us to discussion of a third element in diffusion, time.

3. OVER *TIME*

Time is one of the most important considerations in the process of diffusion. The time dimension is involved (1) in the *innovation-decision process* by which an individual passes from first knowledge of the innovation through its adoption or rejection, (2) in the *innovativeness* of the individual, that is, the relative earliness-lateness with which an individual adopts an innovation when compared with other members of his social system, and (3) in the innovation's *rate of adoption* in a social system, usually measured as the number of members of the system that adopt in a given time period.

The Innovation-decision Process

The *innovation-decision process* is the mental process through which an individual passes from first knowledge of an innovation to a decision to adopt or reject, and to confirmation of this decision. We conceptualize four main functions in the process:

(1) knowledge, (2) persuasion (attitude formation and change), (3) decision (adoption or rejection), and (4) confirmation.

The innovation-decision can take a negative turn; that is, the final decision can be *rejection*, a decision not to adopt an innovation. Or another decision can be made at a later point in time to *discontinue* or cease the use of an innovation, after previously adopting it.

The last function in the innovation-decision process is *confirmation*, a stage at which the receiver seeks reinforcement for the adoption or rejection decision he has made. Occasionally, however, conflicting and contradictory messages reach the receiver about the innovation, and this may lead to discontinuance on one hand, or later adoption (after rejection) on the other.

Innovativeness and Adopter Categories

Innovativeness is the degree to which an adopting unit is relatively earlier in adopting new ideas than the other members of his system. It is handy to refer to a particular individual as in the "late majority" adopter category, or as an "innovator" or "laggard." This shorthand notation saves words and contributes to clearer understanding. For diffusion researches show clearly that those in each of the adopter categories have a great deal in common. If a receiver is like most others in the late majority category, he is below average in social status, has little use of mass media channels, and secures most of his new ideas from peers via interpersonal channels. *Adopter categories* are the classifications of members of a social system on the basis of innovativeness. The five adopter categories are: (1) innovators, (2) early adopters, (3) early majority, (4) late majority, and (5) laggards.

These categories, clearly, have full meaning only when applied to *successful* innovations--those which eventually receive acceptance by most or all of the population. Little research has been done on failed innovations, a broad spectrum indeed, which might extend from the Nehru jacket to the Planning-Programming-Budgeting System (PPBS). For example, we do not know whether the characteristics identified previously as important to the adopter category of an individual in the context of a successful innovation are applicable in the same fashion in the case of failed innovations. Understanding the failure of innovation seems as important as understanding success, since at the moment of initiation of an innovation its eventual success or failure is not known. The fact that little is known about failed innovations reinforces the point that past research about innovations has tended to the normative, dealing only with "good," "desirable" innovations. Although innovativeness is a useful characteristic, it may eventually need to be supplemented with a measure of the success in judging the *value* (for the individual or the organization) of the proposed innovation. Such a concept of "judgmental innovativeness" may be more useful as we seek to develop a general theory which can deal equally well wth successes and failures in the innovation process.

Obviously, the measure of innovativeness and the classification of the system's members into adopter categories are based upon the relative *time* at which an innovation is adopted.

Rate of Adoption

There is yet a third specific way in which the time dimension is involved in the diffusion of innovations. *Rate of adoption* is the relative speed with which an innovation is adopted by members of a social system. This rate of adoption is usually measured by the length of time required for a certain percentage of the members of a system to adopt an innovation. So we see that rate of adoption is measured for an innovation or a system, rather than for an individual. Innovations that are perceived by receivers as possessing greater relative advantage, compatibility, etc., have a more rapid rate of adoption.

There are also differences in the rate of adoption for the same innovation in different social systems. What is a social system?

4. TO MEMBERS OF A *SOCIAL SYSTEM*

A *social system* is defined as a collectivity of individuals, or units, who are functionally differentiated and engaged in collective problem solving with respect to a common goal. The members or units of a social system may be individuals, informal groups, complex organizations, or subsystems. The social system analyzed in a diffusion study may consist of all the peasants in a Latin American village, a mental hospital, farmers of an Ohio county, medical doctors in a large city, or members of an aborigine tribe. Each unit in a social system can be functionally differentiated from every other member.

The defining feature of such a system is the *interaction* between its component elements; it is not necessary for such interactions to be consciously purposive for a system to exist, although the existence of common objectives simplifies the detection of interactions for the analyst. It should be noted that the term "structure" has multiple acceptable meanings; the communication model discussed here emphasizes *communication structure*, which is not necessarily equivalent to either the role structure or the formal authority structure, alhtough it may correlate highly with them. The terminology of "open systems theory," with its emphasis on the permeability of system boundaries, allows the development of very flexible and useful analytical models (Kutz and Kahn 1964).

It is important to remember that diffusion occurs within a social system, since the social structure of the system affects the innovations' diffusion patterns in several ways. Further, the social system constitutes a set of boundaries within which innovations diffuse. In this section, we shall deal with how the social structure affects diffusion, the roles of opinion leaders and change agents, and, lastly, types of innovation-decisions.

Opinion Leaders and Change Agents

Very often the most innovative member of a system is perceived as a deviant from the social system, and he is accorded a somewhat dubious status of low credibility. His role in diffusion (especially in persuading others of the innovation) is therefore likely to be limited. On the other hand, there are members of the system who function in the role of opinion leader. They provide information and advice about innovations to many others in the system.

Opinion leadership is defined as the ability informally to influence the attitudes or behavior of individuals in a desired way with relative frequency. Thus, it is a type of informal leadership, rather than being a function of the individual's formal position or status in the system. Opinion leadership is earned and maintained by the individual's technical competence, social accessibility, and conformity to the system's norms. Several researchers indicate that when the social system is modern, the opinion leaders are quite innovative, but when the norms are traditional, the leaders also reflect this norm in their behavior. By their close conformity to the system's norms, the opinion leaders serve as an apt model for the innovation behavior of their followers.

The direction of causality is not certain. Opinion leaders may be leaders because they conform, or they may conform because they are leaders. We do not clearly understand how much flexibility a leader may have to change his approach without moving from "leader" to "crank" roles. In short, we are not sure how much the normative structure of a given system may be modified with regard to innovation by an individual, or what the correlates of such modification might be.

In any system, naturally, there may be both innovative and also more traditional opinion leaders. So these influentials can lead in the promotion of new ideas, or they can head an active opposition. In general, when leaders are compared with their followers, we find that they: (1) are more exposed to all forms of extra-system communication, (2) are more cosmopolite, (3) have higher social status, and (4) are more innovative (although the exact degree of innovativeness depends, in part, on the system's norms).

Opinion leaders are usually members of the social system in which they exert their influence. In some instances, individuals with influence in the social system are professionals representing change agencies external to the system. A *change agent* is a professional who influences innovation-decisions in a direction deemed desirable by a change agency. He usually seeks to obtain the adoption of new ideas, but he may also attempt to slow down diffusion and prevent the adoption of undesirable innovations. Change agents often use opinion leaders within a given social system to prime the pump of planned change. There is research evidence that opinion leaders can be "worn out" by change agents who overuse them. Opinion leaders then are perceived by their peers as too much like the change agents; thus, the opinion leaders lose their credibility with their former followers.

Types of Innovation Decisions

The social system has yet another important kind of influence on the diffusion of new ideas. Innovations can be adopted or rejected by individual members of the entire social system. The relationships between the social system and the decision to adopt innovations may be described in the following manner:

1. *Optional decisions* are made by an individual regardless of the decisions of other members of the system. Even in this case, the individual's decision is undoubtedly influenced by the norms of his social system and his need to conform to group pressures. The decision of an individual to begin wearing contact lenses instead of eye glasses, an Iowa farmer's decision to adopt hybrid corn, and the adoption of contraceptive pills are examples of optional decisions.

2. *Collective decisions* are those in which individuals in the social system agree to make by consensus. All must conform to the system's decision once it is made. An example is fluoridation of a city's drinking water. Once the community decision is made, the individual has little practical choice but to adopt fluoridated water.

3. *Authority decisions* are those forced upon an individual by someone in a superordinate power position, such as a supervisor in a bureaucratic organization. The individual's attitude toward the innovation is not the prime force in his adoption or rejection; he is simply told of, and expected to comply with, the innovation-decision. Few research studies have yet been conducted of this type of innovation-decision, which must be very common in an organizational society such as the USA today.

Collective decisions are probably much more common than optional decisions in bureaucratic organizations, such as factories, hospitals, schools, or labor unions, in comparison with other fields, like agriculture, where innovation-decisions are usually optional.

In the transference of the innovation diffusion model from the level of the individual to the level of the organization, we have sometimes given in to a too-close parallelism. The question of the definition of "adoption" is subject to this problem. The usual approach has been to make adoption a "0-1" variable--either the innovation is adopted or it is not, and the only problem is the definition of the exact moment when the change from "0" to "1" takes place. Even in the case of the individual, the process consequences of an adoption predicated on "malicious obedience" will be rather different from those based on a wholehearted internalization of the innovation. On the organizational level, the variability of both the degree and content of adoption becomes an important factor.

Moreover, this variability is often weighted. The consequences of lower implementation or internalization of an innovation are more evident and severe in some parts of an organization than in others. For example, the adoption of a new contraceptive method by a family planning clinic clearly depends much more on the physicians than on the receptionists. If we continue to regard adoption for analytical purposes as a dichotomous variable, we are forced to develop some criteria as to the allocation of this weighted variability.

Clearly, it is not usually sufficient to say that adoption takes place at the moment when the leader so decides. Particularly in the cases of collective and authority decisions, where there is an implication of group activity in the adoption process, the simple dichotomous approach which we carry from individual studies tends to obscure some very important features of the process of adoption. In an organizational context, the decision to adopt an innovation, and the implementation of the decision, may be two quite different things.

Generally, the fastest rate of adoption of innovations can be by authority decisions (depending, of course, on whether the authorities are traditional or modern). In turn, optional decisions can be made more rapidly than those of the collective type. Although made most rapidly, authority decisions are more likely to be circumvented, and may eventually lead to a high rate of discontinuance of the innovation.[8] Where change depends upon compliance under public surveillance, it is not likely to remain once the surveillance is removed.

The type of innovation-decision for a given idea may change or be changed over time. For instance, automobile seat belts during the early years of their diffusion were installed in private autos largely as optional decisions. Then in the 1960s many States began to require, by law, installation of seat belts in all new cars. In 1968, a Federal law was passed to this effect. Then in 1974 all new autos were required to have an ignition/seat belt interlock system to prevent starting the motor until the seat belt is fastened. So an optional innovation-decision became a collective (or an authority) decision (later abandoned).

LIMITATIONS AND SHORTCOMINGS OF THE CLASSICAL DIFFUSION MODEL

In the previous section, we summarized the main elements in the classical diffusion model. Here we shall criticize this model from the viewpoint of what types of behavior it has not been able to help us understand, especially from the viewpoint of our central concern in this essay with the process of technological innovation.

Origins and Backgrounds of the Diffusion Model

In any given field of scientific research a major breakthrough occurs from time to time. This "revolutionary paradigm" (or major reconceptualization) constitutes a new way of looking at some phenomenon, and usually sets off a furious amount of intellectual effort (Kuhn 1962). Promising young scientists are attracted to the field, either to advance the revolutionary paradigm with their research, or to disprove certain aspects of it. Gradually a scientific consensus about the field is developed. Eventually, perhaps after several generations of academic scholars, this "invisible college" composed of researchers on a common topic reaches a point at which few findings of an exciting nature emerge, and the field begins to decline in scientific interest (Crane 1972); it splinters into numerous invisible colleges, each of which grows around a new revolutionary paradigm.

That is the normal path of science. Research on the diffusion of innovations has followed these stages rather closely, although the final stage of demise has not yet set in (Crane, 1972). The revolutionary breakthrough occurred in the early 1940s when two rural sociologists, Bryce Ryan and Neal Gross (1943), investigated the diffusion of hybrid seed corn among Iowa farmers.[9] "The hybrid corn study is undoubtedly the most widely known rural sociological inquiry of all time, and even today it ranks as a classical study of diffusion" (Rogers with Shoemaker 1971, p. 54).

The hybrid corn study set forth a new approach to the study of communication and change in human behavior. Within a few years, its leads were being followed up by an increasing number of scholars. By 1950 there were 73 diffusion publications; by 1960, 508; by 1970, 1970; and there are over 2400 today. At first, most diffusion researches were completed by rural sociologists who investigated the diffusion of agricultural innovations in the United States. But soon, the diffusion approach was being pursued in other fields: Education, anthropology, communication, marketing, economics, and medical sociology. Each of these disciplines conducted diffusion research in a somewhat specialized way, without much interchange with other diffusion research traditions. In the early 1960s, each of these diffusion traditions operated as a separate invisible college, and the total field of diffusion was relatively unintegrated. About this time, however, a major intellectual watershed ocurred in diffusion research, and thereafter it proceeded in a different manner.

By the mid-1960s the old boundaries between the diffusion traditions began to break down. It became apparent that the diffusion of innovations was a general process, independent of the various disciplines involved in its study, the different types of innovations being analyzed, or the variety of research methods utilized to investigate diffusion. By the early 1970s Rogers with Shoemaker (1971, p. 47) concluded: "Diffusion research is thus emerging as a single, intergrated body of concepts and generalizations, even though the investigations are conducted by researchers in several scientific disciplines."

So the scientific field of diffusion research emerged as an invisible college. This was not an entirely beneficial development, however; concepts, methods, and interpretations can become stereotyped. Scientific standardization is not always a blessing. We feel that to be especially true in the case of the some 373 diffusion research publications dealing with the adoption of innovations by multimember units such as organizations. The structural distinctiveness of organizations as units of innovation decision making has not been fully realized (as we discuss later in some detail). These organizational innovativeness studies were much (too much) influenced by previous research on single-member units. The reciprocal flow of improvements, modifications, and amplifications in the diffusion model, resulting from diffusion research in organizations, has not yet occured. The present essay, of course, aims to rectify this situation.

Questionable Assumptions of the Classical Diffusion Model

We have already mentioned that diffusion scientists investigating organizational in-innovativeness assumed that an identical approach could be utilized as had been followed in the earlier diffusion studies where individuals (like farmers) were adopting units. We shall amplify on the seriousness of this assumption shortly, after specifying certain other questionable assumptions of diffusion research, as it has developed until today.

Assumption 1: *The innovation is advantageous for all adopters.* The classical model assumed that everyone should adopt, that diffusion rates should be rapidly increased, and that rejection was an undesirable and/or irrational decision.[10] This pro-change bias may have been justified in the case of many of the agricultural innovations that were originally studied, and which strongly flavor our conception of the diffusion process. After all, hybrid corn was a highly profitable innovation for farmers.

But many innovations have no such universal usefulness. Some new ideas are the result of behavioral science research, rather than biological or physical science investigation. The innovation may not even have a physical referent, or object. And such innovations may only be appropriate for adoption by *some* individuals, and by *some* organizations.

The implication for future research is to study the diffusion of a wider range of innovations, including some that should be rejected by some units. Diffusion research would thus shed its pro-change bias. And our focus should be upon how innovation-decisions are made (and by whom), with what criteria, and to what consequences.

Further, we need to understand the motive for adopting an innovation. Among U. S. farmers, an economic motivation is the paramount thrust for adopting innovations. But what benefits are perceived for adoption of an innovation by a hospital administrator? Improved patient care? Prestige? Mohr (1969) concludes from his study: "A great deal of innovation in health organizations, especially large or successful ones, is 'slack' innovation. After solution of immediate problems, the quest for prestige rather than the quest for organizational effectiveness or corporate profit motivates the adoption of most new programs and technologies." Mohr subsequently suggested that the organization's perceptions of "risk" and "threat" may be important in understanding how 'slack' is translated into actual innovative behavior.

The salience of the prestige motive has also been found by Becker (1970) among local health departments, and inferred by Walker (1969) in his investigation of the diffusion of innovations among American state governments.

One important dimension may be the location of the innovation on the "transitive-reflexive" continuum. "Transitive" innovations are those directly related to the provision of services to clients; "reflective" innovations are those related more to the internal operation of the organization, which are not visible to clients. These terms are adapted from Mohr (1973). Part of the effect of this dimension may lie in the degree of threat and of commitment to change (as we noted earlier) implied by the innovation. The effect of incorporating this intervening dimension into an analysis is to allow a more precise understanding of just how an innovation interacts with the needs of an organization.

In suggesting that not all innovations should be adopted by all organizations, we imply that there are some criteria by which this judgment can be made, by either the organization or the research analyst. In turn, this implication raises questions as to the specificity and objectivity of such criteria--as well as their origin.

Assumption 2: *Cross-sectional data gathering at one point in time is sufficient for study of a process like diffusion.* We stress that diffusion is a *process*, meaning that it is continuous through time, never ending and never beginning. Diffusion, like other processes, flows like a temporal stream. Someone has suggested that all processes should always begin and end with the word "and," implying that no process ever begins, as there is always something that precedes it. No experience ever ends; something always follows it.

However, for heuristic purposes of understanding we often find it convenient to "freeze" the action in this continuous "film" and to isolate certain elements or variables in the diffusion process. Most social-science research methods are better suited to obtaining snapshots of behavior, rather than moving pictures, which would be more appropriate for determining the time order of the variables. Correlation analysis of one-shot survey data is overwhelmingly the favorite methodology of diffusion investigators.[11] Essentially, this amounts to making the diffusion process "time-less." It is convenient for the researchers, but intellectually deceitful to the process they are investigating.

Speaking about the studies of organizational diffusion, Hage and Aiken (1970, pp. 28-29) critically note that:

> The current state of knowledge about organizational behavior is based to a great extent on cross-sectional studies, that is, studies which examine the relationships among various organizational characteristics at some given point in time. While such studies are important and have provided a much better understanding of organizational behavior, at the same time they fail to sensitize the student of organizations to the ongoing process of change in organizations. Organizations seldom stand still....A temporal perspective, one in which the focus of attention is not only on the interrelationships of parts of organizations, but more importantly on changes in these interrelationships, is ultimately necessary for a more complete understanding of organizational behavior.

A further problem with the cross-sectional survey data is that they are unable to answer many fo the "why" questions about diffusion.[12] The one-shot survey provides grist for description, of course, and makes a type of correlation analysis possible: various independent variables are associated with the dependent variable (usually innovativeness). But little can be learned from such a correlational approach about *why* a particular independent variable covaries with innovativeness.

"Such factors (as wealth, size, cosmopoliteness, etc.) may be causes of innovation, or effects of innovativeness, or they may be involved with innovation in cycles of reciprocal casuality through time, or both they and the adoption of new ideas may be caused by an outside factor not considered in a given study" (Mohr 1966, p. 20). Future research must be designed so as to probe the time-ordered linkages among the independent and dependent variables. And one-shot surveys cannot tell us much about time-order.

A critic of the diffusion field may be correct when he judged it to be "A mile wide and an inch deep."

Assumption 3: *Innovation decisions are made by individuals, not organizations.*[13] Almost all diffusion research in the past has ignored the effect of organizational structure.[14] Hospitals, schools, and factories, of course, are organizations, even though farmers are not. But even when medical, educational, and industrial diffusion has been investigated, it has not been studied *within* the organization. Instead, the process has been studied as if innovations diffused only from organization to organization. For example, most educational diffusion has been analyzed with the school system as the unit of analysis. So the school is typically considered as a parallel unit to the farmer; and likewise with hospitals, industrial firms, and other organizations. They have all been reduced to "individuals" on the Procrustean bed of diffusion research.

Becker (1970) studied the diffusion of health innovations among 95 local health departments in three states. But the health department was the unit of analysis, and the innovations (measles immunizations and diabetes screening) were studied as they spread from department to department. The data for each organization were gathered from the director of the local department. This research approach, also followed by most other scholars of organizational innovativeness, tells us much about the process of organization-to-organization diffusion, but nothing about the *intra-organizational process*. In essence, by this most typical diffusion research approach, organizations are artificially de-structured or de-organized. It is almost as if, by heuristic means, the health departments and other agencies were removed from their organizational contexts. Thus, the classical diffusion model is made to fit the research task,[15] rather than vice versa.

Needed is another type of research style in which we seek to understand how innovations spread *within* an organizational structure (Table 1). Such a design would call for a more complete merger of organizational theory and the diffusion model. We would study how organizational structure affects the diffusion of innovations. The new ideas would then be something like tracers, coursing through the structure. We predict the results of such study would illuminate new aspects of organizational structure and functioning, as well as highlight needed modifications in the diffusion model. In a later section, we shall describe a possible research design for a study of innovation in organizations.

TABLE 1.--Much research has studied individual innovativeness, but the innovation process within organizations has been almost completely ignored by diffusion researchers.

Process Being Investigated	Unit of Analysis	
	Individual	Organization
1. Innovation process (e.g., stages and role in this process)	About 40 or 50 researchers	Almost none
2. Diffusion process (e.g., correlates of innovativeness, opinion leadership patterns, etc.)	Most of the 2400 diffusion reports	About 373 publications

The open-system models of organizational structure and function we are using allow us to treat the distinction between two organizations as a variable quantity dependent on the purposes of the analysis, rather than as a single-valued either/or function. Thus, the distinction which we have drawn here between inter- and intra-organizational approaches is a matter of degree, not a firm distinction in kind. What is important is the degree to which the system that we make the object of our primary concern is tied to its environment--what Mott (1972) called the "relative degree of system closure"--and the nature of the structures and processes which implement that connection.

We may criticize the approach that treats innovation decisions at the organizational level as simple extensions of the individual-level model, but we should not forget that organizations are composed of individuals, not solely of abstract forces. It is sometimes useful to distinguish between the class of "innovations-*within*-the-organization," those innovations which require primarily the action (or inaction) of individuals acting as individuals (such as the institution of a new incentive-pay system), versus the class of "innovations-of-the-organization," those which require activity on the organization or system level but no direct activity on the part of most members (such as the institution of a computerized billing process). In a sense, this distinction represents an aspect of the "reflective-transitive" dimension noted previously; it should be used as a variable continuum rather than an analytically neat dichotomy. Again, as in the case of the inter/intra distinction, this break is useful only to help us focus our attention at particular aspects of the general system.

Now we turn to an overview, and critique, of some 373 organizational innovativeness researches.

RESEARCH ON ORGANIZATIONAL INNOVATIVENESS

Our concern here is with the 373 publications reporting the results of research on organizational innovativeness, which are a subset of the some 2400 documents dealing with the diffusion of innovations (Fig. 1). Most of these 373 publications are relatively late on the diffusion research scene, and are heavily influenced by their nonorganization intellectual ancestors.[16]

FIG. 1.--The 373 organizational innovativeness publications represent a subset of research on the diffusion of innovations.

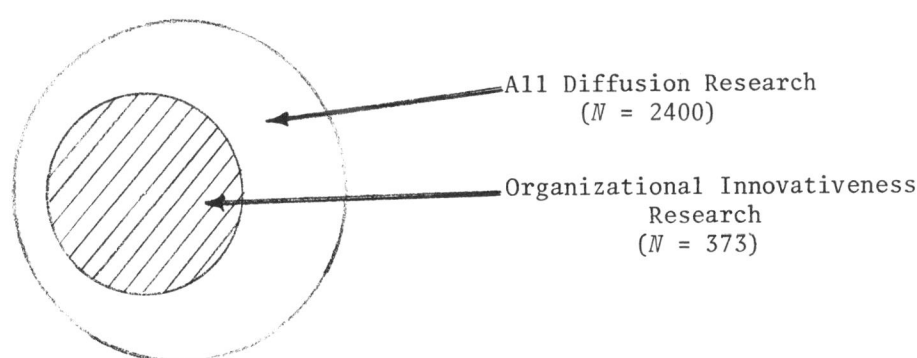

Organizations are obviously important in modern societies. In fact, they are "the *major* mechanisms for achieving man's goals" (Hage and Aiken, 1970, p. 5). So an understanding of the innovation process in organizations is a very fundamental piece of knowledge. And it certainly is a crucial basis for comprehending the nature of technological advance in modern American society. That is why we give so much attention to research on *organizational* diffusion in the present essay.

Lack of Research on Innovation in Organizations

Hage and Aiken (1970, p. xii) list three general research questions to guide the study of innovation in organizations. The first two deal with determining (1) what organizational characteristics, and (2) what environmental factors account for variations in organizational innovativeness. In the main, the 373 organizational diffusion publication (mentioned previously) have tried to answer these two interrelated questions. Hage and Aiken's third question is: "What is the process of adopting new programs or other changes that is, innovations in organizations?" Almost no studies have dealt with the innovation *process* in organizations: "There have been few studies of the actual process of change in organizations" (Hage and Aiken 1970, p. xv). In fact, these authors can only provide one case illustration of this type of research in their book. Actually there are very few others (Table 1).

The large-sample survey researches were actually only studies of organizational *innovativeness*, but not of organizational *innovation*. There is a very important difference. The large-sample surveys mainly compared the various organizational units (that were their respondents) on innovativeness; they searched for correlations between selected (and often rather stereotyped) independent variables with innovativeness. As stated previously, these 373 studies of organizational innovativeness tell us very little, or nothing, about the process of within-organization innovation. As we noted earlier, "innovativeness" when treated as an abstract quality, apart from the context of specific innovations, is not very meaningful because of the number of possible relevant dimensions of innovations and their highly perceptual quality. This problem greatly complicates the comparison of various studies; the detection of general, rather than situation-specific, factors seems to be very much a matter of the particular analyst's perspective.

Diffusion from Organization to Organization

We have been rather critical in this essay of the large-survey studies of organizational innovativeness. They have only provided a general picture of the variables associated with *innovativeness*, and have not really dealt with the problem of how innovations *diffuse* from organization-to-organization. But in the past several years, some beginnings have been made to deal with this issue.

Walker (1969) studied the innovativeness of American States on the basis of their time of adoption of 88 innovative programs--such as having a Board of Health, slaughterhouse inspection, and a gasoline tax. The large-population, industrialized, and urbanized States are most innovative: New York, Massachusetts, California, New Jersey, and Michigan. In each region of the United States certain States emerged as opinion leaders; once they adopted a new program, other States in their "league" followed along. If an innovation was first adopted by a State other than an opinion leader, it spread to other states slowly or not at all. Thus, a communication structure *seemed* to exist for innovation diffusion among the States.

In a further analysis, Walker (1971) utilized sociometric data from personal interviews with State officials in ten States for the actual determination of the communication flows, patterns of sociometric opinion leadership, and clique-like networks among the American States. The innovative States tended to be the "national league" opinion leaders, but State officials also tended to look to their immediate neighbors when searching for guidance about innovations: "State administrators communicate most readily with their counterparts in states that they believe have similar resources, social problems, and administrative styles" (Walker 1971, p. 381). For instance, Iowa officials followed Michigan's and California's lead, although they were much influenced by Wisconsin.

Becker (1970) gathered sociometric data about organization-to-organization diffusion, for a high-risk and a low-risk health innovation from 95 directors of local health departments in three States. He found that "Earliest adopters of the easily accepted program measles immunizations were opinion leaders, while pioneers [of the more risky innovation, diabetes screening] were found to be marginal to their groups." This research result is parallel to comparable findings from diffusion research among individuals (Rogers with Shoemaker 1971).

"Contrary to the current view in diffusion research literature, time of adoption is more likely to be a cause than a result of centrality in information networks" (Becker 1970). So the health professional's innovativeness determined his popularity as an opinion leader, rather than vice versa. "A desire to maintain or increase prestige (tempered by the risks of adoption) motivates the professional to seek...innovations" (Becker 1970). We are not entirely convinced that Becker's survey data allow him to come to these cause-effect conclusions, but in any event his investigation does demonstrate the utility of incorporating sociometric data in studies of organizational innovativeness. This approach begins to allow us some limited insight into the over-time *process* of organization-to-organization diffusion. Unfortunately, Becker was forced to assume the rather important reductionism that the *director* of a local health department somehow "represented" his entire organization in the sociometric relationships that he provided as data. This assumption is doubtful in the case of any organization, other than very small units, and is rather dubious in the case of local health departments which often range in size up to several hundred employees. But nevertheless the potential of sociometric data in providing insight into organization-to-organization diffusion was clearly demonstrated.

Future research is needed to confirm the suspicion suggested by the work of Walker (1969) and Becker (1970), that patterns of opinion leadership exist in the diffusion of innovations among organizations, just as among individuals. Such research should also address the major problem, not directly explored in either of these studies, of possible interactions between leadership and structural variables, which might produce patterns of innovation behavior different from what might be suspected on the basis of additive modeling alone.

Sources of Data

In addition to its lack of concern with (1) the innovation process within organizations, and (2) organization-to-organization diffusion, the studies of organizational innovativeness suffer as a result of the type of data that is typically gathered by the researcher. Usually, he sends a mailed questionnaire to the top man in the organization, or else interviews him personally. All data about the organization's innovativeness, its structural characteristics, and the leader's characteristics are obtained from him. Naturally the researcher only gains a picture of organizational innovativeness as seen from the top. Obviously the perceptions and viewpoints of the various members of the organization cannot be incorporated in the analysis, because they have not been measured. Obviously one cannot adequately study the process of innovation in the organization with just one member's viewpoint of the process.

But the usual studies of organizational innovativeness do enable us to know "a little about a lot." A large sample of organizations can be studied when there is only one respondent for each organization. Such studies are relatively cheap in data-gathering cost per organization.

A very few recent researches have incorporated data from numerous individuals in each organization. For instance, Kaluzny and others (1973) gathered data via personal interviews (with the administrators only), and via mailed questionnaires with the following respondents in each of 68 hospitals in New York state:

1. The admistrator, assistant/associate administrators, and department heads.
2. A sample of the hospital trustees.
3. A sample of all the physicians in each hospital.

Hage and Dewar (1973) also gathered data from several individuals in each of the 16 organizations which they studied.

The usual sources of data (from only the chief executive of the organization) pose special problems for the measurement of innovativeness, which is the most common dependent variable in these studies. Mohr's (1969) approach to operationalization is fairly typical. His 93 directors of local health departments were asked which of a number of innovative programs their agency had adopted; more innovativeness "points" were awarded to health departments which had adopted more innovations from the list.[17]

There is a logical difficulty in using retrospective data about innovativeness in analysis which includes independent variables which are measured cross-sectionally (Kaluzny et al. 1973). *In essence, one is correlating yesterday's innovativeness with today's independent variables.* In many cases, the problem may be even more severe; the correlation is between today's variables and the *post hoc*, selective, congitive-dissonance-reducing *perception* of yesterday's innovativeness. Success is likely to be remembered much more than failure; even the organization's files may be purged or records of innovations that did not work out. The past constantly tends to be reinterpreted in terms of the present or the anticipated future. As we noted earlier, innovation diffusion is a highly time-dependent process, and it is very difficult to analyze it with data gathered at only one point in time. However, the solution is certainly not to measure the dependent variable at t_1 and the independent variables at t_2.[18]

A number of possible sources of data about innovativeness can be tapped:

1. Published records on time of adoption of the innovations, an approach used by Walker (1969).
2. Agency records, such as the notes kept from regular staff meetings, where innovation decisions are discussed.
3. Personal interviews (or mailed questionnaires) with executives in the organization, or with all of the members.
4. Observation and inspection of the innovation's adoption.

The latter sources/methods of data-gathering are probably more accurate, but they are also more costly to obtain. Needed is the incorporation of several, or all, of these four approaches in the same study, so that more precise data will be available on the relative costs and accuracy of each. This is the multi-operational approach (Siber 1973).

Individual Versus Organizational Characteristics and Innovativeness

There are two broad classes of independent variables that have been used to predict organizational innovativeness: (1) individual characteristics of the organization's leader or leaders, and (2) organizational characteristics like size, decentralization, and system openness.

Many studies of innovation have focused on individuals rather than on organizations. However, individuals may have a substantial influence on the adoption of new means and goals by organizations. The influence of top management, for example, was found by Burns and Stalker (1961) and by Mytinger (1968) to be highly significant for the innovativeness of the organizations they led....It is therefore entirely possible that the same factors that seem to cause farmers and doctors to innovate for themselves might also cause executives to innovate, or at least try to innovate, for their organizations (Mohr 1966, pp. 20-21).

The research question that remains to be fully answered in future investigations is: *To what extent is the degree of innovativeness of organizations a function of (1) the personal characteristics of individual executives, versus (2) the structural characteristics of the organization?* Even though a full answer is not yet available, recent research offers some insight.

First of all, we must point out that the question of individual *versus* organizational variables in explaining organizational innovativeness is something of a phony issue. Obviously both sets of independent variables would be expected to co-vary with innovativeness, and the two are interdependent in their relationship with innovativeness. The interaction of structural and individual variables is illustrated by Mohr's (1969) study of local health departments: "When resources are high...a unit increase in health officer motivation an individual variable ...has about 4½ times the effect upon innovation, as it does when resources are low."

So multivariate statistical tools should be utilized in organizational innovativeness studies to entangle the joint effects of individual and structural variables on the dependent variable. And the effects seem to be complementary, rather than competitive.

Hage and Aiken (1970, pp. 122-123) conclude their investigation of innovativeness in 16 rehabilitation agencies by stating that "structural properties were much more highly associated with program change that is, innovativeness than attitudes toward change [of the executives]. This implies that the structure of an organization may be more crucial for the successful implementation of change than the particular blend of personality types in an organization." This conclusion may be somewhat overenthusiastic, considering that the relative contribution of individual and organizational variables to explaining innovativeness depends, in part, on how precisely such individual and organizational variables are measured. Hage and Aiken (1970) express a strong antipsychological bias in their intellectual position, and as both are sociologists, one might surmise that their measurement of structural characteristics would be more thorough than their operationalization of the individual variables.[19]

The question of the relationship between leader variables and structural characteristics is complicated by a serious problem--as yet to be addressed in research--of the possible interaction over time of the leader and his structure. In short, leaders tend to be modified by the organizations in which they function, and in turn shape and modify the structural characteristics of the organizations. We would expect, for example, a leader with democratic beliefs to move the whole organization in more democratic directions, regardless of the specific connection with innovation. In practice, this might lead to a common store of variability which some studies might tap as structural, some as leader--but which should be understood to result from a common phenomenon.

Hage and Dewar (1973) provide a novel test of the importance of leadership versus organizational variables in predicting innovativeness. Using data gathered from 16 rehabilitation agencies in 1964, they predicted innovativeness (measured as the number of innovations adopted from 1964-1967) with (1) the attitudes of the executive director, (2) the attitudes of the executive director and all formal leaders (department heads) in his agency, the formal elite inner circle, (3) the attitudes of the executive director, and all agency members who said they participate in decisions, the informal elite inner circle, and (4) all members of the agency. Correlations with innovativeness were 0.60, 0.45, 0.69, and 0.34, respectively. Structural variables like complexity, decentralization, and formalization were correlated 0.45, 0.33, and -0.07, respectively.[20]

So it appears that (1) leader attitudes toward change predict organizational innovativeness about as well (actually somewhat better) than do organizational structure variables, (2) the executives' attitudes predict innovativeness better than all members' attitudes predict innovativeness better than all members' attitudes, and about as well as the informal elite circles. Perhaps the agency director typically surrounds himself with a set of elite with attitudes toward change rather similar to his own.

When organizational structural variables and size are controlled, the executives' personal characteristics tend to be less important as predictors of organizational innovativeness. For example, Veney et al. (1971) reported that organizational variables accounted for 41 per cent, contextual variables[21] for 31 per cent, and the administrators' personal characteristics for only 5 per cent of the variation in organizational innovativeness in a national sample of 480 hospitals. The hospital administrators' personal variables included formal education, cosmopoliteness, attitude toward change, and tenure. Naturally there are problems of classifying exactly which variables are executives' personal characteristics, and which are organizational structural variables, in these studies. For instance, the executives' cosmopoliteness (in the eyes of one researcher) may be considered the organizations' openness by another scholar.

"Is innovativeness due to the man, the agency, or the place? No one of these can be said to play in singular role" (Mytinger 1968, p. 7). The innovativeness of 40 local health departments in California was related (1) to their size in staff and budget, (2) which in turn rested on the size of the community they served, and (3) the cosmopoliteness, accreditation, and status among his peers of the health officer. Overall, "This study suggests that *size*--size of community and size of the health department--is perhaps the most compelling concomitant to innovativeness" (Mytinger 1968, p. 7). Similar evidence for the importance of size as a predictor of organizational innovativeness is provided by Mohr (1969), Kaluzny et al. (1973), Mansfield (1963b), and several others.

Why do researches consistently find that *size* is one of the best predictors of organizational innovativeness? First, size is a variable that is easily measured, and presumably with a relatively high degree of precision.[22] So size has been included for study in almost every organizational diffusion investigation (and, in fact, in almost every single-member diffusion study). Second, size probably is a surrogate measure of several dimensions that lead to innovation: total resources, slack resources, organizational structure, etc. These intermediate variables have not been clearly identified, nor adequately measured in most researches. Doubtless these unidentified/unmeasured variables are a fundamental, and intellectually deceiving, reason for the size-innovativeness research "finding." Few scholars have much theoretical interest in size as a variable. We urge that once its effects on innovativeness (although the intervening variables) are isolated and understood, size should be dropped from further study, or at most relegated to being only a control variable.[23]

In fact, size was an obvious variable for study in the large-sample survey researches on organizational innovativeness. In the intensive studies of organizational *innovation* that we recommend for future research, which might be conducted in only a very small number of organizations, size could easily (and probably *must*) be used only as a control variable. For example, in a comparative systems analysis of a pair of organizations (one of which is highly innovative, and one of which is noninnovative), they might simply be matched so as to be similar in size (thus removing its effect).

So--good-by, size. Or at least turn it over, and see what's underneath.

Structural Characteristics and Organizational Innovativeness

A recent, and we feel healthy, trend in organizational innovativeness research is the inclusion of independent variables measuring certain dimensions of the organization's structure: centralization, complexity, formalization, openness. Figure 2 shows some of the main structural variables, plus the individual variables, that we expect to be associated with organizational innovativeness. We divide these variables into three classifications: individual characteristics, internal structural characteristics, and

FIG. 2.—Paradigm of independent variables related to organizational innovativeness. *Sources:* Becker and Whisler (1967); Hage and Aiken (1970).

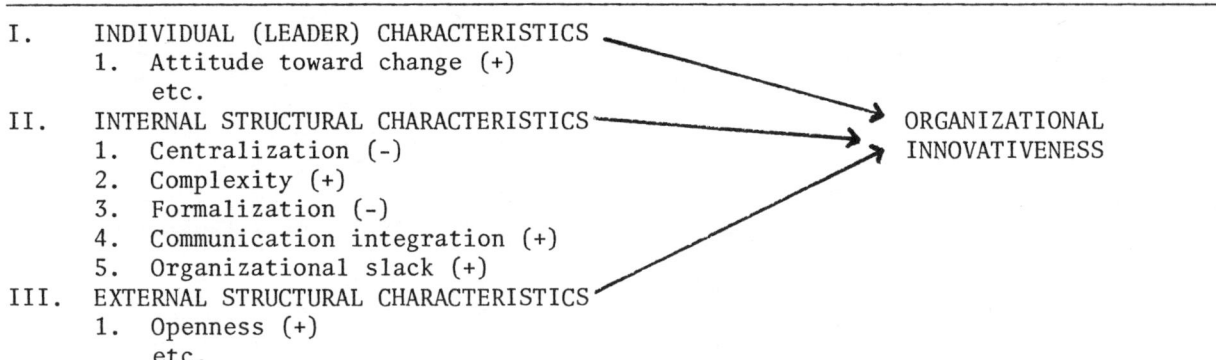

external structural characteristics.

Hage and Aiken (1970) have played a key role in specifying the main internal structural variables that are related to organizational innovativeness, and their concepts have been picked up by other scholars (for example, Kaluzny et al. 1973).

Under open-systems terminology the term "structure" can have many meanings. The "control" or "authority" structure, which is usually operationalized as "the structure of the organization" because of its public, available character, is not necessarily fully equivalent to the "communication structure" or the "role structure," and is quite different from "normative structure." Thus the term, though vital, has an ambiguous set of referents, and it is always necessary to identify which sort of structure is being described. The selection of a structure for study is normally a function of the type of analysis to be undertaken. The specific aspects of structure described here are in theory applicable to almost any notion of structure, although it is only through specific operationalization that they acquire replicable and reliable meaning.

Centralization is the degree to which an organization's power and decision-making are concentrated in relatively fewer individuals (Hage and Aiken 1970, p. 38). Centralization has usually been found to be negatively associated with innovativeness; that is, the more power is concentrated in an organization, the less innovative it tends to be. The range of new ideas in an organization seems to be restricted when a few central leaders dominate the scene. The *initiation* of innovativeness in a centralized organization may actually encourage the *implementation* of innovations, once the innovation decision is made.[24] In a centralized organization, top leaders are poorly positioned to identify operational problems, or to suggest relevant innovations to meet these needs.[25]

A future research direction on innovation in organizations could be to determine how effective participatory management styles (a move toward greater decentralization) are in encouraging organizational innovativeness.

Complexity is the degree to which an organization's members possess a relatively high level of knowledge and expertise, usually measured by the members' range of occupational specialties and the degree of professionalism (training) of each. We generally expect that complexity is positively related to innovativeness, as the members supposedly have a wider grasp of innovations.

Formalization is the degree to which an organization has codified its jobs by means of rules. Such formalization acts to limit the range of latitude available to the organization's staff in deciding about innovations.

Communication integration is the degree to which the members of a system are interconnected by interpersonal communication patterns.[26] New ideas can flow more easily and rapidly among an organization's members if it has higher integration, and so we expect this variable to be positively related to innovativeness.[27] Aiken and Hage (1972) found innovativeness to be moderately related to the number of interorganizational linkages, although there seemed to be no detectable relationship with the number of boundary-spanning roles in the organization. Aiken and Hage (1968) further noted that organizations with larger numbers of joint programs with other organizations tended to be more innovative. The role of boundary maintenance and boundary spanning was noted in particular by Thompson (1962, 1967), although not fully tested.

Organizational slack is the degree to which uncommitted resources are available to the organization. This structure variable is expected to co-vary positively with organizational innovativeness: "Slack provides a source of funds for innovation that would not be approved in the face of scarcity..." (Cyert and March 1963, pp. 278-279). Rasner (1968), in a study of hospitals, and Mohr (1969) in an investigation of health departments, found slack to be associated with innovativeness.

One other structural variable which has not usually been treated along with these others, but which probably has effects, is *efficacy*--the degree to which an organizational process is seen as being under technical control, subject to rational (i.e., cause-and-effect) transformation. This concept has been developed by Scott (1972), and represents a more sophisticated version of the "technology" variable discussed by Thompson (1967). Most industrial production processes are high in efficacy; on the other hand, a process such as teaching reading is faily low. The problem is complicated by the fact that organizations customarily have many processes with varying degrees of efficacy in operation at the same time. It should be possible to include some assessment of efficacy in relation to the processes affected by particular innovations; however, it is reasonable to suspect that efficacy may have some major multiplicative/interactive effects which it is dangerous to discard or handle only through controls.

Among the external structural characteristics of an organization, we suggest openness as a variable likely to be associated with innovativeness. *Openness* is the degree to which an organization engages in an exchange of information with its environment. Individuals who provide an organization with openness are called "cosmopolites." In most structures they are concentrated at the very top and the bottom. At the top, executives have wide travel and other types of contact with other organizations; they are in a position to obtain new ideas from sources external to the organization.

In contrast, individuals near the bottom of the organizational hierarchy also have a certain degree of cosmopoliteness, as they deal with the operational aspects of environmental change. For instance, the lower-level workers deal most directly with customers and clients, incoming materials, and energy.

The openness of a system increases the chances that an innovation, or the need for an innovation, will enter the organization. "The interaction between the organization and its environment is crucial to the innovation process" (Zaltman et al. 1973, p. 120). Utterback (1971a) concluded his study of 32 innovations in the instrument industry by stating that "the greater the degree of communication between the firm and it environment at each stage in the process of innovation, other factors being equal, the more effective the firm will be in generating, developing, and implementing new technology." Similarly, a study of innovativeness in 16 American and European steel companies found a correlation of 0.77 with the frequency of communication with the firm's environment (Miller 1971, p. 114).

Incidentally, Mott (1972) treated system openness as a *dependent* variable, resulting in his model from the degree of routine structure imposed by the organization on its input/output processes. He hypothesized a general trend toward system closure, or structuring of I/O, which is halted at a certain point by the exhaustion of efficacy, or technology (which we discussed previously), which requires the organization to rely on the environment. This formulation is obviously a highly dynamic process, with openness continually adjusting itself in response to organizational and environmental counter pressures. Whenever we use openness in a predictive model, we should be thus aware that it is in turn partly the product of our dependent variables, and subject to continual dynamic adjustment. This dynamism is certainly not confined to system openness, but it may be a more pronounced phenomenon than in the case of some of the more formal system characteristics such as centralization.

Our general paradigm of the main variables related to organizational innovativeness (Fig. 2) raises the issue of the relative importance of the internal versus the external structural variables. As was the case with structural and individual variables, there is probably a synergistic interaction between the internal and external variables. In terms of a research question for future inquiry: *What portion of innovations in an organization is imported from the organization's environment, and what share is invented through creative processes with the organization?*

Our paradigm (Fig. 2) is one attempt, based on past research and on theoretical considerations, to formulate a framework for the main classes of independent variables that are related to organizational innovativeness. "Badly needed is a theoretical framework which brings together the external and internal factors, the structural and psychological factors, and certain factors which have not even received mention up to this point" (Becker and Whisler 1967). We agree. Now it needs to be adequately tested.

THE INNOVATION PROCESS

This essay has questioned the intellectual value of the many past studies of organizational innovativeness, and in the previous section suggested some more profitable directions in which such research should proceed. Mainly, however, we recommend an entirely different approach to innovation in organizations. We refer to this relatively unexplored area as the "innovation process." We define the *innovation process* as the sequence of events from perception of a need in an organization, to consequences from the innovation after it is adopted to meet this need.

We see the innovation process as much larger in scope than the diffusion process; in fact diffusion is but one of the five stages in the innovation process. *One of the main shortcomings of past research on the diffusion of innovations is that it only dealt narrowly with the diffusion stage, and thus only considered a rather limited part of the entire process of innovation.*

Now we move to describing the five stages in the innovation process, showing how they must be modified somewhat in an organizational context, and then describing an approach to future research on the innovation process in organizations.

Stages in the Innovation Process

How is a new technological idea born, developed, diffused, adopted, and eventually replaced by another innovation? This innovation process is usually described as a series of stages.

1. *Need* (and idea generation). This stage consists of the recognition that a need for new information exists, that a perceived problem could be overcome by an innovation, that a "performance gap" exists between actualities and expectations (Zaltman et al. 1973, p. 2).[28]

2. *Invention*. A new idea is created, perhaps by recombining other ideas.

3. *Development*. At this phase, often called R & D, the invention is converted into a form that has greater potential utility.

4. *Diffusion*. The new idea, now called an "innovation," is spread throughout a system of units that adopt it.

5. *Consequences*. The innovation leads to various direct effects, and various indirect effects of these effects, some of which were originally intended and anticipated.

Then, of course, this innovation process may recycle, as new needs arise and are recognized; some of these needs may result from the consequences of the earlier innovation.

Naturally, the delineation of these stages is only heuristic, intended to facilitate descriptive understanding. The sequence is not invariable--in the case of organizations particularly, the actual "invention" may well precede the identification of needs, and in fact partly stimulate it. The states may take place within the organization or partly in its environment. In short, our model requires a good deal of interpretation to be useful in specific investigative cases.

The need-identification phase of the innovative process is particularly important, and very little studied or understood. In a sense, the "decision that a decision is needed" is in itself a process with "stimulation/initiation/etc." phases--although in this case the subject matter of the decision is purely conceptual/abstract rather than concrete/real. The output of *this* process is the organization's frame of reference in which actual innovation proposals are then considered and evaluated against the explicit or subconscious criteria resulting from the "decision-to-decide" phase. The effectiveness of the first phase is crucial to the effectiveness of any subsequent innovation decisions. The evaluation of this process by the researcher is particularly difficult because of the inherent difficulty of working with abstractions and because of the high threat to the organization associated with too-close investigation of their workings. Eventually, an explicit model of this part of the innovation process should be developed and explored in relation to subsequent behavior.

THE INNOVATION PROCESS IN ORGANIZATIONS

When we look at the innovation process in an organizational context, the exact stages from need to consequences must be modified somewhat. Table 2 shows the variety of terms that have been used by various authors for their conceptualized stages in the innovation process. A great deal of consensus about the meaning of each of these stages is evident. As our point of comparison in this section, we utilize Rogers with Shoemaker's (1971, p. 276) conceptualization of the collective innovation-decision process.

TABLE 2.--Comparison of stages in the innovation process in organizations.

Rogers with Shoemaker (1971, p. 306)	Hage and Aiken (1970, p. 113)	Becker and Whisler (1967)	Zaltman et al. (1973, p. 62)
1. Stimulation	1. Evaluation	1. Stimulus	1. Knowledge awareness
2. Initiation	2. Initiation	2. Conception	2. Attitude toward the innovation
3. Legitimation	--	3. Proposal	--
4. Decision	--	--	3. Decision
5. Action/ Implementation	3. Implementation	4. Adoption	4. Initial implementation
Consequences	4. Routinization	--	5. Continued sustained implementation

1. *Stimulation* is the stage in the innovation process at which someone becomes aware that a need exists for a certain innovation in an organization.
2. *Initiation* is the stage in the innovation process at which the new idea receives increased attention from members of the organization and is further adapted to the needs of the system.
3. *Legitimation* is the stage in the innovation process at which an innovation is approved or sanctioned by those who represent the organization's norms and values, and who possess formal and informal power.
4. *Decision* is the choice to act by the members of the organization.
5. *Action/implementation* is the overt behavior involved in effectuating the decision to adopt the innovation.

Certain individuals often play each of the specialized roles that correspond to each stage in the innovation process. For example, stimulators are often very cosmopolite and may only be temporary members of the organization. A different set of individuals usually are initiators; still others, legitimizers. Research might be directed to identifying who plays each of these roles for a particular innovation, and whether an innovation would die in midprocess if one of these roles were not played, and a stage was hence skipped.[29] Identifying these roles is not easy, as Langrish and others (1972, p. 14) warn: "Innovation is almost by definition a corporate and collaborative effort, and it is correspondingly difficult to disentangle the roles played by particular individuals."

Needed Research on the Innovation Process in Organizations

As stated previously, we lack an empirically based understanding of how communication and decision making are involved in the innovation process within organizations. What is needed is a research design different from that utilized in the past research on organizational innovativeness, one in which we learn "more about less," through detailed observation and interviews with personnel at various hierarchical levels in a very small number of organizations. This approach would allow the research to focus on certain of the concepts from the study of organizational behavior, such as organizational slack, formalization, openness, communication integration, etc. These variables are frequently encountered in the field of organizational behavior, but have not often been used as predictors of innovation in an organization. So the proposed research would represent a hybrid merger of two research traditions, both of which have made a partial contribution to understanding the nature of communication and innovation in organizations.

A further major advantage is implicit in the present proposal in that diffusion research in the past has focused on the *decision-making* aspects of innovation adoption, but it has almost totally ignored the consequences/effects of such innovation. Obviously, innovations are adopted in an effort to attain certain desired and expected consequences; adoption is motivated by a need-sensing process by which the organization becomes aware of a performance gap between actuality and what is possible. But past diffusion research has largely stopped at the point of decision, ignoring the *implementation* of the innovation and its consequences. In the case of the innovation process in an organization, implementation is especially important; often the implementers are a different set of individuals from the decision makers. And many innovation decisions, once made, are never fully implemented.

This process emphasis of the study offers a number of special advantages. Not only are we interested in assessing the effectiveness of the actual innovation-decision phases; the effectiveness of the preceding concept-formation and need-identification phases can also be treated. With such a time-process emphasis, it is possible to consider models in which the content of the outcomes of the earlier phases of the process are predictive variables at the later phases. It is also possible to consider that different combinations of structural and leader characteristics act in different ways at different stages--a set of hypotheses testable only through dynamic research approaches.

Thus, the proposed research design is distinctive in two major senses: (1) appropriate independent variables that indicate various dimensions of organizational structure are incorporated in our paradigm (rather than only the personal characteristics of the individual decision maker, as in many individual-unit diffusion researches), and (2) we focus on how innovation decisions are implemented in an organization, and with what consequences for the organization's performance and effectiveness (Fig. 3).

FIG. 3.--Paradigm of the usual diffusion research design (within dotted boundaries) and a proposed design for research on the innovation process in organizations.

INDEPENDENT VARIABLES	IMMEDIATE DEPENDENT VARIABLE	ULTIMATE DEPENDENT VARIABLE
I. Individual (leader) characteristics and communication behavior (education, attitude toward change, etc.)	Decision to adopt an innovation	Implementation of the innovation and its consequences for organizational performance and effectiveness
II. Internal structural characteristics (organizational slack, communication, integration, formalization, etc.)	Definition of situation; formation of criteria of need against which to evaluate innovations	
III. External structural characteristics (openness; efficacy)		

Obviously no one wants to promote the adoption of innovations by organizations just for the sake of innovation. But this is the implicit message of the organizational innovativeness studies, which all use innovativeness as their dependent variable. Seemingly, the more innovativeness, the better.

Instead, we feel that a more appropriate assumption in such researches should be to understand the process of organizational innovation so that eventually organizations can be more responsive to their needs. Rather than focusing on innovativeness per se, these investigations should study how needs are recognized, and transmitted into decisions to adopt appropriate innovations, which have certain consequences in improved organizational performance. So the ultimate dependent variable should be "organizational effectiveness" in its broadest sense (similar to the construct of Mott, 1972), involving elements of production (output), adaptability, and flexibility--in short, the basic purpose for which organizations exist--and "innovativeness" becomes an intermediate dependent variable, meshing with the need-identification process to lead to effectiveness. The innovation process thus can be seen as part of the whole spectrum of organizational activity, rather than an isolated phenomenon out of a system context.

As argued previously, a fruitful type of research for providing understanding of the innovation process in organizations is an in-depth comparative systems analysis of one or two pairs of organizations, one of which is highly innovative and one of which is not. Thus in each pair, the organizations differ widely on the main variable of interest, innovativeness. The basic research approach is then to identify and specify the various independent variables related to innovativeness; by analyzing organizations that are at the extremes on the continuum of innovativeness, isolation and identification of the independent variables is facilitated.

Naturally, each of the two pairs of organizations should be of similar size, engaged in the same function, and otherwise matched on "control" variables that are not of important theoretical interest in explaining innovativeness in organizations. The central research question here is: "What are the main organizational structure variables that lead one organization to be more innovative than another?"

Earlier, we demonstrated why we also intend to investigate the implementation process in organizations by which innovation decisions are put into action, and with what effects/consequences for organization performance (Fig. 3). This focus calls for an in-depth and over-time analysis of how the need for an innovation occurs in an organization, how the innovation is searched for and how it enters the organization, how the innovation decision is made and by whom, how the implementation process occurs and with what consequences for organizational performance and effectiveness. This is essentially a "tracer" type of study of innovation. We suggest utilizing the following sources of data in each organization of study: (1) observation, (2) personal interviews with members of the organization at various levels in the hierarchy, (3) access to records, publications, and other written sources in the organization, and (4) discussions with knowledgeable individuals who are outside the organization but who have contact with it.

There are several possible units of analysis in research on the innovation process in organizations:

1. The *organization*, as when one might compare the process of innovation in one organization with that in another organization.
2. The *roles* played by individuals, or sets of individuals, in the innovation process; for example, stimulators.
3. The *decision* to adopt an innovation.
4. The *innovation*, by tracing its path through the structure of the organization.

We have called for a kind of "micro"-level research on the innovation process in this essay. Research on a fundamental process like innovation is needed at the micro level, so that the corresponding macro level can be more thoroughly understood, and, if desired, altered through the public policy process.

Thus the central question for future research is to determine the nature of communication and innovation in organizations, and in this way learn more about how bureaucratic organizations could become more responsive to appropriate change.

We close with these words from Machiavelli (1947, p. 15): "It must be remembered that there is nothing more difficult to plan, more doubtful of success, nor more dangerous to manage, than the creation of a new system."

NOTES AND REFERENCES

1. We wish to thank the following colleagues in the Innovation Group at the University of Michigan for their critical review of an earlier draft of this paper: Michael Moch, Herbert Schuette, Jack L. Walker, Kenneth Warner, and Sidney Winter.

2. One attempt at intellectual intergration of these two fields has recently been made by Zaltman et al. (1973).

3. These studies are synthesized by Rogers with Shoemaker (1971).

4. To be more specific, we understand something about certain characteristics of organizations that are related to "innovativeness" as a general concept. We have not yet the type of comparative studies that would enable us to draw conclusions about "innovativeness" in relation to specific activities or resources in cross-context organizational settings.

5. The main ideas in the following section are adapted from Rogers with Shoemaker (1971), where a more detailed treatment of this sketchy description of the diffusion model can be found.

6. The wording of this definition implies that the unit of adoption might be an individual (as in most past diffusion research) or an organization (Zaltman et al. 1973, p. 10).

7. Zaltman et al. (1973, pp. 32-50) suggest numerous other attributes of innovations such as point of origin, gateway capacity, etc., when the innovations are diffusing within organizations.

8. So authority innovation-decisions often result in a rapid rate of adoption, but in a relatively low-*quality* decision that cannot effectively be put into action, at least over an extended period of time.

9. Perhaps some question might be raised as to whether formulation of the diffusion approach truly constituted a paradigm, or only a "quasi-paradigm." The conceptualization was rather distinctive for its time in the social sciences, and it certainly set off a great number of following researches.

10. A fundamental reason, we believe, for this pro-change bias in diffusion research is due to the tendency for diffusion researchers to look at the process from the source's viewpoint in turn may stem from the sponsorship of most diffusion research by sources of innovations (that is, by change agencies).

11. There are more appropriate research methods for studying the processual aspects of diffusion: Field experiments, longitudinal panel studies, and simulation. All such approaches rest on the necessity for gathering data from the same (or similar) respondents at several points in time. One of the few organizational diffusion studies to gather data at even *two* points in time is Hage and Dewar (1973).

12. But it may be possible to *infer* certain of the "why" relationships among variables that are measured cross-sectionally.

13. An organization is a social system composed of individuals working toward an objective that they seek to attain through their collective efforts.

14. Except for a recent handful of studies of organizational innovativeness, which incorporate such organizational variables as centralization, formalization, etc., as independent variables; for an illustration, see Hage and Aiken (1970).

15. Availability of a research technique or a conceptual model encourages its further use. This is Kaplan's (1964, p. 28) Law of the Instrument, which we stated at the beginning of this essay: "Give a small boy a hammer, and he will find that everything he encounters needs pounding." Give social scientists the classical diffusion model and they find that it can be applied to organizations as it was to individual units of adoption.

16. It is interesting that some of the early diffusion studies, prior to the Ryan and Gross (1943) study of hybrid corn, were investigations of organizational innovativeness. Examples are McVoy (1940) and Pemberton (1936). Then the organization was largely forgotten as a unit in diffusion research until about the mid-1960s.

17. Mohr also constructed another measure of organizational innovativeness based on the number of man-years (or dollar-equivalents) added to the department for innovative programs in the previous five years. This variable seems to be a measure of "organizational resources devoted to innovativeness."

18. The reverse procedure might occasionally be appropriate: to measure the independent variables at t_1 and the dependent variables at t_2 as in Hage and Dewar (1973). Some understanding of the meaning of the amount of time lapsed would seem to be mandatory in this case.

19. In fact, other researchers like Mytinger (1968) found that individual leader characteristics are of considerable importance in explaining organizational innovativeness. Most researchers report that the individual variables make at least some contribution to predicting organizational innovativeness, the exact contribution depending in part on exactly which leader variables are measured. Doubtless considerable improvement could be made both in the conceptualization and the measurement of individual variables to be related to organizational innovativeness.

20. A serious methodological problem in this study, as in many others, is the question of multiple causation--in short, the degree to which the structural variables of the model are tapping the same basic source of variability (a general case of the problem noted above). This is a common and troublesome problem. Use of part correlations and multiple analyses of variance can help, and stepwise-regression procedures can bring the problem to the surface; Mott (1972) includes a section on this problem. Another problem is the degree to which their findings are generalizable to other types of organizations. Again, this problem is not unique to research on innovation in organizations.

21. The contextual variables included the population size of the community served, its median income, etc.

22. Although if we regard the organization as a sum of *roles* rather than individuals --a point of view consistent with open-system theory--the actual size of the organization is not necessarily equal to the number of individuals in it. Not all individuals have an equal number of roles or interactions with other roles, and thus contribute unequally to the overall dimensions.

23. Hall (1967) showed that size alone was a poor predictor of structural variables but confirmed that it was associated with many other factors, and thus would play a "stand-in" role in any analysis in which it was included.

24. This is the basic structural dilemma posed by Wilson (1966, pp. 193-218): Decentralization and other structural variables that facilitate the initiation of innovation in an organization tend to impede the innovation's implementation. Sapolsky (1967) found evidence to support Wilson's theory in a study of innovation in department stores.

25. Gordon et al. (1974) found that for their sample of hospitals, the average innovativeness scores of the decentralized hospitals were about 2.5 times greater than those of the centralized hospitals; the data are still being examined for specific findings.

26. Communciation integration is usually measured by means of the matrix multiplication of sociometric data about communication flows or patterns among the members of an organization. Detail on computer matrix multiplication procedures for measuring communication integration is provided by Guimarães (1972).

27. We expect vertical communication integration to be especially important in innovation in organizations, as in suggested by Aiken and Hage's (1971) findings on upward communication flows, which can carry feedback to executives from the operational level.

28. Our emphasis on need as a stage in the innovation process illustrates the "demand" aspect in the diffusion of innovations, just as the change agent and other communication channels represent the innovation "supply" aspects. The role of clients in making demands for innovative services and other behaviors in organizations has been noted by other investigators.

29. Note our earlier discussion of the necessity for a theory of innovation to be able to deal with failure as well as success.

13. TECHNOLOGY FORECASTING LITERATURE: EMERGENCE AND IMPACT ON TECHNOLOGICAL INNOVATION
James R. Bright

This chapter examines the relationship of technological forecasting (TF) and its literature to innovation. How has TF literature emerged and how has it contributed to innovation? How might it affect innovation in the future? What further research is needed to develop better technology forecasts? Because there is great public confusion between technological *speculations* and *prophecies* (hopes, fears, entertainment, or advocacy) and technology *forecasting* (based on relatively explicit modes of reasoning), I have tried to illuminate that distinction. Because technology forecasting methods are growing yearly in rigor and richness and summaries are not easily available, I have included an appendix reviewing TF methods in brief.

WHY TECHNOLOGY FORECASTING IS OF GROWING IMPORTANCE

Society long ago learned the value of anticipating the character, intensity, and timing of major environmental forces in social and economic activities. Forecasts of the weather, agricultural production growth, industrial production, markets, sociological change, government spending, economic conditions, political attitudes, and many other attributes of future conditions are regarded as essential to planning wisely or to dealing effectively with coming changes. The wisdom of trying to anticipate things that influence the future is well proven, even though forecasts are almost certain to lack perfection.

It is thus strange that society has been very slow in coming to grips with the forecasting of technology. It has been clear for two decades and quite apparent to thoughtful men for about 100 years that technology is one of the dominant forces in the world. Why have we been so slow to realize the need for systematic ways to think about the technological future? We have the astonishing situation in which the National Science Foundation--the custodian and guide to the nation's technical future--had done little to improve insight and methodology for anticipating that technological future until the early 1970s.

TECHNOLOGY IS OF INCREASING SIGNIFICANCE

Today technology is one of the most powerful forces in our environment, and at times by far the most important for many firms, institutions, nations, and society at large. The power of technology for good and bad is growing. Its increasing impact stems from a number of factors.

• *The degree of advance in the technical capability* of many new devices and materials over their predecessors often is in multiples of improvement, not a few percentage points. These gains in performance are so great that they abruptly and drastically alter the means, effects, time, or costs of doing things. Thus they disturb, for better or worse, existing practices, institutions, and human activities. For example, consider the speed of commercial computation by computer systems (Table 1).[1] Notice that the degree of advance is in multiples of previous performance and not in percentage points. The same phenomenon appears in many other things: speed of transportation, power of explosives, energy storage, miniaturization of electronic circuits, the impact strength of plastics, and the speed of typesetting.

Dr. Bright, formerly professor of technology management and associate dean at the Graduate School of Business Administration of the University of Texas at Austin, has resigned to devote himself to lecturing and consulting, notably with the Industrial Management Center in Austin.

TABLE 1.--Speed of commercial computer computations.

Date	Computer	Operations/second
1944	Harvard Mark 1	0.403
1946	ENIAC	44.65
1951	Univac 1	271.4
1953	IBM 701	615.7
1961	IMB 7074	31 650
1962	Univac 1107	76 050
1963	RCA 601	58 880
1963	CDC 3600	156 375
1965	IBM 360/75	1 437 806

- *The rapidity of introduction of technical successors* seems to be increasing in many fields. Therefore, each technology concept tends to have a shorter marketplace life because of prompt challenge from a new and superior technology. The phenomenon is most apparent in the extremely short marketplace lives (1-2 years) of various solid-state electronic components, but it is also shown in many other products, where a given design has a marketplace life of nearer to five years than the ten years of a decade ago.
- *The size of resources required* often is so great that the R & D funding capacity of individual firms and even an entire industry is exhausted. The American SST is a case in point; the British-French Concorde transport has required two nations to combine their resources. Rolls-Royce and Lockheed represent two firms in which the financial demands of a single technical advance have been staggering.
- *Technology is a national resource:* in the 1960s both society at large and governments began to recognize that they can and must marshal technological resources for national purposes other than war. This possibility had long been recognized in technical circles. Now need to improve environmental qualities, food supply, and the general quality of life, and the need to explore space and the oceans have come to be widely accepted and demanded. Thus technology is being called upon more frequently and more severely to serve public needs.
- *Technology assessment* as a formal, legal requirement has resulted from concern over the negative effects of technology on the environment and society. This concern has grown since the late 1960s. Of course, alarm about men, machines, unemployment, and skills is an old worry, going back more than two hundred years. The new development is intense concern about other effects of a technology, such as the pollution associated with a production process or product use, and about the delayed consequences of technology as it affects the environment years later or as it leads to new social problems.

By 1973 the President's Council for Environmental Quality and the Environmental Protection Agency were established, environmental impact statements were required in many major programs, and new environmental controls had been proposed and adopted by many agencies. Examples are the Occupational Safety and Health Act and consumer protection regulations. In 1969 the first factories were ordered closed for producing noxious fumes, noise, and stream pollution. The Office of Technology Assessment has been staffed to advise Congress. It is clear that society intends technology to be held accountable for its total impact. Technology assessment is thus becoming a requirement of Federal technology proposals. Assessment requires anticipation and hence a need for technology forecasting.

Since governments and firms have been making hundreds of thousands of plans for at least several hundred years without much explicit technology forecasting, why could society not get along as it has in the past? Why, then, should we not make use of the traditional forecasting procedure--the opinion of the technical expert?

THE DIFFICULTIES INHERENT IN EXPERT OPINION

Interaction of Technologies. The use of an expert's opinion may be very logical but is becoming an increasingly dubious procedure. One reason is that the former ability of a competent technical man to assess progress in his special field is disappearing because of the growing interaction of technologies. For instance, if one wanted to forecast the future of computer technology, should his "expert" be an arithmetic logic mathematician, a solid-state physicist, a memory systems designer, a manufacturing process engineer, an electronic circuit engineer, or some other specialist? Obviously, none of these is adequate because single-field knowledge is no longer capable of assessing total improvement in these technical devices. Technical progress is multidisciplined, but technical experts are not, at least not to the same degree.

The obvious improvement to an individual expert is to form a committee. However, even if one assembles a group representing the appropriate technical experts, opinion alone is becoming less satisfactory because much technology change is becoming less autonomous. New technology is stimulated by interactions with social conditions, political actions, economic conditions, and ecological pressures. One can hardly expect the technical experts also to be competent forecasters in all these nontechnology spheres. Consider forecasting the emergence of the SST airplane, of a new detergent, or the antiballistic missile system. Are the controlling forces to be found only from the study of technology? Figure 1 suggests five environments that interact with each other to influence technical progress.[2]

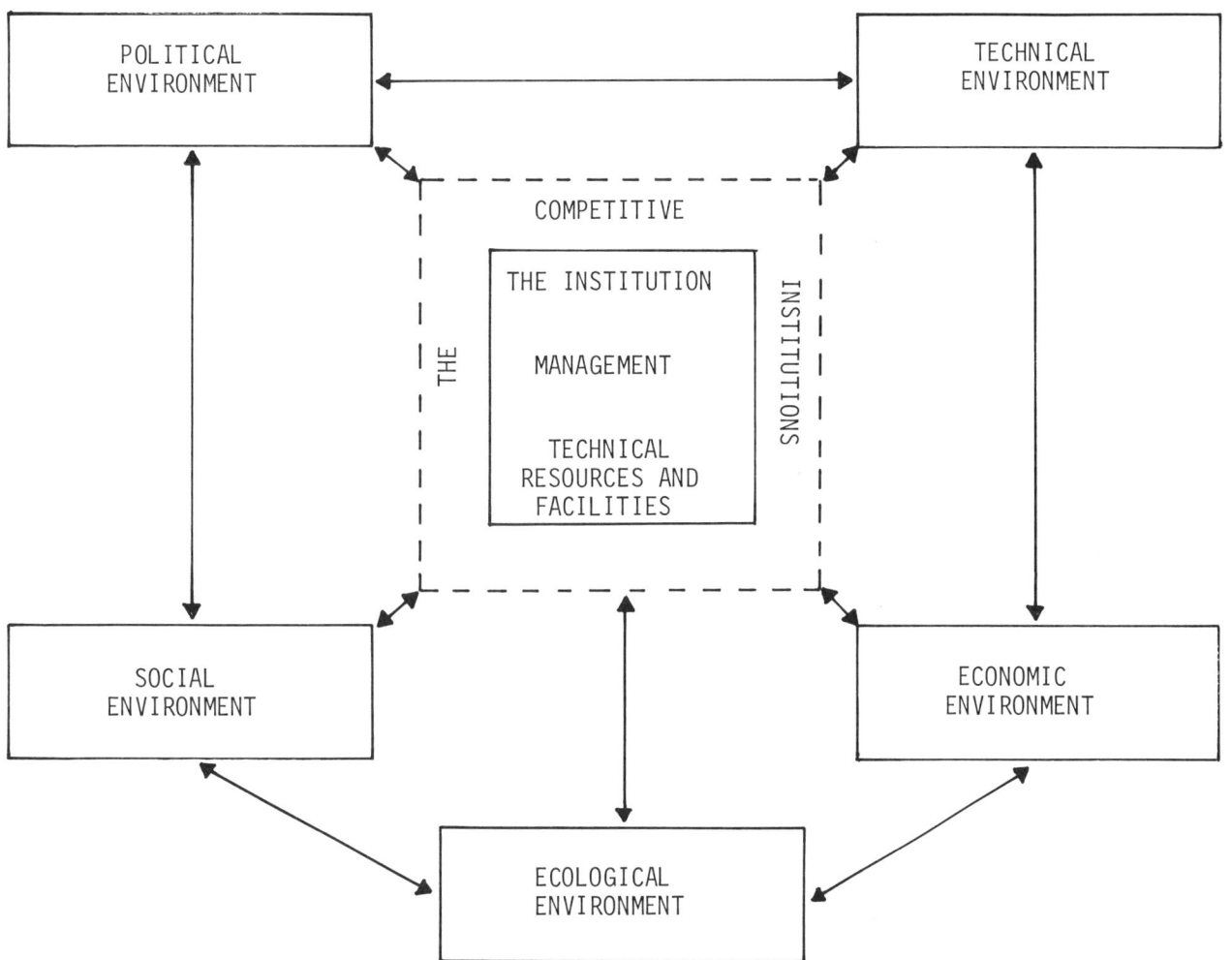

FIG. 1.—Technology and its environmental interactions.

301

Furthermore, there is a vast and often unappreciated distinction between forecasting technical development in the laboratory and forecasting technology in use. Technical knowledge *alone* cannot possibly provide a correct basis for predicting the timing of emergence, diffusion, and impact of new technology.

Value Changes. Behind these environmental relationships is another kind of change only beginning to be dimly understood: the changing value system in our society. Values are virtually a no-man's land to most, and one can hardly expect today's technical expert to predict what value systems will prevail a decade or more from now, when today's laboratory model emerges as a commercial product. Yet these value systems often determine what technological choices society supports.

Methodology for Technology Forecasting. This subject was not taught in engineering schools or scientific courses until about 1969. Past forecasting by expert opinion has largely (but not always) rested on individual knowledge, experience, belief, and intuition. As a result, many scientists, engineers, and inventors of unquestioned technical competence have produced astonishing prediction errors. It is equally true that these experts have had many successes. The significant point is that the forecast user has no way to assess the knowledge, the rationale, the assumptions, and the range of considerations that underlie the technical opinion. Therefore, the prediction that one gets from opinion is of highly uncertain quality, and it cannot be tested and evaluated. It is not a satisfactory basis for decisions committing major resources of an institution.

Given these difficulties, it is clear that society badly needs something better than opinion for forecasting technology.

WHAT IS TECHNOLOGY FORECASTING AND HOW IS IT BEING APPLIED?

Technology forecasting is defined as a quantified prediction of the timing and of the character of the degree to change in attributes of technology associated with the design production, and use of devices, materials, and processes, according to a specified system of data and/or reasoning. The prediction may or may not include an estimate of probability or confidence in the amount of change or its timing. In some cases, relative need, importance, or usefulness can be quantified and used as the basis of the forecasting procedure.[3]

The key distinction of this definition is that the forecast is reproducible through a system of logic; thus it differs from opinion and prophecy in that it rests upon an explicit, stated set of relationships, data, and assumptions. The reasoning procedure yields relatively consistent results independent of the analyst.

Another definition often offered basically reflects the first sentence above but then adds,...*that can be achieved with a specified level of support.*[4] This is a *conditional* forecast; an important philosophical difference is implied. The first definition is a statement of *what will occur.* The latter definition is a statement of what *could occur* under various degrees of effort. Is that more properly planning than forecasting?

The line between forecasting and planning is very difficult to draw, particularly because the ultimate goal of more forecasting is to serve as an aid to planning. Rather than argue over the definition, we should realize that these two concepts are often confused or intertwined--*the prediction of what will materialize and the conditional statement of possibilities*.

What rationales underlie the arguments that technology forecasting is possible by other than opinion? There are at least three that seem to have logical foundations:

1. Technological attributes, as measured by parameters such as strength, pounds per horsepower, speed, abrasion resistance, etc., usually grow in an orderly manner over time or through experience (meaning "cumulative production"). The underlying hypothesis is that there is a consistency or pattern in technological changes and abrupt, major deviations are not common. Consistency provides a basis for trend extrapolation from past data. This is called *exploratory forecasting*.

2. Technology responds to needs, to opportunities, and to the provision of resources. If these supporting forces and definitions of goals can be identified, then the technological progress can be anticipated as a response to that pressure and support. Years ago, this rationale may have seemed quite theoretical. However, it has increasing

validity because of a growing political willingness to use national resources to support the technology leading to desirable social goods. EPA standards on air and water, man on the moon by 1970, and energy independence by 1980 are typical definitions of national goals. Such societal change may justify the *goal-oriented* or *normative forecasting* approach.

3. Forecasting of new technology can be done through monitoring the process of technological innovation, a process that takes years. Signals of this progress can be identified and monitored as the innovation moves through the stages identified in Table 2, which uses a hypothetical cancer cure to illustrate the points.[5]

TABLE 2.--Levels of emergence and impact of technology.

Level 1. That certain knowledge of nature of scientific understanding will be acquired by....*Example:* We shall know the cause of cancer by 1980.

Level 2. That it will be possible to demonstrate a new technical capability on a laboratory basis by....*Example:* We shall have drug to control cancer in mice by 1982.

Level 3. That the new technical capability will be applied to a full-scale prototype or under conditions of field trial....*Example:* We shall have experimental success in curing cancer in people by 1985.

Level 4. That the new technology will be put into first operational use or will be commercially introduced (the first sale) by....*Example:* a drug for cure of cancer will be commercially introduced about 1987.

Level 5. That the new technology will be widely adopted as measured by such things as number of units in use, units of output of the new technology, dollars or sales generated, or any measures expressing the use of the new technology relative to competitive technology by....*Example:* Cancer-curing drugs will be used in 90 per cent of all cancer cases by 1988. Or, sales of cancer-curing drugs will account for 30 per cent of drug industry income by 1988.

Level 6. That certain social (including economic) consequences will follow the use of the new technology by....*Example:* Reduction of death from cancer will add five years to the average life-span by 1990.

Level 7. That future economic, political, social, ecologic, and technical conditions will require the creation or introduction of certain new technical capabilities by.... *Example:* Elimination of cancer as major cause of death will mean that the replacement of deteriorating organs with mechanical substitutes will become the major medical need by the year 2000. The increased elderly population will require vastly expanded retirement communities.

SOURCES OF CONFUSION AND CRITICISM OF TF

People interested in technical progress have various goals and concerns; hence, they look to TF for different information and sometimes for totally illogical things. TF is then criticized for not delivering to expectations! Some expectations commonly encountered are:

• TF can *"tell us what business we should be in in 1985."* Surely such a prediction calls for a decision based on forecasts of many aspects of the future, including assumptions about politics and economics, as well as the firm's resources, its alternatives, and the managerial skills and goals of its leaders 10-15 years hence. A point to stress is that *the output of TF is data about technology--not a decision about management options.*

• *TF attempts to predict profitability of great new inventions.* This commonly repeated accusation simply is not true. To be sure, everyone would like to have such predictions, but these are predictions about technical unknowns *and their economic aspects,* including the general economic climate, markets, and the economics of present and potential technical options. TF *contributes* to such a forecast, but more than technology is involved in reaching a forecast of profits.

• *TF tries to predict "scientific breakthroughs."* Another false accusation. No technology forecasting methodology has been presented with claims of predicting "breakthroughs." Popular and purportedly serious writings may include many predictions about "breakthroughs," but they are based on opinion or prophecy, not on TF methodology. Until now no technology-forecasting methodology has been offered that claim to predict "breakthroughs." However, TF may point to the need for a "breakthrough" or the likelihood of technology reaching a threshold of great technical significance. This was precisely the use of a TA (based on trend extrapolation) at the USAF Avionics Laboratory in 1952. Figure 2 shows the curve that was used to point out to Air Force management that the current 5% failure rate in vacuum-tube circuitry meant that the B-58 bomber "would never fly" (Bright 1972, pp. 2-6). It was statistically impossible to keep so many circuits operational; a new technology was needed. This study led the Air Force to fund $10 million for research on solid-state circuitry. In 1958 Kilby at Texas Instruments produced the first integrated circuit out of this funding effort. Here is a forecast that heavily influenced the emergence of a great contemporary innovation.

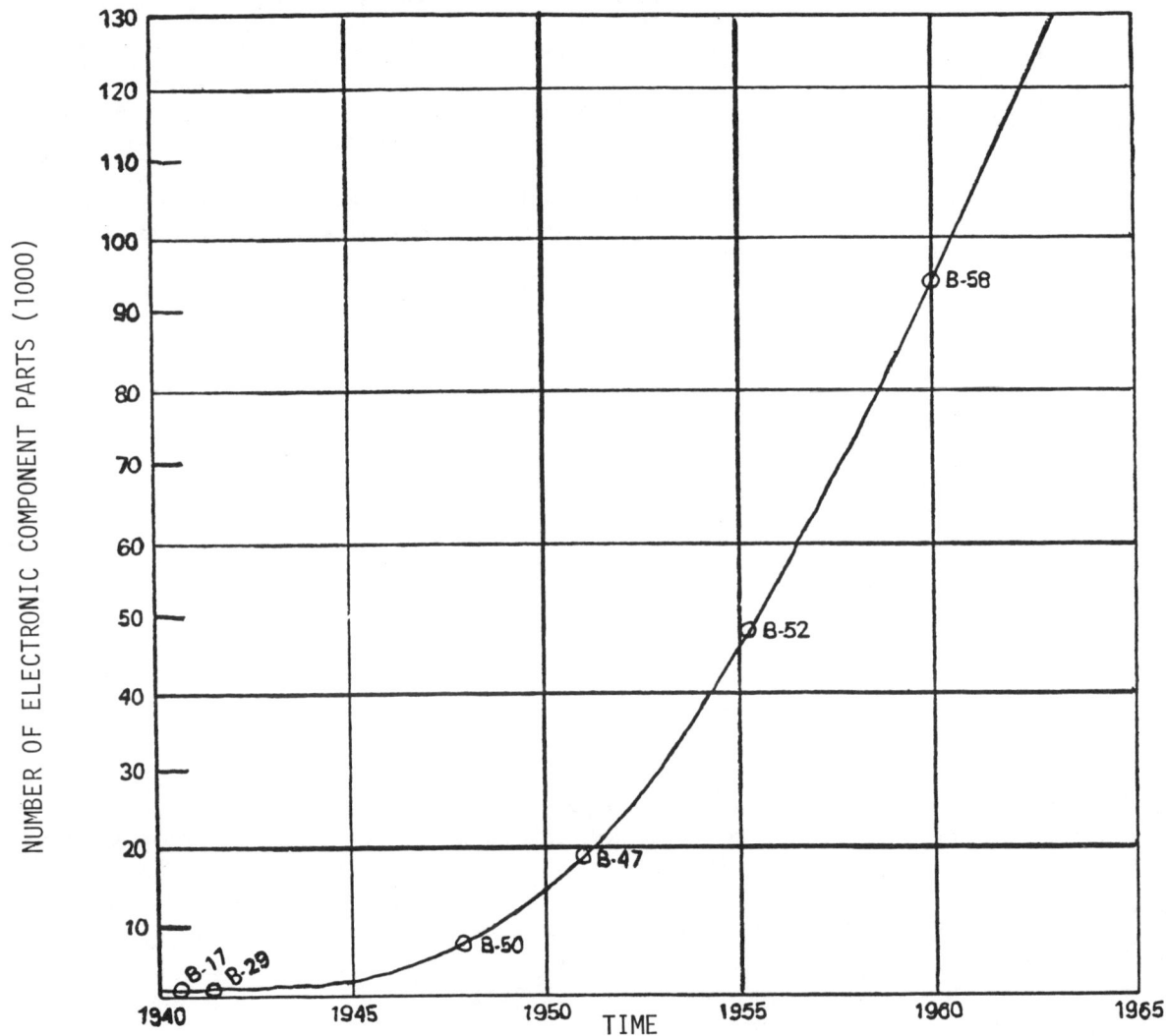

FIG. 2.--Trend of complexity of electronics in USAF weapons systems. (Source: H. V. Noble, USAF Avionics Laboratory, Wright Field; drawn about 1952.)

Martino has explicitly challenged the generally accepted notion that we cannot predict some "breakthroughs." He defines them as "something which appears to transcend prior limitation" and argues that there are precursor events in science which long foreshadow the coming of radical new advances. By using monitoring systems to detect and follow signals of technological change we should be able to avoid many (although not all) technical surprises. To support his idea he developed the "signals" for the coming of atomic power, beginning with Einstein's 1905 paper on mass-energy equivalence and followed by a long and perfectly visible set of events leading to chain reactions in 1942; and similarly for the laser, the field-effect transistor, stereo broadcasting, and space travel. His prediction process rests on the definition of a breakthrough as "technology coming into operational use," as distinct from the discovery of a new scientific principle. Given this caveat, he defends his point soundly (Martino 1972, pp. 210-249). The same monitoring technique was taught in 1964, on the intercontinental ballistic missile (ICBM) as the replacement for the bomber,[6] without however reaching back into science and attacking the "breakthrough" syndrome as does Martino.

- *TF is just another name for something that economists have been doing all along.* Without denigrating the work of economists, we suggest that this is a gross misunderstanding of content and purpose of TF. It is true that economists have been deeply concerned about technology for decades.[7] It is also true that some of the techniques of TF, particularly trend extrapolation, have long been used by the economists. The differences between economists and technology forecasters lie in the parameters used and in the purposes of the forecast. The economist usually works with economic parameters of a technology, such as cost or investment. Dollars do not measure technical capabilities such as the number of electronic circuit functions per square inch. Moreover, it is a fallacy of insidious reasoning (frequently perpetrated by politicians, popular journalists, and vested industrial interests) that technical performance and achievement are necessarily proportional to funding. Doubling investment in research in drilling techniques does not necessarily double the rate at which a drill penetrates the earth's mantle. Dollars are sometimes valid in measuring the rate of emergence of technology from idea to physical reality; nor do they bear systematic relationship to the character or amount of advance in a technical capability.

Quite often, the economist does work with the inputs and/or outputs of a technological device, such as power consumption of tons of product per unit of time. These parameters are of course also useful as technological forecasting data.[8] Whereas the economist's interest are properly directed at economic relationships, the technologist is concerned with technical relationships so as to grasp the significance of future technical states. The manager and government analyst are concerned about both economics and technology, for the technical position ultimately will have economic, institutional, and social impacts.

- *Emotional reactions to the notion of TF also have led to criticism.* Because of personal experiences, hopes, or interests, some moderately to highly charged attitudes have been encountered. The professional scientists or engineer rejects forecasts that violate apparent scientific laws and his own hard-won knowledge of limitations to technical progress--a very understandable reaction.

History also displays a long record of technical speculations by politicians, journalists, consultants, inventors, union leaders, college professors, financial analysts, and social philosophers. Even scientists and engineers are guilty. These popular speculations are often prophecies, hopes, fears, and propaganda. They are usually directed to the social effects or impact on social values by a technical development. They are rarely based on an explicit forecasting procedure. Neither the merits nor demerits of speculation should be charged to TF--but they are!

Confusion also is caused by the present and growing wave of "futurist" writings, meetings, studies, and societies. The "futurists" are largely concerned with "what should be." These well-meaning and sometimes brilliant, useful, appealing, or alarming views argue that (a) man can, to an increasing degree, create the world he wants and should therefore study, choose, and act on desirable goals; (b) technology is the key to solving man's problems and he must support and guide it; and (c) technology is the villain behind man's problem and he must turn it back or at least slow it down.

This type of futurist work includes thoughtful social speculation, "poetry," and even political propaganda about the need for goals, the choice of goals, or the best paths to these goals. Many futurists are amazingly uninformed and disinterested in rigorous forecasting of technology (or economics, political forces, or social forces, for that matter). However, they are strong on speculating and prophesying about parts or all of the future. Many of the present futurist writings seem to be based on personal or group opinion about cause, effect, and desirability, not on methodological forecasting of anything.[9]

"Technology is not the sole determinant of the future." Of course it is not, and TF does not pretend that it is. The goal is to develop understanding of TF in the same way that economic forecasting has acquired depth and usefulness.

"Man does not know how to forecast technology." Wrong. Forecasting technology has had its start, as this literature review will show.

Nothing in the above discussion denies that opinions and speculations may be more accurate than formal forecasting, nor is their usefulness denied as an early-warning system or a framework for societal planning. We should simply realize that prophecy and forecasting are two different things, and should be treated accordingly.

LEVELS OF EMERGENCE AND IMPACT

Much confusion and pointless criticism can be avoided by the recognition that a technology forecast may be addressed to any one or several levels of emergence and impact and should not be unthinkingly applied to other levels. At least seven levels are evident in Table 2.[10]

Levels 1 through 4 are simply predictions of technology moving through different stages of technical achievement. Level 5 implies a prediction about *adoption* of the technology. Adoption depends on all other competing technology, as well as its economic attractiveness, other economic conditions, social acceptability, etc. This is the level on which most economic forecasts about technology operate.

Level 6 may or may not include the assumptions of level 5, depending on whether it concerns micro or macro usage. However, it adds to technological prediction other predictions of the effect on human behavior and of the response of people and institutions.

Level 7 implies a synthesis of forecasts about many attributes of future society plus some value judgments about the goals that could and will be supported by society.

It should be clear that levels 5, 6, and 7 involve much more than technology and probably should not be thought of simply as technology forecasts. It is increasingly erroneous to move into levels 5, 6, and 7 on the assumption that only technology is involved.

The scientist or engineer is perhaps intuitively capable of some sound predictions on levels 2, 3, and possibly 4. However, he moves increasingly farther from his area of expertise as he deliberately or unconsciously translates his technical predictions to level 5--economic growth. Managers and government officials often unwittingly encourage potential error by expecting the technologist to tell them "how important" the new technology will become. The technical man then unwittingly steps into the quagmire of predicting total environmental interactions.

Forecasts for levels 5, 6, and 7 involve a complex set of interactions about which very little is known. Traditionally, estimates of technical progress arose largely from analysis of factors in the technical and economic spheres. Today, much technology progress is controlled by events and trends in other enviornments--the political, social, and ecologic. We have no explicit, disciplined procedure for forecasting these interactions. Systems Dynamics, in the manner of the *Limits to Growth* study, proposes to link these interactions (Meadows et al. 1972). The dynamic modeling procedure is far more "integrative" than any other forecasting concepts, although it is heavily criticized (erroneously) for not embracing "values" and "behavior."

THE USES OF TECHNOLOGY FORECASTING

As a result of the growing forecasting literature, the spread of TF courses (particularly postuniversity short courses for industry), and the number of highly publicized, dramatic techno-economic events since 1970, TF is spreading rapidly into long-range-planning groups. The traumas of the energy crisis, the environmental problems, the troubles of Rolls-Royce (jet engines), Du Pont (Corfam), GE, RCA, and Xerox (computers), Lockheed (Tri-Star), and the Concorde situation on one hand; the dramatic success of Xerox, IBM, Texas Instruments, etc., on the other; and the fantastic advances in electronics, chemicals, medical matters, etc., have sharpened institutional interest in TF throughout the world. The demand for TF education continues to grow. Several specialized consulting firms have mushroomed in response to this demand from industry and government agencies. An analysis of the hundreds of examples in the TF literature (discussed later) shows that TF is being used for the many purposes tabulated in Table 3. More examples can be found scattered throughout the references. One can sum up the general impact of a TF program in the following general terms.

TABLE 3.--Uses of technology forecasting as seen in current TF literature.

1. Identify needs and opportunities.
2. Identify problems and threats.
3. Clarify implications and necessary immediate actions.
4. Set design goals for products.
5. Open the long-range speculative future (legitimizing imagination).
6. Establish relationship to competitors' technology.
7. Establish timing, scale, and nature of facilities.
8. Establish the time at which new product-design specifications will be needed.
9. Estimate rate of market penetration and demand implications.
10. Identify new possibilities in products and processes.
11. Analyze competitive technology approach to achieving a given capability.
12. Detect technical shifts in the firm's activities.
13. Provide a basis for understanding and reflection.
14. Guide to the distribution of research resources.
15. Guide to identifying new research equipment and skill requirements.
16. Explore limiting or influential factors.
17. Communicate technological situations to others.

A TF program helps in the planning and management of technological efforts because it forces explicit consideration of the technological future by management, by functional departments, and by technical personnel. That consideration is based on methods of reasoning that can be analyzed, tested, challenged, and refined, in contrast to opinion, tradition, and superficial assumptions.

It has also been found that a TF program leads to a better understanding of the forces that drive or control technology in one's field. It encourages the forecaster and forecast user to consider interactions between technology and societal changes. Finally, the activity demands that forecast participants and users think beyond their day-to-day activity--something which managers regularly demand and rarely get!

TECHNOLOGY FORECASTING LITERATURE AND ITS INFLUENCE ON THE INNOVATION PROCESS

To grasp the past impact and potential influence of TF literature we must first classify the types of writings that are sometimes included under the TF label. These writings vary not only in objectives and methodology, but they have shifted over time.

Given the earlier definition of TF we must reject the prophecies, fancies, serious speculations, amusing stories, and visionary imaginings about technology and society that extend from the earliest records by man to the present day. Whether correct or incorrect, they do not rest on reproducible logic and data, and one has no way of examining the validity of assumptions and reasoning. Beginning in the late 1700s and appearing strongly by the early 1900s, scientific and engineering knowledge of the day often, but by no means always, becomes part of the basis for the predictions. The "intuition" or subjective experience and subconscious feelings of the technical expert are the basis for the forecast. In the 1920s and 1930s the first glimmerings of forecasting technology by some sort of explicitly expressed logic begin to appear. Table 4 demonstrates the confusing mix of technique and purpose, all still going on simultaneously today!

Examples of all these types of statements about the future (a) have been at times labeled "technology forecasts" and criticized or admired as such; (b) have and have not been influential in creating attitudes and changing actions; and (c) have proved to be correct and incorrect descriptions of the technical future that actually emerged. We are left with a muddled collection of technical prophecies that were right or wrong and based on no expressed reason; expert opinions right and wrong for many reasons; and carefully constructed forecasts that have proved to be brilliantly accurate or far off the mark. We also see heavy use of propaganda as in writings on automation, nuclear warfare,

TABLE 4.—Significant aspects of TF literature.

A. Level of Impact or Emergence Predicted	B. Objective of the Predictor	C. Qualifications of Predictor Relative to the Prediction	D. The Rationale or Methodology Used
1. Acquisition of scientific and technical knowledge	1. To entertain and amuse	1. Technical competence in the predicted area	1. Offhand opinion
2. Emergence of a new capability for societal use	2. Expression of concerned interest	2. Technical competence in un-related technology(ies)	2. Carefully reasoned opinion involving feedback from others
3. Specific technical means of achieving a capability	3. To warn mankind	3. Competence in a discipline interfacing with technology (economics, resources, sociology)	3. Speculation and fantasy
4. How the capability will be used by society	4. To anticipate problems and/or benefits of new technology	4. Novelist	4. Projection of history (apparent trends, analogies, and predecessor events)
5. Impact on one or more societal interfaces (economics, resources, other technology)	5. To prescribe or define utopias	5. Generalist--philosopher	5. Assumption of causality (from current or future developments)
6. Impact on individuals and their behavior	6. To aid management decisions regarding the use of new technology	6. Advocate of a desired future condition	6. Specification of desired goals
7. Impact on social structure and institutional behavior	7. For self-entertainment and reflection	7. Journalist	7. Construction of models to expose future conditions
	8. Propaganda to gain support	8. Spokesman for an institution	8. Parrotry of predictions by others

Note: No judgment of forecasting merit is implied in the order of these factors. It is obvious that the literature of the future, including TF, represents many different combinations of items from all columns.

nuclear power, environmentalism, etc. *Examples of all of them, right and wrong, have occasionally been very influential!*

Literally thousands of "expert opinions," in which accomplished technical men predicted developments with little or limited explanation of rationale, often have been very influential, whether right or wrong. For example, radio came to life as it did because a young Italian student, Marconi, was inspired when he read predictions such as the 1892 prediction by the British physicist William Crookes (in *Fortnightly Review* 51:173-181, 1892):

> Here is unfolded to us a new and astonishing world--one which it is hard to conceive should contain no possibilities of transmitting and receiving intelligence.
>
> Rays of light will not pierce through a wall, nor, as we know only too well, through a London fog. But the electrical vibrations of a yard or more in wavelength...will easily pierce such mediums, which to them will be transparent. Here, then, is revealed the bewildering possibility of telegraphy without wires, posts, cables, or any of our present costly appliances...
>
> This is no mere dream of a visionary philosopher. All the requisites needed to bring it within grasp of daily life are well within the possibilities of discovery, and are so reasonable and so clearly in the path of researches which are now being actively prosecuted in every capital of Europe that we may any day expect to hear that they have emerged from the realms of speculation into those of sober fact.

This examplifies an "expert opinion" technology forecast, based on knowledge of the technological priples. The forecast was correct, and it did influence technological progress.

Contrast that with another "expert opinion" forecast in 1950 by Norbert Wiener, the famous MIT mathematician. In writing about "cybernetics," the control and communication in the animal and the machine, he foresaw the impact of control devices and computers on the factory (in *The Human Use of Human Beings*) as follows:

> I should give a rough estimate that it will take the new tools ten to twenty years to come into their own. A war would change all this overnight....Under these circumstances, the period of about two years which it took for radar to get onto the battlefield with a high degree of effectiveness is scarcely likely to be exceeded by the period of evolution of the automatic factory....Thus a new war will almost inevitably see the automatic age in full swing within less than five years....
>
> Let us remember that the automatic machine, whatever we think of any feelings it may have or may not have, is the precise economic equivalent of slave labor. Any labor which competes with slave labor must accept the economic conditions of slave labor. It is perfectly clear that this will produce an unemployment situation, in comparison with which the present recession and even the depression of the thirties will seem a pleasant joke.

This apparently logical forecast proved to be very wrong in timing and degree of impact. This quotation came to be associated with the automation furor of the mid 1950s. It was cited hundreds of times in speeches and writings by union leaders, especially in the 1954-55 bargaining campaigns of the United Auto Workers. It was also repeatedly used by management, labor economists, sociologists, and the general press. Yet 25 years later, manufacturing labor has not been wiped out, the automatic factory is not a widespread reality, and a depression has not yet been caused by automatic controls despite two wars (including the longest war in our history). Here, then, is an *erroneous forecast that was very influential* in wage negotiations, job classifications, and rate setting.

Another expert opinion forecast typifies influence through a different sort of forecast--the conditional forecast. When Einstein wrote his famous letter to President Roosevelt in 1942, he predicted, in effect, that is is possible to achieve a certain technical capability if the nation will support it. This was a forecast based upon *deep scientific understanding,* given virtually no publicity, yet *incredibly influential* in changing not only technical effort, but the course of world history.

Because of this muddled mix of what is a technological forecast, who made it, how it developed, and what was its purpose, many criticisms have been levelled at TF without justification.

TYPES OF PREDICTION

Technical Prophecy

The earliest literature of this type is found largely in legends and religious writings characterized by soothsaying, entertainment, speculation, and moralizing. There is little scientific reasoning behind them. Serious thinking about the technical future is evidenced in Roger Bacon's statements (1256-1266) about the possibilities of submarines, flying machines, and talking and reasoning automata. The utopias of Francis Bacon include mention of new technology, but without much technical substances. Leonardo da Vinci presents some brilliant technological forecasts that were based on technical insight. It is noteworthy that da Vinci simply described technical possibilities without specifying timing or impact, whereas many of the utopists were concerned with the impact on mankind and society.

Utopian fiction and utopian essays categorize the next emergent literature of future-oriented, quasi-technology predictions. These writings can be defined as the deliberate consideration of the social impact of future technological developments without a tightly reasoned technical logic. In some cases, the predictors do give a technical basis for their speculation, but their technical rationale is of limited depth and breadth. Often it is simply a guess, an opinion, or a hope by an imaginative man who chooses to speculate about the use and impact of technology in the future.

Motives for these writings vary from sheer entertainment and thoughtful speculation to social comment and criticism, and even philosophical views of man's fate. They embrace expressions of hope and despair in technology. This literature begins with the French publicists of the mid-1700s, and includes the French utopists Turgot, Condorcet, and Mercier. In England, Francis Bacon may have been the first utopist to visualize science employed to improve man's future. His *New Atlantis* (1622) was the forerunner of a long string of utopian essays and novels. By the mid-1800s many more utopias have been published, including those of the German Johann K. Friederich, who seemed to have been inspired by the age of steam, and Emilier Souvestre, who in 1945 wrote a satiric criticism of a world in which steam and electric machines did the work and thinking. It is apparent that utopian prophecy is gradually being replaced (but only in part) by technical speculation and science fiction.

Technical Speculation (Beginning in the Early 1800s)

This term includes the predictions of persons who have solid technical competence in some field, but are speculating about the technical future without a strong database or very explicit system of logic. Quite often the topic of speculation lies outside the predictor's technical discipline and the writings often relate to the impact of technology rather than its technical attributes.

Science Fiction and Fiction About Future Social Conditions

Such work includes many predictions about technical possibilities and the impact consequences and social reaction to technology. The works may be fanciful or serious. Motivations behind such writings include entertainment, satire, thoughtful speculation, and social concern. Some science-fiction predictions have been astonishingly accurate; many were based upon a sound extension of technical knowledge.

Some utopian fiction about society and technology has become recognized as outstanding literature, and has been highly influential in affecting societal attitudes. Three outstanding examples are Edward Bellamy's *Looking Backward, 2000-1887*; Aldous Huxley's *Brave New World*; and George Orwell's *Nineteen Eighty-four*.

The foregoing types of literature have been admirably surveyed in *The Tale of the Future, From Start to 1961* by I. F. Clarke. This bibliography annotates nine hundred British books (no articles) on the future. Clarke has since published a series of papers in *Futures* reviewing forecasting since 1763. His "First Forecast of the Future" begins in the June 1969 number and Clarke's papers continue to appear in almost every number of that journal. Clarke does not critically examine the methodology employed over the years. However, his writings are a revelation in showing the enormous amount of technical prophecy and speculation in British literature.

The first serious student of *methods of technical perdiction* is S. Colum Gilfillan. As early as 1911 he proposed (unpublished) the study of successful predictors with view to learning their methodology. He continued these studies to the late 1960s and in 1968 provided us with by far the most comprehensive survey of technology forecasting activity from 1750 to 1967. His summary chapter, "A Sociologist Looks at Technological Prediction," is a masterpiece with almost a hundred references (Bright 1968).

Expert Technical Opinion

The most logical source and common method of TF since the late 1800s has been the opinion of technical experts. Sometimes the prediction was motivated by the imagination, concern, or interest of the individual technical expert. In the early 1900s and just prior to World War II, we find that many journalists interviewed and/or publishers encouraged accomplished scientists, inventors, academicians, and engineers to write about future technology and its impact. Professional societies have also been the platform for many predictions by men of science and engineering. For instance, in 1922 Marconi was honored at a joint meeting of the Institute of Radio Engineers and the American Institute of Electrical Engineers in New York. In his acceptance speech he proposed that radio waves be used to detect the presence and bearing of a ship in bad weather. A. H. Taylor and L. C. Young, from the U.S. Naval Research Laborabory, took their first steps toward the development of radar following this inspiration. This instance exemplifies the use of expert opinion to trigger research and development of a major technological innovation.

The output of such predictive proposals continues to this day. It has been intensified and multiplied through the proliferation of professional societies and trade associations, and their journals, as well as the trade press. A common experience is for such agencies periodically to launch a survey of possibilities and needs in "the next x years in the field of _____."

Formal Technical Committees

Governments frequently convene groups of scientists and engineers to consider technological proposals and possibilities, as well as social situations and technical responses. It has been argued that each major advance in aerospace came about because the U.S. government convened a highly skilled and eminent committee to examine an issue and propose a certain technical advance, such as the ICBM or the establishment of the National Aeronautics and Space Administration (NASA). These committee reports are in fact technical forecasts, but are rarely labeled as such. These reports (a) embody elements of TF; (b) until the late 1960s, have not been labeled or thought of as "technology forecasts;" and (c) often have been highly influential in leading the decision makers to move in certain technological directions.

The Emergence of Formal Technology Forecasts

The concept of explicit technological forecasting designed as an aid to planning and decision making by major institutions apparently begins in 1937 (Inouye and Susskind 1977). Under the aegis of the Department of Interior, the National Resources Committee, through its Science Committee, established a Subcommittee on Technology. The Subcommittee was headed by William F. Ogburn, the distinguished sociologist who was then doubtless the nation's intellectual leader in concern for the study of technology as a force affecting the structure and welfare of society. A 388-page report, *Technological Trends and National Policy*, was published in June 1937. Secretary Ickes's letter of transmittal to President Roosevelt states that the report was "the first major attempt to show the kinds of new inventions which may affect living and working conditions in America in the next 10 to 25 years (U.S. National Resources Committee 1937, p.iii).

The objective of the report was not to encourage technology or identify new technical opportunities to society, but to indicate potential problems associated with the use of the coming inventions and to suggest national policies that would permit adjustment "with the least possible social suffering and loss." The Subcommittee stressed that their concern *was not so much with technology as with planning directed at the social effects of using that technology*. Hence, Part I of the report dealt with social aspects of technology, especially resistance to technological change, unemployment, and productivity. A very brief Part II dealt with "Science and Technology," and Part III consisted of nine sections reviewing individual fields such as agriculture, communication, and power.

Only one chapter in Part I dealt specifically with *methods* of forecasting. Here S. C. Gilfillan discussed "The Prediction of Inventions" and argued that we could learn how to predict better by examining the records of past predictors, noting their methods and areas of success and failure, and deducing some useful insights for the forecaster. His principal conclusion was that the combined efforts of technical men and social scientists need to be united to criticize, refine, and develop better systems of prediction. As specific suggestions for TF methods he argued that:

(a) Inventions form trends, hence allow extrapolation.
(b) Inventions take decades to emerge. The median time from conception to commercial success was 33 years for 19 inventions introduced between 1888 and 1913. Therefore, he argued, one can predict social effects for "inventions already born." This was a clear forerunner of present concept of forecasting by monitoring.
(c) Inventions have "abundant and clear causes" and "it remains to show this wide, causative base can be used for the prediction of an invention. And this we have not learned to do scientifically" (U.S. National Resource Committee 1937, p. 19). We are still applying some techniques such as relevance trees, mission flow analysis, various types of market studies, and scenarios to identify needs, possibilities, and relative importance. This is his "causative" notion embodied in TF.

Gilfillan rejected "optimism," "utopian prophesying" and "wishful thinking" as being highly misleading. He stressed at great length the need to look at technical alternatives for meeting a social need, because many studies showed that functionally equivalent inventions emerged to satisfy a given need. Thus, Gilfillan in 1937 gave us the nucleus of the concept of systematic technology forecasting. He recognized the need to build a database, the use of extrapolation, the principle of monitoring, and the principle of normative (goal-oriented) forecasting.

Since the basic purpose of this Federal report was to assist in developing national policy to ameliorate the impact of new technology, what were the findings? Eleven conclusions about inventions, technology, and unemployment were presented. There were six recommendations.

1. Start planning agencies studies of certain inventions likely to have major social effects, such as the mechanical cotton picker, air-conditioning equipment, plastics, the photoelectric cell, artificial cotton and wool-like fibers, synthetic rubber, prefabricated houses, television, facsimile transmission, the automobile trailer, gasoline produced from coal, steep-flight aircraft, and tray agriculture. (A most interesting list, thought provoking in the different timings and impacts as we are experiencing them now.)
2. Set up a joint committee representing various government agencies to follow and study the potential impact of new technology on unemployment, with dissemination of this advice to industry and labor. (At the President's White House Conference on Automation in 1960, Walter Reuther made the same plea, but it still has not be implemented.)
3. Develop Federal agencies that would continuously study "trends of science and invention" and their possible effects as they affect the work of the departments and agencies.
4. Review the patent system to adapt it better to changing conditions.
5. Investigate the feasibility of collecting data on inventions and discoveries with the goal of gathering information on the more socially significant items in one place.
6. Establish "a permanent over-all planning board" a National Resources Board--to bring the interaction of coming technology and social structure (as well as other factors) before the multitude of Federal, State, county, and city planning groups.

One of the main findings was that "there is usually opportunity to anticipate an invention's impact on society *since it never comes instantaneously without signals*" (U.S. National Resources Committee 1937, p. vii). In other words, some technology forecasting is feasible.

What was the impact of these recommendations? Apparently, the report had very little effect. Only Recommendation 4 on revising the patent system has been implemented, although not for quite the purpose implied in the study.

Recommendation 3 is observed quite widely and rigorously by NASA and by the Departments of Defense, to some extent Agriculture, and Labor. Here and there it appears in some agencies such as the Energy Research and Development Agency and the Federal Aviation Administration. However, by and large there is little continuous study and planning for future technology in most of our government departments. Despite the very vigorous and imaginative efforts of Dr. Herbert Holloman as Assistant Secretary for Research and Development in the Department of Commerce in the late 1960s, this type of effort has so far not come to life in that department.

Recommendation 6 is now coming into being, but only in a very limited way, in the establishment of the Office of Technology Assessment.

Recommendation 2 has been pursued spasmodically by special studies and commissions (on automation) generally in or through the Department of Labor. No advisory service on potential impact has been established.

The vision of Gilfillan and Ogburn in arguing the need for technology forecasting for *policy* purposes was not translated into action at that time, and is only partially implemented today in fragmented form throughout the U.S. government.

TF AS A GUIDE TO MANAGEMENT

The first major organizational effort to forecast technology so as to improve the quality of management decisions was conducted by the U.S. Army Air Corps in 1945. The forecast was the result of the vision of Henry H. Arnold, Commanding General of the Army Air Forces. In 1944 he asked Theodore von Kármán to form a Scientific Advisory Group to give the Air Force scientific guidance in planning their research and development in the 20 years ahead. Arnold's letter of instructions to von Kármán is superb in envisioning the role of technology forecasting. It was included in von Kármán's first report.

HEADQUARTERS, ARMY AIR FORCES, WASHINGTON

7 November 1944

MEMORANDUM FOR DR. VON KARMAN

Subject: AAF Long Range Development Program

1. I believe the security of the United States of America will continue to rest in part in developments instituted by our educational and professional scientists. I am anxious that Air forces postwar and next-war research and development programs be placed on a sound and continuing basis. In addition, I am desirous that these programs be in such form and contain such well thought out, long range thinking that, in addition to guaranteeing the security of our nation and serving as a guide for the next 10-20 year period, that the recommended programs can be used as a basis for adequate Congressional appropriations.

2. To assist you and your associates in our current concepts of war, may I review our principles. The object of total war is to destroy the enemy's will to resist, thereby enabling us to force our will on him. The attainment of war's objective divides itself into three phases: political, strategic and tactical. Political action is directed against resources, and tactical action against his armed forces. Strategical and tactical actions are our main concern and are governed by the principles of objective, surprise, simplicity, mass, offensive, movement, economy of forces, cooperation and security.

3. I believe it is axiomatic that:

a. We as a nation are now one of the predominant powers.

b. We will no doubt have potential enemies that will constitute a continuing threat to the nation.

c. While major wars will continue to be fought principally between the 30th and 60th parallels, north, global war must be contemplated.

d. Our prewar research and development has often been inferior to our enemies.

e. Offensive, not defensive, weapons win wars. Countermeasures are of secondary importance.

f. Our country will not support a large standing army.

g. Peacetime economy requirements indicate that, while the AAF now receives 43% of current War Department appropriations, this allotment or this proportion may not continue.

h. Obsolete equipment, now available in large quantities, may stalemate development and give Congress a false sense of security.

i. While our scientists do not necessarily have the questionable advantage of basic military training, conversely our AAF officers cannot by necessity be professional scientists.

j. Human-sighted (and perhaps radar or television assisted) weapons have more potential efficiency and flexibility than mechanically assisted weapons.

k. It is a fundamental principle of American democracy that personnel casualties are distasteful. We will continue to fight mechanical rather than manpower wars.

l. As yet we have not overcome the problems of great distances, weather and darkness.

m. More potent explosives, supersonic speed, greater mass offensive efficiency, increased weapon flexibility and control, are requirements.

n. The present trend toward terror weapons such as buzz bombs, phosphorus and napalm may further continue toward gas and bacteriological warfare.

4. The possibility of future major wars cannot be overlooked. We, as a nation, may not always have friendly major powers or great oceanic distances as barriers. Likewise, I presume methods of stopping aircraft power plants may soon be available to our enemies. It is not now possible to determine if another totally different weapon will replace the airplane? Are manless remote-controlled radar or television assisted precision military rockets or multiple purpose seekers a possibility? Is atomic propulsion a thought for consideration in future warfare?

5. Except perhaps to review current techniques and research trends, I am asking you and your associates to divorce yourselves from the present war in order to investigate all the possibilities and desirabilities for postwar and future war's development as respects the AAF. Upon completion of your studies, please then give me a report or guide for recommended future AAF research and development programs. May I ask that your final report also include recommendations to the following questions.

a. What assistance should we give or ask from our educational and commercial scientific organizations during peacetime?

b. Is the time approaching when all our scientists and their organizations must give a small portion of their time and resources to assist in avoiding future national peril and winning the next war?

c. What are the best methods of instituting the pilot production of required non-revenue equipments of no commercial value developed exclusively for the postwar period?

d. What proportion of available money should be allocated to research and development?

6. Pending completion of your final report, may I ask that you give me a short monthly written progress report. Meanwhile, I have specifically directed the AC/AS, OC&R (General Wilson) to be responsible for your direct administrative and staff needs. Also, as I have already told you, I welcome you and your associates into my Headquarters. May I again say that the services of the AAF are at your disposal to assist in solving these difficult problems.

/s/H.H. Arnold

The Scientific Advisory Group was established on 1 December 1944 and von Kármán set about selecting permanent staff and part time consultants. The team left for Germany in

May 1945 to study first hand German R & D on high-speed aerodynamics and power. On its return von Kármán wrote a report, "Where We Stand," which became the first of 13 studies making up the series, *Toward New Horizons* (Scientific Advisory Group, U.S. Army Air Force, 1945). A team then went to Japan in September 1945, and on their return the rest of the series of reports was written. One recommendation was that a permanent scientific advisory group be established. In June 1946 a permanent Scientific Advisory Board of 30 members met for the first time, and continues to advise the Air Chief of Staff to this day.

Toward New Horizons was initially classified, and only one study of parts of this report has been available to the public. T. J. O'Connor's doctoral thesis identifies developments in three areas: missiles, jet engines, and jet aircraft.[11] O'Connor found that the recommendations generally were followed, although it was Army Ordnance, not the Air Force, that pursued missile development. The WAC Corporal missile and the Nike missile materialized through this forecast.

The recommendation to fund an ICBM in 1945-1950 was not followed, partly because of funding conditions and partly because of Air Force attitudes. Research on sweptback and delta-winged craft was pursued and produced useful results. Work on tailless aircraft and flying wings was recommended, pursued, but not fruitful. The Arnold Engineering Center was established as a result of the forecast effort in *Toward New Horizons*.

What TF methods were used? That is not very explicit in the study. Two modes of prediction seem dominant. First was the reasoned explanation of significant technical needs, of technical posibilities and progress, and the tradeoffs that favored one line of development over another. It was a kind of goal-oriented, data-and-opinion based forecast. The other mode was simply "expert opinion" without a specified data base or mode of reasoning. Some of these results were quite wrong. For instance, von Kármán said that pilots would utilize the "prone position in very fast interceptors," and his forecast that atomic power would first be applied to propulsion of aircraft rather than stationary power plants was dead wrong. O'Connor concluded that the committee's philosophy was basically, "try everything." Yet they overlooked satellites.

Nevertheless, the significance of this forecasting effort lies in the attempt systematically to examine the technological future and relate possibilities and realities to the goals of a major organization. The effort had major impact on the Air Force technology, but because of secrecy requirements *it had virtually no impact on the methodology of TF or even on the concept of formal technology forecasting.*

After von Kármán's technology forecasting studies, concern for technology accelerated in the military establishment because of international relations. It became painfully obvious that the Cold War was a fact of political life and that the growth of technical capabilities in nuclear weapons, communications systems, aircraft, and ballistic and guided missiles, among other things, was a certainty. The sophistication and cost of new weapons systems absolutely demended better technical planning.

The National Research Council had been making studies (forecasts) prior to World War II, which predicted many scientific advances, although they missed such things as atomic energy, radar, jet propulsion, and antibiotics. After von Kármán's effort, many forecasting studies were launched. *Project Lexington*, made in 1948 for the U.S. Air Force, resulted in the Mariner class of merchant vessels and the atomic depth charge; *Project Charles*, in the establishment of Lincoln Laboratories at MIT. The *Lamp Lighter* report led to the building of the Distant Early Warning (DEW) line, and a 1953 study led to the decision to build the ICBM.

In 1955 the U.S. Army produced a one-time "Technical Capabilities Forecast" through each of its technical services. An official document, "Technical Capabilities Forecast, FY 59-60," was produced in 1957 by the U.S. Army Ordnance Corps. The report outlined ideas for military devices deemed feasible. It was widely circulated and suggestions solicited. Responses were compiled in a document, "Vast and Half-Vast Ideas," and circulated throughout Ordnance Corps laboratories in hopes that it would stimulate more ideas. In 1960 the Army's R & D chief ordered each technical service to prepare a long-range technological forecast; this policy continued annually until 1963. Under reorganization of the technical services the Army Materiel Command took over responsibility of preparing an annual "Long-Range Technological Forecast," which continues.

An Important Change in TF Usage

The character of Army TF changed significantly throughout the 1960s. Initially, the Forecasts predicted types of military devices and dates of availability. However, the forecasters and tactical planners did not always envision the same needs, nor did the technology predicted produce a militarily balanced system. Also, the tactical planners did not anticipate technical possibilities that could affect military tactics. In 1967 the Army therefore began to integrate forecasting and planning. Their TF reports now stressed "what we could do, if we wish," and operational planning documents provided, "what we would like to do." A system of interchange and iteration of these two types of studies merged into the Army's final R & D plans. Their system of planning then led to preparation of three TF volumes, which were updated every two years. Volume 1 covered "Scientific Opportunities," vol. 2, "Technological Capabilities," and vol. 3, "Advanced Systems Concepts" (Bird and Darracott 1968). The combined result has heavily influenced Army R & D work. Methods used include expert opinion, monitoring of worldwide technical developments, special in-depth studies by technical experts, and trend extrapolation.

THE ORIGINS OF TECHNOLOGY FORECASTING METHODOLOGY

TF methodology has its roots in the work of early anthropologists, historians, and sociologists. In 1877 Lewis H. Morgan wrote of the "exponential acceleration" of human progress. In *The Education of Henry Adams* (1918) there is repeated recognition and emphasis on the exponential growth of energy and power.[12]

Here begins formal recognition of exponential technical progress. Apparently, the first attempt to quantify principles of social change (including measures of the impact of certain technologies such as speed of transport, range of missiles, etc.) were presented in 1931 by the sociologist Hornell Hart in *The Technique of Social Progress* (1931). Two TF concepts were formalized: (1) the identification of technical parameters relevant to the purpose of the forecast, and (2) the use of time series of these parameters and their extrapolation. Many citations and charts of the accelerating character of social culture (meaning patterns of behavior) are given in "Acceleration in Social Change," a chapter by Hornell Hart available in two books (1957; 1972). The work of Gilfillan and Ogburn, discussed earlier, comprises the next contribution of sociologists to technology forecasting. However, I cannot find that the TF efforts of these sociologists had any effect in encouraging or delaying innovation. This very perceptive work did not spread into engineering and science education, or into technical planning for the government or industry. Indeed, even most of their own sociologist colleagues have abandoned this field of study and are amazingly ignorant of their work.[13]

The economists have long recognized the interaction of technology and economics. A landmark paper by Siegel appeared in 1953. In the last 20 years or so there has been growing interest and concern among economists over the economics of research and development and technological innovation. Very careful and thorough economic scholars have attempted to develop the relationship of R & D expenditures to technological innovation on national and industry levels. The work of Edwin Mansfield is probably the leading effort in this respect, and many others are making important contributions.[14] The difficulty of such economic work related to technological innovation lies in two aspects: (1) the macro aspects of the available economic data, and (2) in the ambiguity and imperfect reflection of technical progress in the economic data.

Case histories show that some major technological innovations have been brought into being for very little money. To illustrate, consider the following four types of problems with economic data regarding technological innovation. First, beginning in 1934-35, Carlson brought the principle of xerography through the stages of innovation including design concept, patent, laboratory demonstration and crude model (incomplete) of an office copier for roughly $3000 and about 2000 hours of spare time! This was the basis of a business grossing Xerox over $2 billion in 1974. Battelle and Haloid (later Xerox Corporation) carried out further R & D and commercialization of xerography from 1946 to "first profits" in 1953 for an R & D expenditure of perhaps $2 million--a very small expenditure for an enormously profitable innovation.

Second, the correct economic data cannot be ascertained without exhaustive research and adjustment of accounting records. For instance, Xerox chose to include *engineering* expenses for the design and construction of the production equipment along with the conventional R & D expenditures. This approach is contrary to the practice of the majority of firms, I believe. Therefore, Xerox's "R, D, and E" (research, development, and engineering) figures *are inconsistent with much other industry data*. Other firms and government agencies assign similar expenses in different ways and vary them at times.

Third, Xerox since 1960 or so has clearly demonstrated that R & D expenditures do not necessarily produce innovation. They have spent many millions on basic and applied research aimed at producing major new innovations, but are frank to admit (privately, at least) that they have had no success to date. They have, of couse, done a superb job of product refinement and product-line development with xerographic copying. Nevertheless, no great new area of innovation has yet resulted from the millions they have spent on R & D aimed at other subjects. Finally, their financial success in xerography is due very substantially to their mode of charging for the technology. An analysis of three modes of selling (outright sale of machines, leasing by the month, and a lease varying with usage--basically a per copy charge) showed that in the first five years of marketing the Model 914 office copier their "usage" concept produced roughly two times the net income that outright selling would have produced ($660 million vs $330 million) without even considering income flow from depreciation and residual use in leased equipment.[15]

This example is provided only to demonstrate that economic data on R & D activities and on technological innovation is a very limited forecasting tool. That is not the economists' fault, but lies with their materials. Furthermore, economic data does not predict the technical attributes of future technology, and we cannot expect economic forecasting to serve this aspect of technology studies.

THE EMERGENCE OF MODERN TECHNOLOGY FORECASTING METHODOLOGY

About 1959-1960, Ralph C. Lenz Jr., a civilian employee of the U.S. Air Force Systems Command, adapted his Sloan School thesis to an Air Force monograph, *Technological Forecasting*. This booklet rapidly disseminated throughout the military services and some contractor firms. A second edition was distributed in 1962. Lenz's document summed up the state of the art and added some insights of his own. Basically, he proposed six TF methods:

1. Extrapolation of technological trends.
2. Growth analogies, in which it is assumed that a new technology's growth is analogous to biological growth.
3. Trend correlation, then defined as a causal impact of one variable on another.
4. Interdependent relationships, in which the trend of one parameter depends upon the relationship of a number of parameters whose trends must be examined collectively.
5. Trend characteristics, in which one examines the projections of a trend and their possible interactions with "barriers."
6. Dynamic Forecasting, in which a model in the manner of Forrester's industrial dynamics (Forrester 1961, 1969; Chen 1972) is made to represent a future technology situation.

Independently, I launched a course at Harvard Business School in 1960 on "Technological Innovation." This work led to a search for methods of forecasting technological progress in a manner that would be useful to the manager. Early TF materials were reproduced in Harvard Business School case files as "cases" and "notes" in 1960-1962 and were based heavily on Gilfillan and Lenz. One new concept was formalized in this course: forecasting by *monitoring*.

Monitoring is based on the fact that the process of technological innovation takes time measured in decades, and that "signals" of coming technology are evident long before they have societal impact. Forecasting by monitoring was the basis of an exercise on predicting the coming of the missile as a replacement for the bomber. All available TF items were published in a textbook, *Research, Development, and Technological Innovation* (Bright 1964). This text appears to have been the first effort at the teaching of methods of technology forecasting to business (and a few engineering) students. The text also reprinted a basic chapter summing up technical prediction by Gilfillan. Perhaps a dozen companies contacted me about the use of these methods in their work, but no results are known. The influence of this textbook was merely a small stimulus to the teaching of technological innovation; it introduced TF teaching material at perhaps 30 or 40 business schools and a few engineering schools, but probably had little effect on technological innovation.

Another, separate development in the early 1960s was the rise of international interest in TF. Member nations in the Organization for Economic Cooperation and Development (OECD) had, in the 1950s, given strong support to basic research through their respective ministers of science. Investment in basic research was generally applauded as an act of faith. Science was assumed to be the route to economic growth. OECD's Committee for Science Policy therefore supported many international studies of basic research, science, technology, and economic interactions. Through the early 1960s the titles of their publications show a gradually increasing awareness that support of science was not producing all the economic benefits hoped for. As a result, their studies shifted to a closer examination of technological innovation. The OECD Secretariat and staff carefully investigated the literature of invention and innovation. This effort led to European contacts with some American managers and a few academicians who had been concentrating on the "innovation" aspect of technology rather than the "basic research" theme.

Out of this perception and concern the OECD engaged Erich Jantsch as a consultant. His assignment was to prepare a summary of the state of the art of technological forecasting. From October 1965 to May 1966 Jantsch visited twelve OECD countries and Israel. He made approximately 250 contacts and also evaluated 400 literature references. The resulting report, completed in August 1966, was published as the OECD paperback *Technological Forecasting in Perspective* (Jantsch 1967). This monumental survey had great impact. Despite a negligible promotional effort by OECD, the demand for the volume forced two printings of over 8000 copies. The book was widely studied and reviewed with various degrees of understanding, enthusiam, and criticism. Quite apart from the book's great merits, this reaction comprehensively demonstrated the worldwide hunger for and experimentation with ways of thinking vigorously about the technological future. Jantsch's findings were three:

1. TF emerged as a *management* discipline (*not* an engineering or science discipline!) around 1960. It had shifted from the "informed opinion" approach of the mid-1940s to systematic evaluation. TF had been adopted increasingly by military, organizations, research institutes, and industry since the late 1950s.

2. "The value of technological forecasting has been proved," not only by accuracy but by usefulness in defining long-range strategies.

3. TF was "not yet a science, but an art," with gradually evolving techniques, and was undergoing accelerated development as he write. Jantsch identified more than 100 versions of techniques under some 20 approaches.

He estimated that 500 to 600 medium and large-sized American firms had established a TF function as part of their operations, and that they spent approximately $50 million for it in-house and externally--about 1 per cent of their R & D expenditures. Jantsch had to reach these conclusions through a tenuous set of assumptions, and many Americans doubted that 500 U.S. firms had "established a TF function."

However, there was no doubt that since World War II thousands of individuals in science, engineering, and management functions in industry and government had been conducting TF studies whether or not they were so identified. The military underwrote numerous major contracts for long-range technical planning with RAND Corp., Boeing, Lockheed, Honeywell, RCA, General Electric's TEMPO, North American Aviation, and many other firms from the mid-1950s on, involving thousands of investigators who were in fact doing TF work. In addition to a superb international survey of TF activity, Jantsch provided a highly useful summary and annotated bibliography of TF literature and related studies. His book was such an excellent reference souce that it became the major literature landmark itself.

The first U. S. book on *methods* of TF appeared in February 1968. *Technological Forecasting for Industry and Government* was the edited and rewritten proceedings of the first conference for industry on *methods* of TF. It included items by the contributions of twenty-three consultants and authors from industry, government, and academia.[16] In the

Preface I argued that "expert opinion was an unsatisfactory basis on which to make decisions about technology, as evidenced by many past errors." TF should "rest upon an explicit set of quantitative relationships and stated assumptions." The book had several impacts. First, it presented a "how-to-do-it" slant that encouraged industrialists to try TF methods; and second, the wide circulation here and abroad added to general awareness that a common interest in TF was growing in society. The book was the forerunner of a series of books on TF.

In July 1968 the University of Bradford in England held a National Conference on Technological Forecasting. Papers provided the basis for *Technological Forecasting and Corporate Strategy* (Wills et al. 1969).

In June 1968 the University of Strathclyde in Scotland conducted a European conference on Technological Forecasting. Proceedings appeared in 1969 book *Technological Forecasting* (Arnfeld 1969). Robert Ayers produced the first book on TF methodology authored by a single writer: *Technological Forecasting and Long Range Planning* (1969). The American Management Association published a TF summary in 1969 (Lansford 1969). *Technological Forecasting: A Practical Approach* by Marvin Cetron was noteworthy especially for including a much-quoted criticism of TF by MIT's Edward Roberts (Cetron 1969). Two years later a multi-authored volume edited by Marvin Cetron and Christine Ralph appeared, *Industrial Applications of Technological Forecasting* (1971). This book included results of an important survey on TF in industry.

An outstanding volume appeared in 1972. In *Technological Forecasting for Decision-making*, Joseph Martino surveyed technology forecasting and presented theory and practice, with many interesting historical examples and highlights (Martino 1972).

In 1973 an edited *Guide to Practical Technological Forecasting* was published (Bright and Schoeman 1973). This volume presented major papers from some 15 industrial short courses in TF conducted between 1967 and 1972, and attended by more than a thousand industrial and government officials from over 20 countries. It described a range of industrial applications in the late 1960s and reported several new techniques for TF. Among the important new ideas was (1) the first proposal for systematically predicting social resistance to new technology, (2) a proposal for computer generation of future scenarios, (3) a history a morphological analysis (the study of technological form), (4) trend analysis of a very detailed compound parameter (30 variables) for computing power from the first computer through 1967 models, and, perhaps, most important of all, (5) the proposal and demonstration that technology forecasting must include consideration of interfaces with other environments, particularly political forces, ecological interactions, and value changes.

The foregoing books were rather expensive and not well adapted to teaching. The first book of TF exercises was published in 1972. *Brief Introduction to Technology Forecasting: Concepts and Exercises* was designed as a paperback teaching manual (Bright 1972). Within 15 months it was adopted for TF teaching in more than 100 courses in universities and colleges and abroad.

TF has been and is being rapidly improved by the contributions of three new professional journals. In 1968 the British journal of forecasting and planning, *Futures*, was launched by Iliffe. American Elsevier produced the first number of what is now *The Journal of Technological Forecasting and Social Change* in 1969, and the Society for Long Range Planning in Britain introduced a journal that frequently touches on technology forecasting, *Long Range Planning*, in 1968.

One of the most important activities now influencing the field is The World Future Society. Their journal, *The Futurist*, has encouraged forecasting studies by focusing the interest of over 12 000 people on future-oriented topics. However, less than 10 per cent of this journal's content deals with technology forecasting. Most of the publication is directed at social speculation, and much is lamentably weak on methodology.

SEMIPOPULAR TECHNOLOGICAL PREDICTIONS

Ever since the late 1800s there has been a stream of serious and semipopular writings about the future and its interactions with technology. These books arouse interest but do not teach methodology of TF. This enormous literature cannot be reviewed here, but we cite a few examples to recognize and identify its significance. In 1967 *The Year 2000* by Kahn and Wiener captured industrial and public imagination by its wide-ranging set of predictions, woven into an image of possible scenarios. *The Art of Conjecture* (1962) by De Jouvenal suggested a framework for conjecture about the socio-political future. *The Challenge of Man's Future* (1954) by Brown typifies the books that survey man's total situation, outline pitfalls, problems, and possibilities, and suggest certain interactions with technology and possible futures. *The Next Hundred Years* (1936) by Furnas reviews technical developments in a dozen fields and discusses their impact. In the late 1960s the author reviewed his own predictions and found them to have been so conservative as to be amusing. It is also thought-provoking to the forecaster.[17] *Future Shock* by Toffler (1970) typifies the journalistic approach to the future. The author strings together a great many concerns about the future and makes his predictions on the collected opinions that are available in the press. The result is readable, highly appealing, and superficial. Nevertheless, it is influential in creating concern through its enormous circulation and publicity. An annotated bibliography of some 250 scholarly writings, largely sociological, can be found in *The Sociology of the Future* (Bell and Mau 1971).

In addition, there are hundreds of government studies, industry studies, and industrial research efforts about future conditions with respect to national resources. Many of them border on or involve future technology. By and large, the only forecasting techniques are trend extrapolations based on various assumptions. However, dynamic modeling of resources and societal interactions is spreading rapidly, as discussed in the Appendix.

SUMMARY

Neither engineering nor science departments in universities have done much to teach, write, or develop the topic. However, R & D managers in industry and government continue to be deeply interested and are strong in their support and applications of TF techniques to their problems.

From the literature and educational activity by managers and analysts in government and industry, it is evident that TF is becoming more influential in guiding industry and government support. The case of Whirlpool Corp. is an outstanding example of the way in which TF has produced major product changes and anticipatory actions by a firm, moreover one not engaged in "high" technology (Davis 1973).

Present activities suggest that the diffusion of understanding and interest will continue, and that TF will become more influential. Its importance will lie in forcing people concerned with technology into explicit thinking about the future. The future actions of the National Science Foundation on TF research could become an important element for improving technological innovation by sharpening understanding, clarifying issues and alternatives and by providing more lead time.

RECOMMENDATIONS FOR TF RESEARCH

From an examination of experiences with the TF techniques (briefly summarized in Appendix 1) I have identified over two dozen research needs. The principal needs for improving *intuitive forecasting* are:

1. To examine past Delphi studies for causes of accuracy and error with a view to improving the method.

2. To improve the selection and application of expertise to issues.

3. To explore ways other than Delphi techniques for getting more wisdom out of the mind of man. Possibly techniques based on Hegelian philosophy, as suggested by Mason, might have merit.

4. The nominal-group technique and its variations have not been used very much in TF work. Apparently, the technique is almost unknown and unapplied in technology forecasting. It should be studied.

Trend extrapolation. There are major shortcomings in the use of TF trend extrapolation. Here are some needs.

5. Most serious of all is lack of national technological database. Until databases in hundreds of technologies are collected we cannot do very much with trend extrapolation. The National Science Foundation should consider how and where these national

databases should be assembled. They should encourage the collection of the mass of technical data now buried in catalogs, the papers of trade associations, the journals of professional societies, and special studies. The building of this national technical database (possibly at the Bureau of Standards or at the Library of Congress) is a prime and promising task. To demonstrate the significance of this need, one need only to reflect on the quality of forecasting if there were no strong databases in economics, demography, agriculture, or meteorology.

6. Theory or at least an explicit rationale and procedure on how to select significant technical parameters.

7. A theory and procedure on how to combine parameters that reflect design tradeoffs and societal tradeoffs (acceptability).

8. A systematic approach to analyzing the limits of trend extrapolations.

9. Studies to improve curve fitting technical data trend extrapolations.

Structural analysis and normative forecasting. Prime needs are:

10. The decision theorists should be encouraged to apply their skills to goal-oriented methodologies. We especially need to link the definition of goals to the establishment of technical programs.

11. The link between policy making and forecasting should be improved. Procedures for relating policies, plans, and forecasts must be developed.

12. We need much better understanding of how to establish socially desirable and acceptable tradeoffs between technology, social conditions, economics, and the environment. The unbalanced excesses growing out of the well-meaning Environmental Protection Agency regulations and the Department of Transportation automobile safety regulations exemplify the need for greater understanding of system interrelationships, as well as wiser national and international goal-setting procedures.

Monitoring--This procedure requires the following research:

13. Our national government's posture to technological development and the use of technology to serve social needs has been largely reactive rather than anticipatory. The U.S. government should establish a procedure to monitor possible signals of potentially significant technological developments and social changes requiring technological responses.

A system should be devised to feed the resulting analysis to the nation's political leaders and to the scientific establishment, with recommendations for actions that will anticipate problems and minimize them. Such a system would have picked up the potential energy crisis (as was done by many individuals and organizations) and would have presented the evidence to national leaders from a convincing and authoritative platform (as was not done until too late). The monitoring of technical signals should be a major responsibility of the National Science Foundation. Just as the Weather Bureau warns of significant changes in the weather, so NSF should warn the nation of impending technological storms.

14. Historical studies should be undertaken to clarify how signals *could* have been identified and followed. From this analysis, better monitoring procedures might be established.

Dynamic modeling. The violent criticism directed at the *Limits to Growth* study (Cole 1973) should not be allowed to obscure the basic importance of this technique. Our complex society interacts in many nonlinear modes, involving technology, economics, social change, ecology, and politics. We badly need research toward the following imperatives:

15. Improve the models.

16. Develop relevant data and transfer coefficients.

17. Improve the equations governing relationships in parts of the system by involving other disciplines, particularly sociology and economics.

18. Improve the interpretation and exploration of alternatives that affect important systems.

19. The mathematical modeling of complex relationships is only one type of simulation. Research should be undertaken in an attempt to *identify social response* to new technology. By anticipating human reactions we might hope to forestall or minimize some problems and

to improve the quality of what is made available to society. Possibly the Office of Technology Assessment should undertake some such experiments, as well as NSF.

Cross-impact analysis.

20. Further research should be undertaken to improve our understanding of how varieties of possible changes will impact upon each other and upon social response. This work is a kind of "point" analysis of dynamic modeling. No doubt the two should be coupled in research.

Forecasting scientific developments. Only one method of predicting specific scientific advances has been proposed, and the test was very limited (Clingman 1973).

21. Research should be conducted to determine whether scientific progress can be identified and predicted to any extent.

22. Experiments should be designed to see whether and how the awareness of coming scientific knowledge could improve technological innovation and social adaptation.

Study of errors in past forecasts. One way to improve forecasting is to determine why past forecasts were wrong.

23. By considering the nature of these oversights or mistakes, one could, in theory, make sure that he did not fall into the same trap. It should be possible to design forecasting procedures that would at least minimize the number of such errors (Bright 1970; Martino 1972).

Developing and disseminating knowledge. Technology forecasting will not be useful unless we disseminate the techniques to industry and government, and feed the results to planners and policy makers. Therefore, we should:

24. Encourage academic research on TF and the teaching of TF courses in engineering, business, and certain liberal-arts departments such as sociology and political science. NSF should sponsor a national symposium of TF with a view to training and inspiring more competent educators to apply their respective disciplinary skills to improving TF.

25. Encourage the development of special TF courses for industry and government officials, so that the systematic use of TF will become standard practice.

26. Encourage use of explicit TF studies as part of great national issues such as energy, population, ecological conditions, etc.

TF as a national early warning system.

27. The recommendations of the 1937 study on *Technological Trends* should be reviewed and modified appropriately to serve the anticipatory planning needs of the national policy makers and operating heads of departments and agencies. The impact of technological change on labor, education, health, agriculture, transportation, and other areas deserves as much attention as defense and commerce. The Federal government should give formal attention to future technology, just as it does to other key elements of our society.

APPENDIX I. REVIEW OF TF METHODS: CRITIQUE AND RESPONSE

TF methodology today can be classified in perhaps half a dozen categories. Since the field is so new they are briefly reviewed below, with comments and cautions, as a convenience for the reader. Details of the methodology can be obtained from the many books referenced in the footnotes.

INTUITIVE FORECASTING

The improvement of expert opinion has been the goal of much TF research, notably the efforts of Helmer, Dalkey, Enzer, Gordon, Linstone, Turoff, and their colleagues.

The Delphi Technique. In the early 1960s RAND Corp. researchers Helmer, Dalkey, and their colleagues introduced the method to industry. It was intended to improve the use of expert opinion through polling, under conditions intended to minimize the interpersonal bias in committees. These points were anonymity, statistical display of collective predictions, and feedback of reasoning.

Delphi technique rapidly swept throughout the world. By 1974 more than a thousand or so Delphi studies had been conducted by firms, government agencies, trade associations, and consultants. These studies ranged from little internal one-department panels to a Japanese nationwide study involving 4000 predictions. The comments that follow obviously do not apply to every Delphi study.

Experience suggests that Delphi technique appears to be deceptively easy but, in fact, it is exceedingly difficult to do well. Some cautions follow:

A four-round Delphi with more than, say twenty predictions becomes so wearisome and time consuming that participation and thought fall off rapidly.

Many Delphi results are presented as though they were the work of experts, when they are not. They are found to be collections of opinions of persons who have little or no expertise on the *precise technical and economic* issues raised. The technique thus may mesmerize the forecaster and the forecast user into believing that collections of opinions provide valid technological predictions.

The stress on achieving "consensus" is overdone. Some writings imply that consensus means "truth" or "correctness," and seem to suggest that the better the consensus the better the forecast. This proposition is unproven. In fact, one important value of the Delphi statistical display lies in providing the forecast user a picture of the degree of unanimity and the range of opinion. This display of disparity may be extremely useful additional information.

There are other problems. How does one obtain true experts on each topic? Preselection of specialized panels has been tried by TRW and other firms. The unhappy fact is that any preselection process may unwittingly exclude the very interdisciplinary widsom that was sought. Self-rating (sometimes with appropriate weighting) as to degree of expertise on each question has been tried. Another approach is to have a second independent panel repeat the Delphi study and compare the results.

Ambiguity is a problem. Many Delphi predictions prove to be so confusing that the respondents' forecasts are made to different assumptions. The use of qualitative terms like "common use" or "widely accepted" typify this error. The multiple prediction is another source of error as in, "when will electric cars powered by fuel cells be common for suburban use?" Broad and imprecise definitions also confuse experts. Therefore, very careful design of predictions and pretesting are needed. Length must be kept down and repetitions reduced. The mini-Delphi of two rounds, followed by a conference, has been found helpful.

Other Possibilities for Improving Expert Opinion. Interest in the Delphi technique has tended to obscure a basic issue: By what techniques of inquiry can we get improved information out of the minds of knowledgeable people?

The dialectical approach is based on the philosophical concepts of Hegel. The dialectical approach calls for making a forecast and then proposing an extreme opposite or counterforecast. Both forecast positions are then exposed to the strongest possible critical argument. The only known industrial example is described by Mason in "A Dialectical Approach to Strategic Planning" (*Management Science*, April 1969). A model for a dialectical approach was proposed by Ian I. Mitroff in "A Communication Model of Dialectical Inquiring Systems--A Strategy for Strategic Planning" (*ibid.*, June 1971).

A Psycho-heuristic Approach. A psycho-heuristic concept has been suggested by Warren Duff in an unpublished talk to The Southern Sociological Society in 1971. His idea is to have the forecaster make extremely rigorous and critical inquiries of experts in order to elicit sound reasoning to support their forecasts. Obviously, the forecaster must be a technical expert himself in order to ask penetrating and comprehensive questions and to evaluate the validity of answers. This technique borders on the Delphi concept in that the experts do not confront each other. In theory the study director would force the experts to deal with a far more rigorous feedback of reasoning.

A special version of this concept involves querying experts on the nodes and elements of a basically PERT-like network of R & D tasks. Predictive Analytical Forecasting (PAF) was developed by John Vanston of the Department of Nuclear Engineering of the University of Texas at Austin, to predict the emergence of nuclear fusion power under various funding conditions. It has great promise.

TREND EXTRAPOLATION

Trend extrapolation (also called exploratory forecasting) rests on the assumption that technical attributes generally advance in a relatively orderly manner over time, exhibiting patterns of behavior that form fairly well-behaved trends. Therefore, by choosing appropriate parameters, one can develop their past time series and extend them in some manner to predict a future condition or estimate the rate of change. Such projections forecast the status of the measured attributes and so identify future levels of technical performance and functional capability. These projections suggest coming possibilities, potential conflicts, supporting requirements, unacceptable anomalies, and the need for actions and responses on technology-oriented devices and procedures. The forecaster recognizes, of course, that the projections must be studied to identify potential limits and forces for change.

Engineers and scientists have used trend extrapolation for almost 200 years for planning and forecasting. The essential difference from today's practice is that in the past they did not use the term "TF;" there was no formalization of the TF process and there was no attempt to refine the trend-extrapolation techniques and data bases.

It is generally believed that these technical data time series or arithmetic scales produce S-curves, reflecting a slow start, exponential growth, and a leveling off against some limit produced by nature or man. No comprehensive research has been done to prove the universality of this assumption, but almost all data so far plotted show a surprising amount of confirmation. The foregoing S-curve phenomenon comment *refers solely to the achievement of technical capabilities. It does not necessarily apply to the degree of that capability used by society*, which is governed by more than mere availability. This orderly and apparently exponential nature of technical progress may be perhaps explained through a review of the process of technological innovation, as follows.

New technology emerges from an idea, through a struggle to achieve the first crude results, followed by rapid advances to understanding grows and more man-hours, equipment, and funds are committed. Eventually the rate of advance declines as some limit of effectiveness or a technological barrier is reached.

The technology emerges slowly because initially few people are involved, basic scientific knowledge must be gained, and engineering obstacles must be cleared away. There may be lack of scientific and technical understanding, and it may take time to overcome conventional wisdom and erroneous assumptions about the phenomena involved. Funding of the research effort and lack of experimental equipment may also hold back progress.

The advance eventually begins to accelerate because of the rapid commitment of more technical effort and funds once understanding and proof of concept are attained. Then efforts are directed not only to improving the key technical knowledge, but to refining all the facets of technological execution. The spur of competition also inspires rapid improvement through bold new goals and unique combinations of design concepts.

Exponential change is further encouraged by the tendency of technicians and managers to set goals in terms of percentage improvement over present practice. The pattern of action and response between competitors in military hardware also encourages this accelerating advance. Technical progress in a given device usually is the result of accretion of dozens, if not hundreds, of refinements in the component technologies. This accretion leads to exponential gains as well.

Finally, the technical advances level off, for various reasons: technical opportunities for further advance are exhausted; a point of diminishing returns in cost, effort, and usefulness is reached; or some basic limit in nature bars progress. At times, limits are formally imposed by society or informally affected by the climate of the times.

Why should this technical progress be relatively orderly? Because advances in technology, particularly technology that is employed by society, is a compromise among technological possibilities, economics, social conditions, management psychology, skills, resources of the producing group, alternative possibilities and interests, and the reactions of users. This complex mix of offsetting factors seems to moderate great discontinuities and to encourage relatively orderly evolution. While many advances may appear as discontinuities at the moment, in hindsight they are seen as part of an exponential curve or Gompertz curve.

Pros and Cons of Trend Extrapolation. Trend extrapolation has been heavily criticized by some academicians, including some management scientists, sociologists, and "futurists" (Roberts 1969). Arguments (A) are listed below with possible counterarguments (CA) following each:[17]

A. *Simplistic "eyeball" curve fitting in the manner of much TF work today is "naive." Therefore, it is invalid or useless.*
CA. Naive or not, most technological trends appear to have been quite orderly.
CA. Mathematical elegance in curve fitting is of dubious value in many instances because of the crudeness of the data and lack of statistical studies of the phenomena involved.
CA. Useful guidance usually comes from consideration of the approximate conditions forecast, not from the prediction of a precise number and date.

A. *Extending trend lines several decades or centuries leads to ridiculous relationships.*
CA. True enough, so we should not do that. Trend extrapolation loses validity over time. It is a guide to the next 5, 10 and occasionally 20 years, not the next century.
CA. The determination of limits of trends and the manner of approaching those limits is a major part of the forecaster's task. No TF student advocates blind, limitless extrapolation.

A. *There is no proof that past forces will continue to support the trend, so extrapolations are intellectually and philosophically unacceptable.*
CA. On the contrary, the burden of proof is on the critic. The past trend is historical fact. It has resulted from a complex interaction of forces. It is up to the critic (and the forecaster!) to demonstrate what forces will change, when they will affect the projection, and by how much.
CA. Is there evidence that the present is a point of major discontinuity? If that cannot be established, the trend is our best guide to the near future.

A. *We can be certain that future technology will be modified by new developments, attitudes, value systems, and societal choices.*
CA. True. However, this argument applies especially to the technology that is *adopted* by society, but not strongly as to the *creation* of new technical capabilities. Even so, the argument applies only to some technology developments and not to all technical work.
CA. Society is fractionated. Technological activity is international and the technological choices of one nation or group of nations do not necessarily control technology elsewhere. The supersonic transport is a case in point; and the nuclear bomb an unhappy proof of this point. It is by no means proved that the control of a technology in one nation seriously alters the nature of technical advance, except for a few items under agreed international control.

Parameter Selection. Technological attributes (such as "noise") must be converted to measurable parameters (such as "decibel level at 50 feet distance"). This is easier said than done, particularly for historical data. A single parameter rarely encompass all the important attributes of a device or material. Therefore, each situation must be examined to determine the critical parameters. In most case, parameters relevant to performance of the device, including its key parts, materials, and subsystems, and to characteristics that the user desires should be studied. Parameters relevant to related technical and scientific knowledge and even such things as laboratory equipment and measuring devices may be significant in the forecast.

Because of design and operating trade-offs, the parameters or attributes that reveal the full functional capability desired by the user and the compromises inherent in design and operation are needed. This usually requires the construction of compound parameters. Parameters must be applicable to all technical approaches that could provide the capability or service performed by the technology, or the resulting forecast may mislead by omitting an important technical alternative.

There must be an adequate historical database. With too brief a database, the forecaster is in the naive position of extrapolating from a point. This error will be found in an astonishing number of recent speculative technical papers and even official forecasts of government agencies done with the help of TF experts.

Data must be taken from comparable situations. In particular, data from experimental work usually should not be intermixed with data from operationally available technology.

Types of Trend Extrapolations. TF literature identifies these types of technological trend extrapolations. Examples of each can be found throughout the references cited in this chapter.

Single Parameter. Hypothesis: Time series data of a technical attribute that reflects a significant aspect of the technology or its application (e.g. power, speed, cost, strength, hardness) can be plotted and then extrapolated in some manner to predict its future state.

Potential Use: The extrapolation helps to identify the likely degree of change in that attribute at future times. Implications then can be considered, and need for further investigations or actions are indicated.

Cautions: (1) A single parameter often fails to clarify the design trade-offs that are taking place between several attributes; (2) one or a few parameters usually are insufficient to describe all the important changes taking place in a complex technological device, such as an airplane, an oil refinery, or a computer; and (3) no trend goes on unchanged forever--the influential factors must be considered in making the extrapolation.

Compound Parameter. Hypothesis: Combinations of parameters can be developed so that the data will reflect the basic, interrelated changes in the device or service (e.g., power per unit of weight or space; computer operations per dollar). Trends of such factors reflect the technological-economic trade-offs taking place within the device.

Potential Use: Same as previous type. They also may have more relevance to the user because they describe a total service or condition pertinent to the application of the device.

Cautions: (1) More than one compound parameter may be needed; and (2) the compound parameter may still be incomplete and may not fully reflect all the important aspects.

Leading Indicators or Precursors. Hypothesis: Time series data of a single or compound technological attribute (or other phonomenon) that seems to lead the technology in question can be used to forecast a future state (e.g., improved accuracy of measurement might lead to improved accuracy of machine-tool performance).

Potential Use: The rate of change or performance of the precursor and its relationship to the technology under study provide bases for a prediction of future capability.

Cautions: Past relationship between the precursor and the subject technology may not hold for many reasons. For example, different forces may be driving the two phenomena, or certain factors may limit the one parameter but not the other.

Envelope Curves. Hypothesis: A fundamental capability that society desires continues to advance by support of a succession of increasingly effective technologies. As the

S-curves of earlier technologies begin to saturate, a new technology emerges to maintain progress of the fundamental capability. A curve approximately tangent to the tops of the S-curves describes an "envelope" of anticipated change in the basic capability. No definitive theory for drawing or projecting the envelope curve has been suggested, and only piece of research has been done.[18]

Potential Use: S-curves and their envelope curve, even though roughly drawn, suggest the coming of technological change. The S-curves display the growth inability of a given technology to support the advance predicted by the envelope curve. Examination raises questions about the decline and emergence of particular technologies and resulting market impacts.

Cautions: The fact that a given technology no longer supports the envelope curve does not mean either immediate or ultimate economic demise. Many technologies remain effective in certain sectors of society.

Step Functions. Hypothesis: Different industries show patterns of behavior in the scale and timing of technological changes that they introduced into products and facilities, especially as these factors relate to total demand. These patterns are responses to such forces as markets, prices, costs, size of investments, technological considerations, new technology as well as intra-industry relationships, socio-political restraints- and even industry tradition (Simmonds 1973).

Potential Use: By developing these patterns one can estimate the scale and timing of new technology introductions. Impact on the market, prices, and the supplier-customer competitive response cycle may become more clear. An individual form could then make wiser decisions for its own timing and scale of technical facilities through profitability analyses based on various cost-price responses. The technique also is especially useful for studies of intercountry technical competition and economic competition based upon technical facilities.

Cautions: (1) The existence of such patterns must first be established for the particular industry; (2) the future stability of relationships that created these patterns must be carefully considered; (3) there eventually will be limitations to the progressive increases in scale of facilities; (4) new technology and socio-political events can abruptly alter these patterns; and (5) automation, miniaturization, and technological "learning" may occur and so rapidly improve capacity and output as to invalidate the projections at times.

Technological Progress Function. Hypothesis: The technological progress function assumes that a technical parameter improves with cumulative numbers of units produced in the manner exhibited by unit cost and manufacturing time in learning curve theory. Where technological devices are not necessarily produced uniformly over time, a different curve results than from plotting time-related data.[19] The equation:

$$T_i = ai^b$$

where

T_i = technical parameter, characteristic of the ith unit

i = cumulative unit number

b = rate of progress, presumably dependent on environmental considerations such as technical effort measured by rates and levels of investment and man-years

a = constant

Possible justification may lie in the fact that R & D effort often is proportional to production activity and resulting sales income on which R & D budgets are based. Also, technical progress is often the accretion of hundreds of minor improvements, which are "learned" and applied over repetitions of production.

Potential Use: The function provides a basis for technology predictions against growth of technology production or usage instead of time. This basis might be far more appropriate to some technical developments.

Cautions: Same as for other trend extrapolation techniques. The assumption that technical improvement is related to cumulative production may not apply to the particular case.

Substitution Theory (Fisher-Pry). Hypothesis: If a new technology is substituted for an existing technology without a major change in function, it tends to go to completion. Then the time and amount of substitution can be predicted according to a hyperbolic target function based on the average annual rate of displacement. The substitution formula is explained in Fisher and Pry (1971) and in Bright (1972, pp. 6-41).

Potential Use: Predictions of amount and timing and substitution of new technology provide bases for estimates of future market size, production capacity, and supporting activities as well as data for planning phase-out of old technology. Substitution rates in one field or country also might prove to be precursors of substitution rate elsewhere.

Cautions: (1) Accuracy of predictions based on the first 5-10 per cent of displacement data seem to be highly accurate; (2) units of measurement must be carefully chosen to avoid distortion; e.g., the substitution of plastics for steel in automobiles would be badly misrepresented if weight were used as the measure; (3) it has not been explained how and why this displacement theory differs from conventional market product displacement theory; and (4) the very long time spans involved mean that the new technology will be subject to many environmental changes, including competitive technology which presumably reduces reliability. Blackman has argued that this formula is too mechanistic and is unrelated to the forces that cause substitution. He considers Mansfield's approach sounder.

Analogies. Hypothesis: A new technology may emerge in a manner, pattern, and/or rate of change analogous to some predecessor technology or existing natural or social phenomena. One variation is "growth" analogy, which assumes that a technology is analogous to biological activity and "grows" toward a natural limit in the same manner. Another variation is to assume analogy to closely related prior technology.

Potential Use: The analogy provides a pattern that may be a useful guide to a future limited condition. The very evident tapering off of growth has a salutary effect when compared to blind extrapolation of present rates of change.

Cautions: Is the analogy really valid? Consider past socio-economic conditions as well as different uses and needs for the future technology.

Correlation Analysis. Hypothesis: A technological device changes through a blend of many factors of performance, construction, cost, and social usage. Therefore, historical data on the important interrelated parameters should be developed and plotted. Trial extrapolations are then made to test relationships and so to determine the most likely future state or direction. The basic notion is to search for consistency and logical relationships.

Potential Use: By clarification of the interrelationship of elements of the technology correlation analysis may be used to detect the way the factors will influence each other as the technology evolves. It reveals the sensitivity of technological development to future changes in each element and thus suggests the likely trade-offs.

Cautions: This is obviously a complex study that requires very careful consideration of underlying forces that causes changes in each parameter.

NORMATIVE (GOAL-ORIENTED) FORECASTING

This approach to technological prediction rests on the assumption that new technology materializes to fill needs or goals. Therefore, an examination of future needs should reveal major technology that is desirable and probably will be developed in the future. For instance, when President Kennedy announced a program to put man on the moon by 1970, many technical goals were implicit in the effort--guidance, life-support systems, power-storage systems, materials of higher heat resistance, etc. A critical review of this set of future needs could have provided what was, in effect, a forecast of technology. The current definition of energy goals by the Federal government, as well as that funding allocation, is another base for goal-oriented forecasting. The emission standards of the Environmental Protection Agency, with their detailed specifications and time standards, provide an even more precise forecast.

Whether this is forecasting or planning is a question. However, it is obvious that the practice of developing technology to meet defined future needs is a growing factor in the U.S. and world society. Thus, normative forecasting not only has a place in the prediction process, but that place is getting stronger. Methodology in this approach ranges from nothing but a protagonist's view of the future (often resting largely on his own value system) to highly structured reasoning procedures.

Structural analysis has distinct advantages. It orders and systematizes one's thinking. It forces completeness. It attempts to quantify each bit of reasoning. On the other hand, anything not included in the structure will be omitted. It is rigid and so may be misleading. It may give a false impression of validity through the very elegance and quantification of the approach.

Varieties of goal-oriented structural and environmental analysis concepts follow.

Morphological Analysis. A matrix of all theoretically conceivable combinations of technological approaches and configurations is constructed and then inspected to identify untried combinations. One hopes some of them may be promising. The techniques's use is increasing. A few industrial practitioners believe that it has major value in encouraging creativity in technical and managerial thinking. The technique has been studied and perhaps used more extensively in Britain than the U.S. It is a much older technique than generally appreciated.[20]

Relevance Trees. Institutional objectives, then missions, tasks, subsystems, and hardware in turn are identified as levels of the tree. Criteria and their weights are set at each level to support the branch above it. Duplication of systems and hardware appears on lowest levels. A summation of their relevance numbers across any level indicates the relative values of various technologies at that level.

One benefit lies in forcing critical evaluation and definition of alterantives and "relevance" at each level. Obviously, "relevance" must be considered against some agreed-upon future state or set of conditions (a scenario). Also, the analyst must first establish an "objective" against which relevance can be considered. This objective must be clear, minimize confusion of terms, and be acceptable to the decision makers. Each level of the tree must be "closed" (i.e., contain all possible items to support the branch above). The method has been heavily used by Honeywell.[21]

Mission Flow Analysis. Alternative policies and alternative responses and the supporting technology needed for each possible major future development or event are stated. Then by applying probabilities and value, one hopes to identify the most useful technology for the future. This procedure is extremely complex and costly; only a few government-oriented studies exist. It differs importantly from relevance trees in that it begins by recognizing different ways of performing a mission, rather than simply evaluating technology alone (Ginstone 1968, pp. 223-241).

MONITORING

All studies of technological innovation confirm that the process of technological innovation takes time, usually measured in decades. Most technological innovations are visible in society long before they are widely available or applied on a scale having substantial impact. Since new technology emerges from idea into physical reality and then is applied and adopted at some gradual rate rather than instantaneously, it follows that it may be possible to monitor this progress. Then one could predict character and impact (to some extent) and with some lead time.[22]

Phenomena and events outside technology often trigger the development of new technological capabilities or the adoption of technology; or they may alter the relevance of a given technological capability. Therefore, signals of technological change may occur outside technology and "monitoring" must be extended to include these nontechnical environments. Monitoring includes the following activities. (Note their similarity to the economists' study of "indicators.")

1. Searching the environment for signals that may be forerunners of significant technological change;

2. identifying possible alternative consequences if these signals are not spurious and if the trends they suggest continue;

3. choosing the parameters, policies, events, and decisions that should be followed to verify the true speed and direction of technology and the effects of employing that technology; and

4. presenting the data from the foregoing steps in a timely and appropriate manner for management's use in decisions about the organization's reaction.

Monitoring includes much more than simply "scanning" traditional library reference lists. The essence is *evaluation* and *continuous* review. Monitoring is the acceptance of uncertainty as to meaning and rate of change of developments. The forecaster "runs scared"--he tries to stay ahead of developments and their implications--and presents his conclusion relatively late. His contribution lies in providing early warning and in remaining open minded about possible significance. Monitoring includes search, consideration of alternative possibilities and their effects, selection of critical parameters for observation, and a conclusion based on synthesis of progress and implications.

One further dilemma of monitoring is that there are so many apparent signals that information collection, sorting, and evaluating can overwhelm the forecaster. The "noise" in the environment tends to obscure meaning. The only answer to this problem is that *eventually* the true signals, and the meaning of the signals, emerge. With luck the forecaster has gained a little lead time.[23]

SIMULATION OR DYNAMIC MODELING

Forecasting by analyzing the behavior of models has tremendous intellectual appeal. Technologists have long used physical models to predict technical behavior--ship models, airplane wind-tunnel tests, small-scale river flow test basins, etc. Mathematical models to study the behavior of complex systems are well established in operations research literature, systems dynamics, and economics. The idea of using dynamic models to predict future technological conditions has been proposed for at least ten years, but very few examples of such technology forecasts are in print.[24] One of the few industrial firm applications that has been published is by Blackman (1973, pp. 257-267). More probably exist in monographs, government reports, and consulting reports.

The landmark effort on a major scale is doubtless the famous *Limits to Growth* study, in which systems dynamics has been applied to predict possible future conditions (including some technology aspects) on a global scale (Meadows et al. 1972). A further volume, *Toward Global Equilibrium: Collected Papers*, expands the concept (Meadows and Meadows 1973). The study has been violently attacked for many reasons; and the criticisms would provide a list of several pages. A typical short critique is in *Science* (March 1972, p. 1088), and a detailed, scholarly critique was published in two issues of *Futures* (Science Policy Research Unit, Sussex University, Feb. 1973, April 1973). Criticism includes these points:

1. The model is imperfect; therefore, it is useless.
2. The research data are inadequate to substantiate many of the equations and coefficients.
3. The equations and coefficients used are wrong.
4. The mathematics is based on deterministic models when it should be based on probabilistic models.
5. Adequate allowance for man's ingenuity in creating new resources is not made.
6. There are mathematical errors that invalidate the study's conclusions.
7. There is no allowance for behavioral change by mankind and social institutions.

To this Meadows et al. reply that

1. imperfect models are not necessarily useless;
2. more research on basic data is needed, but anyone can change the data and equations to suit his concerns and see what happens--that is, after all, the purpose of the model;
3. the model can indeed allow for behavioral change--simply change the equations, impacts, and responses as desired; and
4. "at least our reasoning is explicit--now make yours explicit."

Regardless of one's opinion on this study, it has provoked worldwide efforts to improve and expand the application of dynamic models to socio-technical systems. Studies have been launched in several European countries, some U.S. regions, and in Japan, South Africa, and elsewhere. These studies will become increasingly influential in setting national and international policies.

Predicting social response to technology through simulation--an imaginative procedure for predicting the behavior of complex technological systems--has been tried by the Institute for the Future (Gordon et al. 1973, pp. 550-572). Their technique is called "simulation gaming" and suggests how people and institutions will respond to technical (and other) possibilities in the future. It can be understood by analogy to physical models. Suppose one created a situation (a "game") involving representative types of people and institutions and then introduced the proposed technology or socio-economic developments into this environment. Perhaps the response of participants would indicate how these social systems will react to these future changes.

One major research experiment attempt was conducted for the state of Connecticut. No claims are made for forecasting precision, but the experimenters concluded that simulation gaming was a very useful device for informing participants about issues and data, improving communications, and becoming more sensitive to aspects of the future.

Perhaps a study of the predictive potential of war games, a method with an accummulated backlog of almost one hundred years of experience, might clarify usefulness and possibilities. Simulation gaming by corporations and government agencies might help the firm anticipate intracompany and industry behavior to potential events. The concept deserves further experimentation.

CROSS-IMPACT ANALYSIS

Future events interact to influence the probability, timing, and impact with other events. By examining these interactions one can develop, according to protagonists, a scenario or description of the future. At least the forecasts will be more internally consistent. Cross-impact analysis is done by assuming that a given prediction is correct; that it "comes true." The forecaster then considers how the occurrence of that event would affect the probability and timing of other predicted events. With each prediction assumed as "true" in turn, a matrix of predictions and their interactions can be systematically explored. Some events obviously will accelarate or inhibit the materialization of other predictions, and impacts will vary. Some predictions will appear to become very tenuous or very certain, and timing may be advanced or delayed.

Cross-impact analysis attempts to do what is implied in all forecasting, to provide a prediction of future conditions based on allowance for all the interacting forces that will shape that future. Cross-impact analysis requires consideration of these interactions. Even in its simplest form it improves the internal consistency of a set of forecasts and clarifies underlying assumptions. It exposes gross errors and inconsistencies.

A more sophisticated approach is proposed by researchers at the Institute for the Future. They suggest that initial probabilities should be applied to each event. By a mathematically development of the potential interactions in terms of mode, strength, and timing, and through use of a randomly selected predicted event as "true," the matrix can be explored on the computer to determine shifts in probabilities and impacts, and eventually a stable condition--a scenario--should result.[25]

Cautions: Cross-impact analysis cannot allow for any future event or condition not included in the matrix. Using one forecast at a time as "true" is clearly erroneous since the future results from the interactions of a number of events, some simultaneous and others at irregular intervals. These events, coming in clusters, often must create a very different impact on other future events than the impact from events taken individually. Finally, the "probability" manipulations proposed are of dubious validity, since the impact of any one event on another surely does not follow the same mathematical relationship of time and force in every instance.

Nevertheless, the cross-impact principle has great merit in forcing systematic and rigorous consideration of interrelationships. Since 1970 extensive efforts have been made to refine the cross-impact technique and to apply it to Delphi forecasts in order to obtain both scenarios and better forecasts (Turoff 1972). In particular, it is argued that cross-impact analysis will refine and improve the Delphi forecasts themselves, as well as increase the perception and depth of understanding of interactive relationships between future trends and events. A cross-impact game technique has been proposed for this purpose (Helmer 1972).

Scenarios. The use of scenarios, meaning descriptions of *possible* futures, is growing rapidly as a planning tool. The scenario itself may be the *result* of a forecast, but it is not a forecasting technique. Almost nothing has been done on *methodology* for developing scenarios. Therefore, we shall omit discussing them in this paper.

OTHER FORECASTING CONCEPTS

Forecasting by Exclusion. It might be possible to improve prediction by systematically excluding unlikely developments. By defining "forbidden regions" one might narrow and better describe the domain of the more probable. Occasionally, one finds this concept applied in engineering and scientific studies, as the boundaries or limits of performance are shown. However, no systematic application to technology forecasting is now known to the author.

Learning from Past Errors. One of the most promising improvements for TF is to study past forecasts and learn from them. This idea was suggested by Gilfillan, although he proposed to study past forecasts to identify reasons for success. From 1969 to 1971 Bright studied dozens of past forecasts to find the reasons for error, with the belief that if reasons for past errors could be identified, it should be possible to improve forecasting by avoiding these errors in the future (Bright 1973). The findings are thought provoking and offer several points worth mentioning here. Major errors occurred because of:

1. Failure to search the early stages of the process of technological innovation and so to detect the coming of technology that was clearly under way.
2. Failure to search widely--particularly, in other nations--for new technology under way.
3. Totally inadequate perception of the influence of nontechnical developments on technology--notably population growth, increased affluence, and changes in laws.
4. Inability or unwillingness to examine the rate of past change in technical parameters and to consider the implications of the exponential progress clearly indicated in that data.
5. Lack of appreciation of the interactive nature of technological developments on each other.

However, the sample of forecasts studied is far too limited and critical review of many more recent forecasts is needed to improve understanding of causes of error. Gordon and colleagues examined the first published Delphi study to check its accuracy five years later. One interesting conclusion was that in many instances the predictions were so fuzzy that it was hard to say whether or not the forecast had materialized! Gordon, too, argues that studies of past forecasts are vital to improve forecasting.

THE DESIGN OF TF STUDIES: WHICH METHODS FOR WHICH PURPOSES?

Designing the TF study for information is only part of the design problem. Good design must be related to the communication and learning needs of the situation. Here are possible uses as drawn from various TF examples in industry and government in recent years.

Delphi Techniques. Variations of Delphi techinques are aimed first at improving the quality of expert opinion and then at presenting the collective conclusions of experts in a more meaningful way. TF Delphi studies have been used to:

1. legitimize imagination throughout a group and direct attention to the long-range future;
2. acquire ideas and speculations beyond the normal bounds and resources of the firm;
3. provide a starting point for in-depth TF studies of potentially significant issues;
4. explore the implications of a prediction developed through other means;
5. detect signals of possible changes that ought to be monitored because of their potential impact.

Technological trend analyses are useful:

1. to determine the rate, direction, and timing of key aspects of a technology;
2. to compare rates of technological advance between technologies;
3. for sensitivity analysis of alterantive changes in technological elements;
4. as a vehicle for studying factors that will influence the limits of a technical trend.

Monitoring systems are useful to:

1. survey the environment for signals of developments that may affect technology;
2. track a faint but potentially important possibility;
3. determine direction and timing of technological progress;
4. determine the true impact of response of a technological development.
5. determine when a threshold of significance has been reached.

Cross-impact analysis has been used:
1. to assure that forecasts are internally consistent;
2. to improve understanding of interactions;
3. to develop and refine scenarios of possible futures.
4. to encourage interdisciplinary and interfunctional consideration of future possibilities.

Structural Analysis (Goal-oriented Techniques). Methods that identify the technology needed to serve future goals have been used to:

1. compare present activities of the firm with what will be needed to meet a future situation;
2. organize thinking so as to display alternatives and assure consideration of each element of complex issues;
3. as a means of developing understanding and agreement on goals and the distribution of effort;
4. to weigh and quantify elements of the decision.

Simulation. Dynamic models enable one to simulate different conditions so as to:

1. gain understanding of the behavior of complex, multifactor, nonlinear systems;
2. study nonlinear interactions over time;
3. examine the sensitivity of a complex system to changes in variables.

Scenarios. Although not forecasting techniques in themselves, alternative possible futures can be used:

1. to analyze the implications of a wide variety of uncertainties in the future;
2. as a basis for a monitoring plan;
3. to identify the need for information and studies in depth;
4. to develop the need for policy decisions;
5. to encourage flexibility in long range planning;
6. to help prepare the firm for contingencies;
7. to sensitize people to the character and influence of possible changes.

NOTES

1. Adapted from Table 1.1 in Bright (1972), pp. 1-2. I draw heavily on this recent condensed textbook because it summarizes and references much existing material.
2. Bright (1972), pp. 1-6. Versions of the chart were published in several papers since 1967. The most widely disseminated version was discussed in Bright (1970).
3. This is a refinement of the definition offered in Bright (1968), p. xi. I am indebted to Ralph C. Lenz Jr. (1962) for the original idea.

4. Marvin Cetron has frequently used this definition because it is especially appropriate to the military TF work, which he pioneered in the U.S. Navy. An interesting but less-well-known definition is discussed and compared with others by Erich Jantsch (1967) in his classic book *Technology Forecasting in Perspective*, pp. 23-28: "Technological forecasting is the probabilistic assessment of future technology transfer."

5. Bright (1968), pp. xi-xii.

6. Bright (1964), pp. 735-736. The same material was used in case form at Harvard Business School in 1960, when TV was first introduced at that institution.

7. See Siegel (1953) for an interesting demonstration of the economists' deep concern for technology and economics. Unfortunately, economic data do not define nor predict technical capabilities.

8. An introductory view of economics and technology appears in Rosenberg (1972). For a wide sampling in economic writing on this subject, see Rosenberg (1971).

9. J. F. Coates (1973). A brilliant attack on the sloppy thinking with which the "futurist" movement abounds. Dror also has made a sharp attack on the World Future Conference along the same lines.

10. From 1968 on this list of stages and discussion appeared in many conference papers here and abroad dealing with technological innovation. In 1972 Martino offered a slightly different version (1972, pp. 5-8).

11. T. J. O'Connor, thesis, University of Denver (1969). This study examines some effects of some of the von Kármán reports.

12. Henry Adams, *The Education of Henry Adams* (New York: The Modern Library, 1931). Adams graduated from Harvard in 1858. The last four chapters express his awareness of the impact of science on society through its exponential growth late in his life.

13. See Bell and Mau (1971). Although a useful annotated bibliography of about 250 items is provided, the omission of the work of the sociologist TF pioneers in the 1920s and 1930s may be indicative of the discipline's attitude toward TF methodology.

14. Mansfield (1968a). Mansfield's many excellent papers and books are the basis source on the link between economics and technology.

15. From an unpublished manuscript by J. R. Bright and data in the Annual Reports of Xerox Corporation (and it predecessors, Haloid and Haloid Xerox), 1948-1965. Also based on many original notes from Battelle and Xerox and a tape recording by the inventor, Chester Carlson, obtained by Prof. Ermenc of Dartmouth College.

16. Bright (1968). The Military Operations Research Society had previously conducted (1966 and 1967) two conferences on TF. Attendance was restricted. I proposed an industrial TF conference to the Harvard Business School Associates. Because of scheduling delays I held the program privately in May 1967 at the Lake Placid Club, N.Y. Approximately 130 persons from six countries attended.

17. Adapted from introduction to Bright (1972).

18. Mohn (1972). Mohn's doctoral dissertation (University of Texas, 1971) is one of the few efforts to explore the evolution of a technical parameter (typesetting speed) over centuries.

19. Fusfeld has convincingly demonstrated that some technology improves in relation to cumulative production. It is as through technological advances were the result of "practice" and reflected a kind of learning curve. See Fusfeld (1973, 1970).

20. Gerardin (1973). References in Gerardin's chapter provide a lengthy guide to the key literature on this type of structural analysis.

21. Esch (1973). Also see Swager (1973).

22. Bright (1973). Davis (1973) describes the most successful TF monitoring program in a nonhigh-technology firm.

23. Texas Instruments also used this technique, limited to technical signals only, in the mid-1960s. See Herbert Sorrows in *Research Management*, March 1967.

24. See Roberts (1964). This is the pioneering book devoted to systems dynamics in a technical area. We do not seem to find much application to technological forecasting.

25. Initial quantitative work was described by Gordon and Haywood (1968). See also Cetron and Ralph (1971, Chap. 5).

14. TECHNOLOGICAL INNOVATIONS AND ECONOMIC GROWTH
Simon Kuznets

I. TECHNOLOGICAL INNOVATION: DEFINITION AND PHASES

Technological innovation is defined here as application of a new element in material technology, i.e., in man's knowledge and capacity to manipulate the natural environment for human ends. Since human ends, and means to attain them, are diverse, we focus our discussion by limiting it to material technology as related to production of economic goods. The new element may be an invention, or a discovery of previously unknown properties of nature, directly usable for economic production. Application may be just sufficient to test economic promise; but we deal here with innovations that had a perceptible economic impact, with application on a fairly wide scale. Finally, we distinguish material technology from social technology, i.e., the capacity of men to organize for various social purposes, economic production among them. Obviously, both social and technological innovations can, independently of each other, contribute to economic product. But, as argued below, such independent contributions of the capacity to control nature and of the capacity to organize society are possible only within narrow limits.

The new element in a technological innovation may yield several types of economic contribution, and of widely different magnitude. In economic analysis, innovations can be classified in various ways: those that yield new consumer goods, in which the basic choices are made by entrepreneurs (private or public); and those that are cost-reducing for established goods (producer or consumer), which can be distinguished from those that yield new products (producer or consumer). The typology can be further refined if we identify the sector of production most directly affected, and recognize the social adjustments required to exploit the economic contribution of specific innovations. A cost-reducing innovation has a different effect in an industry with product of low long-term income and price elasticity of demand from that in a sector producing goods of high demand elasticity. A process innovation that raises the scale of the productive unit (plant) appreciably requires individual and social adjustments different from an innovation that does not affect the scale of production. Innovations can be cost-reducing, but relatively more capital-saving or more labor-saving. The above classifications distinguish innovations by type of economic contribution, by their effects rather than by their causes or sources (the latter are touched upon briefly in a later section); but even this typology is incomplete, and could hardly be exhaustive. The purpose of these brief comments is to suggest the pervasive effects of technological innovations on different industries, different population groups, different requirements for social change. Yet such pervasiveness is combined with unequal impact on the different production sectors: at any given time innovations tend to be concentrated in a few sectors, and the identities of the latter change over the long stretches of historical time.

In addition to diversity in *type*, the diversity in *size* of the economic contribution made by the technological innovations should be stressed. The literature in the field commonly distinguishes between "major" inventions and routine, "improvement" inventions,[1] and one would expect that major inventions would lead to major innovations, and minor inventions and improvements to minor innovations. But the distinction is a rough one, complicated by difficulties in defining the unit of innovation properly. Should a single innovation be associated with a single invention or useful discovery? If so, how does one treat the thousands of inventions (patented) that never reach the application stage; or that having reached it, prove commercial failures; or make a useful but quite minor contribution? How does one treat the major inventions that, in the course of application, generate a host of subsidiary minor inventions, and make a revolutionary contribution to

---

Dr. Kuznets, Nobel laureate in economics, is professor emeritus at Harvard University.

economic production? Can one usefully employ a unit that covers, on the one hand, an invention that improves the method of producing toothpicks and, on the other, an innovation that affects radically the structure of the steam engine (such as that made by James Watt)?

The answers to these questions depend partly on existing tools and knowledge in the field, and partly on the purpose of the analysis for which a unit of technological innovation is defined. Given the extreme difficulty of measuring the specific contribution of a technological innovation, we cannot hope to calibrate innovations by their magnitude, within the wide range that measurement would probably reveal. Nor can we assume a supply of even rough data on innovations that have never reached the commercial stage; or having reached it, had only a minor success. And, in a discussion of the role of technology in economic growth, we must define the relevant unit of technological innovation much more specifically than if the field of study were more broadly defined. We limit our discussion below to successful technological innovations, and to those that may be considered major--in the sense of representing a marked advance and substantial economic contribution. We thus exclude both unsuccessful attempts at innovation, and minor improvements and innovations. It is only for relatively major innovations that we have some qualitative and quantitative information. It is only for them that we can observe the interplay of technological and social factors in the relatively long process of cumulative contribution to economic product and economic growth. This limitation follows a common practice in the field--although the persisting scarcity of quantitative measures for various innovations necessarily means that current classifications are partly arbitrary and it is difficult to secure a base for improving them.[2]

Three comments with respect to this choice of major innovations as the basic unit are relevant here, comments that apply also to the lists of major innovations discussed and analyzed in the literature. First, a successful *major* innovation is defined as major because it makes a perceptible economic contribution; and, if it involves a substantial departure from existing practices, is applied because it has proved to be economically successful. Aborted or unsuccessful innovations must, by definition, be minor--in terms of inputs into them and of effects produced. Second, a major innovation, whether or not associated with a major invention, ordinarily involves a host of minor inventions and improvements; in the course of application and diffusion, it usually provides opportunities for gains from subsidiary inventions, improvements, and the very process of learning. This situation implies that minor inventions and improvements that are part and parcel of the major technological innovations are included in analysis and measurement. Only such minor inventions and improvements as are not associated with, and are not a part of, a major technological innovation, are omitted. Third, although the designation "major" implies a minimum-size contribution to economic growth and often implicitly a minimum-size addition to technological knowledge, the unit employed here is intermediate--between minor inventions and improvements, on the one hand, and large groups of major inventions-innovations closely related to some basic and widespread contribution, on the other. If the definition excludes a minor innovation such as the modification of a plow sulky (for which numerous U. S. patents were issued in the 19th Century), it does include introduction of the condenser (and such related changes as the air pump and cylinder closure) in Watt's stationary engine--separately from the development of high-pressure stationary steam engines, and from the use of such engines in railroad transport. The interdependence of even *major* technological innovations, their coherence into large technologically and economically revolutionary entities (grouped around a single source of industrial power, or a single source of power for communications, or a single industrial material, or a single mechanical manufacturing function) is of key importance in analyzing the role of technological innovations in economic growth.

A major technological innovation implies a substantial new element in technology, and an opportunity for a sustained, substantial contribution to economic product. This means a systematic succession of distinct phases, in which the emergence of one phase depends on the completion of the preceding one. These phases constitute the life cycle of an innovation; though only roughly defined in the literature, they may be combined into a suggestive sequence.

This sequence begins with an open-end, pre-conception phase--in which existing scientific and technological elements can, in retrospect, be seen as having supplied indispensable ingredients for the new technological conception underlying an innovation. The contribution to technological preparedness, not the emergence of demand, is stressed: the demand, eventually satisfied by the innovation, *may* have emerged during, or even before, the pre-conception phase and stimulated the search for a solution; or it may have always been present (e.g., the demand for easier long-range transport or communications, or for labor-saving processes). Without the scientific and technological preconditions, the demand might never be satisfied--at least not as effectively as by the innovation. This phase is open-ended because the process of scientific and technological advance is continuous and it is difficult to establish the breakthrough basic to the innovation. For example, the scientific bases of electrical innovations were laid down in the early 19th Century, but they did not give rise to innovations until late in the century; and even then one may ask whether they are conditioned by still earlier knowledge. In practice, the decision to limit the backward extension of this phase is usually pragmatic. It is assumed that if the period is long, the net contribution of earlier developments to a much later specific technological conception embodied in the innovation cannot be large.

The first complete phase that follows is that between the conception of the technological innovation as a feasible task and the date of its initial, clearly successful, application--what the NSF-sponsored Battelle (1973) study calls the "innovative" phase, and what we designate for convenience, the initial application, or IA, phase. Frank Lynn subdivided this phase into the incubation and the commercial periods, setting the former between the date when "technical feasibility of an innovation is established" and the date "when its commercial potential becomes evident and efforts are made to convert it into a commercial product" (Lynn 1966, II, p. 38). More elaborate subdivisions are discussed in Mansfield and others (1971, Chap. 6, particularly pp. 110-115). In general, it is difficult to date the initial conception of the innovation, as distinct from development to successful initial application; although the period from conception to first successful commercial application averages about 20 years in the Lynn sample, the variance is extremely wide and is associated largely with the incubation period. That is what we would expect, for the dating depends on identifying an adequate perception and judgment of technological feasibility and economic promise; and only rarely is firm evidence available on such perceptions and judgments.

The completion of the IA phase implies, given the successful initial application of the major innovation, that there will be diffusion from the pioneer to other producers--as well as improvement, as the use of the new element in technology becomes more widespread. The beginning of this diffusion (D) phase is set by the date at which the first successful application is completed. To diagnose the *termination* of the D phase, however, requires that a limit be set at which the diffusion process is completed. In the literature, the limit is narrowed for practical, and often analytical, reasons. Thus, the analysis usually deals with diffusion in use in a single country, not in the world at large. If the world were used, the diffusion phase of hardly any modern technological innovation would approach even the halfway mark. Furthermore, if the major innovation is strictly and narrowly defined, it helps to set limits to the diffusion process. The practice is justified if innovations *A* and *B*, defined narrowly, are, in fact, sufficiently different so that the latter is not merely an imitation. (Thus, in Battelle 1973a, hybrid small grains and Green revolutionary wheat were classified as innovations distinct from that of the earlier hybrid corn, not as diffusion of the underlying hybridization process that began with corn.) However, difficulties persist in defining the limit to the diffusion process of a major innovation. They arise partly from the improvement component of innovation and diffusion.

Consider the simple cases in the United States of diffusion of narrowly specific machines or processes among the economic units that could employ them as substitutes for older machines or processes (summarized in Mansfield 1968b, pp. 114-119). The rate of diffusion is measured as the change in the per cent of U.S. firms introducing the new machines, or in the percentage of total acreage of corn planted to hybrid seed. The process is completed when the percentages reach 100, i.e., when all firms in the industry have introduced the new machines, or when all acreage has been planted with new seed. However, the diffusion of the innovation, as far as *economic* contribution is concerned, may be, even at this 100-percent point, far from complete. The new machines, representing a recent innovation, are being continually improved, possibly well beyond the point when the 100-percent level is reached, and the same is true of hybrid seed. The likelihood is that in all such cases the rate of improvement, even beyond the 100-percent mark, will be, at least for a while, greater than that in the sectors of the economy that have not been blessed with recent innovations. Hence, even in these cases the diffusion process is much longer than that suggested by the measures (which averaged about 20 years). The case is even stronger for large unit-cost consumer durables. Even if we could set some absolute limit to the number of cars (or refrigerators, or television sets) per family in a country, diffusion would not be completed when this number is reached, since substantial *quality* and *cost* improvements in the new product would still be possible. This opportunity would mean further diffusion in the sense of a continued spread of the contribution of the innovation. Unless we assume that the net contribution of an innovation, stemming either from quality or cost changes, is fixed when the first successful application is completed--an untenable assumption--diffusion is a function not only of the first successful breakthrough, but also of the improvements that follows; and may extend well beyond the diffusion in mere numbers.

Conceptually, the D phase is completed when some approximate upper limit in numbers (per unit of the universe of users) and some lower limit in the yield of further improvements (quality or cost) are reached, indicating that innovation effects have been exhausted. These upper limits on numbers and lower limits on improvements would ideally have to be determined for each innovation. A clear case of termination of the diffusion phase is one in which a later innovation, yielding products and processes that sharply compete with the older innovation, closes off drastically the economic grounds for further spread of numbers or further improvements of the older product. But in other cases, the closure by obsolescence, introduced by competitive pressure of technological progress, is not as clear. The only reasonable observation at this point is that the diffusion phase of major innovations is fairly long; and that further study is needed to distinguish important subphases within it, since the early periods of the phase may well witness much higher rates of diffusion and improvement than the later periods. Moreover, the pattern of movement within the long diffusion phase may differ among types of innovation, particularly between innovations yielding new specific producer goods, the broader innovations relating to power and materials, and innovations bearing on consumer goods that may affect long-term location and consumption patterns.

The completion of the diffusion phase means, in a sense, the end of a technological innovation as an innovation. But an identifiable component in the economy, usually of substantial magnitude and often far larger than originally expected, remains--and can be associated with the specific innovation. The subsequent development of this component may, in another sense, be viewed as the next phase in the life of a given technological innovation. This phase may give rise to some major economic and social problems; and may be affected not only by policies adopted during the period, but also by policies acted upon in the earlier phases. A change in the diffusion phase may postpone and mitigate the problems of the following phase, which, for obvious reasons, may be designated as that of slowdown and obsolescence (SO). It is thus necessary to stress the inevitability of the (SO) phase in its eventual emergence in the history of any specific technological innovation.

The course of economic change in the SO phase of an innovation differs widely, depending on the nature of the original innovation and the patterns of technological progress elsewhere. To illustrate, in the case of canals and steamboats on internal waterways, the competitive pressure of steam railroads, a later innovation, caused a sharp contraction not only of growth rates but also of absolute volumes of activity. A similar pattern is seen, in recent decades, in the use of steam railroads as passenger and to some extent as freight carriers. Innovations in the textile and older industries caused a marked slowdown, but no absolute contraction. Steel output, a major innovation that is nearly 100 years old, is still rising. But whether reflected in a sharp contraction of volume, or in a slowdown in the rate of technological improvement, diffusion, and growth, the SO phase raises major problems of adjustment--to change in growth potentials, to displacement of previously effectively employed resources, to general loss of position within the economic and social structure. For some innovations, this phase is complete, in the sense that the obsolescence results in contraction of absolute volumes to minimum or negligible levels. For others it remains open-ended--if the end is defined as relatively complete displacement--despite the fact that the pace of growth is far lower than in the preceding phases, and far lower than those of more recent innovations in the same or other fields.

Some aspects of the rough phasing of the life cycle of a major technological innovation is touched upon in later discussion. But at this point, three general comments are relevant. First, the time spans are long: about twenty years for the initial application phase, and much more than twenty years for the properly defined phase of diffusion. Thus, even if we disregard the less determinate time spans of the phases of pre-conception and of slowdown and obsolescence, the periods are of half century or longer. The spans are long not merely because of the time required to solve the purely technological problems in breaking through to an effective invention and resolving the difficulties in development, prototype production, etc., but also because of the time required to make the complementary and other organizational and social adjustments that would assure adequate diffusion and economic success. It is in the interplay of technological advance and organizational, economic, and social adjustments that the crucial feature of the innovation, the *application* of a new technological element, lies.

Second, the duration of the life cycle of an innovation is partly a function of its magnitude, for both the technological task and the process of social adjustment are the more formidable, the greater the departure from established technological practices and the greater the potential diffusion of the innovation. That makes it all the more important to emphasize that even major innovations come in large, coherent groups; and that for both a firm and an economy, the analysis of the innovation process and of the contribution of innovation to economic growth must take account of this *clustering* pattern. For such clusters or groups, the interplay of technological, economic, and social factors covers long periods, and contains multiple inputs and feedbacks among the various aspects. Well over a century elapsed between the time when the first steam engines were shown to constitute a major technological advance and the time when diffusion of steam power was reasonably complete (from the 1770s to well beyond mid-19th Century, in the *developed* countries of the world). The time span was much shorter for narrower and more specialized innovations like the zipper, and even the sewing machine.

Finally, while the definition of phases and the classification of periods within the life cycle of major technological innovation are crude and subject to revision with further study, the concept of the sequence of distinct phases and of a life-cycle pattern seems crucial in the analysis of the contribution of the innovation process to economic and social change. Phase distinction is essential to the study of the emergence, diffusion, and obsolescence of an innovation, in view of the duration, complexity, and cumulative changeability of the process. And any consideration of policy intervention into the process must be oriented, in the first instance, to significant phases. Obviously, policies relating to pre-conception conditions would differ from these applying to the incubation subphase: and the latter would differ from those relevant to the development and initial successful application subphase; and so on along the sequence. Furthermore, consideration of all the phases, including the expected phase of slowdown and obsolescence, is important in that policies affecting one phase, would also have some consequences for the next phase. Thus, prevention of monopolistic ossification in the diffusion phase might yield obvious benefits in both prolonging diffusion and alleviating, in the later stage, the effects of obsolescence.

II. CONTRIBUTION TO ECONOMIC GROWTH

By economic growth we mean a long-term rise in the capacity to supply economic goods, capacity defined within a specified social context and supply evaluated in terms of costs and returns to the human carriers. The growth of a firm, an industry, or a country is subject to constraint by basic rules of social and economic organization. And in the evaluation of such growth, the augmented flow of goods must be offset by all the costs impinging on men as producers and consumers. The observation is particularly important in a discussion of the contribution of technological innovations to economic growth, because the innovations involve changes in the rules of economic and social organization, and may carry with them costs and returns that do not appear in the conventional economic calculus. We shall discuss the contribution of technological innovation as suggested by the long stretch of modern growth in the developed market economies, since late 18th Century--omitting the difficult problems of the peculiar role of technological innovations in the growth of dictatorially planned economies of the Communist type.

One, apparently simple, way to gauge the contribution of technological innovations to a country's economic growth is to identify both new final products (as final products are defined in national economic accounting) and the new processes or tools that affected significantly the costs and supply of older final products; and then to calculate the country's final output that is accounted for by the "new" components. The identification of "new" would require knowledge of the date when the innovation was introduced; and specification of the period back within which the product or process was to have been introduced to count as "new." Definitions of that period, and judgments of the effect of the new tools or processes on the economic base of an old final product (like corn), may differ. But experimenting with periods of different duration might, in itself, prove interesting: it might indicate the variation in the proportion of "new" with the changing duration of the backward extension. And the attempt to ascertain the specific contribution of new technology to old final products would yield insights of its own.

Of course, such a count of "new" components, and the proportion of aggregate national product for which they account, would, if practicable, suggest only the *gross* contribution of technological innovation. It would not reveal the loss involved in the obsolescence of older products and techniques, and the additional costs sustained in introducing the technological innovations (compared with undisturbed continuance of old ways). This point is discussed at length later. Here we should note that the identification of new final products and new production tools and processes may be extremely useful in revealing the effect of technological innovation on patterns of work and life, and implicitly on patterns of social organization. If by new we mean innovations introduced during the past century (i.e., back to the 1870s), removal of these new components would leave the country without electric power, telephones, automobiles, airplanes, movies, radio, and television--let alone the major transformations in educational and medical services reflecting the new components within them. Nor would we have highly organized large plants, automatic controls, computers, and the innovations in the biological processes. The list could easily be extended by scholars more familiar with technological innovations of the past century, and would be sufficiently impressive even for the past half century.

Two broad comments are suggested by this new-components identifying approach. First, it reveals the large contribution of modern technological innovations in the way of new consumer products--as distinct from modification of methods of producing old and established products. Consequently, not only conditions of work, but also conditions of life in general, have been widely affected; and the present generations uses, and is surrounded by, consumer goods quite different from those of the preceding generation. By identifying these new products, as well as the old that they replaced, we secure an insight into the nature of the change, which cannot be conveyed by the anonymity-imposing calculations of inputs and outputs weighted by market prices in the national economic accounts or in the analysis based upon them. Thus, the replacement of candles or oil lamps by electric illumination within the households had consequences over and above the large reduction in cost per illumination unit--some positive, in making population less dependent on nature and its illumination; others negative, in making them more dependent on a single source of power, and the danger of breakdowns. The result was not merely an economic gain in dollars and cents, but a change in the quality of the good and in the corollaries of its wide adoption by households. A similar statement can be made regarding the new production processes--which required different patterns of work. And that means that in addition to the quantitative gap in economic levels among succeeding generations of consumers and workers, there is a qualitative gap of obvious relevance to the problem of measurement and analysis.

Second, since "new" components displace old, whether in consumer goods or in producer tools, and processes, the older options disappear. By and large, the process becomes irreversible, particularly since prolonged success and diffusion of the new component removes, for a long period, incentives for further improvement in the old, displaced component. Consequently, technological innovations may not mean a much wider choice, either for the consumer or the producer. Unless a consumer is exceptionally powerful, in economic and social terms, he is free to choose only what the majority of consumers choose (and thus create an economic base for an effective technological innovation). A consumer who prefers candles for illumination may find it difficult to get an adequate supply of good quality today; and the same may be true of a consumer desiring "natural" ice, or other types of old-fashioned consumer products (not to mention the less developed countries, which would find it difficult to secure an older technology, most suitable to their factor proportions). The disappearance of the old options is a result of various constraints that are typical of modern technology. These and other corollaries must be taken into account in a quantitative approach to the estimation of the *net* contribution of technological innovations to economic growth--which must be net, since the measures of economic growth are also based on the *net* balance of output over input of relevant production factors.

A major technological innovation often requires, for its efficient channeling, a scale of production, a stock of fixed capital, continuity of operation and hence of work, and a relation between the human and other factors of production, that are quite different from those characterizing the earlier technology. The amount of fixed capital, the impersonal character of the operating firm, and the personal relations within the firm needed to operate a steam railroad are vastly different from those of the horse-and-carriage transport of earlier times. The needs of the modern petroleum-refining firm and of a tractor-producing plant are far different from those of the hay and feed, and horse farm, of the past. This means that older patterns of organization of production must be modified--with consequent changes in conditions of participation of men as workers, entrepreneurs, or capital owners. Indeed, much of the diffusion phase of a major innovation can be described in terms of a gradual crystallization of the institutional conditions for increasingly efficient application, even allowing for minor sub-inventions and technical improvements. The inflexible demands of a technological innovation may involve entirely new characteristics of organization of the productive plant and firm, new methods of recruitment of labor force, new conditions for participation of management and of capital owners--in short, a variety of innovations in organizational and social technology. The variety of such innovations is quite extensive--ranging from new legal devices associated with the modern corporation and other modern organizations, and basic shifts in the allocation of jobs among the labor force, to shifts in recognized power relations among the various groups involved in industries affected by technological innovations, with consequent countervailing pressures.[3]

Some of the changes just noted are part and parcel of the technological innovation proper--because many new inventions, tools, and products would not be feasible unless there is a change in scale. Steam power could not be used economically in small doses, and the increasing advantage of high-pressure engines meant a minimum scale of operation that far exceeded that prevailing in the past; and this is true of the innovations in metallurgy and in the fabrication of standardized products. But if large scale had to be introduced (and if, in controlling large charges of power and fixed capital, the envelope had to be massive), the older type of firm unit, an individual or family-entrepreneur owner, could not be preserved. This meant also the legal and institutional changes that brought the modern corporation with its separation of ownership from management; the impersonal character of its organization; and the contrast in power between individual workers and the firm unit, until the workers could find strength in the modern labor union. Thus, a direct thread runs from the major technological innovations, with their scale and other constraints, to a variety of organizational and institutional changes. One major problem in measuring the net contribution of a technological innovation to economic growth lies in identifying the *indispensable* social and institutional changes; distinguishing them from other concomitants that were largely discretionary; and measuring the magnitude of the former, in terms of the costs or returns that may have been involved.

If through their effects on scale and character of the production process, technological innovations changed the conditions of *work*, the latter, in turn, brought about changes in patterns of life required to meet them. If new production processes required scales of output that made urban centers indispensable, migration from the countryside and a shift from country to city patterns of living were called for. If the scale of production meant that employee status would become a dominant feature of the labor force (instead of small-scale entrepreneurship or work on own account) and recruitment would be largely based on overtly tested skills and qualities, the whole life cycle phase of preparation for a job, of the current and next generation, would be affected. Such preparation is clearly a large part of a pattern of consumption and living, consumption as now defined in the national economic accounts and in much of our analysis. And the combination of what might be called production-requirements-of-work-participation, with an existing scale of consumer wants and priorities, may create imbalances that, in turn, would be resolved by technological innovations in consumer goods--which, once made, would further affect patterns of living. Thus the desire to combine participation in work in the high employment-and-income-potential fields of new technology--which meant working in cities-- with living in the more pleasant rural-like surroundings (the desire enhanced by shorter working hours, leaving more time for leisure) had several consequences. It induced the spread of commuting; created a wide demand for automobiles; led to the development of dormitory suburbs; and had other positive and negative consequences. To be sure, the urbanization-compelling change in production scale, necessitated by technological innovations, was only one factor in the combination. Yet it made an important contribution, and this contribution to changes in patterns of consumption and life should be included in evaluating the net yield of technological innovations.

The organizational changes (and those in conditions of work and living) just noted were complementary adjustments, in the sense that they were needed--and some indispensably--for effective channeling of the technological innovation. But the latter may have had also other, negative, impacts: on the capital and labor resources engaged in the old technology displaced by the new; and on natural resources, taken here to include all major environmental aspects of nature. We comment briefly on these *dislocative* effects.[4] Regardless of the rate of adjustment of displaced labor and capital to other opportunities, the displacement and readjustment costs must be debited to the particular technological innovation that is their source. And the impact on natural resources, either by way of enrichment or depletion, must also be taken into account--even though such an impact may not appear directly in the firm's or nation's calculation of costs and returns. At present, neither the deterioration of the environment and the depletion of natural resources, nor the enrichment of these resources by discoveries and additions to our technological knowledge (except to the extent of actual *input* of labor and reproducible capital) enter the calculations in the national economic accounts.

It follows that the net economic yield of a given technological innovation (or of a group of them) must involve not only a comparison of the economic value of output and input as conventionally measured, but also an evaluation of the required changes in conditions of work and life, of both the complementary and the dislocative adjustments. Obviously, the specific contributions of a given technological innovation under the various heads--let alone their combination into a single measure of net yield--cannot be easily ascertained; the difficulties are enormous. Yet it is useful to know what a *full* evaluation of the return from a technological innovation involves. The discussion suggests a few general observations on the evaluation; first, of individual innovations, and then of their combined flow in the course of economic growth.

In evaluating the contribution of a single innovation or coherent group of technological innovations, the phases over the long sequence seem particularly important. The nature and magnitude of the conventional and nonconventional costs and returns would vary greatly among the several phases. The relation of output to input would differ markedly between the IA phase (when the breakthrough occurs but when the total realized contribution is probably far short of the accumulated inputs or costs) and the D phase (when the effects of mass spread of the innovation are realized). Likewise, the requirements, with whatever costs and returns they signify, would have to be met in the early stages in the life-cycle of an innovation--if the latter is to realize its potential. The competitive impact on older components rendered obsolescent by the innovation, presumably moderate at the beginning, would become acute as the innovation diffuses to large market magnitudes, but might cease with the re-absorption of the displaced labor and capital long before the diffusion of the innovation is completed. Some negative externalities, in the way of pollution or impairment of the environment, hardly perceptible in the earlier and middle stages of the life cycle, might become increasingly pressing as the innovation reaches the later subphases of the diffusion process and constitutes an increasingly massive component within production processes and the pattern of life of the country's population. Such variation in the nature and magnitude of costs and returns among the successive phases and subphases would mean a corresponding variation in the magnitude and sources of the *net* contribution to a country's economy. The greatest contribution might well come, not in the early states of an innovation's life cycle, despite rapid increases in productivity; but later, when the innovation has reached large volumes, relative to countrywide product, but is still growing at substantially higher rates than the older sectors. Sustained study of the life cycle of technological innovations is still in the future; and when the results accumulate, the definitions and structures of the sequence of inputs and outputs may have to be revised.

In an evaluation of the type discussed in the preceding paragraph, two questions arise--in addition to that of combining measurable conventional inputs and outputs with the nonconventional ones that are so difficult to measure. How far back do we trace the contributing inputs? How far forward do we trace the possible indirect outputs? The first question is associated with our earlier comments on the open-end pre-conception phase in the life of an innovation. Difficult as it is to set a stopping date in the backward extension in the search for the sources of a given innovation, it is even more difficult to identify the antecedents so as to estimate the costs that should be debited as an input appropriate to the given innovation. Likewise, if (as argued below) a major innovation, in its spread, provides opportunities and pressures not only for related innovations but also for new tools, insights, and puzzles for scientific study, and leads to further advance of science, how far forward should we trace such effects so as to measure the output attributable to them? Given the nature of the questions, the answers must be pragmatic, depending largely on a judgment of the yield, in the way of additional debitable costs and creditable returns, of the pursuit of further antecedents and more indirect consequences. And the problems are aggravated when we shift to the flow of technological innovations in their *aggregate* contribution to a country's economic growth.

Finally, given the long life cycle of a major innovation (and even longer spans for coherent clusters of them), the novel and unforeseen in the new element of technology, the changing interplay with the required social innovations and adjustments, and the dependence of a single major innovation in the longer run on the unfolding of the cluster of which it is a part, it is difficult to discern the eventual shape and estimate the magnitude of an innovation in its positive and negative aspects. The projection and prognostication may become increasingly successful as, with the passage of time and development, the broad contours of the innovation emerge more clearly. But history is full of cases in which even the inventors and the initial entrepreneurs had only a dim notion of the eventual nature and magnitude of the innovation that they were successfully promoting. And there are many cases in which experts and active workers in science and technology were blind to the implications and consequences of research and invention efforts in which they were engaged. To be sure, such errors in prognosis are much larger for the major technological innovations, and particularly clusters of them, than for the minor inventions and improvements that can be diagnosed and projected more successfully, once their market value has been tested. But it is the major innovations and clusters of them that substantially change, both directly and through the opportunities that they provide for minor improvements, a country's capacity for higher levels of economic performance and higher rates of economic growth. Here again the forward evaluation of a major technological innovation will be improved only as further studies of inputs and outputs during the successive phases in the long life cycle yield better understanding of the social and technological interplays in the sequence.

When we consider the contribution of the *whole* flow of technological innovations to the country's product and economic growth, the questions raised in connection with the evaluation of a single innovation assume different aspects and parameters. The requirement to trace the inputs and outputs attributable to a *specific* innovation is no longer operative, but the task of evaluating total inputs and outputs, conventional and nonconventional, remains. And the questions concerning distant antecedents and the more indirect consequences now emerge as part of the problem of allowing for a time lag between all contributing inputs and all resulting outputs. The aggregation of several innovations means that new, recent, and older innovations enter the current flow, adding their contributions in different phases of their life cycles. The life cycle pattern of a single innovation disappears, and is replaced by a potentially more stable and continuous pattern of the absolute or proportional contributions of all technological innovations, major and minor, to the growing product of a country's economy. But this aggregate net contribution, and the underlying inputs and outputs, may be subject to a swing, with a much longer duration and a much narrower proportional amplitude than that of a life cycle of a single innovation or a group of them. This swing might be observed if, in a pioneer or early follower country, the pace of growth accelerates with the initiation or adoption of a broad new technology and then slows down as later follower countries enter the process and reduce the comparative advantage of the pioneer or early follower.

Current aggregative analysis has benefited from the great additions to national economic accounts data in recent decades, and developed as an empirical counterpart to theoretical structure of the economics of allocation of resources, production, and pricing. The comparisons of the combined growth rates of inputs with those of output derived in this analysis leave a positive residual, which is due, largely if not entirely, to the contribution of technological change. This is a net contribution, since all the relevant inputs have presumably been accounted for and since the total output is presumably complete and unduplicated. Even on the assumption of total inclusion of both inputs and outputs, the residual may reflect factors other than technological change; e.g., improvements in the quality of human resources (not due to any inputs) or trends in natural or political conditions having little to do with technology. But it is reasonable to assume that technological change is an important, perhaps the dominant, component in this unaccounted-for residual.

Several groups of problems have emerged in the development of this approach: of inclusiveness of the inputs involved in the production process, with particular reference to possible investments in human capital (now treated as consumption in conventional accounting); to the depletion of natural resources, now not covered; and to the changes in patterns of life and consumption imposed by the new technology that do not represent pure consumption. There is also a problem as to how the inputs are to be combined to derive the proper magnitude of the growth of total inputs, with reference to effects of scale and nonlinearities in the relations. Then there is the question whether the coverage of output is inclusive, either with respect to some positive items (e.g., the value of increased leisure) or to negative items (e.g., pollution and general impairment of the environment). Different solutions to these three groups of problems--inputs, weighting, and outputs--would mean differences in the magnitude and pattern of the unaccounted-for residual. And for the remaining residual there would still be the fourth problem: how to distinguish the contribution of technological innovations from that of other factors (positive and negative). An enormous literature has grown up in recent decades on the first three groups of questions, in a field of national income and productivity that has had a long history going back to the foundation of the discipline.[5] Although it cannot be summarized effectively here, some comments on the recent trends, and on the problems, at least as one investigator sees them, may be appropriate.

If labor inputs are measured in hours, with allowance for age and sex composition of the labor force; if the net stock of reproducible capital is used to gauge capital inputs; and if labor and capital are weighted by the rough proportions of labor and capital shares in total factor income, factor inputs can be estimated. The results, for a number of developed countries over long growth periods, are clear: the combined factor inputs account for a small fraction of per capita product, leaving a large unaccounted-for residual. This residual, well over three-quarters of the growth in per capita product, is presumably due largly to underlying technological innovations, which permit larger outputs with unchanged amounts of labor and reproducible capital. The inclusion of land, or similar stocks of nonreproducible resources, would only raise the proportion of the residual in the growth of per capita product. Revision of the inputs, the weights for combining them, or the output would have to be large, and in the direction of raising the growth rate of input and decreasing that of output, in order to account for the unaccounted-for residual.

The recent efforts at amplification and revision of the productivity calculations, and the implicit revisions of input and output, give the impression of two tendencies. The first is to reclassify some consumption items as prerequisites of higher productivity, either as investment of human capital or as changes in consumption required by the new technology and yet representing no additional consumer welfare. The second is a bias, in considering the social and other adjustments to technological innovation, toward emphasizing the negative or problem aspects and neglecting the positive aspects (also excluded from conventional measurement). This is an individual observer's reaction; it can only be illustrated, not proved. Yet one *should* start with a natural suspicion. If aggregative analysis and the underlying theory yield a residual in the form of unexplained total factor productivity, it is a natural inclination to reduce the residual by *ad hoc* reclassifications of older components, before considering whether the underlying analytical system should be revised to permit integration of a novel element, such as technological innovation, so far exogenous to the analysis. Also, if with rapid growth, as conventionally measured, there are associated discomforts and negative aspects, it is a natural inclination to concentrate on them and introduce them as improper omission, without allowing at the same time for the omission of *positive* elements.

The comment just made is not to condemn critical consideration of inputs and outputs in conventional economic accounts. Consumption, as now defined, should be critically examined and any capital investment or hidden cost identified and excluded; outputs should be examined and both the positive and the negative byproducts included. The argument is rather for a broader view of the issues. Thus, if formal education is found to have increased and to be positively associated with incomes of individuals, informal education, on the job or through some apprentice system, may have declined in relative proportion. Can we say with assurance that fewer years were spent on education (formal and informal) by a farmer in medieval times than by a farmer in modern times? Or can we say that the gross capital stock, including land, per farmer, was much lower in medieval than in modern times--and sufficiently so as to explain the vast differential in agricultural productivity between the two epochs? Similar questions can be asked concerning a medieval craftsman, who presumably went through over eleven years of training as apprentice and journeyman, when compared with a modern blue-collar worker. To be sure, the former's capital endowment was lower than that of a modern worker, but the difference would hardly account, in conventional economic terms, for the immense difference in productivity. We put these questions to suggest that it is not education, and not material capital as such, but accession of useful knowledge and its innovative application that is crucial; and the *indispensable* inputs in modern education, to use one component in which the consumption element can be quite large, may not, on a per capita basis, much exceed that in earlier times.

Similar questions can be asked about changes in conditions of life and negative externalities. Assuming that we find urbanization raising the cost of living as compared with that in the countryside, can we also assume that there was no consumer preference for the cities that warranted greater outlays without allowing for romantic notions about the countryside (in its "pure" form, unameliorated by proximity of cities)? If we find pollution and other forms of depletion of the environment resulting from some components of economic growth, do we also recognize the enormous positive effects of modern technology on health and longevity (not included now as positive components in conventional economic accounting)? If we identify the pressures of modern life as a negative externality, do we also recognize the enormous increase in leisure made possible by the advanced technology underlying modern economic growth?

To be sure, emphasis on neglected negative byproducts is a useful goad to reform and change, and much of analysis in the economic and social literature appears to be aimed at diagnosing and remedying problems, not at a balanced presentation. But if we are to consider the *net* contribution of technological innovations to economic growth, it is a balanced view, including both positive and negative aspects, that we must seek. Attaining such a balance is difficult because measurable items, like the components of inputs and outputs conventionally defined, cannot be readily combined with evaluation of major changes in conditions of life, work, and in organizational structure of society and man. For this reason, it is all too easy to advance a selected, yet unbalanced view--with resulting controversies for which there is no simple resolution.

Under the circumstances, no fully acceptable measure of the *net* contribution of technological innovations to modern economic growth is at hand. However, it is possible to attempt a variety of experimental adjustments to current conventional measures, in which additional costs and returns are estimated, along the lines illustrated by the recent Nordhaus-Tobin Study (1972). Further work in the same direction will at least enhance our appreciation of the magnitudes involved, and may limit the range of disagreement in the field. But for our purposes, a simple summary of the contribution of technological innovations to modern economic growth can be suggested. First, technological innovations produced a thoroughgoing transformation of conditions of work and of life, while increasing enormously the volume of final goods and changing substantially some of their characteristics. Indeed, the novel elements in modern economic growth that distinguish it qualitatively from the past are largely due to technological innovations and adjustments to them. Second, mass application of technological innovations had, inevitably, some negative consequences. All material technology is a disturbance of the natural ecological balance; all technological innovations also have elements of the unexpected, some negative; and, given the complex sequential interplay of technological and social innovations, timely prevention of these negative consequences is unlikely. The current critical reaction to economic growth and technological innovation is a natural response to the negative consequences, the adjustment or amelioration of which has been delayed. Third, although it is difficult to reduce the contribution to a single quantitative estimate of all the costs involved in conventional inputs, nonconventional requirements, and negative by-products, one cannot avoid the conclusion that the net contribution of technological innovations must have been positive and large. The nearest definition of economic product--flow of goods to households minus any components that may be suspect as hidden capital goods (e.g., technical education) and with an allowance for offsets to additional costs or discomforts of living imposed by technology--would still show a rapid increase per head, against fewer working hours. This increase cannot be at the expense of capital stock in the aggregate, even including natural resources (which has in fact grown on a per capita basis). The finding suggested by aggregative economic measures is amply supported by related measures of health, longevity, nutrition, supply of mechanical energy, and of a variety of consumer goods, on per capita basis, for the economically developed countries. Such great increase can be only due largely to the application of new knowledge, technological innovations, and their social complementarities.

That the net contribution is large also explains the pressure and striving, against obstacles, for the useful application of new knowledge. Inference of a negative or minor contribution would mean that innovating societies were foolishly misled, time after time, as to the ultimate consequences of the innovations that they encouraged. To be sure, given the organization of the modern world, societies were compelled to modernize and adopt technological innovations, if they were to retain independence. But judged by current experience, this element of compulsion was of secondary importance. The basic datum is that the modernizing societies wanted modern economic growth and the technological innovations that were its substance.

Underlying the broad conclusion above is the assumption of a social consensus within modern society, favorable to economic growth. The net contribution of technology to economic growth would have been evaluated quite differently by early Christian ascetics, who considered only a *minimum* supply of economic goods needed. That group's estimate of economic growth and product would be far smaller than our conventional one; and its judgment of the effect of technological innovation would be negative, for its contribution to economic goods would be viewed as unnecessary and its influence on holy life pernicious. A similar judgment might prevail in a society that puts a low valuation on material welfare. Some groups, even within modern society, may deviate from the dominant notions, in their judgment of the value of material welfare in the specific forms contributed by modern technology; or in their appraisal of effects of modern technology on capital stock, viewed broadly as capacity for desirable growth in the future. But if we consider the effective consensus of a modern society, through its successive generations, the estimate of the net contribution of technological innovation is positive and high, and likely to remain so with better balancing of conventional and nonconventional costs and returns.

III. CONTRIBUTION OF ECONOMIC GROWTH TO TECHNOLOGICAL INNOVATION

The discussion above centered on the contribution of technological innovations to economic growth and stressed their dominant role in the increase of per capita product, in the major transformation in conditions of work and of life, and in the institutional structure of modern societies. But the association is two-directional: economic growth contributes to a greater flow of technological innovation. The combination of a contribution of growth to further innovation with the contribution of past innovations to growth forms a continuous and self-reinforcing mechanism--subject to limits, but sustainable by continuous feedback between growth and technological change. In this section we comment on the contribution of growth to the flow of technology, the latter again largely viewed as the flow of major innovations.

In its IA phase, a major technological innovation is initiated by a fairly clear perception of a need requiring some new element of technology, and of the capacity to supply this element effectively. It is marked by a subphase in which the technological task posed is worked out. It terminates with the first economically successful application. The ingredients, present in different proportions throughout the IA phase, are technological knowledge servicing inventive capacity, an overt or potentially large need or demand, and entrepreneurial ability to mobilize and manage the economic resources involved. It is in the diffusion phase that many of the necessary social and institutional adjustments are made, a host of minor improvements and inventions emerge, and other effects assume substantial magnitudes, with their specific characteristics more fully revealed. In dealing with the contribution of economic growth to the enhanced flow of major technological innovations, we shall be concerned with the impact of the former on additions to knowledge in the pre-conception phase, and on the supply of technological capacity, perception of need, and supply of entrepreneurial talent in the IA phase--not with the diffusion phase.

In a discussion of the effects of economic growth on further technological innovation, the initial argument may be put simply. If modern economic growth is due, in good part, to the productivity-raising effects of technological innovation, and if the societies involved recognize this connection and consider its results desirable, they will presumably use resources, augmented by past growth, to induce further technological innovation. The societies will base these uses on several judgments. Given ever-present large or potential demand for more goods, they will consider the value of enhancing the stock of basic knowledge and thus the broader technological base from which inventions and improvements can be made. They will judge the value of increasing the supply, and strengthening the motivation, of would-be inventors and discoverers. They will weigh the value of stimulating the supply and facilitating the emergence of necessary entrepreneurship. Both the pioneer and follower countries in modern technology and modern economic growth would certainly be inclined toward such uses of resources--the follower countries after modifying the available technology to suit their own resource endowment, and then attempting to pioneer in new fields to catch up with or surpass the leader. Assuming that such uses of resources by modern societies, and the policies adopted, are well conceived, the resulting enhancement of the rate of additions to the stock of useful knowledge, to inventive capacity, and to entrepreneurial talent should, all other conditions being equal, augment the flow of further innovations.

This two-way connection between the contributions of technological innovation and economic growth is obvious enough. All societies that recognize a high positive yield of any given use of resources ordinarily attempt to employ part of the proceeds to enhance such advantageous uses. The literature of economic history, and of history of science and technology, is full of references to attempts by modern societies, since the 17th Century, to further the growth of natural sciences and useful arts. They have encouraged academies and scientific organizations, established new universities for the natural sciences, rewarded inventors with prizes and patents, and instituted a host of measures based on a recognition of the connection between economic growth and the widening application of a growing stock of useful knowledge to production problems. There has been a long evolution in the social recognition of this connection between technological innovation and economic growth, and in the changing institutional and ideological framework that conditioned social attitudes to science, technology, and innovations. But practical problems of optimal selection of means of enhancing further technological innovation aside, the linkage between the growth-promoting experience of past innovations and the innovation-promoting consequences of past growth is quite obvious in modern societies.

A related general argument would stress the effect of accumulated *experience*, and institutional innovations, through which modern economic growth may stimulate further technological innovations. The experience with past innovations has presumably yielded new ways of handling not only the organizational problems involved in the movement from conception to first successful application, but also those involved in the development of basic science and its connections with applied science. Hence, all other conditions being equal, the experience yield of technological innovation-based economic growth should facilitate a greater flow of further technological innovations in the future. If the obstacles to such further innovation that may also be the result of past economic growth are not increasing, the observation provides another general argument for expecting that economic growth will enhance the further flow of technological innovation--even apart from the greater volume of resources for that purpose and the public policy designed for that end.

But modern economic growth influences further technological innovation in other ways. A somewhat less obvious but crucial connection is indicated by the major role that wide application of new technology plays in yielding a variety of tested knowledge and new tools not previously available, relating to the particular aspect of nature that underlies the new technology. Modern science has characteristically always been stimulated by new data and new tools. If we recognize that applications of new technology (or of old in new combinations) provide science with new data, new tools, and new puzzles; that, in response to these stimuli, science may generate new discoveries and new insights; and that, on the basis of the latter, new technological opportunities relevant to potentially large demands may be glimpsed, to be realized in new inventions and technological innovations--we observe another connection between past technological changes in their introduction and diffusion in economic growth, and a flow of new technological innovations. That this later flow is likely to be at a greater rate than the earlier does not follow directly from the preceding argument. But it would follow if we argue further that, in modern economic growth, cumulative advance of applied technology means a greater flow of new data, new tools, and new puzzles--and thus a greater stimulus to the advance of science and to the related further advances in invention and technological innovation.

In view of the importance of scientific knowledge in modern technology and some tendency in the literature to draw a sharp line of division between science and technology and to assume separate lines of cumulation in the two, it may be useful to restate the argument just advanced.[6] The argument consists of the following theses. First, modern science, in its attention to data and puzzles yielded by practical production problems, is distinctive; and a connection may be drawn between the observation of production practices and some major scientific discoveries. To cite an early illustration: observation of the limited capacity of pumps led (in the 17th Century) to the discovery of the earth's atmosphere and its pressure. Second, whatever science discovers or generalizes about (and it does so about tested aspects of behavior of nature) is grist to the mill of technology, for the latter is essentially manipulation of the forces of nature for man's purposes. Hence, no matter how variable the gap in time may be between scientific discovery and technological application, greater knowledge about behavior of nature will eventually be found relevant to some practical application. Third, once successful application in the form of a major technological innovation is made, it is usually found that the knowledge of the process and material involved in the earlier science discovery is *incomplete* from the technological standpoint, although no additional knowledge was needed in establishing the original discovery or finding. Fourth, such additional knowledge, and new tools made economically possible through mass application of the innovation, may (and usually do) provide new data, new insights, new puzzles. Thus, the use of the Savery and particularly Newcomen atmospheric engines, a technological application of the discovery of the atmosphere in the preceding century, led to the basic improvements by Watt, which replaced the atmospheric with a true steam engine. The development of the latter to a high-pressure type and the observation of increased efficiency at higher pressures led to puzzles, which in their scientific resolution yielded the thermodynamics branch of modern physics (with its laws of conservation of energy and of entropy). A similar learning process in the course of mass application can be illustrated in other fields by the Edison effect, or the discovery of the way short rays are transmitted via the ionosphere--with obvious stimuli for modern physics and astronomy. Fifth, the further advance of science provides bases for new technologies; and so the chain continues to mass applications, more data, more tools, more puzzles, and so on.

The argument above is not in contradiction with the observation of differences in characteristics between scientists and technologists; or with the historical fact that some countries, like the United States in the 19th and early 20th Centuries, made no great contributions to basic science and yet were leaders in invention and technological advance; while others, like France, were much greater contributors to basic science than to new technology. The differences in the approach to problems between scientists and technologist-inventors, and the corresponding differences in their personal characteristics and backgrounds, are, if anything, in partial support of the proposition above that mass application results in much new knowledge. Apparently the setting of the problem in basic science often does not require fully tested knowledge about *technologically* important characteristics--even though the latter, when established, do provide food for further scientific thought. Thus Hertz, establishing the existence of short rays, in accordance with the prediction of Maxwell's theory, was not induced to inquire precisely how these rays were transmitted through space. Nor is the difference in national identity of the locus of advance in basic science and in technology surprising: given the transnational character of science, any nation that is capable of understanding it can build on it; and the new technology, even if it is based on new science, may well be more easily developed in a country other than the pioneer in the scientific advance. The same difference may occur between the locus of technological advance and of science based on the knowledge-yield of the former; thermodynamics was formulated in France (by Carnot), in Germany (by Clausius), and only partly in Britain (by Joule and Kelvin)--and yet Britain was the major developer of the steam engine. The argument is simply that in the development of modern technology, the common base of the economic growth of developed countries, an important feedback relation exists between development of science and technology and economic growth, and between economic growth and further development of science and technological innovation. The argument holds on an international scale, with allowance for time lags and for differences in national locus between the several links in the connection.

The hypothesis just advanced bears also upon the question whether, with enlargement of scale, the activity-yielding additions to new knowledge and new technology is subject to diminishing or increasing returns. Evidence is available showing that modern societies devote increasing proportions of their resources to augmenting the supply of scientists and technologists. The growth in numbers, and in level of advanced training, of scientists, engineers, and high-level technologists--the group capable of adding both to the growth of natural science and to the flow of inventions, improvements, and other ways of advancing material technology--has been rapid. Indeed, this growth has been greater than that of almost any other occupation (except that directly connected with a completely new branch of production, e.g., computer specialists). If these augmented talents were engaged, in increasing density, in a circumscribed field of science or technology, diminishing returns could be expected. But scientific discovery moves from one field to another, and it is induced to shift its focus and sometimes revise its theoretical structure by an accumulation of new data, new tools, or new puzzles provided by the unexpected lessons of mass application of major technological innovations, themselves perhaps resulting from an earlier scientific advance. Invention and innovation also tend to move from sector to sector, not only because of stimuli from the new scientific discoveries but also because of the exhaustion of cost-reducing opportunities and of demand responsiveness in sectors in which technological innovations of earlier days had time to diffuse widely. The richer promises are elsewhere and are often provided by new types of demand connected with changed conditions of life and work. These points, relating to the need or demand component in the triad involved in major technological innovation, are discussed below; but here, in dealing with the rate of return to an increased supply of providers of new knowledge and new technology, we note the tendency, and inducement, to shift from one field to another. This shift should help to avoid diminishing returns that could be expected with increasing supplies in a narrowly circumscribed field of science or technology.

The answer to the question of the dynamics of returns (diminishing, constant, increasing) to increased supply of participants in the process of adding to basic and applied knowledge and to technological advance, and to a greatly increased flow of material resource input into the tasks, depends partly on the tests one would apply. The *impression* is one of great acceleration in the 19th and 20th Centuries in the rate of discoveries in natural sciences, in the marked multiplication of disciplines and cross disciplines, and in the flow of major technological innovations. Furthermore, the greater density of the network of science and modern technology provides more favorable conditions for the generation of new ideas and inventions more rapidly. The closer relation between science and technology, reflected in the technological breakthroughs by scientists addressing themselves to technological tasks, can be illustrated by the development of nylon and of the transistor. But all these are impressions, and no proper measure of output and input, particularly of the former, are available. Such data as counts of scientific papers, or lists of major discoveries, inventions, or patents, are all limited by the wide diversity in the magnitude of the unit and the difficulty of interpreting a consensus on "major" in terms of a narrow range of magnitude. Such measures as numbers of persons engaged, with adequate levels of training, and even of material resources input, run into difficulties because of possible variance in selectivity and innate capacity of the people involved—surely an important aspect in a production process in which creativity is a major component.

No firm answer can be therefore given. Still, a moderate statement can be made: despite the large increase in the supply of people and resources to the knowledge-augmenting and technology-enhancing activities, diminishing returns, if present, did not prevent the continuance of a vigorous rate of flow of discoveries, inventions, and technological innovations. This observation applies in the international field of modern science and technology—not necessarily to any one country. As already indicated, in some countries the flow of technological innovations did accelerate with entry into modern economic growth; in others, the rate of technological innovation, compared with earlier phases of their economic growth in modern times, may have declined.

What are the effects of modern economic growth on further technological innovation, insofar as demand for its products and supply of requisite entrepreneurship are concerned? An initial examination of this question suggests several impeding effects. As already noted, reduction in costs of a final consumer goods, through preceding innovation, may be such that even marked cuts through further innovation cannot be expected to stimulate greater demand. Thus the need, or market, for further innovations may be damped by effects of past growth. Or, second, economic growth based on past technological innovations and resulting in marked enlargement of the scale of the plant and the firm, may make for entrepreneurship more conservative with respect to further innovations—particularly if the industries are oligopolistic or monopolistic. Finally, modern economic growth, in market economies, has resulted in a substantial rise in the share of the governmental, public sector in total product and resources; and the responsiveness of the public entrepreneur to further technological innovations may be more limited than that of entrepreneurs in the private sector. The three observations just made cover wide and complex topics, on which substantial literature exists. We merely offer a few comments to stress that, as usual, the process even of modern growth has not only self-sustaining but also self-curbing elements.

To illustrate the first observation, consider the case of cotton textiles. The technological innovations, which extended from mechanization of spinning in the 1770s to mechanization of weaving in the 1830s (the two most labor-consuming operations), combined with various auxiliary and complementary innovations, resulted in the conversion of cotton textiles from what was a high-priced luxury to a low-unit-price, mass-use commodity. Once the new technology of producing cotton textiles was established and improved, further technological innovations could reduce the low cost further—and particularly provide some competitive advantage to the pioneering firms. But, by and large, opportunities for substantial gains from further technological innovations, in *that* field, were decisively narrowed; and were certainly much narrower than opportunities for technological advance and innovation in other fields, in which current technology and the high costs of final product left much room for gain from an innovation. This did not mean that technological advance in the light-textiles industry ceased. But much of it came as diffusion of technological change in other fields (e.g., automatic controls over the spindles and looms) or in the form of different textile products, such as rayon and later nylon. There was no pressure for further major technological innovations within the cotton-textile industry proper.

Several aspects of this illustration deserve note, because it typifies the slowing down of incentives to further innovation in industries that reach technological maturity within a given framework of current technology. First, the loss of incentives to innovate was *relative*, not absolute--in the sense that potential returns from further innovation became more promising in other sectors. Thus, assuming a stock of uncommitted inventive and capital resources seeking promising opportunities, the comparative advantages of using it in the mature industry would be far lower than in the past when it was growing. Second, relevant technological innovations that do occur, may originate elsewhere and be introduced into the industry; or may provide a base for the production of a different product, serving the same purpose, thus setting up a new, different, and competing industry. Third, the relevant innovations that occur at this stage are not usually managed by the entrepreneurial talent within the mature industry. The developers of rayon and nylon were not the entrepreneurs in the cotton-textile industry. With the latter reaching maturity, and lack of great potential returns from technological innovations within the old technological framework, the successful entrepreneurs might seek other sources of greater returns--either in the protected markets of the colonies, or by raising the protective barriers of the large domestic market, or by shifting the industry to regions with appreciably lower labor costs and easier labor supply. The last statement illustrates an important point. Maturity of an industry generally means large economic magnitudes, and the power to exert political pressure within the economy or to exploit differentials in supplies of productive factors. These powers are not operative, or are of lesser importance, in the early phases of major rapid technological changes within a new industry.

As already indicated, economic growth means large scale of plant and firm, with the structure of fixed costs and marginal returns leading to monopoly of oligopoly in many modern industries and of public regulation of some. What are the consequences for entrepreneurial response to invention and innovation? As Maclaurin (1949, p. 251) indicated, in a truly competitive industry there would be no research, for there would be only windfall profits; and returns from an innovation could be derived by a pioneer only if protected by a patent (or secret), i.e., by monopolistic power. And it is true that a large-scale firm and even monopoly provide opportunities for internal research (e.g., DuPont work on nylon) that are not available to smaller and competitively pressed units. But in the case of monopoly, or even oligopoly, technological innovation can be slowed down by absence of competitive pressure. Monopolistic concerns, unlike others, can delay innovation even when its technological and long-term economic feasibility is established. Such a delay avoids the destructive effects on capital and labor in uses that would be displaced, but it does represent a slowing down in the rate of technological *innovation*. Finally, the rigid rules under which the regulated public utilities must operate may well inhibit adventurous entrepreneurship that is often required for particular types of innovation, or at least inhibit it more than in industries not subject to this special type of regulation.

The comments above should have been put in the form of questions, for no clear answer can be provided here. But one may suggest a similarity to the effects of slowing down of demand for a consumer good that is the product of past innovations. In the latter case we observed that a loss of incentives for further innovation in the *given*, "mature," field, results in the focus of vigorous technological innovation shifting to another field. That may be true also in the slowing down of technological innovation in a given field because of the existence of a monopoly or oligopoly in it. Unless they are strictly regulated, the dominance of the large concerns in a given field need not prevent them from technological innovation and from research and development outlays in *other*, possibly (but not necessarily) related field. The two sets of cases are parallel in the sense that relaxation of pressure for technological innovation--either because of slowing of demand, or because of attainment of a relatively protected position--may retard the flow of technological innovation in a specific field. But with resources available for innovation, the focus may shift, with accelerated growth following in another field. One may still ask whether this combination of retardation due to dominance in a given field, perhaps before the maximum reduction of real costs through technological innovation has been reached, with a shift of resources for innovation to another field, is an optimal outcome. Policy remedies would depend upon the answer. But the answer would involve difficult comparisons of lost or delayed opportunities in one field with accelerated opportunities in others.

The growth in the share of government--in product, labor force, capital--in the market economies in the course of modern economic growth was due to several factors. Urbanization created situations in which basic services to urban population--in much greater volume than required by the more thinly spread rural population--had to be provided by the public sector rather than sold on the market (police, sanitation, health, etc.). Certain types of capital investment, which could not be handled by, or entrusted to, the private sector, had to be taken over by government (highways, port facilities, and in many countries the railroads, shipping, and airlines, atomic energy, space exploration). Education and health services, not geared to the paying capacity of the recipients, were an ever-growing responsibility of government in a modern society. Finally, the participation of *developed* countries in acute armed conflicts meant devotion of large material resources to defense. The question posed here is whether an increasing proportion of resources in the public sector, without the profit motive and without competitive pressures for efficiency, has not meant a lack of stimuli for significant technological advance.

The question is strictly relevant only to the civilian part of the government sector. One may assume that international competition and tension provide sufficient pressure so that the incentive for technological innovation in weaponry and other components of defense production has not diminished, and is probably not inferior to the competitive pressures within the private sector. And the evidence on the major technological breakthroughs associated with the war and defense industries, some of large value for civilian production in the longer run, is impressive. It is hardly an accident that World War I accelerated the spread of the internal combustion engine in automobile and airplane transportation; or that World War II is associated with the development of atomic power, radar communications, electronic controls and computers, and the space industry. Two sets of questions may still be raised concerning the war- and defense-related industries. The first is as to the level of efficiency, in the absence of cost controls and economic pressures as great as those in the private sector. The second is in regard to the *net* contribution of an advance in military fields to technological innovations relevant to peace-type production and needs. We are in no position to examine the first set of questions, which would involve detailed quantitative analysis under properly set control conditions. The second should be examined in light of the fact that defense purposes emphasize technology of transport, communication, power, and strength of materials--all areas of technology potentially and directly relevant to civilian use. One last point: it is distinctive, and indispensably typical, of modern economic growth that the developed countries are ready to devote large *material* resources to defense purposes, substituting the material for human resources as much as possible. That explains the extremely high technological level sought under the competitive pressures of international strain among the *developed* countries. (The pressure would not be as great in the strain between developed and less developed countries.)

The question about the effect of the increasing share of government on incentives for further technological change relates more properly to the civilian components of government product. The answer would involve an examination of the functioning of the civilian government, and the inducements to technological entrepreneurship that are provided. One would also have to examine both the relatively limited segments of the public non-military activities that are specifically directed at basic and technological research--as a common service to society; and those that encourage higher education, a prerequisite for much of research and technological advance. The question is raised to indicate its relevance in the wider field of inquiry that would deal with feedback effects of economic growth, stemming largely from past technological innovations, on further flow of the latter.

In discussing the effects of economic growth on further technological innovation, we stressed the positive feedback of what might be broadly called *learning* aspects of economic growth--learning about benefits of past innovations as a rationale for more resources and better policy to encourage them; learning, by experience, better ways of generating and successfully channeling technological innovations; and particularly important, learning from the new data, new tools, new puzzles generated in the mass application of innovations. In contrast, in dealing with the effects via demand and entrepreneurial responses, we tended to stress only the inhibiting ones. However, we also indicated that the common response in the private sector to such limiting consequences of economic growth on further technological innovation was a shift in focus. Even in the public sector, within the nonmilitary component, there has been a movement toward greater technological innovation in fields in which the new technology, either because of its dangers or because of its greater social rather than private return, had to be handled by the government (as with atomic energy, space exploration, or geophysical research).

In connection with shifts of technological innovation to new fields of consumer demand, we have already noted that economic growth did have some effects that constituted a positive feedback. But we should add a further note about several aspects of economic growth that constitute positive contributions to further technological innovation of the type that yields new consumer goods. Thus, for one thing, the concentration of population in urban centers, one of the corollaries of modern economic growth, exposed consumers to a demonstration effect not found in the countryside. Also, the anonymity in the cities, contrasted with personal knowledge and status identification in the more settled society of the countryside, permitted consumer choices uninhibited by traditional status restraints. Furthermore, the availability within modern cities (and eventually in the countryside) of sources of industrial power for consumer purposes encouraged the invention of new consumer goods that would use such power. A clear illustration is the proliferation of new consumer durable goods dependent on a continuous supply of electric power, originally provided almost exclusively for illumination purposes. Thus, the shortages of domestic labor, associated with cessation of migration from the countryside, created a demand for substitutes for cooking, cleaning, washing, and similar purposes. This demand, which might otherwise have been met (and was partly met) by development of commercial cooking and baking, commercial laundries, commercial cleaning establishments, etc., stimulated technological innovation that supplied fractional-power electric equipment for many domestic services (and later even for recreation devices, such as the phonograph, radio, and television). Finally, one should note that it was part and parcel of the social and ideological corollaries of modern economic growth that material welfare was assigned a high priority; and that consequently the consumer was quite responsive in recognizing the advantages of a new consumer good when it emerged in the course of a technological innovation. Demand was thus created by supply without a substantial lag; and often with a speed far exceeding the expectations of the innovators.

IV. CONCLUDING COMMENTS

This paper dealt with relations between technological innovations and modern economic growth, against the background of the historical experience of non-Communist developed countries. By technological innovation we mean application of new elements in material technology; i.e., in the knowledge and capacity to manipulate nature for human ends--more specifically here, for economic production. By economic growth we mean the sustained rise in the supply and diversity of economic goods to the population, under acceptable rules of social and economic organization. Because of the wide scope of the subject, the difficulties in establishing quantitative weights, and the limitations of space and competence, the discussion was perforce too brief; and at some points only raised questions, or merely noted a particular aspect of the problem. Far from attempting to represent an adequate summary of the state of knowledge as found in the literature, the discussion was limited to reflections by an observer who has worked on quantitative aspects of modern economic growth, while entertaining a strong interest in the contributions of technology to it.

Given the brevity and uneven character of the discussion, a detailed summary is hardly necessary. But it might help to list the major observations.

(1) Technological innovation played a key role in the rise of product and productivity in modern economic growth; and also induced major transformations of conditions of work and life.

(2) These transformations were required to channel new technology effectively, and meant organizational changes in the earlier institutions that governed production. The resulting changes in conditions of work for the active participants were a major element in changing conditions of life. Thus, technological innovations required innovations in social structure and even in prevailing attitudes. They also required adjustment to resulting displacement of resources in earlier, and obsolete, uses.

(3) A major technological innovation--i.e., one with a substantial economic impact-- involves a sequence of phases stretching over a long life cycle. It takes time to pass from the perception of a new technological possibility to economic mass production, requiring economic and social changes; and to reach the period of slowdown and obsolescence. Of the four phases suggested--pre-conception, initial application (IA), diffusion (D), and slowdown and obsolescence (SO), the IA and D phases alone account, on the average, for half a century. The phase sequence, the differences in input-output relations in the several phases, and the phase differences in the interplays between the technological, economic, and social adjustments, bear clearly upon analysis, policy consideration, and prognosis.

(4) The clustering of even major technological innovations into groups of related changes (stemming from exploitation of one source of power, or from a new industrial material, or from the interrelation of functions within a production process), combined with the interplay between innovations and the social and institutional adjustments to them, lengthens the sequence of distinctive phases and adds to their complexity.

(5) The focus of technological innovations shifts over time from one sector of the economy to another, and creates new sectors. Their immediate impact is always unequal among sectors, and hence among social groups in the economy. This inequality of impact is itself a social and economic problem that requires adjustments.

(6) Because of the combination of conventional economic inputs with required changes in conditions of work and life, and because of the combination of conventional economic outputs with possible nonconventional byproducts of technologically induced economic growth, an adequate quantitative gauge of the *net* contribution of technological innovations to economic growth is still to be established. Current measures of total factor productivity, though possibly reflecting largely technological changes, are either limited to conventional input and output, or involve a variety of *ad hoc*, and not fully tested, assumptions as bases for inclusion of nonconventional inputs or byproducts.

(7) Such a net measure may not be of much value, since a variety of elements, in both inputs and outputs, is needed to give meaning to the comparison. Yet the search for such a net measure helps to focus the analysis. Provisionally, one may justifiably argue that the social valuation of technologically facilitated, modern, economic growth is high and positive, with the critical reactions reflecting responses to temporary lags in adjustment.

(8) Technologically induced economic growth, having been attained, stimulates further technological innovation. Society is stimulated through the learning of past benefits and through the effort made to allocate more resources and provide more favorable institutional conditions for further innovation. Past experience with technological innovations may make for better ways of developing successful innovations in the future. A particularly important stimulus is the learning that takes place through mass application of recent new technology and yields new data, new tools, new insights and puzzles to natural science, and helps widen the base provided by the latter to further technological breakthroughs and innovations.

(9) Economic growth hastens the maturity of the older fields by slowing down growth of final demand for their products. It may also limit the conditions for responsive innovative entrepreneurship in the established and modernized fields because of the large scale of the firm, and possible dominance of a few in an oligopolistic or monopolistic situation. Furthermore, the rise in the share of the public sector is a factor, since in its nonmilitary areas it may not be easily responsive to technological innovation. The slowing down of the older sectors, once modernized through technological

innovation, helps to shift the focus of innovation to other sectors. And these shifts to new sources of power, new materials, new types of producer equipment, and, in an important way, new types of consumer goods, help to maintain a high or increasing pace of technological innovation, and a high or increasing pace of economic growth.

We conclude the paper with brief, tentative impressions of the current state of the field, insofar as it bears on the study of economic consequences and origins of technological innovation; and even more tentative suggestions regarding directions of further work.

The importance of both science and technological innovation for modern economic growth must have become apparent by the middle of the 19th Century (if not earlier), when the development of steam power and its application to railroads, besides other less important technological innovations, were seen to provide new bases for rapid growth and transformation of modern societies. The policies of some governments in planting technological institutes and otherwise encouraging science and technology testify to this recognition. Yet, in both the Classical and (implicitly) the Marxian schools within the economic discipline, technological progress was viewed as a feeble barrier to the expected exhaustion of natural, nonreproducible resources (acting via pressure on profits and capital accumulation, and via limitation on reduction of labor inputs into required means of subsistence). The failure of the resulting long-term prognoses of both schools, apparent by the last quarter of the 19th Century, only led to extruding the subject of economic growth (and the related topic of technology) from the accepted corpus of the discipline. Economics concentrated on the problems of limited means to various ends, within a short-term framework in which technology, institutions, and tastes of consumers were all supposed to be given, i.e., fixed. With few exceptions, of which Schumpeter's monographs just before World War I were most conspicuous, the problems of economic growth were neglected, until their reintroduction largely after World War II. And the short-term orientation of the economic discipline led to a neglect of empirical and quantitative work on long-term trends and macro-quantities. Such work was kept partly alive in the concern with business cycles and then revived, with impressively explosive growth, in the late 1930s, and again largely after World War II.

The explanation of these past trends in the treatment of problems of economic growth, and the associated problems of technological change, is less important here than the consequence for the field of our interest. The consequence was obviously a failure to accumulate economic data relevant to the effects and origins of technological innovation and a limited stock of analytical studies, most of them recent. In particular, there has been no attempt to link technology to the basic output-and-input-statistics of the developed countries. Subclassifications relating output, labor force, capital, etc. to technological innovation are missing from our aggregate measures. The number of economists engaged in the study of technological aspects of economic growth has been, and is, quite small. Even the studies in economic history that would provide systematic accounts of the emergence, development, and phasing in the spread of innovations (single, or in clusters) are few. If there is now a really serious concern with the contributions and sources of technological innovation, with the prospects of specific policy action in the field, then much systematic work must be initiated on the quantitative aspects of the problem, and implicitly on the analytical formulations that would be needed for, and then modified by, such quantitative analysis.

In this development of further quantitative work, linkages with the already accumulated and established measures, countrywide or industry wide, would be particularly important. The possibility of distinguishing within total output, labor force, capital, etc. of the country or of an industry, components associated with recent, not so recent, and much older technology, would add immensely to the quantitative base for the study of the interrelations between technological innovation and modern economic growth. A more systematic analysis of the accumulated experience with R & D outlays and of the costs of innovations would be possible. And systematic historical accounts of the emergence and spread of clusters of major innovations should be encouraged. Such an account of the spread of steam, or of electric power, or of the automobile, each of which meant a major addition to the productive capacity of modern societies, major transformations of conditions of work and life, and much social innovation in the process of adjustment, would illustrate the complex interrelations and the sequential phases involved.

The list could easily be expanded, but would then run the danger of conveying, unwarrantedly, an impression of careful selection and testing. These are only *ad hoc* suggestions, can be no more than that, and are hardly worth multiplication. The main burden of these comments is to stress the long neglect of the field, and the consequences for the current state of our knowledge. The illuminating and intriguing monographs of the handful of economists who have worked on technological innovation are only a beginning. A more sustained volume of systematic quantitative analysis, closely linked to the flow of basic data, and of monographic work on the complex interrelations susceptible of historical study, would add to our understanding, perspective, and stock of empirical findings needed for adequate analysis and policy consideration.

REFERENCES

1. See, for example, Schmookler (1966, particularly pp. 18ff, and the lists in App. C-F).

2. In addition to Schmookler's monograph cited in Ref. 1, see also Schmookler (1972); Jewkes et al. (1969); National Bureau of Economic Research (1962, particularly Part III, pp. 279-360); Mansfield (1968b); and the recent National Science Foundation studies, particularly Myers and Marquis (1969) and Battelle (1973).

3. On organizational change see Berle and Means (1968), particularly book II, Chap. 1; and Boulding (1953). On the labor force see Hoselitz (1953, pp. 52-77), reprinted in Hoselitz (1960, pp. 23-51).

4. For more detailed discussion see Kuznets (1972, pp. 431-461), reprinted in Kuznets (1973, pp. 185-211), which covers several of the points touched on here.

5. It is practicable to give here only a few selected references. The collections of papers in three volumes of the Conference on Research in Income and Wealth (1961, 1967, 1973) are useful, as are those in *The Review of Income and Wealth*, Ser. 18, No. 1 and 2, for March and June 1972, dealing with factor input and productivity. A comprehensive review is presented in Nadiri (1970). The empirical studies by E. F. Denison and J. W. Kendrick are repeatedly referred to in these sources.

6. These comments on effects of the wide application of technological innovations on further growth of science (furthering, in turn, more technological innovation) are the reflections of an amateur: no economist is competent to deal with such connections. They are based on general reading, no expert knowledge. Yet in this important, still to be cultivated, field, even amateur conjectures may be useful in suggesting intriguing connections and perhaps stimulating more thorough consideration.

In addition to many of the books and studies mentioned in Refs. 1 and 2, and general references in the history of science and technology, I was much influenced by three monographs not mentioned so far: Cohen (1948); Maclaurin, (1949), particularly Chaps. 1-4 dealing with the movement from scientific discovery to technological application (and effects of the latter on increased knowledge); and Cardwell (1971). No claim to even a moderately wide coverage of the field is intended.

APPENDIX: SOME PROBLEMS OF ECONOMIC MEASUREMENT

I. MEASUREMENT IN THE ECOLOGICAL SYSTEM OF INNOVATION

The purpose of this brief appendix is to review some of the measurement problems associated with our knowledge or lack thereof of innovation. Measurement is confined to the present and past; although the future is worth knowing, knowledge of the future is gained by forecasting, not by measuring.

This study treats the innovation process as an ecological system. This ecological framework was found to provide a fecund structure for organizing and integrating the diverse parts of innovation into a coherent whole. Likewise, it will be argued that this ecological perspective is useful in organizing and highlighting the problems associated with measuring various aspects of innovation.

For example, the ecological perspective also argues for the dynamic nature of the innovation process. Just as organisms evolve and change in relation to subtle changes in other organisms and in the environment, so do structure and function change and interact with each other and with the larger environments in terms of the innovative process. It is not surpirsing that measurement in such a dynamic, complex, and interdependent world is far from easy.

There are two main groups of researchers who use measurement to study innovation: economists and management scientists. The latter, together with psychologists and social scientists, use behavioral measurements involving a large number of diverse parameters. Methodological problems associated with this type of measurement for innovation are dealt with by Douds and Rubenstein in Chap. 9 above, so we shall not deal with them in this appendix. Here we shall review the problems of economic measurement of innovation.

This review will be illustrative rather than exhaustive for two reasons. First, the literature treating technological growth is but a part of the extensive econometric literature dealing with economic growth. It would be impossible to treat technological change in detail without becoming embroiled in the complexity of the larger literature. Second, recent literature reviews of the problems associated with the measurement of technological change are available and will be referred to here (Nadiri 1970; Mansfield 1972a and b; Stewart 1972; Griliches 1973). These reviews should provide the interested reader with an introit into the literature.

II. TECHNOLOGICAL CHANGE AND GROWTH OF THE ECONOMY

One important measurement concern of economists has been the relationship between technological change and economic growth. In our ecological context this relation is examined at a rather high level of aggregation, usually the nation-state or some large part thereof. In the mid and late 1950s Abramowitz (1956), Fabricant (1959), and Solow (1957) attempted to relate technological change to overall economic growth. They made certain simplifying assumptions about the nature of the production process and incorporated them into an aggregate production function. They next attempted to measure the inputs of resources, mainly labor and capital, and related them to the output of products and services. They found that they were unable to explain a large part of the increase in output in terms of increases is labor and capital. This "residual" was felt to contain a component which could be labeled "technological advance." In one of the earliest studies Solow (1957) found that this "residual" was quite high, and indeed that 87 per cent of the increase in output per man-hour was a consequence of influences other than increments of capital and labor. However, Solow made no allowance for change in the quality of either capital or labor.

Although this "residual" surely contained elements of technological change, it was also likely that it masked other aspects which contributed to economic growth. Denison (1962) tried to take into account the effects of changes in labor quality on the size of this "residual." He specifically identified such aspects of labor quality as reduction in hours worked, the age-sex composition of the labor force, and the level of educational attainment. He found that the last item was especially important in terms of its influence on economic growth. Denison was able significantly to reduce the "residual" by taking into account the effects of increases in labor quality, especially education, on increases in production. The effect of Denison's inclusion of characteristics of labor quality in his model was to reduce the "residual," which among other things contains the effect of technological change, to account for 40 per cent of overall economic growth or less than one-half of Solow's result.

Denison's work, by taking into specific account changes in labor quality, had reduced the number of items included in the "residual" labeled "technological change." Jorgenson and Griliches (1967) extended the work of Denison. They accepted the adjustment that Denison had made in terms of the labor input and attempted to make similar adjustments for capital inputs. They felt that if the effects of improvements in the capital inputs were explicitly accounted for, an additional portion of the "residual" could be explained. They attempted to make adjustments for aggregation and measurement errors in the area of capital inputs. They also hypothesized that if inputs and outputs were correctly measured and if certain biases were eliminated, then the "residual" itself would be significantly reduced, if not eliminated. And indeed they found the "residual" to be very small.

Denison's response (1969) to the revisions of Jorgenson and Griliches brought about further revisions in these estimates (Christensen and Jorgenson 1969; Jorgenson and Griliches 1970). These revisions took the form of including rates of utilization for capital inputs and including in outputs the items which are excluded or under-counted in national income data (Jorgenson and Griliches 1970). As a result of these revisions, the differences between the estimates of Denison and those of Jorgenson and Griliches were reduced.

Three basic sets of problems are encountered in the measurement of the relationship between technological change and the growth of the economy. These sets of problems relate to specification of the model, measure of outputs, and measure of inputs. The problem of developing a model that differentiates the effects on productivity and growth of many interdependent forces is the basic problem of measurement in an ecological system. The complexity of this process and its dynamic nature, coupled with this interdependence, makes it difficult to specify a production function that can separate out the various technical aspects of the production function and their interaction with one another and with changes in factor prices. This is an especially difficult problem in terms of building models capable of empirical investigation, given data sources that are now available. Along the same lines, some of the assumptions that have proved to be necessary to permit an empirical look at this relationship are of questionable validity and may impose unrealistic constraints on the models. Examples of questionable assumptions are: perfect substitution between old and new capital, constant rate of depreciation, and the use of income shares as proxies for the contribution of various inputs to the growth of output.

Difficulties are also encountered in measuring the output of the economy. Some are of the type that have been encountered before in measuring gross national product and national income, whereas others pertain more specifically to the measurement of technological change and growth. One significant difficulty is in treating improvements in the quality of final goods and services. Price indices often do not account for quality changes, or at best underestimate them (Fellner 1972, p. 45; Griliches 1973, p. 67; and Mansfield 1972b, p. 479). Problems are also encountered in handling new products. Present modes of measurement often understate the importance of new products as vehicles of technological change (Nadiri 1970). Existing methods are much better at measuring changes in technology that reduce the cost of products already in existence. Nor is the increase in consumer welfare associated with a larger range of choice included under present means of measurement (Mansfield 1972b, p. 478).

There are also certain conceptual problems of defining and measuring the output of the government and service sectors of the economy. Government output is valued at cost; therefore existing measures of government output are in fact measures of input, so that for measurement purposes there is an implicit assumption of zero productivity change. This assumption is especially important because a large portion of R & D output, as well as a significant portion of goods and services, is sold to the public sector (Griliches 1973, p. 67; Stewart 1972, p. 15; and Nadiri 1970, p. 1170). Moreover, most empirical studies have dealt with private benefits, not social benefits. To the extent that these two series differ significantly, economic growth and the contribution of technology to economic growth is misstated (Nadiri 1970, p. 1170).

The third set of measurement problems relates to factor inputs. As mentioned previously, the quality of labor varies through time. More serious problems of input measurement have been found in attempts to calculate aggregate capital. Some of these problems relate to quality improvements of new capital. One particularly thorny problem is whether to treat capital as a stock or whether to estimate the flow of capital services. The second method is clearly preferable in terms of theoretical validity, but it is extremely difficult empirically. The consumer of capital services is usually also the supplier, and the "purchase" is an internal transaction to the firm and, as such, very difficult to estimate. This feature has led to a rather roundabout method of estimation of capital services, which is explained in some detail by Jorgenson and Griliches (1970). Additional problems are encountered in the use of price indices for investment goods. In many cases these price indices are in fact cost indices (Jorgenson and Griliches 1970).

Not only are theoretical and measurement difficulties encountered in establishing the relationship between technological change and growth, but even greater difficulties present themselves when the relationship between R & D activity and technological change is examined. It is evident that only a portion of the contribution to economic growth is due to formalized R & D expenditures and that many other activities contribute to technological progress. Given these problems, it is not surprising that there have been but few attempts to measure the relationship of one component, R & D expenditures.

Denison (1962, p. 245) made a rather rough estimate of the contribution of organized R & D to growth. By making assumptions of the relative importance of technological progress versus improvements in other types of knowledge (e.g., improved business organization), of the importance of foreign sources of technological knowledge, and of the failure of R & D expenditure figures to include all the activities which lead to technical progress, he arrived at the conclusion that no more than one-fifth of the "residual" can be explained by organized R & D. This translates to a 0.12- to 0.15-percent increase in national product.

Fellner (1970), in his presidential address to the American Economic Association, attempted to estimate the costs and benefits of technological progress. Beginning with a "residual" as a measure of technological progress, he then attempted to estimate a range of rates of return to what he called "activities generating technological progress," which included but were not limited to R & D expenditures. R & D expenditure estimates range from 1/6 to 1/2 of the total cost of productivity gains. He included in his calculations the "unrecorded costs" associated with technological progress, such as obsolescence of durable goods, technological unemployment, etc. Fellner found a social rate of return on "activities generating technological progress" of 13 to 55 per cent. The large range in his estimates was caused by variations of the "unrecorded costs" of technical progress. He concluded that the marginal rate of return is higher for these activities than for physical capital.

Griliches also attempted to estimate the contribution of R & D to economic growth. Making what he called "arbitrary but perhaps not unreasonable assumptions" relating to rates of return on private and public R & D investments and externalities associated with these investments, he calculated the contribution of R & D to economic growth at between 1/3 and 1/2 per cent per year. He then proceeded to explain why this estimate may in fact be too high (Griliches 1973, pp. 77-79).

The attempts to gauge the effect of R & D expenditures or investments on overall economic growth are in a preliminary stage. In addition to the problem of determining the relationship between technological change and economic growth, there is the added problem of ascertaining the relationship between R & D expenditures and technological change, with R & D expenditures being only one factor affecting technological change. Given these difficulties, it is not surprising that attempts to relate R & D expenditures to technological and economic growth have been tentative. Improvements in this area await a clearer understanding of the nature of the interrelated components of technological growth as well as better and more disaggregated data (Nadiri 1970, p. 1169; Stewart 1972, p. 17).

III. R & D EXPENDITURES AND INDUSTRY PRODUCTIVITY

In the late 1950s work was undertaken to investigate technological change and productivity growth at the industry level. The most comprehensive of these studies was conducted by Kendrick at the National Bureau of Economic Research. Kendrick calculated average annual rates of change in total factor productivity by 33 industry groups for the period 1899-1954 (Kendrick 1961, pp. 136-137). Using Kendrick's basic data, Terleckyj was able to relate average annual growth of factor productivity to research intensity for 19 industrial sectors classified according to the two-digit code of the U. S. Department of Commerce's Standard Industrial Classification (SIC) system. Using his concept of research intensity, which he defined as the ratio of research inputs to total input, Terleckyj found that a "tenfold difference in research intensity was associated with a difference of roughly one percentage point in the annual rate of growth of industry productivity" (Terleckyj 1967, p. 376).

Brown and Conrad (1967) treated research and education as fundamental variables in trying to explain productivity. Using the 1958 input-output table in an attempt to trace the interindustry flow of knowledge, they found that there was a significant and positive correlation between R & D expenditures and productivity. Their results indicated that a 1-percent increase in R & D expenditures was associated with a 0.2- to 0.3-percent increase in productivity, depending upon the industry. Unfortunately, the measures used by Brown and Conrad are not directly comparable to the measure used by Terleckyj. But Terleckyj (1967), in reviewing the Brown and Conrad work, indicated that their estimate of the effect of R & D expenditures on productivity was somewhat higher than the results he obtained.

Mansfield (1965), in a study of ten firms in the chemical and petroleum industries and ten two-digit SIC industries, found that the rate of productivity change was related to the rate of growth of accumulated R & D (the sum of all R & D expenditures over time) for both the firms and the industries. The marginal rates of return depended in part on whether technological change was embodied (included in new capital goods) or disembodied (such things as improved organization, which affects both old and new capital). In the petroleum industry the marginal rate of return was about 40 to 60 per cent regardless of whether technological change was embodied or not. In the chemical industry embodied technological change yielded a marginal rate of return of 30 per cent and embodied change only 7 per cent. High rates of return (of at least 15 per cent) were found for the ten industry groups. Minasian (1969) studied seventeen firms in the chemical industry and also found a positive and significant relationship between value added and accumulated R & D. He found a gross rate of return on investment in R & D of 54 per cent, compared with 9 per cent for physical capital.

The above-mentioned studies, though differing in time periods, industries covered, and methodology, yielded a fairly high and consistent relationship between R & D expenditures and increases in productivity. The rate of return on research was often in the 30- to 50-percent range (Griliches 1973). This high figure is impressive since industry and firm level studies have the advantage that R & D expenditures can be treated as an explicit variable rather than being treated as part of a residual, as was the case in studies of the economy (Mansfield 1972a and b).

But even at the industry and firm level, difficulties in measuring the relationship between R & D expenditures and increases in productivity occur. Many of the problems that were encountered at the level of the economy remain. There are still problems of measuring outputs and inputs and problems in model specification (Mansfield 1972a and b). Changes in product and input quality also present problems at the industry level. It is still difficult, if not impossible, to gauge the relative contributions of the many determinants of productivity increases. Also, the necessity of securing adequate price and cost deflators remains troublesome.

A major problem associated with industry level studies relates to the fact that much of the R & D activity is directed at improving the productivity of products that will be used in an industry other than the industry doing the R & D (Mansfield 1972a and b). Improvements in farm tractors may not show up as productivity increases in the tractor industry, but as increases in productivity in the farming industry (Griliches 1973, p. 68). Although attempts have been made to trace interindustry or explicit interfirm flows of technology by means of input and output tables, by Brown and Conrad (1967), much still remains to be done. The problem is made more difficult by the fact that time lags exist between the time R & D expenditures are made and increases in productivity take place both within the same firm or industry or in some other industry.

IV. PATENT STATISTICS

One measure frequently used in studies relating to innovation and economic growth is the counting of patents in technological areas of particular interest. Perhaps the most significant use of patent statistics are the time-series and cross-sectional studies of Jacob Schmookler (1966), which were summarized in Chap. 2 above.

In brief, he found that changes in investment in an industry were strongly correlated with changes in the number of capital-goods patents in that industry. This correlation was interpreted as meaning that invention was stimulated by economic demand. Rosenberg (1974) has criticized Schmookler's neglect of supply factors. Our interest in this section, however, is in Schmookler's use of patent statistics as a measure of innovation.

Schmookler (1966, p. 1) argued that understanding invention is important to understanding technical progress generally. However, invention is but one component of technical progress. The linkage between patented inventions and technological progress is complicated by a number of factors. First, not all inventions are patented. Reasons for not patenting include antitrust problems, the ability of firms to exploit an invention before a competitor can imitate it, and the secrecy involved in classified defense work. Second, not all patented inventions are applied commercially (i.e., become innovations). Third, inventions, when applied, differ considerably among themselves in the technical progress they effect (Kuznets, 1962). Thus, patenting is a possible intermediate state in innovation; it is not a final output.

On the input side Sanders (1962) argued that if patents are to be a measure of inventive activity,

1. the proportion of inventive activity resulting in patented inventions must have remained invariant over the span of time during which patents are deemed to serve as a useful index; and
2. the inputs per average patent must have remained similarly invariant.

The nonpatenting phenomenon, which was especially important after the middle 1930s (Schmookler 1966, Chap. 2), introduces an uncertainty in the proportion of inventive activity resulting in patents. Likewise the input per patent altered drastically after World War II because of the dramatic increase in industrial R & D and the increasing complexity of inventions (Schmookler 1966, p. 39).

Thus, although patents statistics may be the most useful data available in periods before industrial R & D became important, they are not very desirable as a final measure; and if Schmookler's results give strong qualitative indications of the importance of demand, it is important to remember that they are based on pre-1950 data. The problems mentioned earlier, such as nonpatenting, make patent statistics questionable measures for later periods.

V. THE DATABASE

We have argued, and we hope convincingly, for viewing innovation as being a dynamic and ecological process with complex interaction between function and structure. This kind of structure, though it may be meaningful in terms of understanding the process, assured that difficulties would be encountered in measuring the magnitude of the components of the system. In order for measurement to take place, a structure must be enumerated and measurements taken in relation to that structure. There is a fundamental conflict between the requirements for orderly measurement (that the structure remain unchanged so that results are comparable) and the dynamic nature of the innovative process.

The basic data source concerning formalized R & D expenditures is a time series that has been collected by the Bureau of the Census for the National Science Foundation since 1953. In conducting this survey the National Science Foundation attempts to delineate what activities should be included in R & D expenditures and what should be excluded

(NSF 1973, pp. 84-85). However, the quality of the received data is limited by the fact that except for a very small section of the survey, the NSF has to depend on the voluntary compliance of those being surveyed, and has to rely on their estimates of the magnitude of the expenditures. The NSF survey breaks down R & D activity by industry, in most cases two-digit SIC classifications and in some cases three- and four-industry SIC classifications. Certain difficulties arise when Standard Industrial Classifications are used. The categories are determined in large part by existing industry structure and to facilitate ease of reporting. These categories are often based on particular existing technology, materials, or past competitive relationships. As such, they are often poor instruments for measuring changes in technology that have the effect of changing industry structure, including material substitutability and changing competitive relationships (Chap. 2 above).

The National Science Foundation data also classify firms according to size. The basic size variable used here is number of employees. Some difficulties are encountered in using the number of employees as the major determinant of size classification, especially when the largest-size category is 10 000 employees and over.

The NSF data also relate R & D expenditures by activity (basic research, applied research, or development) and distinguish between Federally sponsored and company-sponsored R & D. The number of scientists and engineers employed in R & D activity by industry, firm size, etc. is also given. The presumption is that this extensive data collection does catch most of the formalized R & D as defined by the National Science Foundation, but as we have seen not all the firm's productivity-increasing activities are a result of R & D expenditures by any means.

The last important type of data is patent statistics. There are several problems in interpreting these data. The classifications of the patent office are designed to facilitate patent searches. Schmookler took a year to convert the data he needed from the patent office classification to the Standard Industrial Classification scheme, which he did by assigning relatively homogeneous patent office subclasses to particular industries in which they were used (Schmookler 1966, pp. 20-23). Subclasses that were not homogeneous or could be used in many industries were ignored. Thus Schmookler's railroad patent data did not contain such pertinent inventions as diesel engines and bearings! Although other researchers may need the data in other forms (e.g., by general areas of technology), it does appear difficult to process these data because of their original classification scheme.

Measurement in the innovation field certainly remains fraught with problems. In the areas dealt with by economists, the data are not only limited, but their linkage to what they purport to measure continues to change with time. Finally, much of what we would like to know, such as the changes caused by an innovation, lies in the future, outside the scope of measurement, and subject to the yet primitive techniques of forecasting.

BIBLIOGRAPHY

NOTE: The Bibliography is arranged alphabetically, by first-named author. Most entries are followed (in brackets) by the numbers of the pages on which the reference is made. For the meaning of the abbreviations ASQ, IRE, and IEEE, see note at the beginning of the Chap. 10 Bibliography on p. 259.

Abernathy, W. J., and R. R. Rosenbloom, 1968, "Parallel and sequential R & D strategies: Application of a simple model," *Trans. IEEE* EM-15: 2-10.

Abramovitz, M., 1956, "Resource and output trends in the U.S. since 1870," *Am. Econ. Rev.* 46: 5-23 [357].

Achiadelis, B., P. Jervis, and A. Robertson, *Project Sappho: A Study of Success and Failure in Industrial Innovations*, Report to the Science Research Council, University of Sussex (Science Policy Research Unit) [9, 25, 218].

Adelman, M.A., 1954, "Comments on books by Galbraith and Lilienthal," *Northwestern Univ. Law Rev.* 49: 157-163 [32].

Aiken, M., and L. Hage, 1968, "Organizational interdependence and intra-organizational structure," *Am. Sociol. Rev.* 33: 912-930 [290].

———, 1971, "The organic organization and innovation," *Sociology* 5: 63-82 [297].

———, 1972, "Organizational permeability, boundary spanners, and organizational structure," paper presented to the American Sociological Association [290].

Alboosta, C. A., and A. G. Holzman, 1970, "Optimal funding of an R & D project portfolio," presented at the 11th Institute of Management Science meeting, Los Angeles [104, 108, 118].

Allen, T. J., 1966, "Performance of information channels in the transfer of technology," paper read at the MIT Conference on the Transfer of Technology [213].

———, 1967a, "Communication in the research and development laboratory," *Technological Rev.* 70: 31-37 [117, 118].

———, 1967b, "The diffusion of technological information," in *Proc. 20th National Conf. Administration of Research*, Denver, Colo.: Denver Research Institute, pp. 111-120 [118].

———, 1969, "Information needs and uses," *Ann. Rev. Information Science and Technology* 4: 3-29 [118].

———, 1971, "Communications networks in R & D laboratories," *R & D Management* 1: 14-21 [117].

Allen, T. J., and S. I. Cohen, 1966, "Information flow in an R & D laboratory," working paper, MIT (Sloan School of Management) [81-84, 95, 117, 144].

———, 1969, "Information flow in an R & D laboratory," *ASQ* 14: 12-19 [95].

Allen, T. J., and D. G. Marquis, 1964, "Positive and negative biasing sets: The effects of prior experience on research performance," *Trans. IEEE* EM-11: 158-162 [85-87, 256].

Allison, D., 1965, "The university and regional prosperity," *International Science and Technology* 40: 22-31 [117].

Anderson, H. H., Ed., 1959, *Creativity and Its Cultivation*, New York: Harper [52].

Andrew, G. H. L., 1954, "Assessing priorities for technical effort," *Operational Res. Quarterly* 5: 67-80 [110, 118].

Andrews, F. M., and G. F. Farris, 1967, "Supervisory practices and innovation in scientific teams," *Personnel Psychology* 20: 497-516 [79, 80, 251].

———, 1972, "Time pressure and performance of scientists and engineers: A five year panel study," *Organizational Behavior and Human Performance* 8: 185-200 [109].

Andrews, K., 1971, *The Concept of Corporate Strategy*, Chicago: Dow-Jones-Irwin [224, 225].

Ansoff, H. I., 1962, "Evaluation of applied research in a business firm," in J. R. Bright, Ed., *Technological Planning on the Corporate Level*, Cambridge, Mass.: Harvard University (Graduate School of Business Administration), pp. 209-224.

———, 1965, *Corporate Strategy*, New York: McGraw-Hill [230].

Anthony, R. N., 1965, *Planning and Control*, Cambridge, Mass.: Harvard University Press.

Apstein, M., 1965, "Effectiveness of military laboratories as a function of contract activity," *Trans. IEEE* EM-12: 44-50 [246, 251].

Aram, J. D., and Setrak Javian, 1973, "Correlates of success on customer-initiated R & D projects," *Trans. IEEE* EM-20: 108-113, 1973.

Argyris, C., 1965a, *Integrating the Individual and the Organization*, New York: Wiley [117].

———, 1965b, *Organization and Innovation*, Homewood, Ill.: Irwin-Dorsey Press [228].

Arnfeld, R. V., Ed., 1969, *Technological Forecasting*, Edinburgh: Edinburgh University Press [319].

Arrow, K., 1962a, "Economic welfare and the allocation of resources for invention," in R. R. Nelson, Ed., *The Rate and Direction of Inventive Activity*, Princeton, N.J.: Princeton University Press [204].

———, 1962b, "The economic implications of learning by doing," *Rev. Economic Studies* 29: 155-173 [214].

Asher, D. I., 1962, "A linear programming model for the allocation of R & D efforts," *Trans. IRE* EM-9: 154-157 [105, 118].

Atkinson, A. C., and A. H. Bobis, 1969, "A mathematical basis for the selection of research projects," *Trans. IEEE* EM-16: 2-8 [105, 108, 118].

Averch, H., and L. Johnson, 1962, "Behavior of the firm under regulatory constraint," *Am. Economic Rev.* 52: 1052-1069 [44].

Avery, R. W., 1959, "Technical objectives and the production of ideas," presented at MIT Industrial Liaison Program seminar on the Organization of R & D in Decentralized Companies [98, 117].

———, 1960, "Enculturation in industrial research," *Trans. IRE* EM-7: 20-24 [98, 117].

Ayers, R. U., 1969, *Technological Forecasting and Long Range Planning*, New York: McGraw-Hill [118, 319].

Baier, K., 1969, "What is value? An analysis of the concept," in K. Baier and N. Rescher, Eds., New York: The Free Press [20].

Baker, G. L., et al., 1963, *A Century of Service: The First 100 Years of the United States Department of Agriculture*, Washington, D.C. [40, 116].

Baker, N. R., 1965, "The influence of several organizational factors on the idea generation and submission behavior of industrial researchers and technicians," doctoral dissertation, Northwestern University, Evanston, Ill. [98, 99, 117, 138].

Baker, N. R., and J. R. Freeland, 1972a, "Recent advances in R & D benefit measurement," presented at the 41st National Meeting of ORSA, New Orleans, La. [62, 104, 105].

———, 1972b, "Structuring informational flow to enhance innovation," *Management Science* 19: 105-115 [61, 62, 85, 88, 99-100, 117, 118, 128, 139].

Baker, N. R., and W. H. Pound, 1964, "R & D project selection: Where we stand," *Trans. IEEE* EM-11: 124-134 [100, 104, 105, 118].

Baker, N. R., J. Siegman, and J. Larson, 1971, "The relationship between certain characteristics of industrial research proposals and their subsequent disposition," *Trans. IEEE* EM-18: 118-123 [69, 101, 105, 117, 251].

Baker, N. R., J. Siegman, and A. H. Rubenstein, 1967, "The effects of perceived needs and means on the generation of ideas for industrial R & D projects," *Trans. IEEE* EM-14: 156-163 [46, 90, 96, 98, 117, 202, 251].

Baker, N. R., C. Shumway, D. M. Maher, W. E. Souder, and A. H. Rubenstein, "An R & D project selection and budget allocation model for large, decentralized organizations," Evanston, Ill.: Northwestern University (Program of Research on the Management of R & D) [102, 106, 108].

Barnett, H. G., 1953, *Innovation: The Basis of Cultural Change*, New York: McGraw-Hill [20, 127].

Barnett, H., and C. Morse, 1963, *Scarcity and Growth*, Baltimore, Md.: The Johns Hopkins Press [182, 188, 190].

Barth, R. T., 1970, "The relationship of intergroup organizational climate with communication and joint decision making between task-interdependent R & D groups," doctoral dissertation, Northwestern University, Evanston, Ill.

Bar-Zakay, S. N., 1970, *Technology Transfer Model*, Santa Monica, Calif.: RAND Corp., Publication 4509 [119].

Bass, L. W., 1962, "Historical development of industrial research in the United States,"

Chemistry and Industry 62: 1000-1003 [56].

Battelle-Columbus Laboratories, 1973, *Interaction of Science and Technology in the Innovative Process: Some Case Studies*, Columbus, Ohio: Report prepared for the National Science Foundation, NSF-C667 [12, 24, 25, 117, 217, 337, 356].

Baty, G., 1964, *Initial Financing of the New Research-based Enterprise in New England*, Boston: Federal Reserve Bank [117].

Bean, A. S., 1968, "A short report on the product manager concept," submitted to A. H. Rubenstein [112].

Becker, M. H., 1970, "Sociometric location and innovativeness: Reformulation and extension of the diffusion model," *Am. Sociological Rev.* 35: 267-282 [135, 136, 148, 282, 286].

Becker, S. W., and T. L. Whisler, 1967, "The innovative organization: A selective view of current theory and research," *J. Business* 40: 462-469 [290, 292, 293].

Beged-Dov, A. G., 1965, "Optimal assignment of R & D projects in a large company using an integer programming model," *Trans. IEEE* EM-12: 138-142 [105].

Bell, D., J. Chilcott, A. Read, and R. Salway, 1967, "Linear programming for resource allocation," London: Central Electricity Generating Board, Rep. RD/H/R2-3.

Bell, W., and J. A. Mau, Eds., 1971, *The Sociology of the Future*, New York: Russel Sage Foundation [320, 324].

Ben-David, J., 1971, *The Scientist's Role in Society*, Englewood Cliffs, N.J.: Prentice-Hall [8].

Bender, A. D., and E. B. Pyle, 1972, "Planning, control, and resource allocation models in R & D: A performance review by a pharmaceutical manufacturer," TIMS 19th International Meeting, Houston, Tex.

Bennis, W. G., 1969, *Organization Development: Its Nature, Origins, and Prospects*, Reading, Mass.: Addison Nelson [59, 73].

Berle, A. A., and G. Means, 1968, *Modern Corporation and Private Property*, New York: Harcourt, Brace and World, rev. ed. [356].

Bernal, J. D., 1953, *Science and Industry in the Nineteenth Century*, London: Rutledge and Kegan Paul [23].

Binder, A., 1964, "Statistical theory," in P. R. Farnsworth, D. McNear, and Q. McNear, Eds., *Annual Rev. Psychology* 15: 277-310 [249].

Birch, H. G., and Rabinowitz, H. S., 1951, "The negative effect of previous experience on productive thinking," *J. Experimental Psychology* 41: 121-126 [86].

Bird, J. R., and H. T. Darracott, 1968, "Developing and using the U.S. Army long-range technological forecast," in J. R. Bird, Ed., *Technological Forecasting*, Englewood Cliffs, N.J.: Prentice-Hall [316].

Blackman, A. W., 1973, "Forecasting through dynamic modeling," in J. R. Bright and M. E. Schoeman, Eds., *A Guide to Practical Technological Forecasting*, Englewood Cliffs, N.J.: Prentice-Hall [330].

Bolton, J., 1972, "Small firms," presented at Management Forum, University of Manchester [117].

Booz, Allen and Hamilton, Inc., 1960, *Management of New Products*, New York: Booz, Allen and Hamilton, Inc.

Boretsky, M., 1971, "Concerns about the present American position in international trade," in *Technology and International Trade*, Washington, D.C.: National Academy of Engineering [205].

Boulding, K. E., 1953, *The Organizational Revolution*, New York: Harper [356].

Bower, J. L., 1967, "Strategy as a problem solving theory of business planning," working paper, Boston: Harvard Business School [224].

———, 1970, *Managing the Resource Allocation Process: A Study of Corporate Planning and Investment*, Boston: Harvard Business School [227].

Boyd, R., 1972, "World dynamics: A note," *Science* 177: 516-518.

Bralley, J. A., 1960, "Project selection in industrial R & D: Problems and decision processes," in M. C. Yovits, Ed., *Research Program Effectiveness*, New York: Gordon and Breach [100, 101, 117, 118].

Bright, J. R., 1964, *Research, Development, and Technological Forecasting*, Homewood, Ill.: R. D. Irwin [20, 103, 127, 145, 317, 334].

———, Ed., 1968, *Technological Forecasting for Industry and Government*, Englewood Cliffs, N.J.: Prentice Hall [211, 311, 333, 334].

Bright, J. R., 1970, "Evaluating signals of technological change," *Harvard Business Rev.* 48: 62-70 [322, 323].

———, 1972, *A Brief Introduction to Technology Forecasting*, Austin, Tex.: Pemaquid Press, 2nd ed. [27, 304, 319, 333, 334].

———, 1973, "Forecasting by monitoring signals of technological change," in J. R. Bright and M. E. Schoeman, Eds., *A Guide to Practical Technological Forecasting*, Englewood Cliffs, N.J.: Prentice-Hall, pp. 238-256 [332].

Bright, J. R., and M. E. Schoeman, Eds., 1970, *Technological Forecasting: An Academic Inquiry*, Canoga Park, Calif.: Xyyyx Information Corp.

———, 1973, *A Guide to Practical Technological Forecasting*, Englewood Cliffs, N.J.: Prentice-Hall [319, 334].

Brittain, J. E., 1970, "The introduction of the loading coil: George A. Campbell and Michael I. Pupin," *Technology and Culture* 11: 36-57 [25].

Brooks, H., 1966, "National science policy and technology transfer," in *Proc. Conf. on Technology Transfer and Innovation*, Washington, D.C.: National Science Foundation (NSF 67-5) [120].

Brown, G., 1957, "Characteristics of new enterprises," *New England Business Rev.*, June: 1-4 [37].

Brown, H., 1954, *The Challenge of Man's Future*, New York: Viking Press [320].

Brown, L. A., 1968, *Diffusion Dynamics: A Review and Revision of the Quantitative Theory of the Spatial Diffusion of Innovation*, Lund [142].

———, 1969, "Diffusion of innovation: A macro view," *Economic Development and Cultural Change* 17: 189-209.

———, 1972, "Diffusion processes: Recent developments and their relevance to growth pole," presented at the International Geographical Union Conference on Regional Aspects of Economic Development [122, 142].

Brown, L. A., and K. R. Cox, 1971, "Empirical regularities in the diffusion of innovation," *Ann. Assoc. Am. Geographers* 61: 551-559 [122].

Brown, M., and A. Conrad, 1967, "The influence of research and education of CES production relations," in M. Brown, Ed., *The Theory and Empirical Analysis of Production*, New York: Columbia University Press, pp. 341-372.

Brown, Y., 1965, "R & D differences among industries," in R. Tybout, Ed., *Economics of Research and Development*, Columbus, Ohio: Ohio State University Press.

Bruner, J., and L. Postman, 1947, "Emotional selectivity in perception and reaction," *J. Personality* 16: 69-77 [128].

Burkholder, D., 1963, "The role of pharmaceutical detailmen in a large teaching hospital," *Am. J. Hospital Pharmacy* 20: 274-285 [146].

Burns, T., 1961, "Research, development, and production," *Trans. IRE* EM-8: 15-23 [251].

———, 1969, "Models, images, and myths," in W. H. Gruber and D. G. Marquis, Eds., *Factors in the Transfer of Technology*, Cambridge, Mass.: MIT Press [119].

Burns, T., and G. Stalker, 1961, *The Management of Innovation*, London: Tavistock [8, 72, 96, 112, 117, 207, 221, 228, 288].

———, 1966, *The Management of Innovation*, London: Tavistock, rev. ed.

Butrick, F. M., 1971, *How to Develop New Products for Sale to Industry*, Englewood Cliffs, N.J.: Prentice-Hall.

Campbell, D. T., 1968, "Quasi-experimental design," in J. L. Sills, Ed., *International Encyclopedia of the Social Sciences*, New York: Macmillan, vol. 5, pp. 259-263 [254, 255].

Campbell, D. T., and D. W. Fiske, 1959, "Convergent and discriminant validation by the multitrait-multimethod matrix," *Psychological Bull.* 56: 81-105 [248, 249].

Campbell, D. T., and J. C. Stanley, 1966, *Experimental and Quasi-experimental Designs for Research*, Chicago: Rand McNally [240, 254, 255].

Capron, W. F., 1966, "Comments," *Am. Economic Rev.* 56: 508-511.

———, Ed., 1971, *Technological Change in Regulated Industries*, Washington, D.C.: Brookings Institute [44, 45].

Cardwell, D. S. L., 1971, *From Watt to Clausius: The Rise of Thermodynamics in the Early Industrial Age*, London: Heinemann; New York: Cornell University Press [23, 356].

Carlson, R. O., 1968, "Summary and critique of educational diffusion research," paper presented at the National Conference on the Diffusion of Educational Ideas, East Lansing, Mich. [150].

Carnegie Corporation of New York *Quarterly* 9 (July 1960) [117].

Carter, A., 1970, "Technological forecasting and input-output analysis," *Technological Forecasting and Social Change* 1: 287-300 [188, 212].

Carter, C. F., and B. R. Williams, 1957, *Industry and Technical Progress: Factors Governing the Speed of Application of Science*, London: Oxford University Press [46, 90, 103, 202].

Cattell, J. M., 1915, "Families of American men of science," *Popular Science Monthly* 86: 404-417 [52].

Cetron, M. J., Ed., 1969, *Techological Forecasting: A Practical Approach*, New York: Gordon and Breach [319].

Cetron, M. J., and C. Ralph, Eds., 1971, *Industrial Applications of Technological Forecasting*, New York: Wiley [211, 212, 319, 334].

Cetron, M. J., J. Martino, and L. Roepcke, 1967, "The selection of R & D program content: Survey of quantitative methods," *Trans. IEEE* EM-14: 4-13 [104, 106, 118].

Chakrabarti, A. K., 1972, "A review of the models on technology transfer," Working paper Ser. No. 29, Evanston, Ill.: East-West Center, Technology and Development Institute [119, 120].

Chandler, A. D., Jr., 1962, *Strategy of Structure*, Cambridge, Mass., MIT Press [230].

Charnes, A., R. W. Clower, and K. O. Kortanek, 1967, "Effective control through coherent decentralization with preemptive goals," *Econometrica* 35: 294-319 [118].

Charnes, A., and A. C. Stedry, 1964, "Chance-constrained model for real-time control in research and development management," *Management Science* 12: B352-362 [105, 108].

Charpie, R. L., 1967, *Technological Innovation: Its Environment and Management*, Washington, D.C.: Department of Commerce, Report 0-242-736 [7, 9, 27, 42, 43, 51, 60, 208].

———, 1972, Testimony before U.S. Congress, House Subcommittee on Science, Research, and Development [214].

Chen, K., Ed., *Urban Dynamics: Extensions and Reflections*, San Francisco: San Francisco Press [317].

Chollar, R. G., G. J. Wilson, and B. K. Green, 1958, "Creativity techniques in action," *Research Management* 1: 5-22 [117].

Christensen, L., and D. Jorgenson, 1969, "The measurement of U.S. real capital input, 1929-1957," *Rev. Income and Wealth* 15: 293-320 [358].

Churchman, C. W., and R. L. Ackoff, 1954, "An approximate measure of value," *Operations Research* 2: 172-180 [118].

Churchman, C.W., R. L. Ackoff, and E.L. Arnoff, 1957, *Introduction to Operations Research*, New York: Wiley.

Clingman, W. H., 1973, "Predicting scientific and exploratory development," in J. R. Bright and M. E. Schoeman, Ed., *A Guide to Practical Technology Forecasting*, Englewood Cliffs, N.J.: Prentice-Hall, pp. 69-113 [322].

Coates, J. F., 1973, "Anti-intellectualism and other plagues on managing the future," *Technological Forecasting and Social Change* 4: 243-262 [334].

Cochran, M. A., E. B. Pyle III, L. C. Greene, H. A. Clymer, and D. Bender, 1971, "Investment model for R & D project evaluation and selection," *Trans. IEEE* EM-18: 89-100 [100, 108, 118].

Cohen, I. B., 1948, *Science, Servant of Man*, Boston: Little, Brown [356].

Cohen, K. J., and R. M. Cyert, 1973, "Strategy: Formulation, implementation, and monitoring," *J. Business* 46: 349-367 [228].

Cole, H. S. D., Ed., 1973, *Models of Doom: A Critique of The Limits to Growth*, New York: Universe Books [321].

Coleman, J. E. Katz, and H. Menzel, 1957, "The diffusion of an innovation among physicians," *Sociometry* 20: 253-270 [129, 135, 145].

Coler, M. A., Ed., 1963, *Essays on Creativity in the Sciences*, New York: New York University Press [53].

Collier, B. D., G. W. Cox, A. W. Johnson, and P. C. Miller, 1973, *Dynamic Ecology*, Englewood Cliffs, N.J.: Prentice-Hall.

Comanor, W. S., 1965, "Research and technical change in the pharmaceutical industry," *Rev. Economics and Statistics* 47: 182-190 [37, 214].

Comanor, W. S., 1967, "Market structure, product differentiation and industrial research," *Quarterly J. Economics* 81: 639-657 [36].

Comanor, W. S., and F. M. Scherer, 1969, "Patent statistics as a measure of technical change," *J. Political Economy* 77: 392-398 [34].

Conference on Research in Income and Wealth, 1961, *Studies in Income and Wealth*, vol. 25, Princeton, N.J.: Princeton University Press [356].

———, 1967, *Studies in Income and Wealth*, vol. 31, New York: Columbia University Press.

———, 1973, *Studies in Income and Wealth*, vol. 38, New York: Columbia University Press.

Connolly, T., 1972, "The diffuse decision: An integrating focus," presented at joint TIMS/ORSA/AIIE Meeting, Atlantic City, N.J. [102].

Cooper, A. C., 1972, "The founding of technologically based firms," in A. Cooper and J. Komives, Eds., *Technical Entrepreneurship: A Symposium*, Milwaukee, Wis.: Center for Venture Management [57, 58, 117].

———, 1973, "Technical entrepreneurship: What do we know?" *R & D Management* 3: 59-64 [57].

Cooper, R., 1971, "Technology and U.S. trade: A historical view," in *Technology and International Trade*, Washington, D.C.: National Academy of Engineering [57, 213, 251].

Cordtz, D., 1971, "Bringing the laboratory down to earth," *Fortune* 83: 106-122 [112, 115].

Corey, E. R., 1956, *The Development of Markets for New Materials*, Boston: Harvard Business School [223].

Cox, C. M., 1926, "Early mental traits of 300 geniuses," in L. M. Terman, Ed., *Genetic Studies of Genius*, Stanford, Calif.: Stanford University Press, vol. 2 [52].

Cramer, R. H., and B. E. Smith, 1964, "Decision models for the selection of research projects," *Engineering Economist* 9: 1-20 [103, 105].

Crane, D., 1972, *Invisible Colleges: Diffusion of Knowledge in Scientific Communities*, Chicago: University of Chicago Press [8, 281].

Crombie, A. C., Ed., 1963, *Scientific Change*, New York: Basic Books [17].

Cronbach, L. J., 1951, "Coefficient 'alpha' and the internal structure of tests," *Psychometrica* 16: 297-334 [247].

Cronbach, L. J., and P. E. Mehl, 1955, "Construct validity in psychological tests," *Psychological Bull.* 52: 281-302 [248].

Cyert, R. M., and J. G. March, 1963, *A Behavioral Theory of the Firm*, Englewood Cliffs, N.J.: Prentice-Hall [291].

Dalby, J. F., 1971, "Practical refinements to the cross-impact matrix technique of technological forecasting," in M. J. Cetron and C. A. Ralph, Eds., *Industrial Applications of Technological Forecasting*, New York: Wiley.

Damon, P. E., 1973, "Modeling the world," *Science* 180: 1236-1237 [30].

Dantzig, G. B., and P. Wolfe, 1961, "The decomposition algorithm for linear programming," *Econometrica* 29: 767-778 [108].

David, P. A., 1975, *Technical Choice, Innovation, and Economic Growth*, London: Cambridge University Press.

Davis, K., 1965, "Mutuality in understanding of the program manager's management role, *Trans. IEEE* EM-12: 117-123 [251].

Davis, R. C., 1973, "Organizing and conducting technological forecasting in the consumer goods firm," in J. R. Bright and M. E. Schoeman, Eds., *A Guide to Practical Technological Forecasting*, Englewood Cliffs, N.J.: Prentice-Hall [320, 334].

Dean, B. V., and I. E. Houser, 1967, "Advanced materiel systems planning," *IEEE Trans.* EM-14: 21-43 [108].

Dean, B. V., and M. J. Nishry, "Scoring and profitability models for evaluating and selecting engineering projects," *Operations Research* 13: 550-570 [118].

Demaree, A. T., 1970, "GE's costly ventures into the future," *Fortune* 82: 88-93.

Denison, E., 1962, *The Sources of Economic Growth in the United States and the Alternatives Before Us*, New York: Committee for Economic Development [357, 359].

———, 1969, "Some major issues in productivity: An examination of estimates by Jorgenson and Griliches," *Survey of Current Business* 49: 1-28 [358].

Department of Commerce, 1967, *Technological Innovation: Its Environment and Management*, Washington, D.C.

Deutermann, E., 1966, "Seeding science-based industry," *New England Business Rev.*, pp. 7-15 [117].

Diamond, S., 1959, *Information and Error*, New York: Basic Books [238].

Diebold Institute, 1973, *Private Enterprise in a Post-industrial Society*, No. 2, New York [147].

Disman, S., 1962, "Selecting R & D projects for profit," *Chemical Engineering* 69: 87-90 [118].

Dixon, R. B., 1928, *The Building of Cultures*, New York: Scribner [6].

Doctors, S. I., 1969, *The Role of the Federal Agencies in Technology Transfer*, Cambridge, Mass.: MIT Press [40, 119, 146, 147].

Douds, C. F., 1970, "The effects of work-related values on communication between R & D groups," doctoral dissertation, Northwestern University, Evanston, Ill. [249, 258].

———, 1971, "The state of the art in the study of technology transfer: A brief study," *R & D Management* 1: 125-131 [233].

———, 1972, "Experiment sub-n: Varieties of meaning of the concept," Northwestern University (Program of Research on the Management of Research and Development, IE/MS Department), paper 72/59 [240].

———, 1973, "Natural experiments for technology transfer from national laboratories: Specification of independent variables and parameters," Northwestern University (Program of Research on the Management of Research and Development, IE/MS Department), paper 73/95 [240].

Douds, C. F., and Rubenstein, A. H., 1966, "Some models of organizational interfaces in the R & D process," Northwestern University (Program of Research on the Management of Research and Development IE/MS Department).

Draheim, K., 1972, "Factors influencing the rate of formation of technical companies," in A. Cooper and J. Komives, Eds., *Technical Entrepreneurship: A Symposium*, Milwaukee, Wis.: Center for Venture Management [57].

Draheim, K., R. P. Howell, and A. Shapero, 1966, *The Development of a Potential R & D Complex*, Menlo Park, Calif.: Stanford Research Institute [57, 58].

Drucker, P. F., 1976a, "Technological trends in the twentieth century," in M. Kranzberg and C. W. Pursell, Eds., *Technology in Western Civilization*, New York: Oxford University Press, Vol. 2, Chap. 2, pp. 10-22 [4].

———, 1967b, "Technology and society in the twentieth century," in M. Kranzberg and C. W. Pursell, Eds., *Technology in Western Civilization*, New York: Oxford University Press, Vol. 2, Chap. 3, pp. 22-33 [4].

Dubin, R., 1958, *The World of Work*, Englewood Cliffs, N.J.: Prentice-Hall [8].

Dunlop, R. A., 1963, "Developing an experience register for engineers," *Trans. IEEE* EM-10: 6-15.

Eckenrode, R. T., 1965, "Weighting multiple criteria," *Management Science* 12: 180-192 [117, 118].

Eiduson, B. T., 1966, "Productivity rate in research scientists," *Am. Scientist* 64: 57-63.

Einhorn, H. J., 1970, "The use of non-linear, non-compensatory models in decision making," *Psychological Bull.* 73: 221-230 [118].

Enos, J. L., 1962a, *Petroleum Progress and Profits*, Cambridge, Mass.: MIT Press [181, 230].

———, 1962b, "Invention and innovations in the petroleum industry," in R. R. Nelson, Ed., *The Rate and Direction of Inventive Activity*, Princeton, N.J.: Princeton University Press, pp. 229-321 [217].

Erickson, C. E., B. S. Gantz, and R. W. Stephenson, 1968, "Development of the job description form: Its use in the functional categorization of professional jobs," *Trans. IEEE* EM-15: 119-129 [251].

Esch, M. E., 1973, "Honeywell's pattern: Planning assistance through technical evaluation of relevance numbers," in J. R. Bright and M. E. Schoeman, Eds., *A Guide to Practical Technological Forecasting*, Englewood Cliffs, N.J.: Prentice-Hall, pp. 147-163 [330, 334].

Etzioni, A., 1961, *A Comparative Analysis of Complex Organizations*, New York: Free Press [8].

Eyring, H. B., 1966, "Some sources of uncertainty and their consequences in engineering design projects," *Trans. IEEE* EM-13: 167-181 [251].

Fabricant, S., 1959, *Basic Facts on Productivity Change*, New York: Columbia University Press [357].

Farrington, B., 1949, *Greek Science*, Harmondsworth (England): Penguin Books [19, 22].

Farris, G. F., 1969a, "Some antecedents and consequences of scientific performance, *Trans. IEEE* EM-16: 6-16 [109, 252].

———, 1969b, "Organizational influence and individual performance: A longitudinal study," *J. Appl. Psychology* 53: 87-92 [109, 111].

———, 1972, "The effect of individual roles performance in innovative groups," *R & D Management* 3: 23-28 [110].

Fellner, W., 1951, "The influence of market structure on technological progress," *Quarterly J. Economics* 65: 556-577 [214].

———, 1970, "Trends in the activities generating technological progress," *Am. Economic Rev.* 60: 1-29 [202, 359].

———, 1972, "The progress-generating sector's claim to high priority," in *Research and Development and Economic Growth/Productivity*, Washington, D.C.: National Science Foundation, NSF 72-303, pp. 37-46 [358].

Fishburn, P. C., 1964, *Decision and Value Theory*, New York: Wiley.

Fisher, R. A., 1925, *Statistical Methods for Research Workers*, London: Oliver and Boyd [240].

———, 1935, *The Design of Experiments*, London: Oliver and Boyd [240].

Fisher, J. C., and R. H. Pry, 1971, "A simple substitution model of technological change," in M. Cetron and C. Ralph, Eds., *Industrial Applications of Technological Forecasting*, New York: Wiley, pp. 290-307.

Fliegel, F. C., and J. E. Kivlin, 1966, "Farmer's perception of farm practice attributes," *Rural Sociology* 31: 197-206 [124].

———, 1968, "A cross-national comparison of farmers' perceptions of innovations as related to adoption behavior," *Rural Sociology* 33: 437-449 [124].

Folger, A., and G. Gordon, 1962, "Scientific accomplishment and social organization: A review of the literature," *Am. Behavioral Scientist* 6: 51-58 [236].

Forbes, R. J., 1958, *Man the Maker*, New York: Abelard-Schuman [2].

Fornas, C. C., 1936, *The Next Hundred Years*, New York: Reynal and Hitchcock.

Forrester, J. W., 1961, *Industrial Dynamics*, Cambridge, Mass.: MIT Press [317].

———, 1969, *Urban Dynamics*, Cambridge, Mass.: MIT Press [317].

———, 1971, *World Dynamics*, Cambridge, Mass.: Wright Allen [30, 192].

Forseth, D. A., 1965, "The role of government-sponsored research laboratories in the generation of new enterprises: A comparative analysis," S. M. thesis, MIT (Sloan School of Management), Cambridge, Mass.

Francis, D. G., and E. M. Rogers, 1962, "Adoption of a nonrecommended innovation: The grass incubator," presented at Rural Sociology Society, University Park, Pa. [125].

Freeland, J. R., and N. R. Baker, 1972, "Mathematical models of resource allocation decision making in hierarchical organizations," presented at joint ORSA/TIMS AIIE meeting, Atlantic City, N.J. [108, 118].

———, 1973, "A goal partitioning procedure for modeling coordination activities in a hierarchical organization," Research Paper No. 184, Stanford University (Graduate School of Business), Stanford, Calif.

Freeman, C., 1965, "Research and development in electronic capital goods," *National Institute Economic Rev.* 60 [34, 46, 211].

———, 1973, "A study of success and failure in industrial innovation," in B. R. Williams, Ed., *Science and Technology in Economic Growth*, London: Macmillan [206].

Frischmuth, C. S., and T. J. Allen, 1969, "A model for the evaluation and description of technical problem solving," *Trans. IEEE* EM-16: 58-64 [251].

Fusfeld, A. R., 1970, "The technological progress function: A new technique for forecasting," *Technological Forecasting and Social Change* 1: 301-312 [334].

———, 1973, "The technological progress function," in J. R. Bright and M. E. Schoeman, Eds., *A Guide to Practical Technological Forecasting*, Englewood Cliffs, N.J.: Prentice-Hall [334].

Galbraith, J., 1970, "Environmental and technological determinants of organizational design," in J. W. Lorsch and P. R. Lawrence, Eds., *Studies in Organizational Design*, Homewood, Ill.: R. D. Irwin, pp. 113-139 [92].

Galbraith, J. K., 1952, *American Capitalism*, Boston: Houghton Mifflin [33, 207, 209].

Galton, F., 1870, *Hereditary Genius*, New York: Macmillan [52].

Gargiulo, G. R., J. Hannoch, D. B. Hertz, and T. Zang, 1961, "Developing systematic procedures for directing research programs," *Trans. IRE* EM-8: 24-29 [118].

Gellman, A., 1971, "Surface freight transportation," in W. F. Capron, Ed., *Technological Change in Regulated Industries*, Washington, D.C.: Brookings Institute, pp. 166-196 [95].

———, 1973, "innovation as a process," paper delivered at NATO-Advanced Study Institute on Technology Transfer, Evry, France [13].

Gellman Research Associates, 1974, "Economic regulation and technological innovation: A cross-national survey of literature and analysis," prepared for the National R & D Assessment Program, Washington, D.C.: National Science Foundation [45].

Gemmill, G. R., and H. J. Thamhain, 1973, "The effectiveness of different power styles of project managers in gaining project support," *Trans. IEEE* EM-20: 38-44 [251].

Geoffrion, A.M., J.S. Dyer, and A. Feinberg, 1971, "An interactive approach for multi-criterion optimization with an application to the operation of an academic department," working paper No. 176, Western Management Science Institute, UCLA, Los Angeles, Calif. [108].

Gerardin, L., 1973, "Morphological analysis: A method for creativity," in J. R. Bright and M. E. Schoeman, Eds., *A Guide to Practical Technological Forecasting*, Englewood Cliffs, N.J.: Prentice-Hall [334].

Gershinowitz, H., 1960, "Sustaining creativity against organizational pressures," *Research Management* 3: 49-56 [117, 118].

Gibbons, M., and R. D. Johnston, 1970, "Relationship between science and technology," *Nature* 227: 125-127 [25].

Gilfillan, S. C., 1935, *The Sociology of Invention*, Chicago: Follett Publishing Co. [8, 56, 145, 168, 181].

———, 1960, "An attempt to measure the rise of American inventing and the decline of patenting," *Technology and Culture* 1: 201-214, 227-234 [117].

———, 1968, "A sociologist looks at technical prediction," in J. R. Bright, Ed., *Technological Forecasting*, Englewood Cliffs, N.J.: Prentice-Hall, pp. 3-24.

———, 1969, "Some racial comparisons of inventiveness," *Mankind Quarterly* 9: 120-129 [17].

———, 1971, *Supplement to the Sociology of Invention*, San Francisco: San Francisco Press [181].

Gille, B., 1967, *Engineers of the Renaissance*, Cambridge, Mass.: MIT Press [22].

Gillette, R., 1972, "The limits to growth: Hard sell for a computer view of Doomsday," *Science* 175: 1088-1092.

Gisser, P., 1972, *Launching the New Industrial Product*, New York: American Management Association.

Glaser, B. G., 1963: "The impact of differential promotion systems on careers," *Trans. IEEE* EM-20: 38-44, 1973.

Gloskey, C. R., 1960, "Research on a research department: An analysis of economic decisions on projects," *Trans. IRE* EM-7: 166-173 [251].

Gold, B., 1969, "The framework of decisions for major technological innovations," in K. Baier and N. Rescher, Eds., *Values and the Future*, New York: Free Press.

Goldhar, J. D., 1970, "An exploratory study of technological innovation," doctoral dissertation, George Washington University, Washington, D.C. [46, 90, 202].

Goodman, R. A., "Organization and manpower utilization in research and development," *Trans. IEEE* EM-15: 198-204, 1968 [251].

Goodwin, P. G., 1972, "A method for evaluation of subsystem alternative designs," *Trans. IEEE* EM-19: 12-21 [118].

Gordon, G., 1963, "The problem of assessing scientific accomplishment: A potential solution," *Trans. IEEE* EM-10: 192-197 [251].

———, 1964, "The organization designed to produce change," presented at Seminar of Innovative Organization, University of Chicago, Chicago, Ill. [69].

———, 1967, "A paradox of research administration," *Proc. 20th Nat. Conf. Administra-*

tion of Research, Denver, Colo.: Denver Research Institute, pp. 15-25.

Gordon, G., and S. Marquis, 1964, "Resources and scientific innovation," presented at American Sociological Association, Montreal.

——, 1966, "Freedom, visibility of consequences, and scientific innovation," Am. J. Sociology 72: 195-202 [78].

Gordon, G., et al., 1974, "Organizational structure, environmental diversity, and hospital adoption of medical innovations," in A. D. Kaluzny, Ed., Innovation of Health Care Organizations: An Issue in Organizational Change, Chapel Hill, N.C. [297].

Gordon, T. J., and H. S. Becker, 1972, "The cross-impact approach to technology assessment," Research Management 15: 73-80 [207-214].

Gordon, T. J., and H. Haywood, 1968, "Initial experiments with the cross-impact matrix method of forecasting," Futures 1: 100-116 [334].

Gordon, T. J., S. Enzer, and R. Rochberg, 1973, "Experiment in simulation gaming for social policy studies," in J. R. Bright and M. E. Schoeman, Eds., A Guide to Practical Technological Forecasting, Englewood Cliffs, N.J.: Prentice-Hall, pp. 550-572.

Grabowski, H., 1968, "The determinants of industrial R & D," J. Political Economy 76: 292-306 [209].

Grayson, R., 1968, "The effect of formal organizational structure on new product development for branded consumer package goods," doctoral dissertation, New York University, New York [114].

The Green Revolution, 1970, Subcommittee on National Security Policy and Scientific Developments, 91st Congress, Washington, D.C.: Government Printing Office, pp. 38-612 [117].

Griliches, Z., 1957, "Hybrid corn: An exploration in the economics of technological change," Econometrica 25: 501-522 [20, 142].

——, 1958, "Research costs and social returns: Hybrid corn and related innovations," J. Political Economy 66: 419-431 [186, 187, 202, 204, 205].

——, 1964, "Research expenditures, education, and the aggregate agricultural production function," Am. Economic Rev. 54: 962-974 [202, 204].

——, 1973, "Notes on the role of education in production functions and growth accounting," in B. Williams, Ed., Science and Technology in Economic Growth, London: Macmillan, pp. 59-83 [357-360].

Gross, N. C., J. B. Giacquinta, and M. Bernstein, 1971, Implementing Organizational Innovations: A Sociological Analysis of Planned Educational Change, New York: Basic Books [275].

Gruber, W., D. Mehta, and R. Vernon, 1967, "The R & D factor in international trade and international investment of U.S. industries," J. Political Economy 75: 20-37 [214].

Guimarães, L., 1972, "Communication integration in modern and traditional social systems: A comparative analysis across twenty communities of Minas Gerais, Brazil," doctoral dissertation, Michigan State University, East Lansing, Mich. [297].

Habakkuk, H. J., 1962, American and British Technology in the Nineteenth Century, Cambridge: Cambridge Univesity Press [206].

Haefle, J. W., 1962, Creativity and Innovation, New York: Reinhold [117].

Hage, J., and M. Aiken, 1970, Social Change in Complex Organizations, New York: Random House [68, 71, 112, 117, 283, 285, 288, 290, 293, 296].

Hage, J., and R. Dewar, 1973, "Elite values versus organizational structure in predicting innovation," Administrative Science Quarterly 18: 279-290 [287, 288, 296, 297].

Hagen, E. E., 1971, "How economic growth begins: A theory of social change," in P. Kilby, Ed., Entrepreneurship and Economic Development New York: Free Press.

Hagerstrand, T., 1952, Innovation Diffusion as a Spatial Process, Chicago: University of Chicago Press.

——, 1965a, "A monte Carlo approach to diffusion," European J. Sociology 6: 43-67 [112].

——, 1965b, "Quantitative techniques for analysis of the spread of information and technology," in C. A. Anderson and M. J. Bowman, Eds., Education and Economic Development, Chicago: Aldine Publishing Co. [122].

——, 1968, "Diffusion: The diffusion of innovations," in D. L. Sills, Ed., International Encyclopedia of the Social Sciences, New York: Macmillan, 4: 174-177 [121, 122, 135].

Hake, B., 1971, New Product Strategy, New York: Pitman.

Hall, A. R., 1963, "Merton revisited, or science and society in the seventeenth century," *History of Science* 2: 1-16 [23].

Hall, R. H., 1967, "Organizational size, complexity, and formalization," *Am. Sociological Rev.* 32: 903-912 [297].

Hamberg, D., 1966, *R & D: Essays on the Economics of Research and Development*, New York: Random House [34, 42, 49, 61, 69, 117, 209].

Hanneman, G. J., 1971, "A computer simulation of information diffusion in a peasant community," M.A. thesis, Michigan State University, East Lansing, Mich. [122].

Hardin, C. M., Ed., 1969, *Overcoming World Hunger*, Englewood Cliffs, N.J.: Prentice-Hall [117].

Harding, T. S., 1947, *Two Blades of Grass: A History of Scientific Development in the U.S. Department of Agriculture*, Norman, Okla.: University of Oklahoma Press [116].

Harris, H., 1965, Testimony to the U.S. Senate Subcommittee on Antitrust and Monopoly of the Committee on the Judiciary, Part 1, 88th Congress, 2nd Session, p. 176 [39].

Harrold, R. W., 1969, "An evaluation of measurable characteristics within Army laboratories," *Trans. IEEE* EM-16: 16-23 [251].

Hart, H., 1931, *The Technique of Social Progress*, New York: Holt [316].

———, 1957, "Acceleration in social change," in F. R. Allen, H. Hart, D. C. Miller, W. F. Ogburn, and M. F. Nimkoff, Eds., *Technology and Social Change*, New York: Appleton-Century-Crofts [316].

———, 1959, "Social theory and social change," in L. Gross, Ed., *Symposium on Sociological Theory*, White Plains, N.Y.: Row, Peterson.

———, 1972, "Acceleration in soical change," in N. de Nevers, Ed., *Technology and Society*, Reading, Mass.: Addison-Wesley, pp. 51-78 [316].

Harvey, E., and R. Mills, 1970, "Patterns of organizational adaptation: A political perspective," in M. N. Bald, Ed., *Power in Organizations*, Nashville, Tenn.: Vanderbilt University [117].

Havelock, R. G., 1969, *Planning for Innovation Through Dissemination and Utilization of Knowledge*, Ann Arbor, Mich.: Center for Utilization of Knowledge, Institute for Social Research.

Hayami, Y., and V. W. Ruttan, 1971, *Agricultural Development: An International Perspective*, Baltimore, Md.: Johns Hopkins Press [21, 128, 187, 188].

Helmer, O., 1972, "Cross impact gaming," *Futures* 4: 149-167 [331].

Helmer, O., and T. Gordon, 1964, "Report on a long-range forecasting study," Santa Monica, Calif.: RAND Corp. [211].

Hennipman, R., 1954, "Monopoly: Impediment or stimulus to economic progress?" in *Monopoly and Competition and Their Regulation*, International Economic Association, New York: Macmillan [214].

Hertz, D. G., 1964, "Risk analysis in capital investment," *Harvard Business Rev.* 42: 95-106 [103, 118].

Hespos, R. E., and P. E. Strassman, 1965, "Stochastic decision trees for the analysis of investment decisions," *Management Science* 11: 244-259 [103].

Hess, S. W., 1962, "A dynamic programming approach to R & D budgeting and project selection," *Trans. IRE* EM-9: 170-179 [105].

Hetman, F., 1973, *Society and the Assessment of Technology*, Paris: Organisation for Economic Cooperation and Development [4, 21].

Hewlett, R. G., and O. E. Anderson, 1962, *The New World, 1939/1946: A History of the United States Atomic Energy Commission*, University Park, Pa.: Penn State University Press [181].

Hicks, J. R., 1963, *The Theory of Wages*, New York: St. Martin's Press, 2nd ed. [24, 206].

Hill, R., and Hlavacek, J., 1972, "The venture team: A new concept in marketing organization," *J. Marketing* 36: 44-50 [115].

Hill, S. C., 1970, "A natural experiment on the influence of leadership behavior patterns on scientific productivity," *Trans. IEEE* EM-17: 10-20 [29, 138, 254, 255].

Hillier, J., 1960, "A theory of communication in a research laboratory," *Research Management* 3: 255-270 [117, 118].

———, 1968, "Venture activities in a large corporation," *Trans. IEEE* EM-15: 65-70.

Hodgen, M. T., 1952, *Change and History*, New York: Wenner-Gren Foundation [4].

Hodgins, J. W., 1972, "Management challenges to the entrepreneur," *Business Quarterly* 37: 42-50 [117].

Holland, W. E., 1972, "Characteristics of individuals with high information potential in government research and development organizations," *Trans. IEEE* EM-19: 38-44 [25].

Hollander, S., 1965, *The Sources of Increased Efficiency: A Study of Du Pont Rayon Plants*, Cambridge, Mass.: MIT Press [29, 67, 73, 117, 214, 230].

Hoselitz, B. F., 1953, "Social structure and economic growth," *Economia Internazionale* 6: 52-77 [356].

―――, 1960, *Sociological Aspects of Economic Growth*, Glencoe, Ill.: Free Press [356].

Hough, G. W., 1974, *Technology Diffusion: Federal Programs and Procedures*, Mt. Airy, Md.: Lomond Systems.

Houton, F. W., 1963, "Work assignment and interpersonal relations in a research organization: Some participant observations," *ASQ* 7: 502-521 [98, 109].

Howell, R. P., 1972, "Comparative profiles--entrepreneurs versus the hired executive: San Francisco Peninsula semiconductor industry," in A. Cooper and J. Komives, Eds., *Technical Entrepreneurship: A Symposium*, Milwaukee, Wis.: Center for Venture Management [57].

Hruschka, E., and H. Rheinwald, 1965, "The effectiveness of German pilot farms," *Sociologia Ruralis* 5: 101-111 [124].

Hufbauer, G., 1966, *Synthetic Materials and the Theory of International Trade*, Boston: Harvard University Press [214].

Hughes, T. P., 1966, *Lives of the Engineers: Selections from Samuel Smiles*, Cambridge, Mass.: MIT Press [52, 181].

―――, 1971, *Elmer Sperry: Inventor and Engineer*, Baltimore, Md.: Johns Hopkins Press [5, 26, 54, 168, 180, 181].

Hunter, L., 1949, *Steamboats on the Western Rivers: An Economic and Technological History*, New York: Octagon Books [181].

Hyman, R., 1964, "Creativity and the prepared mind: The role of information and induced attitudes," in C. W. Taylor, Ed., *Widening Horizons in Creativity*, New York: Wiley [80].

Illinois Institute of Technology Research Institute, 1968, *Technology in Retrospect and Critical Events in Science*, Report NSF-C535, Chicago, Ill. [10, 12, 24, 214, 234, 235].

Inouye, A., and C. Süsskind, "'Technological Trends and National Policy,' 1937: The first modern technology assessment," *Technology and Culture* 18 (No. 4), 1977 [311].

Isenson, R. S., 1967, *Project Hindsight: Final Report, Task I*, Office of the Director of Defense Research and Engineering, Washington, D.C. [12, 24].

Jantsch, E., 1967, *Technological Forecasting in Perspective*, Paris: Organisation for Economic Cooperation and Development [318, 334].

Jewkes, J., D. Sawers, and R. Stillerman, 1958, *The Sources of Invention*, New York: Macmillan [51, 168, 217].

―――, 1969, *The Sources of Invention*, New York: Macmillan, rev. ed. [12, 23, 25, 209].

Jones, S. L., and J. E. Arnold, 1962, "The creative individual in industrial research," *Trans. IRE* EM-9: 51-55 [98, 117, 118].

Jones, S. V., 1973, *Inventions Necessity Is Not the Mother of: Patents Ridiculous and Sublime*, New York: Quadrangle Books [17].

Jorgenson, D., and Z. Griliches, 1967, "The explanation of productivity change," *Rev. Economic Studies* 34: 249-284 [358].

―――, 1970, "The sources of economic growth: A reply to Edward F. Denison," discussion paper No. 131, Cambridge, Mass.: Harvard Institute of Economic Research [358, 359].

Josephson, M., 1959, *Edison*, New York: McGraw-Hill [5, 54, 180-182].

Jouvenal, B. de, 1967, *The Art of Conjecture*, New York: Basic Books [320].

Kahn, A. E., 1971, *Economics of Regulation: Principles and Institutions*, New York: Wiley [44].

Kahn, H., undated, "If the rich stop aiding the poor...," United Nations Centre for Economic and Social Information, Development Forum.

Kahn, H., and A. Weiner, 1967, *The Year 2000*, New York: Macmillan [320].

Kaluzny, A. D., et al., 1973, "Predicting two types of program innovation from organizational characteristics: The case of hospitals," Chapel Hill, N.C.: University of North Carolina (School of Public Health) [287, 289, 290].

Kaplan, A., 1964, *The Conduct of Inquiry*, San Francisco: Chandler [275, 296].

Kaplan, A. D. H., 1954, *Big Enterprise in a Competitive System*, Washington, D.C.: Brookings Institute [33].

Kaplan, N., "Some organizational factors affecting creativity," *Trans. IRE* EM-7: 24-30 [98. 117. 118. 251].

Karger, D., and R. Murdick, 1972, *New Products Venture Management*, New York: Gordon and Breach.

Katz, E., 1960, "The two-step flow of communication," in W. Schram, Ed., *Mass Communications*, Urbana: University of Illinois [117].

Katz, E., and P. F. Lazarsfeld, 1955, *Personal Influence*, Glencoe, Ill.: Free Press [117].

Kauper, T. E., 1973, "Patent law revision," testimony before the U.S. Senate Subcommittee on Patents, Trademarks, and Copyrights, of the Committee of the Judiciary, 93rd Congress, first session, 291-304 [34].

Keesing, D., 1967, "The impact of research and development on U.S. trade," *J. Political Exonomy* 75: 38-45 [214].

Kegan, D. L., 1969, "Measures of the usefulness of written technical information to chemical researchers," M.S. thesis, Northwestern University, Evanston, Ill. [257].

Kendrick, J., 1954, "National productivity and its long-range projection," in *Long-range Economic Projection*, National Bureau of Economic Research [61, 212].

———, 1961, *Productivity Trends in the United States*, Princeton, N.J.: Princeton University Press [360].

Kerker, M., 1961, "Science and the steam engine," *Technology and Culture* 2: 381-390 [23].

Kerlinger, F. N., 1965, *Foundations of Behavioral Research*, New York: Holt, Rinehart and Winston [245, 247].

Khandwalla, P. N., 1973, "Mass output orientation of operations technology and organizational structure," working paper, Faculty of Management McGill University [229].

King, S., 1973, *Developing New Brands*, New York: Wiley.

Kivlin, J. E., 1960, "Characteristics of farm practices associated with rate of adoption," doctoral dissertation, Pennsylvania State University, University Park, Pa. [124].

Klein, B., 1965, "Policy issues involved in the conduct of military development programs," in *Economics of Research and Development*, Columbus: Ohio State University Press [207].

Klein, B., and W. Meckling, 1958, "Application of operations research to development decisions," *Operations Research* 6: 352-363 [102].

Knight, K. E., 1965, "Some general organizational factors that influence innovative behavior," *Trans. IEEE* EM-12: 2-9 [251].

Knoblauch, H.C., et al., 1962, *State Agricultural Experiment Stations: A History of Research Policy and Procedures*, USDA Miscellaneous Report No. 904 [40, 116].

Komives, J. L., 1972, "A preliminary study of the personal values of high technology entrepreneurs," in A. Cooper and J. Komives, Eds., *Technical Entrepreneurship: A Symposium*, Milwaukee, Wis.: Center for Venture Management [57].

Kornhauser, W., 1962, *Scientists in Industry: Conflict and Accomodation*, Berkeley: University of California Press [8, 98, 117].

Kottenstette, J. P., and Rusnak, J. J., 1973, "A new perspective on the inter-sectoral movement of new technology," *IEEE Trans.* EM-20: 102-107 [141, 142, 145, 150].

Kranshar, P. M., 1969, *New Products and Diversification*, London: Business Books.

Kranzberg, M., 1963, "The technical act," in C. F. Stoner, Ed., *The Technical Order*, Detroit, Mich.: Wayne State University Press, pp. 135-139 [12].

———, 1967a, "Retrospect and Prospect: Comments," in D. L. Spencer and A. Woroniak, Eds., *The Transfer of Technology to Developing Countries*, New York: Praeger [8, 120].

———, 1967b, "The spectrum of science-technology," *J. Sciential Laboratories* 48: 47-58 [23].

———, 1967c, "The unity of science-technology," *Am. Scientist* 55: 48-66 [23].

———, 1968, "The disunity of science-technology," *Am. Scientist* 56: 21-34 [23].

———, 1971, "Science, technology, and warfare: Action, reaction, and interaction in the post-World War II era," in M. D. Wright and L. J. Paseek, Eds., *Science, Technology, and Warfare*, Washington, D.C.: Office of Air Force History and U.S. Air Force Academy

(GPO No. 0-409-818) [23].

Kranzberg, M., 1973, "Can technological progress continue to provide for the future?" in A. Weintraub, E. Schwartz, and J. R. Aronson, Eds., *The Economic Growth Controversy*, White Plains, N.Y.: International Arts and Sciences Press [30].

Kroh, R. G., 1961, "The institutional location of the scientist and his scientific values," *Trans. IRE* EM-8: 133-139 [251].

Kruybosch, C. E., 1972, "Management styles and social structure in 'identical' engineering groups," *Trans. IEEE* EM-19: 92-102 [243].

Kubie, L. S., 1958, *Neurotic Distortion of the Creative Process*, Lawrence: University of Kansas Press [53].

Kuhn, T. S., 1962, *The Structure of Scientific Revolutions*, Chicago: University of Chicago Press [17, 120, 281].

Kuhn, T. S., and N. Kaplan, 1959, "Environmental conditions affecting creativity," in C. W. Taylor, Ed., *Third University of Utah Research Conference on the Identification of Creative Scientific Talent*, Salt Lake City, Utah: University of Utah Press, pp. 313-316 [97, 117].

Kunik, I. J., 1960, "A patent attorney takes issue," *Technology and Culture* 1: 221-227 [117].

Kutz, D., and R. Kahn, 1964, *The Social Psychology of Organizations*, New York: Wiley.

Kuznets, S., 1930, *Secular Movements in Production and Prices*, Clifton, N.J.: Kelley.

———, 1962, "Inventive activity: Problems of definition and measurement," in R. R. Nelson, Ed., *The Rate and Direction of Inventive Activity: Economic and Social Factors*, Princeton, N.J.: Princeton University Press, pp. 19-43 [3, 214, 361].

———, 1966, *Modern Economic Growth*, New Haven, Conn.: Yale University Press.

———, 1972, "Innovations and adjustments in economic growth," *Swedish J. Economics* 74: 431-461 [356].

———, 1973, *Population, Capital and Growth*, New York: Norton [356].

Lambert, D., 1971, "The role of climate in the economic development of nations," *Land Economics* 47: 339-344 [17].

Lamont, L. M., 1971, *Technology Transfer, Innovation, and Marketing in Science-oriented Spin-off Firms*, Ann Arbor, Mich.: University of Michigan [57].

———, 1972, "The role of marketing in technical entrepreneurship," in A. Cooper and J. Komives, Eds., *Technical Entrepreneurship: A Symposium*, Milwaukee, Wis.: Center for Venture Management [117].

Langrish, J., 1971, "Technology transfer: Some British data," *R & D Management* 1: 133-136 [46, 90, 117].

Langrish, J., M. Gibbons, W. G. Evans, and F. R. Jevons, 1972, *Wealth from Knowledge*, London: Macmillan [25, 47, 48, 84, 89, 94, 96, 200, 202, 204, 293].

Lansford, J., 1969, *Technological Forecasting Methodologies*, New York: American Management Center [319].

LaPorte, T., and D. Metlay, 1974, "Technology observed: Attitudes of a wary public," working paper No. 9, Univeristy of California, Berkeley (Institute of Governmental Studies) [20, 21].

Lawrence, P., and J. Lorsch, 1967, "Differentiation and integration in complex organizations," *ASQ* 12: 1-47 [207].

Layton, C., 1969, *European Advanced Technology*, London: Allen and Unwin.

Layton, E. T., 1971, "Mirror-image twins: The communities of science and technology in nineteenth century America," *Technology and Culture* 12: 562-580 [23].

———, 1974, "Technology of knowledge," *Technology and Culture* 15: 31-41.

Lazarsfeld, P. F., B. Berelson, and H. Gaudet, 1948, *The People's Choice*, New York: Columbia University Press [117].

Lehman, H. C., 1953a, *Age and Achievement*, Princeton, N.J.: Princeton University Press.

———, 1953b, "The chemist's most creative years," *Science* 127: 1213-1222 [117].

Lenz, R. C., Jr., 1962, *Technological Forecasting* (ASD-TDR-62-414), Wright-Patterson Air Force Base, Ohio: Aeronautical Systems Division, U.S. Air Force [333].

———, 1970, "Practical application of technical trend forecasting," in M. Cetron and J. Goldhar, Eds., *The Science of Managing Organized Technology*, London: Gordon and Breach [212].

Lieberman, H. R., 1968, "Technology: Alchemists of Route 128," *New York Times*, 8 January 1968 [56].

Linstone, H. A., 1968, "On mirages," in J. R. Bright, Ed., *Technological Forecasting for Industry and Government*, Englewood Cliffs, N.J.: Prentice Hall [329].

Lipetz, B., 1965, *A Guide to Case Studies of Scientific Activity*, Carlisle, Mass.: Intermedia, Inc. [234].

Little, Arthur D., Inc., 1965, *Management Factors Affecting Research and Exploratory Development*, Cambridge, Mass.: Arthur D. Little, Inc. [6].

———, 1969, "Program evaluation of the Office of State Technical Services," prepared for U.S. Department of Commerce (Office of Program Planning), Cambridge, Mass: Arthur D. Little, Inc. [120, 150].

Litvak, I. A., and C. J. Maule, 1972, "Managing the entrepreneurial enterprise," *Business Quarterly* 37: 42-50 [117].

———, 1973, "Some characteristics of successful entrepreneurs in Canada," *Trans. IEEE* EM-20: 62-68 [241].

Livingston, R. B., 1974, letter to *Science* 184: 849-850 [117].

Luce, R. D., and H. Raiffa, 1957, *Games and Decisions*, New York: Wiley [117].

Luchins, A. S., 1942, "Mechanization in problem solving," *Psychological Monographs* 54 (No. 248) [86, 87].

Lynn, F., 1966, "An investigation of the rate of development and diffusion of technology in our modern industrial society," report of the President's Commission on Technology Automation and Economic Progress, *Technology and the American Economy*, Washington, D.C., App. to vol. 2 [202, 217, 337].

McClelland, D. C., 1971, *The Achieving Society*, Princeton, N.J.: Van Nostrand [57].

Maclaurin, W. R., 1949, *Invention and Innovation in the Radio Industry*, New York: Macmillan [181, 351, 356].

———, 1954, "Technological progress in some American industries," *Am. Economic Rev.* 44: 178-200 [214].

———, 1955, "New products innovation and introduction: Their broad implications," in A. H. Rubenstein, Ed., *Coordination, Control, and Financing of Industrial Research*, New York: King Crown Press [117, 118].

McPherson, J. H., 1965, "How to manage creative engineers," *Mechanical Engineering* 87: 32-36 [97, 117].

McVoy, E. C., 1940, "Patterns of diffusion in the United States," *Am. Sociological Rev.* 5: 219-227 [296].

MAB (Materials Advisory Board), 1966, Report of the ad hoc committee on Principles of Research-engineering Interaction, Washington, D.C.: NAS-NRC (AD 636529)[255].

Machiavelli, N., 1947, T. B. Bergin, Ed., *The Prince*, New York: Appleton-Century-Crofts [296].

Machlup, F., 1962, *The Production and Distribution of Knowledge in the United States*, Princeton, N.J.: Princeton University Press [56].

Mahdavi, K. B., 1972, *Technological Innovation: An Efficiency Investigation*, Stockholm: Beckmans [13].

Maher, P. M., 1972, "Results from an experiment with a computer based R & D project selection technique," 19th International Meeting of TIMS, Houston, Tex. [107, 118].

Maher, P. M., N. R. Baker, W. E. Souder, A. H. Rubenstein, A. R. Gallant, and C. R. Shumway, 1975, "Diffuse decision making in a hierarchical organization: An empirical examination," *Management Science* 21: 697-707.

Maier, N. R. G., 1961, "Reasoning in humans," in T. L. Harris and W. E. Schonn, Eds., *Selected Readings on the Learning Process*, New York: Oxford University Press [87].

Mansfield, E., 1961, "Technical change and rate of innovation," *Econometrica* 29: 741-746 [123-126].

———, 1963a, "Intrafirm rates of diffusion of an innovation," *Rev. Economics and Statistics* 45: 348-359 [125, 136].

———, 1963b, "Size of firm, market structure, and innovation, *J. Political Economy* 71: 556-576 [126].

———, 1965, "Rates of return from industrial research and development," *Am. Economic Rev.* 55: 310-322 [360].

Mansfield, E., 1966, "Technological change: Measurements, determinants, and diffusion," report to the President of the National Committee on Technology, Automation, and Economic Progress [123, 124, 140].

———, 1968a, *Industrial Research and Technological Innovation*, New York: Norton [2, 61, 101, 134, 202, 209, 218, 334].

———, 1968b, *The Economics of Technological Change*, New York: Norton [20, 39, 59, 61, 68, 103, 126, 135, 136, 204, 205, 209-211, 217, 229, 337, 356].

———, 1971, *Technological Change*, New York: Norton [7, 20, 61, 68, 102, 103, 105, 118, 136].

———, 1972a, "The multinational firm and technological change," paper read at the Rockefeller Foundation; also in J. Dunning, Ed., *Economic Analysis and the Multinational Enterprise*, New York: George Allen and Unwin, 1974 [357, 360].

———, 1972b, "Contribution of R & D to economic growth in the United States, *Science* 175: 477-486 [9, 61, 69, 202, 204, 207, 357, 358, 360].

———, 1973a, "Determinants of the speed of application of new technology," in B. Williams, Ed., *Science and Technology in Economic Growth*, New York: Wiley [212].

———, 1973b, "Technological forecasting," paper read at meeting of International Economic Association, Moscow [214].

———, 1974, "Technology and technological change," in J. Dunning, Ed., *Economic Analysis and the Multinational Enterprise*, New York: Allen and Unwin.

Mansfield, E., F. Husic, J. Rapoport, A. Romeo, E. Villani, and S. Wagner, 1976, *The Production and Diffusion of New Industrial Technology*, New York: Norton [200, 202, 205, 210, 214].

Mansfield, E., J. Rapoport, J. Schnee, S. Wagner, and M. Hamburger, 1971, *Research and Innovation in the Modern Corporation*, New York: Norton [19, 37, 49, 59, 60, 68, 103, 112, 117, 126-128, 200-202, 207, 209, 337].

Mansfield, E., and S. Wagner, 1975, "Organizational and strategic factors associated with probabilities of success in industrial R & D," *J. Business* 48: 179-198 [204, 207, 208].

Mantoux, P., 1961, *The Industrial Revolution in the Eighteenth Century*, New York: Harper and Row, rev. ed. [26].

March, J. G., and H. A. Simon, 1958, *Organizations*, New York: Wiley [98].

Marcson, S., 1960, *The Scientist in American Industry*, New York: Harper [8, 117, 118, 251].

———, 1960a, "Role concept of engineering managers," and "Role adaptation of scientists in industrial research," *Trans. IRE* EM-7: 3-34 and 159-166, 1960 [260].

Markham, J. W., 1962, "Inventive activity: Government controls and the legal environment,"in R. R. Nelson, Ed., *The Rate and Direction of Inventive Activity*, Princeton, N.J.: Princeton University Press, pp. 587-608 [39, 40, 42].

———, 1965, "Market structure, business conduct, and innovation," *Am. Economic Rev.* 60: 323-332 [33, 34, 38, 220, 230].

———, 1973, "Market concentration and innovation," unpublished draft.

Marquis, D. G., 1969a, "The anatomy of successful innovations," *Innovation* 7: 28-37.

———, 1969b, "Factors in the transfer of technology," *Innovation* 7: 289-301 [226].

Marquis, D. G., and T. A. Allen, 1967, "Communication patterns in applied psychology," *Am. Psychologist* 21: 1052-1060 [93, 95, 117, 142].

Marschak, T. A., 1962, "Strategy and organization in a system development project," in R. R. Nelson, Ed., *The Rate and Direction of Inventive Activity*, Princeton, N.J.: Princeton University Press [23, 235].

Marschak, T. A., T. K. Glennan Jr., and R. Summers, 1967, *Strategy for R & D: Studies in Microeconomics of Development*, New York: Springer [61, 102, 103, 117, 235, 242].

Martino, J. P., 1972, *Technological Forecasting for Decision Making*, New York: American Elsevier [305, 319, 322, 334].

———, 1973, "A survey of behavioral science contributions to laboratory management," *Trans. IEEE* EM-20: 68-75 [111].

Mason, R. G., 1964, "The use of information sources in the process of adoption," *Rural Sociology* 29: 40-52 [125].

———, 1969, "A dialectical approach to strategic planning," *Management Science* (April) [323].

Meadows, D. L., and D. H. Meadows, 1973, *Toward Global Equilibrium: Collected Papers*, Cambridge, Mass.: Wright-Allen Press [330].

Meadows, D. H., D. L. Meadows, J. Randers, and W. W. Behrens III, 1972, *The Limits to Growth*, Washington, D.C.: Potomac Associates [30, 195, 306, 330].

Meltzer, L., 1956, "Scientific productivity in organizational settings," *J. Social Issues* 12: 32-40.

Meltzer, L., and J. Salter, 1962, "Organizational structure and the performance and job satisfaction of physiologists," *Am. Sociological Rev.* 27: 351-362.

Menzel, H., and E. Katz, 1956, "Social relations and innovations in the medical profession: The epidemiology of a new drug," *Public Opinion Quarterly* 19: 337-352.

Merton, R. K., 1935, "Fluctuations in the rate of industrial invention," *Quarterly J. Economics* 49: 454-474.

———, 1938, "Science, technology, and society in seventeenth century England," *Osiris* 4 (Pt. 2); reprinted in book form 1970, New York: Howard Fertig [23].

———, 1973, *The Sociology of Science*, Chicago: University of Chicago Press [215].

Merton, R. K., and E. Barber, 1961, "Singles and multiples in scientific discovery: A chapter in the sociology of science," *Proc. Am. Philosophical Society* 105: 470-486 [17].

Metlay, D., 1971, "Public attitudes toward technology," in T. A. La Porte et al., Eds., *A Perspective in the Assessment of Large Scale Technology: The Case of the STOL Aircraft Transport System*, Berkeley, Calif.: University of California (Institute of Government Studies), pp. 91-102 [1, 20, 21].

———, 1972, "Public attitudes toward technology: A preliminary report," in T. A. La Porte et al., Eds., *Social Change, Public Response, and the Regulation of Large Scale Technology*, Berkeley, Calif.: University of California (Institute of Government Studies), pp. 71-123 [1, 20, 21].

Meyer, H., 1969, "If people fear to fail, can organizations ever succeed?" *Innovation* 7: 58-63 [258].

Miller, J. R., 1970, *Professional Decision Making*, New York: Praeger [118].

Miller, R. E., 1971, *Innovation, Organization and Environment: A Study of Sixteen American and West European Steel Firms*, Sherbrooke, Canada: University of Sherbrooke (Institute for Research on the Improvement of Administration) [291].

Minasian, J., 1969, "Research and development, production functions, and rates of return," *Am. Economic Rev.* 59: 80-85 [202, 204, 360].

Mishan, E., 1967, *The Costs of Economic Growth*, New York: Praeger [204].

Mitroff, I. I., 1971, "A communication model of dialectical inquiring systems," *Management Science* 17: B634-648 [323].

Mohn, N. C., Jr., 1972, "Application of trend concepts in forecasting typesetting technology," *Technological Forecasting and Social Change* 3(No. 2) [334].

Mohr, L. B., 1966, "Determinants of innovation in organizations," doctoral dissertation, University of Michigan, Ann Arbor [283, 288].

———, 1969, "Determinants of innovation in organizations," *Am. Political Science Rev.* 63: 111-126 [282, 287-289, 291].

———, 1973, "The concept of organizational goal," *Am. Political Science Rev.* 67: 470-481 [282].

Molella, A. P., and N. Reingold, 1973, "Theorists and ingenious mechanics: Joseph Henry defies science," *Science Studies* 3: 323-351 [23].

Moor, W. C., 1972, "The development and preliminary test of behavior related dimensions of information systems," *J. Am. Society Information Science* 23: 50-57 [257].

Moore, J. R., and N. R. Baker, 1969a, "An analytical approach to scoring model design-application to research and development project selection," *Trans. IEEE* EM-16: 90-98 [104, 118].

———, 1969b, "Computational analysis of scoring models for R & D project selection," *Management Science* 16: B212-232.

Mordka, I., "A comparison of a research and development laboratory's organization structures," *Trans. IEEE* EM-14: 170-177, 1967.

Morison, E. E., 1966, *Men, Machines, and Modern Times*, Cambridge, Mass.: MIT Press [181, 214].

Morley, J., 1968, *Launching a New Product*, London: Business Books.

Morris, J. M., 1962, "Administration of research in industry," *Research Management* 5: 237-247 [98, 117].

Morse, E. V., and B. Gordon, 1968, "Creative potentials and organizational structure," *Proc. Academy Management* 37-48 [84].

Morton, J. A., 1964, "From research to technology," *International Science and Technology* 29: 82-104 [112].

———, 1968, "The manager's changing role in technological innovation," *Bell Telephone Mag.* 47: 8-15 [112, 116, 161].

———, 1971, *Organizing for Innovation*, New York: McGraw-Hill [6].

Motley, C. M., and R. D. Newton, 1959, "The selection of projects for industrial research," *Operational Research* 7: 740-451 [118].

Mott, P. E., 1972, *The Characteristics of Effective Organizations*, New York: Harper and Row [283, 291, 295, 297].

Mottur, E., 1968, *The Processes of Technological Innovation: A Conceptual Systems Model*, Washington, D.C.: National Bureau of Standards, Report 9689 [13].

Mueller, R. K., 1971, *The Innovation Ethic*, New York: American Management Association, Inc. [27].

Mueller, W. F., 1962, "The origins of the basic inventions underlying Du Pont's major product and process innovations, 1920-1950," in R. R. Nelson, Ed., *The Rate and Direction of Inventive Activity: Economic and Social Factors*, a report of the National Bureau of Economic Research, Princeton, N.J.: Princeton University Press [12, 25, 29, 207, 209, 217].

Mullins, G. J., 1963, "Prediction of creativity in a sample of research scientists," *Trans. IEEE* EM-10: 52-57.

Musson, A. E., and E. Robinson, 1969, *Science and Technology in the Industrial Revolution*, Toronto: University of Toronto Press [23].

Myers, S., 1965, "Attitude and innovation," *International Science and Technology*, October: 91-96 [8].

———, 1966, "Industrial innovations and the utilization of research output," *Proc. 20th Nat. Conf. Administration of Research*, Miami, Fla.

Myers, S., and D. Marquis, 1969, "Successful industrial innovations," Washington, D.C.: National Science Foundation, report NSF 69-17 [12, 25, 46, 61, 90, 117, 118, 123, 142, 200, 202, 215, 218, 236, 356].

Myers, S., et al., 1973, "Notes of a panel presentation on institutional barriers to innovation and diffusion in the service sector," Engineering Foundation Conferences, Henniker, N.H. [112].

Mytinger, R. E., 1968, *Innovation in Local Health Services: A Study of the Adoption of New Programs by Local Health Departments with Particular Reference to Newer Medical Care Activities*, U.S. Public Health Service publication 1664-2 [288, 289, 297].

Nasbeth, L., and G. F. Ray, 1974, *The Diffusion of New Industrial Processes*, Cambridge: Cambridge University Press [125, 126, 134, 150].

Nadiri, M. I., 1970, "Some approaches to the theory and measurement of total factor productivity: A survey," *J. Economic Literature* 8: 1137-1177 [356-359].

National Academy of Engineering, 1971, *Technology and International Trade*, Washington, D.C. [214].

National Academy of Sciences Materials Advisory Board, 1966, *Report of the Ad Hoc Committee on Principles of Research-Engineering Interaction*, Washington, D.C.: National Research Council (AD-636529) [202].

National Academy of Sciences, 1967, *Applied Science and Technological Progress*, Washington, D.C. [207, 214].

———, 1973, *U.S. International Firms and R, D & E in Developing Countries*, Report of the Ad Hoc Panel on Science and Technology for International Development, Washington, D.C. [147].

National Bureau of Economic Research, 1962, *The Rate and Direction of Inventive Activity* (R. R. Nelson, Ed.), Princeton, N.J.: Princeton University Press [214, 356].

———, 1967, *The Theory and Empirical Analysis of Production*, New York.

———, 1970, *The Technology Factor in International Trade*, New York.

National Bureau of Standards, 1966, *Technology and World Trade*, Washington, D.C. [214].

National Science Foundation, 1967, *Proc. Conf. Technology Transfer and Innovation*,

National Science Foundation, 1968, *Research and Development in Industry*, Washington, D.C.

———, 1971, *Research and Development in Industry*, Washington, D.C. [220].

———, 1972, *Research and Development and Economic Growth-Productivity*, Washington, D.C. [219].

———, 1973, "Public attitudes toward science and technology," in *Science Indicators*, report of the National Science Board, Washington, D.C. [20, 34, 39, 40].

Nelkin, D., 1971, *The Politics of Housing Innovation: The Fate of the Civilian Industrial Technology Program*, Ithaca, N.Y.: Cornell University Press [146, 269].

Nelson, R. R., 1959, "The simple economics of basic scientific research," *J. Political Economy* 67: 297-306 [204].

———, 1962, "The link between science and invention: The case of the transistor," in R. R. Nelson, Ed., *The Rate and Direction of Inventive Activity*, Princeton, N.J.: Princeton University Press [12, 25, 29, 235].

Nelson, R. R., and E. S. Phelps, 1966, "Investment in humans, technological diffusion, and economic growth," *Am. Economic Assoc. Papers and Proc.* 56: 69-75 [128].

Nelson, R., M. Peck, and E. Kalachek, 1967, *Technology, Economic Growth, and Public Policy*, Washington, D.C.: Brookings Institute [37, 204].

Nordhaus, W., 1969, *Invention, Growth, and Welfare*, Cambridge, Mass.: MIT Press.

Nordhaus, W., and J. Tobin, 1972, "Is growth obsolete?" in National Bureau of Economic Research, *Economic Growth, Colloquium V of Economic Research: Retrospect and Prospect*, New York: Columbia University Press [346].

Normann, R., 1971, "Organizational innovativeness: Product variation and reorientation," *ASQ* 16: 203-215 [222].

Northwestern University, 1964, Annual report, Program of Research on the Management of Research and Development, Department of Industrial Engineering and Management Sciences, Evanston, Ill. [117].

———, 1970, Army Materiel Command Seminar on Value Measurement, AMC Value Methodology Experiments, session at Airlie House, Va., Department of Industrial Engineering and Management Sciences, Evanston, Ill. [118].

Nutt, A. B., 1965, "An approach to research and development effectiveness," *Trans. IEEE* EM-12: 103-112 [118].

———, 1972, "Some considerations in implementing an R & D resources allocation system," 19th International TIMS meeting, Houston, Texas.

Odum, E. P., 1963, *Ecology*, New York: Holt, Reinhart and Winston [13].

OECD, 1967, E. Jantsch, *Technological Forecasting in Perspective*, Paris [211, 214].

———, 1969, *Gaps in Technology*, a series of reports, Paris [214].

Oppenheim, S. C., 1957, "Comments," *Patents, Trademark, and Copyright J.* 52: 135-136 [43].

Parker, W., and J. Klein, 1966, "Productivity growth in grain production in the United States, 1840-60 and 1900-10," in D. Brady, Ed., *Output, Employment, and Productivity in the United States after 1800*, New York: Columbia University Press [187].

Parnes, S. J., 1964, "Research on developing creative behavior," in C. W. Taylor, Ed., *Widening Horizons in Creativity*, New York: Wiley [88].

Parnes, S. J., and H. F. Harding, 1962, *A Source Book for Creative Engineering*, New York: Scribner [117].

Parnes, S. J., and A. Meadow, 1963, "Development of individual creative talent," in C. W. Taylor, Ed., *Scientific Creativity: Its Recognition and Development.* New York: Wiley [88].

Passer, H. C., 1953, *The Electrical Manufacturers, 1875-1900*, Cambridge, Mass.: Harvard University Press [181, 182].

Pavitt, K., and S. Wald, 1971, *The Conditions for Success in Technological Innovation*, Paris: OECD [205, 208, 211, 214].

Peak, H., 1953, "Problems of objective observation," in L. Festinger and E. Katz, Eds., *Research Methods in the Behavioral Sciences*, New York: Holt, Rinehart and Winston, pp. 441-465 [247].

Peck, M. J., 1962, "Inventions in the post-war American aluminum industry," in R. R.

Nelson, Ed., *The Rate and Direction of Inventive Activity*, National Bureau of Economic Research, Princeton, N.J.: Princeton University Press, pp. 279-298 [42, 214].

Peck, M. J., and F. M. Scherer, 1962, *The Weapons Acquisition Process: An Economic Analysis*, Boston: Harvard University Press [23].

Pelz, D. C., 1955, "Leadership within a hierarchical organization," *J. Social Issues* 7: 49-55.

———, 1956, "Some social factors related to performance in a research organization," *ASQ* 1: 310-325.

Pelz, D. C., and F. M. Andrews, 1966, *Scientists in Organizations: Productive Climates for Research and Development*, New York: Wiley [79, 116, 207, 251, 253].

Pemberton, H. E., 1936, "The curve of culture diffusion rate," *Am. Sociological Rev.* 1: 547-556 [296].

Pessemier, E. A., 1966, *New Product Decisions: An Analytical Approach*, New York: McGraw-Hill [115].

Pessemier, E. A., and N. R. Baker, 1971, "Project and program decisions in research and development," *R & D Management* 2: 75-83 [105, 118].

Pessemier, E., and H. Root, 1973, "The dimensions of new product planning," *J. Marketing* 37: 10-18 [115].

Pessemier, E. A., and R. Teach, 1966, "A single subject scaling model using judged distances between pairs of stimuli," Papers 143 and 282, Purdue University (Institute for Research in Behavioral Economic and Management Sciences) [118].

Peters, D. H., 1968, "Incidence and exploitation of commercial ideas in university departments and laboratories," doctoral dissertation, MIT Sloan School of Management [84].

Peters, D. H., and E. B. Roberts, 1967, "Unutilized ideas in university laboratories," unpublished paper, MIT Sloan School of Management [99, 118].

Peterson, R. W., 1967, "New venture management in a large company," *Harvard Business Rev.* 45: 68-76 [115, 207].

Petrini, F., 1966, "The rate of adoption of selected agricultural innovations," Report 53, Agricultural College of Sweden, Uppsala [124].

Phillips, A., 1956, "Concentration, scale, and technological change in selected manufacturing industries, 1899-1939," *J. Industrial Economics* 4: 1179-1193 [36].

Pigou, A. C., 1928, *The Economics of Welfare*, London: Macmillan, 3rd ed. [204].

Polanyi, M., 1944, "Patent reform," *Rev. Economic Studies* 11: 61-76 [42].

President's Commission on Technology, Automation, and Economic Progress, 1966, *Technology and the American Economy* (with Appendixes), Washington, D.C. [204].

Price, D. J. de S., 1965, "Is technology historically independent of science? A study in statistical historiography," *Technology and Culture* 6: 553-568 [22, 94].

Price, W. J., and L. W. Bass, 1969, "Scientific research and the innovative process," *Science* 164: 802-809 [23, 214, 235].

Quinn, J. B., 1959, *Yardsticks for Industrial Research*, New York: Ronald Press [117].

———, 1960, "How to evaluate research output," *Harvard Business Rev.* 38: 69-80.

———, 1967, "Technological forecasting," *Harvard Business Rev.* 45: 89-106 [214].

Quinn, J. B., and J. A. Mueller, 1963, "Transferring research results to operations," *Harvard Business Rev.* 41: 49-66 [112, 115, 116, 161].

Rae, J. B., 1967, "The invention of invention," in M. Kranzberg and C. W. Pursell, Eds., *Technology in Western Civilization*, New York: Oxford University Press, vol. 1, chap. 19 [4].

Rasner, M. M., 1968, "Economic determinants of organizational innovation," *ASQ* 12: 614-625 [291].

Rescher, N., 1969, "What is a value change?" in K. Baier and N. Rescher, Eds., *Values and the Future*, New York: Free Press [20, 22].

Roberts, E. B., 1964, *The Dynamics of Research and Development*, New York: Harper and Row [334].

———, 1968, "Exploratory and normative technological forecasting: A critical appraisal," paper read at NATO Seminar on Technological Forecasting [212].

———, 1969, "Exploratory and normative technological forecasting: A critical appraisal," in M. J. Cetron, Ed., *Technological Forecasting: A Practical Approach*, New York: Gordon and Breach [325].

Roberts, E. B., 1970, "Entrepreneurship and technology," in M. J. Cetron and J. D. Goldhar, Eds., *The Science of Managing Organized Technology*, New York: Gordon and Breach, pp. 485-509 [57].

―――, 1972, "Influences upon performance of new technical enterprises," in A. Cooper and J. Komives, Eds., *Technical Entrepreneurship: A Symposium*, Milwaukee, Wis.: Center for Venture Management [117].

Roberts, E. B., and H. A. Wainer, 1971, "Some characteristics of technical entrepreneurs," *Trans. IEEE* EM-13: 181-187 [57, 251].

Roe, A., 1953, *The Making of a Scientist*, New York: Dodd, Mead [52].

―――, 1964, "The psychology of scientists," in K. Hill, Ed., *The Management of Scientists*, Boston: Beacon [117].

Roepcke, L. A., W. E. Rafert, and R. P. Benedict, 1968, "Army Materiel Command: Planning principles and philosophy," *Trans. IEEE* EM-15: 150-178.

Rogers, C. R., 1964, "Toward a theory of creativity," *ETC: Rev. General Semantics* 2: 249-256 [53].

Rogers, E. M., 1962, *Diffusion of Innovations*, New York: Free Press [128, 142].

Rogers, E. M., with F. F. Shoemaker, 1971. *Communication of Innovations: A Cross-cultural Approach*, New York: Free Press [19, 20, 120, 121, 123-129, 135, 136, 140, 142-144, 148-150, 161, 229, 281, 286, 293, 296].

Rokeach, M., 1960, *The Open and Closed Mind*, New York: Basic Books [127].

Ronken, H. O., and P. R. Lawrence, 1952, "Administering changes," working paper, Harvard University (Graduate School of Business), Boston [79].

Root, H. P., "New product investment decisions: The process and procedures," in F. C. Allvine, Ed., *American Marketing Association Combined Proceedings, Spring and Fall Conferences*.

Rosen, E. M., and W. E. Souder, 1965, "A method for allocating R & D expenditures," *Trans. IEEE* EM-12: 87-93 [105, 108, 118].

Rosenberg, N., Ed., 1969, *The American System of Manufactures*, Edinburgh: Edinburgh University Press [144].

―――, 1970, "Economic development and the transfer of technology: Some historical perspectives," *Technology and Culture* 11: 550-575 [8].

―――, 1971, *Economics of Technological Change*, Baltimore: Penguin Books [334].

―――, 1972, *Technology and American Economic Growth*, New York: Harper and Row [191, 215, 216, 334].

―――, 1973, "Innovative responses to materials shortages," *Annual J. Am. Economics Association* 63: 111-127 [30].

―――, 1974, "Science, invention, and economic growth," *Economic J.* 84: 90-108 [25, 29, 361].

Rosenbloom, R., and F. Wolek, 1970, *Technology and Information Transfer*, Cambridge, Mass.: Harvard University Press.

Rossman, J., 1964, *Industrial Creativity: The Psychology of the Inventor*, New Hyde Park, N.Y.: University Books [48].

Rubenstein, A. H., 1957, "Setting criteria for R & D," *Harvard Business Rev.* 35: 95-104 [49, 59, 100, 101, 103, 117].

―――, 1962, "The constraints of decentralization," *Chemical Engineering Progress* 58: 11-15.

―――, 1964, "Organizational factors affecting research and development decision-making in large decentralized companies," *Management Science* 10: 618-634 [60, 61, 67, 117].

―――, 1968, "Research-on-research: The state of the art in 1968," *Research Management* 11: 279-304 [61, 117].

―――, 1971, "A longitudinal study of the development of information style," *Management Information Systems*, Program of Research on the Management of Research and Development, Northwestern University, Evanston, Ill. [252].

―――, 1973, "Basic research on technology transfer," *Proc. NATO Conference on Technology Transfer*, Paris [232].

Rubenstein, A. H., and R. Hannenberg, 1962, "Idea flow and project selection in several industrial research and development laboratories," in *Proc. Conf. Economic and Social Factors in Technological Research and Development*, Ohio State University [61, 117].

Rubenstein, A. H., and M. Radnor, 1963, "Top management's role in research planning in large decentralized companies," in G. Krewers and G. Morlat, Eds., *Proc. 3rd Intern. Conf. Operational Research*, Paris [70, 160].

Rubenstein, A. H., R. T. Barth, and C. F. Douds, 1969, "Coupling relations in product and systems development," *Proc. National Electronics Conf.* 25: 893-898 [112].

Rubenstein, A. H., G. J. Rath, R. D. O'Keefe, J. A. Kernaghan, W. C. Moore, and D. J. Werner, 1973, "Behavioral factors influencing the adoption of an experimental information system," *Hospital Administration*, Fall: 27-43 [256].

Rubin, I. M., and W. Seelig, 1967, "Experience as a factor in the selection and performance of project managers," *Trans. IEEE* EM-14: 131-135 [251].

Rubin, I. M., A. C. Stedry, and R. D. Willits, 1965, "Influences related to time allocation of R & D supervisors," *Trans. IEEE* EM-12: 70-79 [251].

Rudelius, W., 1969, "Methods of analyzing the impact of program stretch-outs," *Trans. IEEE* EM-16: 23-25 [251].

Ruefli, T. W., 1971, "A generalized goal decomposition model," *Management Science* 17: B505-518 [118].

Ruttan, V., 1959, "Usher and Schumpeter on invention, innovation, and technological change," *Quarterly J. Economics* 73: 596-606 [213].

Ryan, B., and N. C. Gross, 1943, "The diffusion of hybrid seed corn in two Iowa communities," *Rural Sociology* 8: 15-24 [281, 296].

Salter, W. E. G., 1960, *Productivity and Technical Change*, Cambridge: Cambridge University Press [29].

Samuelson, P., 1947, *Foundations of Economic Analysis*, Cambridge, Mass.: Harvard University Press [15].

Sanders, B. S., 1962, "Some difficulties in measuring inventive activity," in R. R. Nelson, Ed., *The Rate and Direction of Inventive Activity*, National Bureau for Economic Research, Princeton, N.J.: Princeton University Press, pp. 53-77 [361].

Sanders, B. S., J. Rossman, and L. J. Harris, 1959, "Patent acquisition by corporations," *Patent, Trademark, and Coypright J.* 54 [35].

Sapolsky, H. M., 1967, "Organizational structure and innovation," *J. Business* 40: 497-510 [297].

Scherer, F. M., 1958, *Patents and the Corporation*, Boston: J. J. Galvin [43].

———, 1965a, Testimony before the U.S. Senate Subcommittee on Antitrust and Monopoly of the Committee on the Judiciary, Part 3: Concentration, Invention, and Innovation, 89th Congress, First Session, pp. 1188-1206 [34, 46].

———, 1965b, "Invention and innovation in the Watt-Boulton steam engine venture," *Technology and Culture* 6: 165-187 [53].

———, 1965c, "Firm size, market structure, opportunity, and the output of patented inventions," *Am. Economic Rev.* 55: 1097-1125 [103, 209].

———, 1965d, "Government research and development programs," in R. Dorfman, Ed., *Measuring Benefits of Government Investments*, Washington, D.C.: Brookings Institute.

———, 1967a, "Market structure and employment of scientists and engineers," *Am. Economic Rev.* 57: 524-531 [36].

———, 1967b, "Research and development resource allocation under rivalry," *Quarterly J. Economics* 81: 359-394 [36].

———, 1970, *Industrial Market Structure and Economic Performance*, Chicago: Rand McNally [36, 37, 210, 220, 229].

Schmookler, J., 1960, "An economist takes issue," *Technology and Culture* 1: 214-220 [117].

———, 1965a, "Technological change and economic theory," *Am. Economic Rev.* 55: 333-341 [32].

———, 1965b, Testimony before the U.S. Senate Subcommittee on Antitrust and Monopoly of the Committee on the Judiciary, Part 3: Concentration, Invention, and Innovation, 89th Congress, First Session, pp. 1257-1269.

———, 1966, *Invention and Economic Growth*, Cambridge, Mass.: Harvard University Press [9, 43, 202, 214, 216, 356, 361, 362].

———, 1972, *Patents, Invention, and Economic Change*, Cambridge, Mass.: Harvard University Press [356].

Schon, D. A., 1967, *Technology and Change*, New York: Delacorte Press [8, 37, 207, 214].

Schumpeter, J. A., 1939, *Business Cycles*, New York: McGraw-Hill [2, 5, 117, 125].

———, 1943, *Capitalism, Socialism, and Democracy*, London: Allen and Unwin [327].

———, 1949, *The Theory of Economic Development*, Cambridge, Mass.: Harvard University Press.

———, 1950, *Capitalism, Socialism and Democracy*, New York: Harper, rev. ed. [209].

Schwartz, J. J., 1973, "The decision to innovate," DBA dissertation, Harvard University [67, 68, 79, 117].

Science Policy Research Unit, 1973, "The limits of growth controversy," *Futures* 5: 4-152 [330].

Scientific Advisory Group, U.S. Army Air Force, 1945, *Where We Stand* (the "Von Kármán Reports"), unpublished, classified [315].

Scott, A., 1962, "The development of extractive industries," *Canadian J. Economics and Political Science* [30, 189].

Scott, B. R., 1973, "The industrial state: Old myths and new realities," *Harvard Business Rev.* 51: 133-148 [60, 70, 112, 229].

Scott, W. R., 1972, "Professionals in hospitals: Technology and the organization of work," in B. S. Georgopoulos, Ed., *Organization Research on Health Institutions*, Ann Arbor, Mich.: University of Michigan (Institute for Social Research) [291].

Scriven, M., 1974, "The exact role of value judgments in science," in K. F. Schaffner and R. S. Cohen, Eds., *Philosophy of Science Association 1972*, Dordrecht (Netherlands): Reidel, pp. 219-247 [20].

Seiler, R., 1965, *Improving the Effectiveness of Research and Development*, New York: McGraw-Hill.

Sellitz, C., M. Jahoda, M. Deutsch, and S. W. Cook, 1965, *Research Methods in Social Relations*, New York: Holt, Rinehart and Winston, rev. ed. [247, 248].

Shapero, A., 1971, *An Action Program for Entrepreneurship*, Austin, Tex.: Multidisciplinary Research [57].

Sharlin, H. I., 1967, "Applications of electricity," in M. Kranzberg and C. W. Pursell Jr., Eds., *Technology and Western Civilization*, New York: Oxford University Press, vol. 1, chap. 34 [54].

Shepard, H. A., 1959, "Major researches in creativity," *Research Management* 2: 203-220 [117].

Shepart, W. G., 1971, "The competitive margin in communications," in W. F. Capron, Ed., *Technological Change in Regulated Industries*, Washington, D.C.: Brookings Institute, pp. 86-112 [44].

Sherwin, C. W., and R. S. Isenson, 1966, *First Interim Report on Project Hindsight: Summary*, Washington, D.C.: Office of the Director of Defense Research and Engineering (AD 642-200) [12, 24, 46, 90, 214, 234, 235].

———, 1967, "Project Hindsight: A Defense Department study of the utility of research," *Science* 156: 1571-1577.

Shumway, C. R., P. M. Maher, N. R. Baker, W. E. Souder, A. H. Rubenstein, and A. R. Gallant, 1975, "Diffuse decision making in hierarchical organizations," *Management Science* 21: 697-707 [102].

Sieber, S. D., 1973, "The integration of fieldwork and survey methods," *Am. J. Sociology* 78: 1335-1359 [287].

Siegel, I. H., 1953, "Technological change and long run forecasting," *J. Business of the University of Chicago* 26: 141-156 [334].

Siegman, J., N. R. Baker, and A. H. Rubenstein, 1969, "Control mechanisms in the R & D idea flow process: Model and behavioral study," unpublished paper, POMRAD, Northwestern University [98, 117].

Sigford, J. V., and R. H. Parvin, 1965, "Project PATTERN: A methodology for determining relevance in complex decision making," *Trans. IEEE* EM-12: 9-13 [118].

Simmonds, W. H. C., 1973, "Analysis of industrial behavior and its use," in J. R. Bright and M. E. Schoeman, Eds., *A Guide to Practical Technological Forecasting*, Englewood Cliffs, N.J.: Prentice-Hall, pp. 215-237 [327].

Sinclair, D. L., 1972, "Management challenges to the entrepreneur," *Business Quarterly* 37: 60-63.

Singh, R. N., 1966, "Characteristics of farm innovations associated with the rate of adoption," Report 14, Agricultural Extension Education, Ontario [124].

Slocum, D. H., 1972, *New Venture Methodology*, New York: American Management Association.

Smith, C. S., 1960, *A History of Metallography*, Chicago: University of Chicago Press [22].

Smith, W. R., 1959, "Favorable and unfavorable working conditions reported by scientists at two research centers," in C. W. Taylor, Ed., *Third University of Utah Research Conference on the Identification of Creative Scientific Talent*, Salt Lake City: University of Utah Press [117].

Solo, R. A., 1975, *Organizing Science for Technology Transfer in Economic Development*, East Lansing: Michigan State University Press.

Solow, R., 1957, "Technical change and the aggregate production function," *Rev. Economics and Statistics* 39: 312-320 [8, 357].

———, 1973, "Is the end of the world at hand?" *Challenge* 16: 39-50 [214].

Souder, W. E., 1972a, "Comparative analysis of R & D investment models," *AIIE Trans.* 4: 57-64 [104, 106, 107, 118].

———, 1972b, "A scoring methodology for assessing the suitability of management science models," *Management Science* 18: B526-545 [106, 118].

———, 1973, "Utility and perceived acceptability of R & D project selection models," *Management Science* 19: 1384-1394 [106].

Spencer, D. L., and A. Woroniak, Eds., 1967, *The Transfer of Technology to Developing Countries*, New York: Praeger [119].

Spicer, E. H., Ed., 1953, *Human Problems in Technological Change*, New York: Russell Sage Foundation [20, 127, 145].

Stakman, E. C., R. Bradfield, and P. C. Mangelsdorf, 1967, *Campaigns Against Hunger*, Cambridge, Mass.: Belknap Press [117].

Stein, M. I., and S. J. Heinze, 1960, *Creativity and the Individual: Summaries of Selected Literature in Psychology and Psychiatry*, Glencoe, Ill.: Free Press [117].

Stephenson, R. W., B. S. Gantz, and C. E. Erickson, 1971, "Development of organizational climate inventories for use in R & D organizations," *Trans. IEEE* EM-18: 38-50 [251].

Stewart, C., 1972, "A summary of the state of the art on the relationship between R & D and economic growth/productivity," in National Science Foundation, *Research and Development and Economic Growth/Productibity*, NSF 72-303, Washington, D.C., pp. 11-19 [357-359].

Stigler, G. J., 1952, "The case against big business," *Fortune* 45 (May): 123 [32].

———, 1956, "Industrial organization and economic progress," in L. D. White, Ed., *The State of the Social Sciences*, Chicago: University of Chicago Press, pp. 269-282 [36].

Storer, N. W., 1962, "Research orientations and attitudes toward teamwork," *Trans. IRE* EM-9: 29-33 [118, 251].

Strassmann, W. P., 1959, *Risk and Technological Innovation*, Ithaca, N.Y.: Cornell University Press.

Susbauer, J. C., 1967, "The science entrepreneur," *Industrial Research* 9: 23-30.

———, 1969, "The technical company formation process: A particular aspect of entrepreneurship," doctoral dissertation, University of Texas, Austin [57].

———, 1972, "The technical entrepreneurship process in Austin, Texas," in A. Cooper and J. Komives, Eds., *Technical Entrepreneurship: A Symposium*, Milwaukee, Wis.: Center for Venture Management [58, 117].

Susskind, Charles, 1967, "Conference on Electron Device Research: A Pattern to Be Copied?" *IEEE Spectrum* 4(No. 2): 100-102 [94].

Swager, W. L., 1973, "Perspective trees: A method for creatively using forecasts," in J. R. Bright and ,. E. Schoeman, Eds., *A Guide to Practical Technological Forecasting*, Englewood Cliffs, N.J.: Prentice-Hall, pp. 164-190 [334].

Tannenbaum, M., **1970**, "A booming technology, a better environment: Can we have both?" *Bell Telephone Magazine*, May-June: 25-26 [4, 90].

Tannenbaum, M., et al., 1966, Report of the ad-hoc committee on principles of research-engineering interaction, National Academy of Science (Materials Advisory Board), Washington, D.C.: National Research Council, MAB 222-M [46].

Taviss, I., 1972, "A survey of popular attitudes toward technology," *Technology and Culture* 13: 606-621 [20].

———, 1973, "Futurology and the study of values," in F. Tugwell, Ed., *Search for Al-*

ternatives, Cambridge, Mass.: Winthrop [22].

Taylor, C. W., 1956, *First Conference on the Identification of Creative Scientific Talent*, Salt Lake City, Utah, New York: McGraw-Hill [52].

———, Ed., 1964, *Creativity: Progress and Potential*, New York: McGraw-Hill [117].

Taylor, C. W., W. R. Smith, and B. Ghiselin, 1961, "A study of the multiple contributions of scientists at one research organization," *Trans. IRE* EM-8: 194-201 [251].

Taylor, R. L., and J. M. Utterback, 1975, "A longitudinal study of communications in research: Technical and managerial influences," *Trans. IEEE* EM-22: 80-87.

Teich, A. H., Ed., 1972, *Technology and Man's Future*, New York: St. Martin's Press [21].

Terleckij, N., 1959, "Sources of productivity change," doctoral dissertation, Columbia University.

———, 1967, "The influence of research and education on CES production relations: Comment," in M. Brown, Ed., *The Theory and Empirical Analysis of Production*, New York: Columbia University Press, pp. 372-379 [360].

Thompson, C. W. N., 1969, "Management research methods for engineering managers," *Society of Automotive Engineers Monograph* 690362 [240, 257].

———, 1970, "Study of the user's 'natural queue' of documents," *Trans. IEEE* EWS-13: 73-78 [257].

———, 1974, "Administrative experiments: The experience of fifty-eight engineers and engineering managers," *Trans. IEEE* EM-21: 42-50 [240].

Thompson, J. D., 1962, "Organizations and output transactions," *Am. J. Sociology* 68: 309-324 [291].

———, 1967, *Organizations in Action*, New York: McGraw-Hill [291].

Toffler, A., 1970, *Future shock*, New York: Random House.

Trueswell, R. W., G. J. Rath, and A. H. Rubenstein, 1968, "A study of the information searching behavior of X-ray crystallographers," Program of Research on the Management of Research and Development, paper 68/16, Northwestern University, Evanston, Ill. [257].

Turoff, M., 1972, "An alternative approach to cross-impact literature," *Technological Forecasting and Social Change* 3: 309-399 [320, 330].

Uman, D. B., 1969, *New Product Programs*, New York: American Management Association.

UNESCO, 1971, *International Aspects of Technological Innovation: Proceedings of a Science Policy Symposium*, Paris, 1970, Paris: UNESCO.

United Nations Secretariat, 1973, Department of Economic and Social Affairs, Document ST/GCA/190 [147].

U.S. Department of Commerce, *Technological Innovation: Its Environment and Management*, GPO: 0-242-736, Washington, D.C. [200].

U.S. National Resources Committee, 1937, *Technological Trends and National Policy*, Washington, D.C.: Government Printing Office [311, 312].

U.S. Senate, 1973, Patent law revision, hearings before the Subcommittee on Patents, Trademarks, and Copyrights, Committee on the Judiciary, 93rd Congress, First Session [46].

Usher, A. P., 1954, *A History of Mechanical Inventions*, Boston, Harvard University Press [168, 181, 213].

Utterback, J. M., 1964, "Summary of variables and propositions concerning idea flow," working paper, Department of Industrial Engineering and Management Sciences, Northwestern University, Evanston, Ill. [98, 117].

———, 1969, "The process of innovation in instrument firms," doctoral dissertation, MIT School of Management, Cambridge, Mass. [46, 90, 98, 117, 118].

———, 1971a, "The process of technological innovation within the firm," *Academy of Management J.* 14: 75-88 [111, 251].

———, 1971b, "The process of innovation: A study of the origination and development of ideas for new scientific instruments," *Trans. IEEE* EM-18: 124-131 [83, 90, 117].

———, 1973, "Innovation in industry, and diffusion of technology," working paper, Graduate School of Business, University of Indiana, Bloomington, Ind. [61, 117, 118].

———, 1974, "Innovation in industry and diffusion of technology," *Science* 183: 620-626 [48, 90, 92, 94, 97, 117, 142, 217, 229].

Vaughan, F. L., 1956, *The U.S. Patent System*, Norman: University of Oklahoma Press [46].

Veney, J.E., et al., 1971, "Implementation of health programs in hospitals," *Health*

Services Research 6: 350-361 [289].

Vernon, R., 1966, "International investment and international trade in the product cycle," *Quarterly J. Economics* 80: 190-207 [214].

Villard, H. H., 1958, "Competition, oligopoly and research," *J. Political Economy* 66: 483-497 [33].

Vollmer, H. M., 1964, "Application of the behavioral sciences to research management," Stanford Research Institute report AFOSR 64-2555.

Vroom, V. H., Ed., 1967, *Methods of Organizational Research*, Pittsburgh: University of Pittsburgh Press [237].

Walker, J. L., 1969, "The diffusion of innovations among the American states," *Am. Political Science Rev.* 63: 880-899 [135, 282, 286, 287].

———, 1971, "Innovations in state politics," in H. Jacob and K. N. Vines, Eds., *Politics in the American States: A Comparative Analysis*, Boston: Little, Brown [286].

Wallmark, J. T., and B. Sellerberg, 1966, "Efficiency vs. size of research teams," *Trans. IEEE* EM-13: 137-142.

Walton, E., 1960, "What is the role of the government laboratory: A questionnaire study in one government laboratory," *Trans. IRE* EM-7: 114-117, 1960.

Watanabe, M. 1974, "The conception of nature in Japanese culture," *Science* 183: 279-282 [117].

Waterman, A., 1965, "The changing environment of science," *Science* 147: 13-18 [22].

Watt, K. F. E., 1966, *System Analysis in Ecology*, New York: Academic Press [14].

Webb, E., D. T. Campbell, R. D. Schwartz, and L. Sechrist, 1966, *Unobtrusive Measures*, Chicago: Rand McNally [248, 249].

Weber, M., 1930, *Protestant Ethic and the Spirit of Capitalism*, New York: Scribner [19].

Weiner, C., 1973, "How the transistor emerged," *IEEE Spectrum* 10: 24-33 [9, 12, 25].

Weitzman, M., 1970, "Iterative multi-level planning with production targets," *Econometrica* 38: 50-65 [118].

Werner, D. J., 1969, "A theoretical and empirical investigation of the relationships between some measures of information-related behavior and characteristics of the individual's task, person, organizational environment, and professional environment," Ph.D. dissertation, Northwestern University, Evanston, Ill. [257].

Wertime, T. A., 1962, *The Coming of the Age of Steel*, Chicago: University of Chicago Press [22].

West, S. S., 1960, "The ideology of academic scientists," *Trans. IEEE* EM-19: 53-60 [251].

White, F., 1961, *American Industrial Research Laboratories*, Washington, D.C.: Public Affairs Press [56].

White, L., jr., 1962, *Medieval Technology and Social Change*, Oxford: Oxford University Press [17, 19].

———, 1967, "The historical roots of our ecological crisis," *Science* 155: 1203-1207 [19].

White, L. J., 1971, *The Automobile Industry Since 1945*, Cambridge, Mass.: Harvard University Press [223].

Whitehead, A. N., 1925, *Science and the Modern World*, New York: Macmillan [48].

Whitman, A., 1974, "Inventors invent, but the question is, how?" *New York Times*, February 24 [17].

Wiener, N., 1950, *The Human Use of Human Beings*, Boston: Houghton Mifflin [309].

Wik, R. M., 1966, "Science and American agriculture," in D. D. Van Tassel and M. G. Hall, *Science and Society in the United States*, Homewood, Ill.: Dorsey Press [116].

Williams, B., Ed., 1973, *Science and Technology in Economic Growth*, New York: Macmillan.

Williams, R. M., Jr., 1967, "Recent developments in research on social institutions," *Ann. Am. Academy Political and Social Science* 374: 171-184 [20].

Williamson, M., 1960, "Role of the technical staff in product innovations," *Research/Development* 11: 91-95 [117, 118].

Wills, G., D. Ashton, and B. Taylor, Eds., *Technological Forecasting and Corporate Strategy*, London: Crosby Lockwood & Son [319].

Wilson, J. Q., 1966, "Innovation in organization: Notes toward a theory," in J. D.

Thompson, Ed., *Approaches to Organizational Design*, Pittsburgh: University of Pittsburgh Press [70, 112, 117, 297].

Winch, R. F., and D. T. Campbell, 1969, "Proof? No. Evidence? Yes. The significance of tests of significance," *Am. Sociologist* 4: 140-143 [244].

Wolek, F. W., and B. C. Griffith, 1974, "Policy and informal communications in applied science and technology," working paper 133, University of Pennsylvania (Wharton School of Management) [251].

Woodward, J., 1965, *Industrial Organization: Theory and Practice*, London: Oxford University Press [112, 221].

Young, H. C., 1973, "Some effects of the product development setting, information exchange, and marketing-R & D coupling on product development," doctoral dissertation, Northwestern University, Evanston, Ill. [112, 114].

Zaltman, G., R. Duncan, and J. Holbek, 1973, *Innovations and Organizations*, New York: Wiley [60, 70-72, 112, 117, 229, 275, 291-293, 296].

Zilsel, E., 1945, "The genesis of the idea of scientific progress," *J. History of Ideas* 6: 325-349 [23].

Zvorykin, A. A., I. I. Osimova, V. I. Chernyshev, and S. V. Shukhardin, 1962, *Istoria tekhniki*, Moscow: Izdatelstvo Sotsialno-Ekonomicheskoi Literatury [26].

INDEX TO CONTRIBUTORS

Baker, Norman R., *1*
Bright, James R., *299*
Douds, Charles T., *231*
Eveland, John Dudley, *275*
Hughes, Thomas P., *166*
Kelly, Patrick, *iii*, *1*
Kranzberg, Melvin, *iii*, *1*
Kuznets, Simon, *335*
Mansfield, Edwin, *199*
Mitzner, Morris, *1*
Rogers, Everett M., *275*
Rosenberg, Nathan, *185*
Rosenbloom, Richard S., *215*
Rossini, Frederick A., *1*
Rubenstein, Albert H., *231*
Strassmann, W. Paul, *263*
Susskind, Charles, *ix*
Tarpley, Fred A. Jr., *1*
Zybkow, Martha, *ix*